2802754746
WITHDRAWN
AF841470

CEPHALOPOD NEUROBIOLOGY

Neuroscience studies in squid, octopus, and cuttlefish

EDITED BY

N. JOAN ABBOTT

King's College, London

RODDY WILLIAMSON

and

LINDA MADDOCK

Marine Biological Association, Plymouth

OXFORD NEW YORK TOKYO

OXFORD UNIVERSITY PRESS

1995

Oxford University Press, Walton Street, Oxford OX2 6DP

Oxford New York
Athens Auckland Bangkok Bombay
Calcutta Cape Town Dar es Salaam Delhi
Florence Hong Kong Istanbul Karachi
Kuala Lumpur Madras Madrid Melbourne
Mexico City Nairobi Paris Singapore
Taipei Tokyo Toronto

and associated companies in
Berlin Ibadan

Oxford is a trade mark of Oxford University Press

Published in the United States
by Oxford University Press Inc., New York

© Oxford University Press, 1995

All rights reserved. No part of this publication may be reproduced, stored in a retrieval system, or transmitted, in any form or by any means, without the prior permission in writing of Oxford University Press. Within the UK, exceptions are allowed in respect of any fair dealing for the purpose of research or private study, or criticism or review, as permitted under the Copyright, Designs and Patents Act, 1988, or in the case of reprographic reproduction in accordance with the terms of the licences issued by the Copyright Licensing Agency. Enquiries concerning reproduction outside these terms and in other countries should be sent to the Rights Department, Oxford University Press, at the address above.

This book is sold subject to the condition that it shall not, by way of trade or otherwise, be lent, re-sold, hired out, or otherwise circulated without the publisher's prior consent in any form of binding or cover other than that in which it is published and without a similar condition including this condition being imposed on the subsequent purchaser.

A catalogue record for this book is available from the British Library

Library of Congress Cataloging in Publication Data
(Data available on request)

ISBN 0 19 854790 0

Typeset by Footnote Graphics, Warminster, Wiltshire
Printed in Great Britain by
St Edmundsbury Press, Bury St Edmunds, Suffolk

UNIVERSITY
COLLEGE LONDON
LIBRARY

Preface

Cephalopods are among the most intelligent invertebrates, able to compete successfully with vertebrates in several marine habitats. They have highly developed sensory and motor systems, and large multilobed brains, capable of complex analysis, learning, and social behaviour. Certain cephalopod nerves, synapses, and muscles have particular features that make them outstanding preparations for investigating basic questions in neuroscience, from the mechanisms underlying neural excitability and synaptic transmission, to the role of glial cells, to the way muscle activity is controlled. Within the last five years, many of the powerful new techniques of molecular biology and electrophysiology have been applied to advantageous cephalopod preparations, and the exciting new work deserves a wider audience. This volume brings together 32 chapters written by experts in each topic, and illustrates the many links between cellular physiology and the function of the animal as a whole.

The volume reviews work on particular cephalopod preparations that have contributed to recent developments. The squid axon has given information on the axonal cytoskeleton, axon transport, and the ionic mechanisms underlying action potential generation and conduction. Studies on the squid giant synapse give information about the molecular events involved in neurotransmission both pre- and postsynaptically. Studies on the glial (Schwann) cells around the squid giant axon provide evidence for axon–Schwann cell signalling. Studies on the morphology and innervation of cephalopod muscles show how they are used in ventilation, swimming and escape responses. The chromatophore muscles of the skin enable cephalopods to adopt a variety of patterns for display and camouflage, and the mechanisms for their control are beginning to be understood. The brains of several cephalopod species have been mapped in elegant morphological studies, and new electrical recording methods are making possible the correlation of structure and function. The sensory systems of cephalopods resemble their vertebrate counterparts in important respects, but show subtle differences that highlight basic principles of sensory function.

This is a book to be enjoyed by both vertebrate and invertebrate neuroscientists, and by anyone interested in the basic principles that control neural function.

London
Plymouth
August 1994

N.J.A.
R.W.
L.M.

Acknowledgements

This volume had its origin in an International Symposium organized in Cambridge, England, organized by the Centre for Cephalopod Neurobiology at the Marine Biological Association Laboratory (MBA), Plymouth, England, sponsored by the MBA, and with financial support from the MBA, the Wellcome Trust and the UK Physiological Society. We are grateful to the Director of the MBA, Dr Michael Whitfield, for his interest and support.

Contents

PART VI: CENTRAL NERVOUS SYSTEM

PART VII: SENSORY FUNCTION

Contributors

Editors:

Dr N. Joan Abbott
Physiology Group, Biomedical Sciences Division, King's College, University of London, Strand, London WC2R 2LS, UK.

Dr Roddy Williamson
Marine Biological Association, The Laboratory, Citadel Hill, Plymouth PL1 2PB, UK.

Dr Linda Maddock
Marine Biological Association, The Laboratory, Citadel Hill, Plymouth PL1 2PB, UK.

Contributors:

Dr N. Joan Abbott
Physiology Group, Biomedical Sciences Division, King's College, University of London, Strand, London WC2R 2LS, UK.

J. Adjaye, Biophysics Group, King's College, University of London, 26/29 Drury Lane, London WC2B 5RL, UK.

Dr T. J. A. Allen
Department of Physiology, School of Veterinary Science, Park Row, Bristol BS1 5LS, UK.

Dr Francisco Bezanilla
Department of Physiology, UCLA School of Medicine, Center for Health Sciences, 10833 Le Conte Avenue, Los Angeles, CA 90024-1751, USA.

Dr Q. Bone, FRS
Marine Biological Association, The Laboratory, Citadel Hill, Plymouth PL1 2PB, UK.

Dr Scott T. Brady
Department of Cell Biology and Neuroscience, University of Texas Southwestern Medical Center, 5323 Harry Hines Boulevard, Dallas, TX 75235-9039, USA.

Dr E. R. Brown
Marine Biological Association, The Laboratory, Citadel Hill, Plymoth PL1 2PB, UK.

Dr M. Bundgaard
Department of General Physiology and Biophysics, Panum Institute, University of Copenhagen, Blegdamsvej 3C, DK-2200 Copenhagen N, Denmark.

Dr B. U. Budelmann
Marine Biomedical Institute and Department of Otolaryngology, University of Texas Medical Branch, Galveston, TX 77555-0843, USA.

Professor T. J. Bullock
Department of Neurosciences A-001, University of California San Diego, La Jolla, CA 92093, USA.

Ana M. Correa, Department of Physiology, UCLA School of Medicine, Center for Health Sciences, 10833 Le Conte Avenue, Los Angeles, CA 90024-1751, USA.

Dr C. J. Cornwell
c/o Dr J. B. Messenger, Department of Animal and Plant Sciences, School of Biological Sciences, Sheffield University, POB 601, Sheffield S10 3UQ, UK.

Dr Joachim Deitmer
Department of Allgemeine Zoologie, University of Kaiserlautern, Kaiserlautern, Germany.

Dr A. De Santis
Stazione Zoologica 'A. Dohrn', Villa Comunale, Napoli 80121, Italy.

Dr P. A. M. Eagles
Biophysics Group, King's College, University of London, 26/29 Drury Lane, London WC2B 5RL, UK.

Dr Peter D. Evans
AFRC Laboratory of Molecular Signalling, Department of Zoology, University of Cambridge, Downing Street, Cambridge CB2 3EJ, UK.

Dr Ian C. Forster
Physiologisches Institut, Universität Zürich, Winterthurerstrasse 190, CH-8057 Zürich, Switzerland.

Dr Kishio Furuya
Department of Cell Physiology, National Institute for Physiological Sciences, Myodaiji, Okazaki, 444 Japan.

Dr W. F. Gilly
Hopkins Marine Station, Stanford University, Department of Biological Sciences, Pacific Grove, CA 93950-3094, USA.

Dr Nikolaus G. Greeff
Physiologisches Institut, Universität Zürich, Winterthurerstrasse 190, CH-8057 Zürich, Switzerland.

Dr Michael Hans
Department of Membrane Biophysics, Max Planck Institute for Biophysical Chemistry, Am Fassberg, D-3400 Goettingen, Germany.

Professor B. M. Hendry
The Renal Unit, Dulwich Hospital, East Dulwich Grove, London SE22 8PT, UK.

Dr Michinori Ichikawa
Electrotechnical Laboratory, 1-1-4 Umezono, Tsukuba-shi, Ibaraki, 305 Japan.

Dr Isao Inoue
Institute for Enzyme Research, Tokushima University, Tokushima, 770 Japan.

Professor Richard D. Keynes, FRS
Physiological Laboratory, University of Cambridge, Downing Street, Cambridge CB2 3EG, UK.

Professor U. Kishimoto
Department of Biology, College of General Education, Osaka University, Toyonaka, 560 Japan.

Dr Y. Kito
Department of Biology, Faculty of Science, Osaka University, Machikaneyama 1-1, Toyonaka, Osaka, 560 Japan.

Dr Fumio Kukita
Department of Cell Physiology, National Institute for Physiological Sciences, Myodaiji, Okazaki, 444 Japan.

Dr N. J. Lane
Department of Zoology, University of Cambridge, Downing Street, Cambridge CB2 3EJ, UK.

Dr K. A. Langmack
Department of Medical Physics, Addenbrookes Hospital, Hills Road, Cambridge, UK.

Dr Philip L. Leopold
Department of Pathology, College of Physicians and Surgeons, Columbia University, 630 West 168th Street, New York, NY 10032.

Dr Jen-Wei Lin
Department of Physiology and Biophysics, New York University School of Medicine, New York, NY 10016, USA.

Professor Rodolfo Llinás
Department of Physiology and Biophysics, New York University Medical Center, 550 First Avenue, New York, NY 10016, USA.

Dr M. T. Lucero
Department of Physiology, 410 Chipeta Way, University of Utah, Salt Lake City, UT 84108, USA.

Professor Gen Matsumoto
Electrotechnical Laboratory, 1-1-4 Umezono, Tsukuba-shi, Ibaraki, 305 Japan.

Dr Rosa Maria Merzon
Centro de Neurociencias, Instituto Internacional de Estudios Avanzados (IDEA), Caracas, Venezuela.

Dr J. B. Messenger
Department of Animal and Plant Sciences, School of Biological Sciences, University of Sheffield, POB 601, Sheffield S10 3UQ, UK.

Dr M. Michinomae
Department of Biology, Faculty of Science, Konan University, Okamoto 8-9-1, Higashinada, Kobe, 658 Japan.

Dr J. A. Miyan
Department of Physiological Sciences, University of Manchester, Stopford Building, Oxford Road, Manchester M13 9PT, UK.

Dr K. Narita
Department of Biology, Faculty of Science, Osaka University, Machikaneyama 1-1, Toyonaka, Osaka, 560 Japan.

Professor C. Nicholson
Department of Physiology and Biophysics, New York University Medical Center, 550 First Avenue, New York, NY 10016, USA.

Dr D. C. Ogden
National Institute for Medical Research, The Ridgeway, Mill Hill, London NW7 1AA, UK.

T. Ohkawa
Department of Biology, College of General Education, Osaka University, Toyonaka, 560 Japan.

Professor A. Packard
Dipartimento di Zoologia, Universita di Napoli, Via Mezzocannone 8, 80134, Napoli, Italy.
and Stazione Zoologica, 80121 Napoli, Italy.

Dr M. Perri
Hopkins Marine Station, Stanford University, Department of Biological Sciences, Pacific Grove, CA 93950-3094, USA.

Dr Y. Pichon
Equipe de Neurobiologie, Neuropharmacologie Moléculaire et Ecotoxicologie, CNRS Université de Rennes I, Bâtiment 13, Complexe Universitaire de Beaulieu, 35042-Rennes Cedex, France.

Ms K. W. T. Potter
Department of Neurosurgery, University of Wisconsin, 4620 MSC, 1300 University Avenue, Madison, WI 53706, USA.

Dr Vincenzina Reale
AFRC Laboratory of Molecular Signalling, Department of Zoology, University of Cambridge, Downing Street, Cambridge CB2 3EJ, UK.

Dr Patricia A. Revest
Physiology Group, Biomedical Sciences Division, King's College, Strand, London WC2R 2LS, UK.

Mr J. Rosenthal
Hopkins Marine Station, Stanford University, Department of Biological Sciences, Pacific Grove, CA 93950-3094, USA.

Dr B. Rouot
INSERM u.65, Université Montpellier II, C.P. 100, Place Eugène Bataillon, 34095 Montpellier Cedex 05, France.

Dr H. R. Saibil
Department of Crystallography, Birkbeck College, University of London, Malet Street, London WC1E 7HX, UK.

Dr M. Seidou
The Aichi Prefectural University of Fine Arts 1-1 Sagamine, Yazako, Nagakute-cho, Aichi, Japan.

Dr Mutsuyuki Sugimori
Department of Physiology and Biophysics, New York University Medical Center, 550 First Avenue, New York, NY 10016, USA.

Dr Dieter Swandulla
Department of Membrane Biophysics, Max Planck Institute for Biophysical Chemistry, Am Fassberg, D-3400 Goettingen, Germany.

Dr I. Tsutsui
National Institute for Physiological Sciences, Myodaiji, Okazaki, 444 Japan.

Dr M. Usher
Department of Physiology, University College, University of London, Gower St, London WC1E 6BT, UK.

Dr C. Venien-Bryan
Department of Biochemistry, Oxford University, UK.

Dr Jorge Villegas
Centro de Neurociencias, Instituto Internacional de Estudios Avanzados (IDEA), Caracas, Venezuela.

Dr J. R. Wilkinson
Department of Crystallography, Birkbeck College, University of London, Malet Street, London WC1E 7HX, UK.

Dr R. Williamson
Marine Biological Association, The Laboratory, Citadel Hill, Plymouth PL1 2PB, UK.

Professor Shunichi Yamagishi
Department of Cell Physiology, National Institute for Physiological Sciences, Myodaiji, Okazaki, 444 Japan.

Professor J. Z. Young, FRS
Department of Experimental Psychology, University of Oxford, South Parks Road, Oxford OX1 3UD, UK.

Dr Karl Zipser
Department of Brain and Cognitive Sciences, Massachusetts Institute of Technology, Cambridge, MA 02139, USA.

Abbreviations

[^{14}C]DG	[^{14}C]deoxyglucose
α-GPDH	α-glycerophosphate dehydrogenase
γ-DGG	γ-D-glutamylglycine
4-AP	4-aminopyridine
5-HT	5-hydroxytryptamine, serotonin
ABRM	anterior byssus retractor muscle, mollusc
ACh	acetylcholine
ACL	anterior chromatophore lobe
AMP-PNP	adenylyl-5′-imidodiphosphate
AMPA	α-amino-3-hydroxyl-5-methyl-4-isoxazole propionic acid
APB (= AP-4)	DL-2-amino-4-phosphonobutyric acid
APV (= AP-5)	DL-2-amino-5-phosphonopentanoic acid
ARF	ADP-ribosylation factor
ASW	artificial sea water
BBB	blood–brain barrier
BSA	bovine serum albumin
BTX	batrachotoxin
cAMP	cyclic adenosine monophosphate
CANP	calcium-activated neutral protease
cGMP	cyclic guanosine monophosphate
cis,2,3-PDA	*cis*-2,3-piperidine dicarboxylic acid
CNQX	6-cyano-7-nitroquinoxaline-2,3-dione
CNS	central nervous system
CPP	3-((R)-2-carboxypiperazine-4-yl)-propyl-1-phosphonic acid
CTX	*cholera* toxin
DAB	diaminobenzidine
DNQX	6,7-dinitroquinoxaline-2,3-dione
DPAT (= 8-OH-DPAT)	(±)-8-hydroxy-dipropylaminotetralin HBr
DTT	dithiothreitol
ED_{50}	effective dose for 50% inhibition
EDRF	endothelium-derived relaxing factor
EEG	electroencephalogram
EM	electron microscope
EPSP	excitatory post-synaptic potential
F	Faraday
FH space	Frankenhaeuser–Hodgkin space

G-protein	GTP-binding protein
GAMS	γ-D-glutamylaminomethylsulphonic acid
GDEE	glutamate diethyl ester
GDP	guanosine diphosphate
GFL	giant fibre lobe
g_{ion}	ionic conductance, e.g. g_K, K^+ conductance
GppNHp	(non-hydrolysable GTP analogue)
GTP	guanosine triphosphate
GTPγS	(non-hydrolysable GTP analogue)
HH model/equations	Hodgkin–Huxley model/equations
HPLC	high performance liquid chromatography
HRP	horseradish peroxidase
I–V curve	current–voltage curve
IP_3	inositol 1,4,5-trisphosphate
ISM	ion-selective microelectrode
JSTX	Joro spider toxin
kDa	kilodaltons, unit of molecular weight
K_m	Michaelis constant
L-ODAP	β-*N*-oxalyl-L-α,β-diamino propionic acid
LDH	lactate dehydrogenase
MAPs	microtubule-associated proteins
NAAG	*N*-acetyl aspartyl glutamate
NAD	nicotine adenine dinucleotide
NADPH	reduced nicotinamide adenine dinucleotide phosphate
NFL	low molecular weight triplet component of mammalian neurofilaments
NFM	middle molecular weight component of mammalian neurofilaments
NMDA	*N*-methyl-D-aspartic acid
NO	nitric oxide
NTPs	nucleoside triphosphates
ODH	octopine dehydrogenase
PCL	posterior chromatophore lobe
PCMS	*p*-chloromercuriphenylsulfonic acid
PDE	phosphodiesterase
PHI	peptide histine isoleucine
PMA	paramedian area of skin
PTX	*pertussis* toxin
QED	quantum emission domain
R	gas constant
SDS-PAGE	sodium dodecyl sulfate polyacrylamide gel electrophoresis
SEM	standard error of the mean

SNFL	squid neurofilament low molecular weight polypeptide
SR	sarcoplasmic reticulum
STX	saxitoxin
T	absolute temperature
T system	transverse tubular system
TBA	tetrabutylammonium
TCA	trichloroacetic acid
TEA	tetraethylammonium
TMA	tetramethylammonium
TRSV	Texas red-labelled synaptic vesicles
TTX	tetrodotoxin
VIP	vasoactive intestinal peptide

PART I

Squid axon structure, biochemistry, and transport

1

The cytoskeleton of the squid giant axon

J. ADJAYE and P. A. M. EAGLES

Summary

The squid giant axon has given invaluable information about the axonal cytoskeleton. About 6 per cent of the axoplasmic protein is actin, half polymerized as actin filaments, under the axolemma and in the core axoplasm, where they play a role in organelle mobility. Actin-binding proteins like spectrin are present. Around 20 per cent axoplasmic protein is tubulin, and 75 per cent of this is polymeric, forming microtubules similar to those of flagellae. X-ray diffraction studies give information about microtubular substructure, indicating an 8 nm repeat equivalent to the size of the α,β tubulin dimer. Microtubule-associated proteins include the motor proteins kinesin and the cortical protein axolinin. The neurofilament is the most abundant cytoskeletal component in squid axoplasm, accounting for 13 per cent axoplasmic protein, and being 95 per cent polymerized. The polypeptide structure has been analysed, and the filaments shown to belong to the family of intermediate filaments. They are altered by two post-translational events: proteolysis and phosphorylation. Progress on cloning squid neurofilament genes is described.

For decades, the squid giant axon has provided an unparalleled model for studies on the cytoskeleton. A major feature of the preparations is that the cytoplasm of the axon can be easily removed from the surrounding membrane (Bear *et al.* 1937; Baker *et al.* 1962; Brown and Lasek 1990), and thus pure neuronal axoplasm can be studied, free of contamination from the axolemma and from other cell types. Hundreds of milligrams of axoplasm can be quickly extracted from a single axon providing enough material in a near native state for conventional biochemistry. It is not surprising therefore that studies on the squid giant axon have been pivotal in our understanding of the structure and function of neurofilaments and microtubules, major elements of this cytoskeleton. There is also little doubt that our knowledge about microtubules in organelle motility has developed rapidly because of the ease and speed with which the squid axonal cytoskeleton can be extracted and analysed (see Chapter 2).

Actin

About 6 per cent of the total axoplasmic protein is composed of actin, of which about half is in the polymer state (Morris and Lasek 1984). A cortical layer of actin filaments lies beneath the surrounding axolemma of the giant axon, although this accounts for no more than about 8 per cent of the actin (Fath and Lasek 1988).

Most of the actin filaments are still attached to the column of axoplasm after extraction and can be gently removed for analysis. Electron microscopy shows that most of these filaments are short in length, generally about 0.4 μm, although there is a population of filaments that are twice this length (Fath and Lasek 1988). Microfilaments have been observed to be closely associated with microtubule bundles along their long axis, although the reason for this is unclear (Fath and Lasek 1988). In addition to actin there are reports of actin-binding proteins being present in axoplasm. Myosin, for example, has been extracted from squid optic lobe (See and Metuzals 1976) and spectrin is also present (see review by Brown and Lasek 1990).

There has been much speculation regarding the function of the microfilament in axoplasm. It may provide some part of the mechanism for slow axonal transport although definitive evidence is lacking. A role for actin has recently been demonstrated in organelle motility within axoplasm and thus some of these short filaments are presumably engaged in this activity (Kuznetsov *et al.* 1992). The relationships between actin-based and microtubule-based motility are intriguing and should provide exciting prospects for further studies.

Microtubules

Squid axoplasm contains abundant microtubules, and this situation can be compared with that in *Myxicola*, the marine fanworm, where microtubules are relatively sparse. The profuse traffic of organelles seen in squid axoplasm, in contrast to that seen in *Myxicola*, no doubt accounts for some of this difference.

About 20 per cent of the axoplasmic protein is tubulin and nearly 75 per cent of this is polymeric. Many of the microtubules are stable polymers that do not show the rapid length changes characteristic of dynamic instability (Weiss *et al.* 1988).

Structural studies on axoplasmic microtubules are important because here one has a chance of studying native tubules essentially free from polymerization artefacts that result from purification of tubulin. Electron microscopy, using the negative staining technique, reveals the basic microtubule structure (Fig. 1.1). Such microtubules are often well preserved with little flattening of

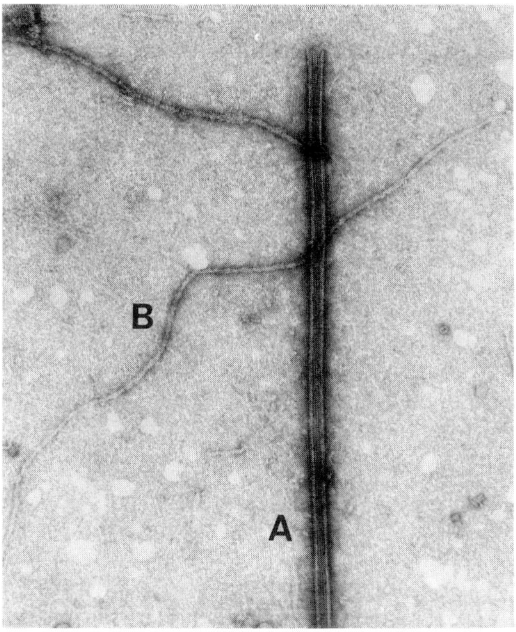

Fig. 1.1. Electron micrograph of a negatively stained sample of axoplasm. The microtubule (A) is well stained and shows protofilaments and tubulin subunits. The neurofilaments (B) appear to be less rigid and show a wavy profile. No subunit detail is seen in neurofilaments that have been negatively stained. This probably reflects the disordered peripheral domains of P200, which lie near to the surface of the filament.

the wall. Analysis of micrographs and of their diffraction patterns (Amos and Eagles 1987; Wais-Steider *et al.* 1988) shows that axoplasmic microtubules are similar in many structural respects to flagellar tubules, with the most likely number of protofilaments being 13.

The majority of microtubules in squid axoplasm run with their axes roughly parallel to the sides of the axon. They are therefore already well orientated and remain so even after the axoplasm has been extruded. This situation has considerable advantage because axoplasmic fibres can be easily pulled out that have tubules so well ordered that they produce interpretable X-ray diffraction patterns. The pictures (Fig. 1.2) show layer lines at a spacing corresponding to a 4-nm repeat arising from the axial repeat of the tubulin molecule. Weaker reflections are also present between these strong layer lines, indicating that the true repeat is 8 nm, which corresponds to the size of the α, β-tubulin dimer.

The reflections in the X-ray pattern vary in intensity depending on the hydration state of the axoplasm, and a detailed analysis of the changes has been performed (Wais-Steider *et al.* 1988). The simplest interpretation of the changes seen is that the tubulin subunit exhibits some flexibility in its

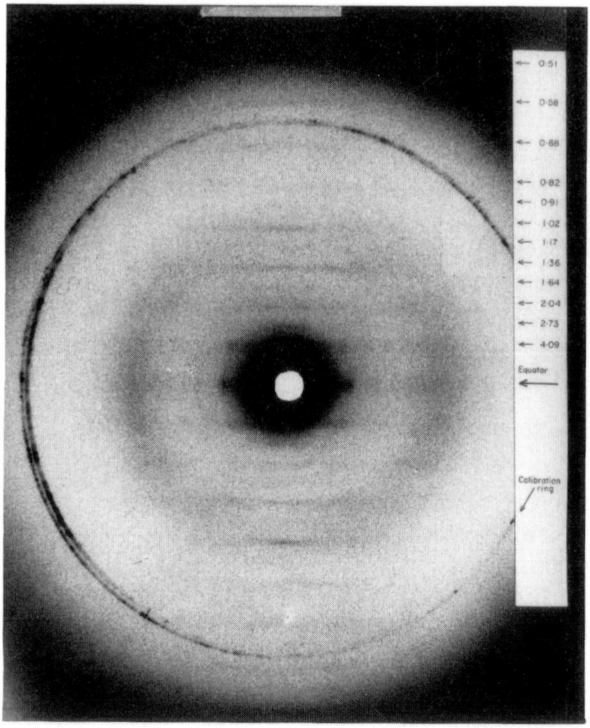

Fig. 1.2. X-ray diffraction pattern from a fibre pulled from squid axoplasm. The pattern seen here arises from microtubules. The layer-lines can be indexed on a basic repeat of 8.18 nm, corresponding to the repeat of the tubulin dimer. The equivalent spacing (in nm) of the layer lines is indicated by arrows above the equator. (Reproduced with permission from Wais-Steider *et al.* (1987).)

associations with neighbouring molecules, enabling subunits to tilt or rotate within the wall of the microtubule. This ability of the subunit to alter its orientation may have implications for the functioning of the microtubule although further work is needed to clarify this area.

Another issue that has remained unresolved and that is potentially solvable by X-ray diffraction techniques is the nature of the microtubule lattice in normal cytoplasmic microtubules. The A lattice allows for a closed tube of 13 protofilaments, where neighbouring tubulin dimers make consistent bonds. If the B lattice exists, however, the tube will have an open seam running along its length that is presumably filled by other proteins (Amos and Eagles 1987). Axoplasmic microtubules are ideal for answering this question and current work is aimed at elucidating it.

Microtubule-associated proteins (MAPs) can be classified according to whether they are motor proteins like kinesin or dynein or whether they act so as to stabilize the microtubule lattice as do the high-molecular-weight MAPs and tau molecules.

Kinesin was first isolated from squid axoplasm (Vale *et al.* 1985*a*,*b*) and much of the early work concerning the characterization of the motor proteins and the role of microtubules benefited from this source of material (Allen *et al.* 1982). A number of other MAPs have also been identified in squid axoplasm (Sakai and Matsumoto 1978; Gilbert and Sloboda 1986; Pratt 1986), although their precise function is still unclear. One of these proteins, axolinin, is present in large amounts in the cortical layer and may modify microtubules in this region for particular roles associated with the membrane (Sakai *et al.* 1985).

Neurofilaments

The most abundant cytoskeletal component in squid axoplasm is the neurofilament. Neurofilament polypeptides constitute about 13 per cent of the total axoplasmic protein, and most of these molecules (95 per cent) are polymerized into stable structures that do not readily exchange subunits with the environment.

Electron microscopy of negatively stained preparations (Fig. 1.1) shows that the filaments have diameters of around 9 nm with lengths of many microns (Wais-Steider *et al.* 1983). They course in a helical path within the squid axon (Metuzals 1969; Metuzals *et al.* 1981). Paracrystals can be formed from squid neurofilament polypeptides, and these paracrystals have helped in the structural analysis of the proteins (Metuzals *et al.* 1988).

The polypeptide composition of squid neurofilaments is shown in Fig. 1.3. The major polypeptides have apparent molecular masses of 200 kDa (P200) and 60 kDa (P60), although there is also present, a protein of mass approximately 400 kDa (reviewed in Eagles *et al.* 1990). Studies show additional axoplasmic proteins around 65 kDa, and higher bands below P200 (Cohen *et al.* 1987). Some of the bands may be non-phosphorylated precursors of the mature P200, because it is well known that posttranslational phosphorylation of neurofilament polypeptides greatly reduces their mobility on sodium dodecyl sulfate polyacrylamide gel electrophoresis (SDS-PAGE). All of these polypeptides react with a monoclonal antibody, anti-IFA, on Western blotting, indicating that they contain the consensus sequence YRKLLEGEE that is a diagnostic feature of intermediate filaments.

X-ray diffraction analysis of purified squid neurofilaments shows that they contain coiled coils of α-helices (Wais-Steider *et al.* 1983). It is believed that the coiled coils are formed by both of the major polypeptides, P200 and P60. The genes for these proteins have recently been sequenced (Szaro *et al.* 1991; Way *et al.* 1992) and the deduced amino acid sequence shows that the proteins have α-helical rod regions, characteristic of intermediate filament polypeptides, that bear heptad repeats.

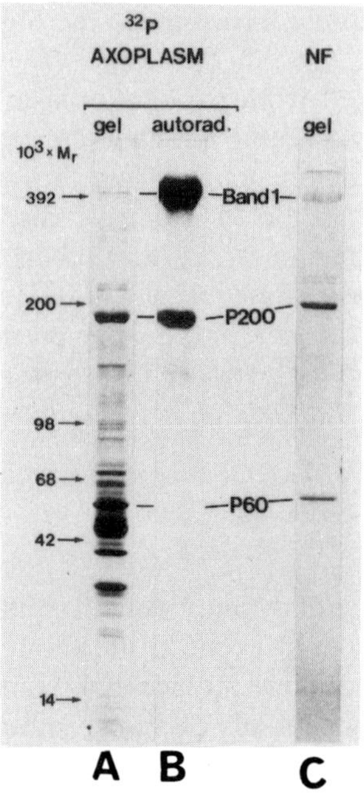

Fig. 1.3. Analysis by SDS-PAGE of (C) a purified preparation of squid neurofilaments and (A) a sample of whole squid axoplasm that had been incubated with [^{32}P]ATP. (B) The corresponding autoradiograph of the sample incubated with radioactivity is shown in the centre. Purified squid neurofilaments consist of two major polypeptides, P60 and P200; a minor high-molecular-weight band, band 1, which is probably a cross-linked dimer of P200; and a number of faintly staining associated polypeptides. If axoplasm is incubated with [^{32}P]ATP, only P200 and band 1 become labelled. The label incorporated into P200 is located on the peripheral domain of the polypeptide. (Reproduced with permission from Brown and Eagles (1986).)

The arrangement of the polypeptides within the neurofilament is thought to be like that depicted in Fig. 1.4, which is an amalgam of many types of information. P60 is thought to reside primarily in the core of the filament, which is similar to the disposition of the protein NFL, the low-molecular-weight triplet component of mammalian neurofilaments. P200 has both a core domain and a more peripheral domain, again similar to the arrangement of NFM (the middle component) and NFL in the mammalian counterpart (Chin *et al.* 1983).

Squid neurofilaments are altered by two major posttranslational events: proteolysis and phosphorylation (reviewed in Eagles *et al.* 1990). Axoplasm contains a calcium-activated neutral protease (CANP) that, on activation, readily cleaves P200, roughly in the middle of the molecule (Fig. 1.4). The enzyme seems to be specific for this protein. Cleavage, at least in the early

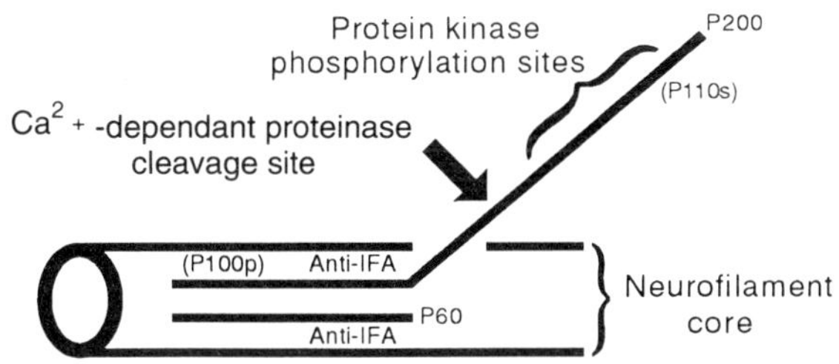

Fig. 1.4. Model for the squid neurofilament. The proposed sites of phosphorylation, calcium-dependent proteolytic cleavage, and anti-IFA binding are indicated. The enzyme calcium-activated neutral protease (CANP) splits P200 into a fragment that binds anti-IFA (P100p), which resides in the core of the filament, and a fragment corresponding to the peripheral domain of the polypeptide (P110s), which contains the phosphorylation sites and the KSP repeats. After cleavage, P100s is soluble and can be easily removed from the filamentous fraction by sedimentation. (Reproduced with permission from Brown and Eagles (1986).)

stages of digestion, results in a neurofilament, recognizable as such under the electron microscope, although lacking the peripheral domain associated with P200. The filaments stay assembled, presumably because the interactions forming the coiled coils, which stabilize the backbone, are not affected by proteolysis.

The phosphorylation of these molecules is carried out by an axonal kinase, specific for neurofilament polypeptides. P200 is again the major participant in the process, although band 1 is also phosphorylated (Fig. 1.3). If axoplasm is incubated with radioactive [^{32}P]ATP, P200 and band 1 are the main products of labelling. The peripheral domain of P200 is where the labelling occurs (Brown and Eagles 1986) and this region contains many putative phosphorylation sites based on the sequence KSP. The arrangement of phosphorylated regions can be well seen using electron microscopy (Martin *et al.* 1990) where a punctate distribution can be demonstrated. This arrangement probably results from a clustering of the peripheral domains of P200 molecules which repeat every 20–40 nm.

The reasons for the two enzymatic events are still unclear. Proteolysis has been implicated in neurofilament removal, although if this were its primary task it would be more sensible for the enzyme to cut at the backbone of the filament rather than in the peripheral regions in order to destroy filament integrity quickly. The process of phosphorylation at the periphery of the filament might indicate some change of conformational state that is needed in this vicinity. Phosphorylation is thought to be developmentally regulated, the polypeptides becoming modified as they move through the initial segment of the axon, the axon hillock (Eagles *et al.* 1990). If phosphorylation were to alter the conformation of the peripheral domain of P200, then the elaboration of the wispy side-arm structures might result, leading to mutual repulsion of neighbouring filaments and an increase in axonal girth.

Cloning of squid neurofilament genes

Recently, we have been involved with the cloning of neurofilament genes from the squid *Loligo forbesi*. mRNA was isolated from squid optic lobes and a cDNA library was constructed from it in the expression vector λgt11. The library was screened with antibodies generated against purified squid neurofilaments that had been previously dephosphorylated. A number of positive clones were isolated; the λgt11 recombinants were purified and their inserts released. The largest insert had a size of 3.5 kb, and was selected for sequencing. The corresponding gene was subsequently shown to produce a single mRNA species (2.6 kb), encoding a protein of mass 60 kDa. We have called this protein 'squid neurofilament lower-molecular-weight polypeptide' (SNFL) (Adjaye *et al.* 1993).

The use of λgt11 allows the production of a lac Z-directed fusion protein incorporating part of the β-galactosidase molecule. The fusion protein, β-gal-SNFL, has a mass of 180 kDa and is recognized by the antibody anti-IFA, the epitope of which is a diagnostic feature of intermediate filaments. The fusion protein is therefore immunologically related to intermediate filaments.

The fusion protein was isolated in bulk and used for the production of a polyclonal serum. The serum, after absorption to remove β-galactosidase antibodies, identified a 60-kDa neurofilament protein on Western blots of crude samples of squid optic lobe and axoplasm, but no reaction was observed with skin, muscle, liver, or heart. Immunocytochemistry with the polyclonal serum showed that it stained filamentous structures inside squid neurones, indicating that it binds to neurofilaments. The serum cross-reacted with neurofilament polypeptides from the marine worm *Myxicola*, fish, and *Xenopus* but not mammals (Adjaye *et al.* 1993). These studies suggest that SNFL is a squid neurofilament molecule that has immunological similarities with other neurofilaments from invertebrates and lower vertebrates.

The predicted protein sequence of the gene (Fig. 1.5) showed that there was an open reading frame covering 557 amino acids, and that the secondary structure of the protein shared features with intermediate filament polypeptides. There are distinct N-terminal, C-terminal, and rod regions. The rod region has classical heptad repeats, indicating coiled coil forming ability, and the predicted length of the coils are similar in length to coils 1a, 1b, 2a, and 2b of intermediate filaments. Furthermore, the rod region has part of the consensus sequence RLNN at its N-terminal end (LNFN), and, at its C-terminal end, it has SRRLLEGSG, similar to the IFA epitope now identified with YRKLLEGEE. In all, the SNFL molecule has all the hallmarks of a classical intermediate filament polypeptide.

While our work was in progress, we learned about the sequence of another neurofilament gene from squid, although from a different species (*Loligo*

```
        Head
1       MEDSSSEVTVNREVTTDNPYTISVTNKTLSAIRIKMFMPRAYELKVTVIK
                                 > Rod          Coil 1A
51      MAFELSTRCSKRDGVVETVLTDVIEGKTMSGYDRSGRVNLLNFNNQVIFR
                                                   L1
101     VVRKTLDSNDSNVVDAIQVRSYAEVIDAKFRYPLTGLLFVELDSKMFPNQ
          Coil 1B
151     LLTISIRKRWKIVNVRSNYDHESRTYNGNWDGTFKKAWTNNRAWVLYDLM

201     INQRYGLDQKELGISVDKWALYEAAQYCDQMVLTAKAARNLATFVTCKLV
                  L1,2              Coil 2
251     IIQSQTDAYKVVRDICSIFRGMSFWNGESISVIIDRTRELAYIFTNDNVV

301     NGDFSYTSQARRACTRRATGCLMMSKTCINRTLSQYSIVEGHSTVRNNVT

351     SITAIGAHVEAKANRRGRWILKTNLRSTTVNFATGLEGMIRTIGDVVAIS
                                     > tail
401     DNFWSSNLTMNLSRRLLEGSGSQIFLPLVDARAGDFIIVNNADGKPVKRT

451     ISSVSADGKTIELTSGFGFPVKPNTVFAIDRTDLRLQQYVVTKIDKADDD

501     EEFTSKSRRWSTILISTMKLITGLTSTTDRQASLDQIRCPKPENVKCLRA

551     RESSGRA*
```

Fig. 1.5. Predicted protein sequence of SNFL. The one-letter amino acid code is used and the proposed rod region together with heptad repeats are marked. The anti-IFA binding site is also indicated.

pealei) (Szaro *et al.* 1991). The gene that we have isolated is completely different to that reported for *L. pealei* in terms of its sequence and the predicted amino acid sequence of the protein. So we conclude that in squid neurones there are two distinct genes encoding two types of low-molecular-weight neurofilament polypeptide.

A novel feature of our protein is that, in the rod region, coil 1b is two heptads longer than that found in mammalian neurofilaments but shorter than that seen for the invertebrate intermediate filaments. Thus it has a peculiar and somewhat atypical structure.

Acknowledgements

Much of the work discussed here was supported by grants from the MRC, the Wellcome Trust, and the SERC. We also are most thankful to the Director and staff of the Marine Biological Laboratory, Plymouth, for the use of their facilities.

References

Adjaye, J., Marsh, P., and Eagles, P. A. M. (1993). Immunological properties and cDNA sequence analysis of an intermediate-filament-protein from squid neuronal tissue. *Journal of Cell Science*, **106**, 1283–90.

Allen, R. D., Metuzals J., Tasaki, I., Brady, S. T., and Gilbert, S. P. (1982). Fast axonal transport in squid giant axon. *Science*, **218**, 1127–8.

Amos, L. and Eagles, P. A. M. (1987). Microtubules. In *Fibrous protein structure* (ed. J. M. Squire and P. J. Vibert), pp. 215–16. Academic Press, London.

Baker, P. F. Hodgkin, A. L., and Shaw, T. I. (1962). Replacement of the axoplasm of giant nerve fibres with artificial solutions. *Journal of Physiology*, **180**, 424–38.

Bear, R. S., Schmitt, F. O., and Young, J. Z. (1937). The ultrastructure of nerve axoplasm. *Proceedings of the Royal Society of London, Series B*, **123**, 505–19.

Brown, A. and Eagles, P. A. M. (1986). Squid neurofilaments: phosphorylation and calcium dependent proteolysis in situ. *Biochemical Journal*, **239**, 191–7.

Brown, A. and Lasek, R. J. (1990). The cytoskeleton of the squid giant axon. In *Squid as experimental animals* (ed. D. L. Gilbert and W. J. Adelman Jr), pp. 235–302. Plenum Publishing Co, New York.

Chin, T. K., Eagles, P. A. M., and Maggs, A. (1983). The proteolytic digestion of ox neurofilaments with trypsin and alpha-chymotrypsin. *Biochemical Journal*, **215**, 239–52.

Cohen, R. S., Pant, H. C., House, S., and Gainer, H. (1987). Biochemical and immunocytochemical characterization and distribution of phosphorylated and non phosphorylated subunits of neurofilaments in squid giant axon and stellate ganglion. *Journal of Neuroscience*, **7**, 2056–74.

Eagles, P. A. M., Pant, H. C., and Gainer, H. (1990). Neurofilaments. In *Cellular and molecular biology of intermediate filaments* (ed R. D. Goldman and P. M. Steinert), pp. 37–94. Plenum Publishing Co, New York.

Fath, K. R. and Lasek, R. J. (1988). Two classes of actin microfilament are associated with the inner cytoskeleton of axons. *Journal of Cell Biology*, **107**, 613–21.

Gilbert, S. P. and Sloboda, R. D. (1986). Identification of a MAP-2-like ATP-binding protein associated with axoplasmic vesicles that translocate on isolated microtubules. *Journal of Cell Biology*, **103**, 947–56.

Kuznetsov, S. A., Langford, G. M., and Weiss, D. G. (1992). Actin-dependent organelle movement in squid axoplasm. *Nature*, **356**, 722–5.

Martin, R. M., Schilling, K., Fritz, W., and Giuditta, A. (1990). Visualization of differential neurofilament phosphorylation in the pre- and postsynaptic axoplasm of the squid giant synapse: an electron spectroscopic study. *Neuroscience*, **37**, 553–62.

Metuzals, J. (1969). Configuration of a filamentous network in the axoplasm of the squid (*Loligo pealei*) giant nerve fibre. *Journal of Cell Biology*, **43**, 480–505.

Metuzals J., Montpetit, V., and Chapin, D. F. (1981). Organization of the neurofilamentous network. *Cell and Tissue Research*, **214**, 455–82.

Metuzals, J., Pant, H., Gainer, H., Eagles, P. A. M., White, N. S., and Houghton, S. (1988). In vitro polymorphism and phase transitions of the neurofilamentous network isolated from the giant axon of the squid (*Loligo pealei*). *Cell and Tissue Research*, **252**, 249–62.

Morris, J. R. and Lasek, R. J. (1984). Monomer–polymer equilibria in the axon: direct measurement of tubulin and actin as polymer and monomer in axoplasm. *Journal of Cell Biology*, **98**, 2064–76.

Pratt, M. M. (1986). Stable complexes of axoplasmic vesicles and microtubles: protein composition and ATPase activity. *Journal of Cell Biology*, **103**, 957–68.

Sakai, H. and Matsumoto, G. (1978). Tubulin and other proteins from squid giant axon. *Journal of Biochemistry*, **83**, 1413–22.

Sakai, H., Matsumoto, G., and Murofushi, H. (1985). Role of microtubules and axolinin in membrane excitation of the squid giant axon. *Advances in Biophysics*, **19**, 43–89.

See, Y. P. and Metuzals, J. (1976). Purification and characterization of squid brain myosin. *Journal of Biological Chemistry*, **251**, 7682–9.

Szaro, B. G., Pant, H. C., Way, J., and Battey, J. (1991). cDNA cloning and characterization of two low molecular weight neurofilament proteins of the squid *Loligo pealei*. *Journal of Biological Chemistry*, **266**, 15035–41.

Vale, R. D., Reese, T. S., and Sheetz, M. P. (1985*a*). Identification of a novel force-generating protein, kinesin, involved in microtubule-based motility. *Cell*, **42**, 39–50.

Vale, R. D., Schnapp, B. J., Reese, T. S., and Sheetz, M. P. (1985*b*). Organelle, bead and microtubule translocations promoted by soluble factors from the squid giant axon. *Cell*, **40**, 559–69.

Wais-Steider, C., Eagles, P. A. M., Gilbert, D. S., and Hopkins, J. (1983). Structure similarities and differences amongst neurofilaments. *Journal of Molecular Biology*, **165**, 393–400.

Wais-Steider, C., White, N. S., Gilbert, D. S., and Eagles, P. A. M. (1987). X-ray diffraction patterns from microtubules and neurofilaments in axoplasm. *Journal of Molecular Biology*, **197**, 205–18.

Way, J., Hellmich, M. R., Jaffe, H., Szaro, B., Pant, H. C., Gainer, H., and Battey, J. (1992). A high-molecular-weight squid neurofilament protein contains a lamin-like rod domain and a tail domain with Lys-Ser-Pro repeats. *Proceeding of the National Academy of Sciences, USA*, **89**, 6963–7.

Weiss, D. G., Langford, G., Seitz-Tutler, D., and Keller, F. (1988). Dynamic instability and motile events of native microtubules from squid axoplasm. *Cell Motility and the Cytoskeleton*, **10**, 1–11.

2

The nervous system of *Loligo pealei* provides multiple models for analysis of organelle motility

PHILIP L. LEOPOLD, JEN-WEI LIN, MUTSUYUKI SUGIMORI, RODOLFO LLINÁS, and SCOTT T. BRADY

Summary

The squid, *Loligo*, has come to be synonymous with the study of electrophysiology. However, *L. pealei* has also made significant contributions in a number of other fields of neurobiology including intracellular transport and targeting. This chapter discusses the use of the squid giant axon and giant synapse in the study of organelle motility. Special attention is paid to the construction of the model systems, and each preparation is illustrated with an example of the model's utility in research on translocating organelles. The giant axon and the giant synapse have taught us principles regarding the enzymology, pharmacology and biochemistry of organelle transport and are likely to increase our understanding of motility mechanisms in the future.

An excellent correlation exists between the seasonal migration of scientists to the Marine Biological Laboratory at Woods Hole, Massachusetts, and the availability of the squid, *Loligo pealei*. This phenomenon can be attributed in large part to the work of Cole and Curtis at Woods Hole and Hodgkin and Huxley at Plymouth, who introduced the use of the squid giant axon for electrophysiological studies (Cole and Curtis 1939; Hodgkin and Huxley 1952; reviewed by Adelman and Gilbert 1990). The popularity of the squid was derived mainly from unique features of the squid nervous system (Fig. 2.1). The absence of myelin in cephalopods led to the development of unusually large axons in the highly mobile squid. In turn, these giant axons required synaptic contact of similar magnitude to assure proper coupling of pre- and postsynaptic action potentials at a synapse.

The large diameter of the squid axon (0.3–0.8 mm in *L. pealei*) permitted early electrophysiologists to place electrodes inside the axon to record changes in the electrical potential across the axolemma. The possibility of making intracellular recordings in the squid giant axon was critical to the study of membrane currents and potentials, as well as other key features of axonal electrophysiology (reviewed by Adelman and Gilbert 1990). The first

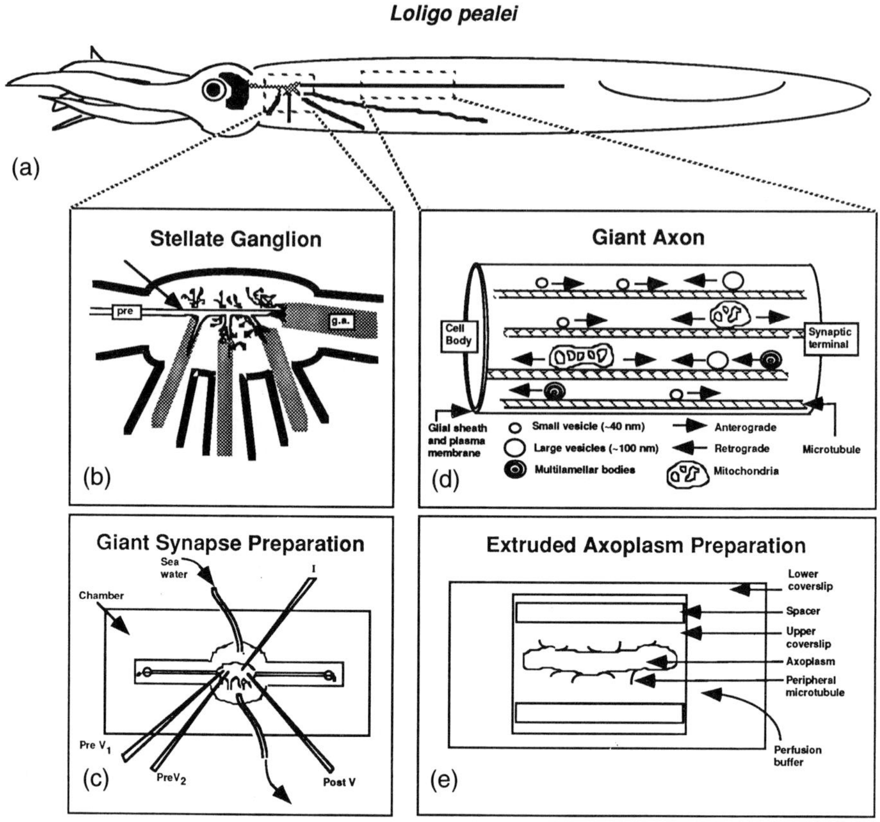

formal mathematical descriptions of the action potential derive directly from studies on the squid giant axon.

The squid nervous system has continued to provide insights into neuronal cell biology, in part through development of novel preparations for dissection of cellular mechanisms. Two of these preparations have proven to be invaluable for the study of motility in the axon and presynaptic terminal. Isolated axoplasm from the squid giant axon may be experimentally manipulated by perfusion with suitable buffers, and materials can be microinjected into the presynaptic terminal of the giant synapse during intracellular recording of electrical properties. Both experimental models are discussed in this chapter with reference to their utility in studying the mechanism of organelle transport.

Replacement of axoplasm in the giant axon was first performed by Oikawa *et al.* (1961) and Baker *et al.* (1961) in order to manipulate intracellular ionic constituents while conducting electrophysiological experiments (reviewed by Adelman and Gilbert 1990). Initially, interest focused on the axonal membrane rather than on properties of the axoplasm. A few investigators recognized

Fig. 2.1. Selected anatomy of the squid nervous system. (**a**) Outline of the optic lobes (black), secondary nerve fibres (grey), and tertiary nerve fibres (black) of *Loligo pealei*. Secondary fibres synapse on to axons of tertiary fibres. Tertiary fibres control muscle contraction in the mantle. This nerve pathway allows the squid to respond rapidly to visual stimuli. (**b**) Detail of the stellate ganglion showing synapse of secondary fibres on axons of tertiary fibres (redrawn from Young 1936; not to scale). g.a., Giant axon. The arrow marks the site of microinjection of fluorescently labelled synaptic vesicles. (**c**) Giant synapse preparation. A combination of presynaptic (PreV1, PreV2), extracellular (I), and postsynaptic electrodes (Post) allows the observer to simultaneously record from presynaptic and postsynaptic cells, inject current, and microinject effectors such as channel blockers (e.g. tetraethyl ammonium), proteins (e.g. synapsin), dyes (e.g. fura 2), or organelles (e.g. synaptic vesicles). Inclusion of other channel blockers and ionic constituents in the perfusion buffer allows isolation of currents due to single classes of ion channels. A continuously flowing bath of sea water maintains the supply of oxygen to the synapse (redrawn from Llinás *et al.* 1981*a*). (**d**) Detail of giant axon (not to scale). This schematic representation of the giant axon shows major classes of organelles which move along microtubules in fast axonal transport. Small vesicles (40 nm diameter) make up the majority of organelles moving toward the synaptic terminal (anterograde) while larger vesicles and multi-lamellar bodies move toward the cell body (retrograde). Healthy mitochondria can change direction, but degenerating mitochondria are found moving in the retrograde direction (summarized from Tsukita and Ishikawa 1980; Fahim *et al.* 1985; Miller and Lasek 1985; Allen *et al.* 1982; and Martz *et al.* 1984). (**e**) The cytoplasm of the giant axon can be squeezed out of the plasma membrane and surrounding connective tissue using a technique that preserves the structure of the axoplasm (Morris and Lasek 1982). The lack of plasma membrane and connective tissue gives the extruded axoplasm several advantages over an intact axon for the study of organelle transport, including decreased light scattering and biochemical accessibility for perfusion of effector molecules. Microscopy and perfusion of the axoplasm are facilitated through the use of a perfusion chamber (redrawn from Brady *et al.* 1985).

---

that the axoplasm was also of interest (Bear *et al.* 1937; Lasek 1984; Brown and Lasek 1990). The large diameter of the axon provided biologists with a unique model that offered the ability to obtain and study a large volume of cytoplasm free of the optical and biochemical barrier presented by the axolemma, glial sheath, and connective tissue. More refined methods were developed for removing the intact axoplasm (extrusion), which preserved the internal structure of the axoplasm (Morris and Lasek 1982, 1984; Lasek 1984). The first studies focused on the structure and composition of the axoplasm, but eventually questions arose about the viability of the axoplasm after extrusion.

The fortuitous development of video-enhanced contrast, differential interference contrast microscopy by Robert Day Allen provided a method for detecting individual microtubules and synaptic vesicle-sized organelles in living preparations (Allen *et al.* 1981). Application of video microscopic techniques first on the intact giant axon (Allen *et al.* 1982) and then on the isolated axoplasm (Brady *et al.* 1982) revealed that an extraordinary amount of membrane organelle transport continued in the axoplasm and offered the first opportunity to view movement of small vesicles along individual microtubules in the axoplasm (Fig. 2.1). Subsequent applications of video microscopy to the field of organelle transport have revolutionized our understanding of the mechanisms for organelle transport in both neurones and non-neuronal cells.

The absence of the normal permeability barriers (axolemma and glial sheath) in the isolated axoplasm has facilitated the characterization of membrane organelle traffic. The axoplasm may be perfused with defined buffers that alter the biochemical microenvironment of moving organelles. One of the most striking results obtained in such perfusion studies came from the use of ATP analogues to characterize the ATPases involved in transport. Perfusion with the ATP analogue, adenylyl-5′-imidodiphosphate (AMP-PNP), caused organelle translocations to stop and left organelles attached to microtubules (Lasek and Brady 1984). The effect of AMP-PNP indicated that a new class of molecular motor was involved in organelle transport (Lasek and Brady 1985). Reasoning that the molecular motor for fast axonal transport of organelles must be attached to microtubules along with the organelles in the presence of AMP-PNP, several laboratories identified a new class of mechanochemical ATPase, the kinesins (Brady 1985; Vale *et al.* 1985), with properties suitable for a molecular motor for organelle transport. This represents only one example of how observations made in isolated axoplasm have yielded insights into common cellular processes in neurones and other cell types.

As with the giant axon, interest in the giant synapse of the squid was initially driven by an interest in electrophysiological properties of the neurone (Bullock 1948). The giant synapse is found in the stellate ganglion, which also gives rise to the giant axons of the squid (Fig. 2.1). Afferent axons from the squid brain make contact with the several efferent giant fibres in order to provide co-ordinated firing during contraction of the mantle. A single afferent secondary fibre branches into a series of digits, like fingers on a hand, after entry into the stellate ganglion. Each digit makes contact with a giant axon and forms a functional synaptic specialization. For electrophysiological studies, the ‘giant synapse’ is generally taken to mean the synapse between the most distal digit of the secondary (presynaptic) fibre as it contacts the largest and most medial efferent fibre, the ‘giant axon’ (Martin and Miledi 1986).

The giant synapse has proven to be an excellent model for the study of neurotransmitter release due to the large diameter of the pre- and postsynaptic cells (Young 1939). The large diameter of the preterminal axon and terminal proper allows intracellular recording to be conducted simultaneously from both pre- and postsynaptic cells (Llinás *et al.* 1981*a,b*). Equally important, either pre- or postsynaptic cells may be microinjected with suitable agents during intracellular recording (Llinás *et al.* 1985). This combination allows investigators to monitor the structural and electrical integrity of the synaptic terminal while introducing pharmacological or biochemically active reagents into the terminal. Through the use of suitable agents, the molecular machinery for neurotransmitter release, endogenous organelle translocation and targeting, or membrane properties can be characterized (Llinás *et al.* 1985, 1989; Lin *et al.* 1990). Polypeptides, enzymes, and even intact membrane-bounded organelles may be introduced into the presynaptic terminal by this method and then analysed by electrophysiology and by video microscopy.

For the purpose of studying organelle transport, the extruded axoplasm and the giant synapse offer complementary information. The extruded axoplasm offers the opportunity to manipulate the biochemical environment of organelle transport and observe the effect of that manipulation on a population of endogenous organelles. In contrast, the giant synapse offers the opportunity to introduce biochemically modified exogenous organelles *in situ* in order to observe how those modifications influence the interaction of organelles with endogenous translocation and targeting machinery. The experiments described in this chapter will illustrate these two models for the use of the squid nervous system in the study of organelle motility.

Extruded squid axoplasm: pharmacology of endogenous organelle transport

The extruded axoplasm model

Axoplasm may be isolated from the squid giant axon simply and efficiently through an extrusion process that maintains the structural integrity of the axoplasm (Bear *et al.* 1937; Morris and Lasek 1982). Extrusion is a mechanical method of removing the permeability barriers that normally surround the axoplasm. In this way, the axoplasm is spared the shock of treatments with detergents or other chaotropic agents that would normally be required to permeabilize the axon. The ultrastructural and biochemical organization of the axoplasm is well preserved following extrusion (Morris and Lasek 1982). Most importantly, membrane-bounded organelles will continue to move toward the synaptic terminal (anterograde) or toward the cell body (retrograde) for hours after extrusion allowing a comparison of motility in the two different directions (Brady *et al.* 1982, 1985).

Following extrusion, the axoplasm becomes accessible to diffusible molecules that would not normally cross the plasma membrane. Components of the perfusion buffer, ranging in size from ions to antibodies, rapidly equilibrate with the interior of the axoplasm. Several key elements of the perfusion protocol ensure that viability of the axoplasm is maintained for several hours. First, the composition of perfusion buffer should approximate the ionic profile and osmolarity of squid axoplasm (Brady *et al.* 1985). Our standard buffer, referred to as buffer X (Brady *et al.* 1982), was derived from descriptions of the small-molecular-weight components of squid axoplasm (Deffner 1960) as well as from 'intracellular' buffers described by Morris and Lasek (1982), Rubinson and Baker (1979), and Adams (1982). The buffer contains low-molecular-weight constituents including aspartate, glycine, betaine, taurine, potassium, and magnesium with a minimal chloride concentration. HEPES is included to enhance the pH buffer capacity, while a Ca^{2+} EGTA buffer system keeps Ca^{2+} levels at approximately 0.1 μM. This buffer is either used at full strength (*ca.* 1.2 osmolarity) or at half strength (0.6 osmolarity). The half-strength buffer X ('buffer X/2') facilitates the dispersion of microtubules and organelles from the body of the axoplasm. Single vesicles can be followed moving along single microtubules at the periphery of the axon in either buffer, although individual microtubules are more readily located when buffer X/2 is employed (Brady *et al.* 1982).

In all cases, the axoplasm is perfused with a volume of buffer equivalent to four volumes of axoplasm (typically about 20 μl). This volume limits the extraction of essential elements of the translocation apparatus following dilution and permits precise determination of final concentrations for all constituents of axoplasm and perfusion buffer. As a result, the isolated axoplasm model can be utilized for analytical studies of organelle transport including comparisons with enzyme kinetic studies. When 1 mM ATP and 1 mM glucose are included in the perfusion buffer, organelle translocation continues for hours at room temperature with minimal decrement in motility.

Characterization of organelle transport

By modifying the composition of the perfusion buffer, the biochemical environment of the axoplasm may be adjusted in order to analyse the influence of enzymatic substrates or modulators on organelle motility. Such experiments with isolated axoplasm can address many questions regarding the mechanisms of organelle motility. As described above, observations in the axoplasm led directly to the identification of a new class of molecular motors, the kinesins, which are involved in the movement of membrane organelles in fast axonal transport. Other studies have demonstrated metabolic and structural requirements for efficient organelle motility. For example, experiments performed by Brady *et al.* (1982, 1985) demonstrated that organelle transport in isolated squid axoplasm required ATP. When endogenous ATP production was eliminated through addition of dinitrophenol (an uncoupler of ATP production by mitochondria) to perfusion buffer, the velocity of organelle motility rapidly declined. However, if the perfusion buffer containing dinitrophenol was supplemented with exogenous ATP, organelle movements continued unabated. Regulation of organelle transport by interaction with the cytoskeleton was demonstrated by McGuiness *et al.* (1989) who showed that perfusion of the extruded axoplasm with dephosphorylated synapsin I inhibited axonal transport of neuronal organelles (McGuiness *et al.* 1989). This observation lent support to a proposal that phosphorylation and dephosphorylation of synapsin I regulates the movement of synaptic vesicles to active zones within the presynaptic terminal (Llinás *et al.* 1985, 1991).

Isolated axoplasm also provides a means of investigating the enzymology and kinetics of fast axonal transport at the level of individual organelles. All of these studies rely on the introduction of specific probes into the axoplasm at precisely defined concentrations. Equally importantly, both anterograde and retrograde organelle motility can be evaluated simultaneously in isolated axoplasm, permitting quantitative comparison of the two. In the following illustration of the isolated axoplasm model, the biochemical environment of organelle transport was tailored to address questions about the nucleoside triphosphate requirements for organelle translocation in both anterograde and retrograde directions.

Application of the model

Application of the extruded axoplasm model is best illustrated by describing strategies that employ the isolated axoplasm to answer specific questions about organelle transport. The design and implications of selected experiments presented here exemplify the utility of this model. In the process, practical considerations emerge that may have general applicability.

Neurones move axonal organelles toward the synaptic terminal (anterograde) as well as toward the cell body (retrograde). Since the populations of organelles moving in these two directions are morphologically (Smith 1980; Tsukita and Ishikawa 1980; Fahim *et al.* 1985; Miller and Lasek 1985) and biochemically distinct (Bisby 1982; Martz *et al.* 1989), mechanisms must exist in the neurone for determining the direction of movement for a particular organelle. One model requires separate force-generating enzymes for each direction of transport (Vale *et al.* 1985). Other models, which may not be mutually exclusive, involve modification of the transport organelle that influences the direction of the motor or a modification of the motor itself. Each of these models makes specific predictions. For example, if two different motors drive anterograde and retrograde motility, one would expect the motors to differ in their pharmacology and biochemistry. To determine whether anterograde and retrograde organelle translocations have different characteristics, the perfusion buffer was modified by the addition of alternative nucleoside triphosphates, and both directions of transport were analysed to identify any differential effects. These experiments define one enzymatic property of anterograde and retrograde transport, namely, the nucleotide specificity associated with the ATPase activity of the organelle motor.

The ability of different nucleoside triphosphates (NTPs) to support bidirectional organelle transport could not be determined until two experimental problems were solved. Endogenous ATP had to be removed without altering levels of the substituted NTP, and synthesis of additional ATP by oxidative phosphorylation or transphosphorylation in the axoplasm had to be blocked. These problems were solved using the following methods. Hexokinase was added to the perfusion buffer to hydrolyse endogenous ATP. Hexokinase was the enzyme of choice because of a high specificity for ATP over other NTPs (Darrow and Colowick 1962), as opposed to enzymes such as apyrase which are more promiscuous. Since hexokinase hydrolyses ATP by transferring a phosphate group to a hexose sugar, a suitable phosphate acceptor was included. 2-Deoxyglucose (5 mM) was used as the phosphate acceptor instead of glucose because it cannot be further metabolized. Use of 2-deoxyglucose prevented an increase in glucose-6-phosphate concentration and subsequent ATP production through glycolysis. Endogenous ATP is typically present at approximately 1 mM in the axoplasm and is diluted to approximately 0.25 mM ATP by perfusion, so excess 2-deoxyglucose was added to ensure efficient depletion of endogenous ATP. Synthesis of new ATP was inhibited by a series of pharmacological agents: dinitrophenol (200 μM) blocks ATP production by mitochondria by uncoupling oxidative phosphorylation; diadenosine pentaphosphate (25 μM) inhibits transphosphorylation of ADP by adenylate kinase in which 2ADP = ATP + AMP; and theophylline inhibits nucleoside diphosphokinase (Lam and Packham 1986), an enzyme that regenerates ATP by catalysing a reaction in which ATP is transphosphorylated by NTPs (NTP + ADP = NDP + ATP; included in selected experiments).

Perfusion of axoplasm with a physiological buffer (buffer X/2) supplemented with 2 mM ATP supports organelle motility for over 40 min at velocities which approximate to anterograde and retrograde transport in freshly extruded, unperfused axoplasm (Fig. 2.2). The ATP depletion buffer described above was sufficient to eliminate all endogenous ATP as well as at least 2 mM additional ATP (Fig. 2.2). However, inclusion of 20 mM ATP in the ATP depletion buffer swamped the phosphate acceptor leaving excess ATP to drive organelle transport at normal rates. Perfusion with ATP depletion buffer thus removed endogenous ATP and prevented synthesis of new ATP, but was not itself deleterious to organelle motility in either the anterograde or retrograde direction.

To determine the ability of NTPs to support organelle transport, various NTPs were included in the ATP depletion buffer. UTP and GTP at concentrations of 10 or 20 mM were capable of supporting organelle transport at 20–60 per cent of velocities generated by similar concentrations of ATP (Fig. 2.3; Leopold *et al.* 1990). Maximum anterograde and retrograde velocities were similar when generated by UTP or GTP hydrolysis and continued for periods exceeding 40 min.

In the original experiments (Leopold *et al.* 1990), no provision was made for blocking conversion of ADP to ATP by nucleoside diphosphokinase. To confirm that motility was due to hydrolysis of the substituted NTP rather than by regeneration of ATP through transphosphorylation of ADP, theophylline was added to the ATP depletion buffer to inhibit nucleoside diphosphokinase.

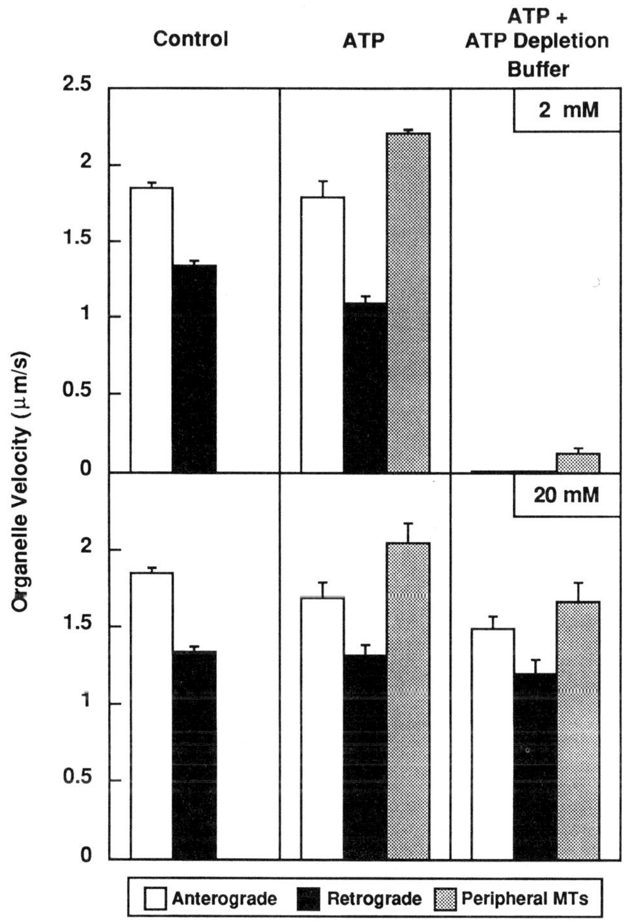

Fig. 2.2. ATP depletion buffer is capable of depleting endogenous ATP as well as supplementary ATP up to a concentration of 2 mM. Allen-video enhanced contrast, differential interference contrast microscopy is capable of detecting all axonally transported organelles including the smallest class of organelles, 40-nm vesicles that are rapidly moved in the anterograde direction (Allen *et al.* 1982; Brady *et al.* 1982). Prior to perfusion, anterograde transport has a significantly higher velocity than retrograde transport (Control). The difference in velocity may reflect the differing abilities of small vesicles (anterograde) and larger vesicles (retrograde) to negotiate the cytoskeletal meshwork in the axoplasm. This hypothesis is supported by the fact that, following perfusion with physiological buffer supplemented with 2 mM ATP (ATP), organelle translocation velocity on microtubules that have been extracted from the periphery of the axoplasm (peripheral MTs) is higher than either direction of transport in the interior of the axoplasm. ATP depletion buffer stops motility in both the anterograde and retrograde directions, even when 2 mM ATP is included in the perfusion buffer. Perfusion of 20 mM ATP overcomes the ability of the perfusion buffer to deplete ATP (see text) and maintains organelle transport at normal levels.

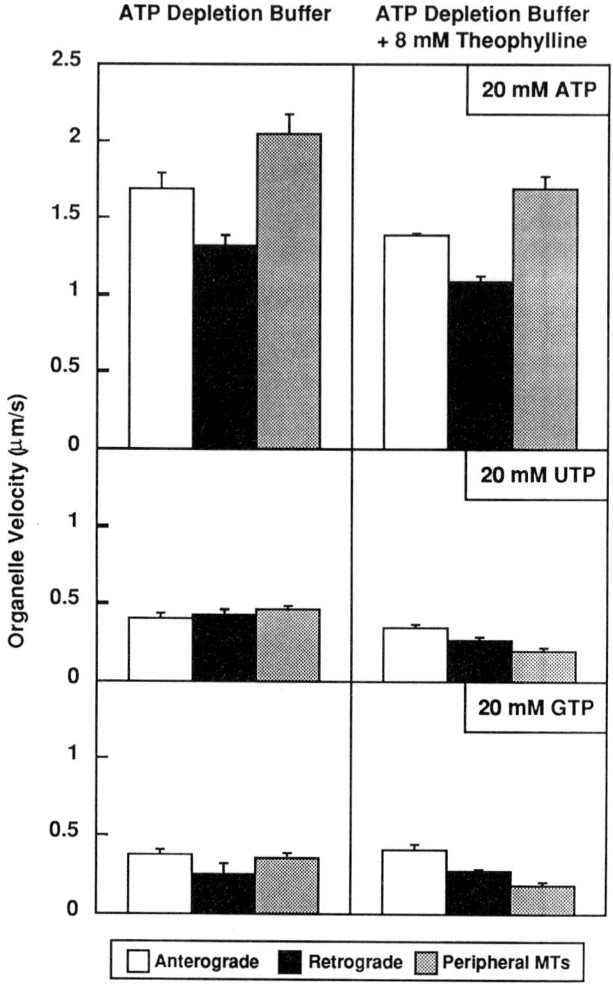

Fig. 2.3. Both UTP and GTP can replace ATP as the energy source for anterograde and retrograde organelle transport in extruded squid axoplasm. Under conditions that deplete endogenous ATP (see text), UTP and GTP at concentrations of 20 mM and 10 mM (Leopold *et al.* 1990) cause organelle motility to continue, but at slower rates than normal movement. Organelle velocities do not change in the presence or absence of theophylline. This result indicates either that nucleoside diphosphokinase is not present in significant amounts in squid axoplasm or, more probably, that diadenosine pentaphosphate, which was included in the ATP depletion buffer to inhibit adenylate kinase, also inhibits nucleoside diphosphokinase. However, the authors could not find precedent for this inhibition in the literature.

The ability of 20 mM UTP or GTP to support bidirectional organelle motility was not affected by the presence of 8 mM theophylline (Fig. 2.3), ruling out the possibility that NTPs supported transport through nucleoside diphosphokinase rephosphorylation of the ADP created when ATP was hydrolysed by hexokinase.

The fact that both anterograde-directed and retrograde-directed organelles responded in a similar manner to UTP and GTP (as well as to ITP, CTP, and TTP; Leopold *et al.* 1990), indicates that motors that drive organelle motility in each direction have a comparable specificity for a spectrum of NTPs. However, the *in vitro* motile activities and ATPases of kinesin and cytoplasmic dynein (the two motors included in most current models of fast axonal transport as anterograde and retrograde organelle motors, respectively) have distinct specificities for various NTPs. For example, kinesin can cause microtubule gliding using a wide variety of NTPs (Porter *et al.* 1987; Cohn *et al.* 1989) while cytoplasmic dynein can only use ATP to induce microtubule gliding (Paschal and Vallee 1987).

Interestingly, squid axoplasm is not the only system in which both directions of organelle motility have similar nucleotide specificities. Schliwa *et al.* (1991) reported that anterograde and retrograde organelle transport in the giant amoeba, *Reticulomyxa*, are indistinguishable by nucleotide specificity. In this system, bidirectional organelle motility has been attributed to a single molecular motor (Euteneuer *et al.* 1988). The nucleotide specificity of organelle motility and several other experimental results are not readily explained by the current kinesin/cytoplasmic dynein model for organelle motility. A complete discussion of this question and the roles of various organelle motors can be found in Brady (1991).

Squid giant synapse: biochemical modification of exogenous organelles

The giant synapse model

Work on the squid giant synapse has defined electrophysiological properties of neurotransmitter release. Early experiments determined that neurotransmitter release depended on an inward calcium current rather than on sodium or potassium currents (Katz and Miledi 1969; Llinás *et al.* 1972; Miledi 1973; Llinás and Nicholson 1975). Subsequently, techniques for recording presynaptic calcium currents in the squid giant synapse along with evoked postsynaptic potentials (EPSPs) generated by the release of neurotransmitters were described (Llinás *et al.* 1981*b*). Therefore, the squid giant synapse can be used to assay directly neurotransmitter release through fusion of endogenous synaptic vesicles with the plasma membrane by recording the associated EPSPs.

Besides providing an assay for exocytosis of endogenous vesicles, the giant synapse offers the ability to introduce foreign organelles into the presynaptic terminal. The presynaptic recording electrode can be used to pressure-inject solutions into the presynaptic cytoplasm. Objects as large as proteins (Llinás *et al.* 1985) or synaptic vesicles (Llinás *et al.* 1989) can be introduced into the presynaptic terminal cytoplasm while recording intracellularly. If the micro-injected vesicles are fluorescently labelled, changes in the distribution of microinjected organelles can be followed (Llinás *et al.* 1989). Furthermore, microinjected vesicles can be biochemically modified prior to introduction into the terminal and the effects of modifications evaluated. These methods make the squid giant synapse a versatile model for studying the complex combination of organelle targeting mechanisms and neurotransmitter release through the use of microinjection experiments.

One ubiquitous criticism of microinjection experiments is that physical insult to the cell may change its physiology. In the giant synapse, the membrane potentials of pre- and postsynaptic membranes may be monitored to ensure that the integrity of the membranes is maintained throughout the experiment. The ability to generate action potentials and observe the post-synaptic response to these stimuli provides a functional gauge for viability of the terminal. Therefore, the presumption of damage to the microinjected cell can be experimentally tested and ruled out as a source of error or misinterpretation.

The two most critical elements of microinjection experiments in the squid giant synapse are the preparation of the recording/microinjection electrode and the choice of giant synapse. The electrode must have an opening large enough to allow microinjection of proteins or organelles while having a small enough diameter to prevent influx of sea water when the presynaptic cell is impaled. The electrode is bevelled to create a large opening without increasing the diameter of the electrode (Llinás *et al.* 1981*a*). Stellate ganglia are selected based on the clarity and regularity of morphology. In particular, the digit of the presynaptic fibre must lie on top of, or next to, the postsynaptic fibre to allow visualization of the electrode while the cell is being pierced. The presynaptic terminal should not contain constrictions or contortions that obscure the exact location of the tip of electrode.

Progress during microinjection is monitored both microscopically and electrophysiologically. The location of the fluorescent dye (whether unconjugated, conjugated to proteins, or embedded in a lipid bilayer) is followed using epifluorescence video microscopy and an image analysis system (Llinás *et al.* 1989). Direct observation of the microinjection via a fluorescent label verifies that the electrode is not clogged and that sufficient experimental solution is dispensed to provide an adequate signal. After impaling the presynaptic cell, the resting potential is checked to confirm that the terminal has not become depolarized. The resting potential is typically monitored throughout the pressure-injection to ensure that the electrode does not slip out of the cell and that the terminal remains viable.

Targeting signals on organelles

The synaptic terminal represents a site at which many controls over vesicle motility converge. While organelle transport in the axon is characterized by long, uninterrupted translocations, organelle motility in the presynaptic terminal must be highly regulated. Vesicles must be targeted to the terminal, then mobilized and placed in apposition with the plasma membrane at the site of the active zone before fusion with the membrane upon stimulation. After fusion, the appropriate membrane components must be capable of reforming by endocytosis from the plasma membrane and either re-enter the cycle or are repackaged for return to the cell body in retrograde axonal transport. Each of these events must be appropriately regulated by proteins of the synaptic vesicle and/or presynaptic terminal.

Synaptic vesicle targeting is probably controlled by a combination of proteins. One protein that is known to regulate synaptic vesicle mobility is synapsin I. Synapsin I interacts with synaptic vesicles and microfilaments in a phosphorylation-dependent manner (reviewed by De Camilli *et al.* 1990). Microinjection of synapsin I into the squid giant synapse demonstrated that dephosphorylated synapsin I, but not tail phosphorylated synapsin I, would inhibit neurotransmitter release, presumably by crosslinking synaptic vesicles to microfilaments (Llinás *et al.* 1985). After synapsin I releases synaptic vesicles from microfilaments, movement of organelles through the microfilament meshwork to the active zone may require the assistance of a class of molecular motors that are related to myosin and that have been proposed to fulfil that role (Adams and Pollard 1986; Pollard *et al.* 1991). Proteins involved in controlling exocytosis have not yet been identified, although several candidates are currently under investigation (reviewed by Pevsner and Scheller (1994) and Ferro-Novick and Jahn (1994)). Still other proteins may regulate anterograde versus retrograde transport of organelles in the synaptic terminal. The giant synapse is particularly well suited for examining mechanisms by which synaptic vesicles are targeted at the synaptic terminal. By fluorescently labelling synaptic vesicles, a cloud of organelles can be visualized at the tip of the electrode during microinjection. Once the borders of the initially injected vesicles are carefully noted, the rate and direction of vesicle migration can be determined by following changes in the fluorescence pattern (Llinás *et al.* 1989).

The following section illustrates the use of the giant synapse in addressing one problem in organelle targeting, the anterograde/retrograde switch. As discussed above, the direction of organelle motility is specifically regulated. The fact that distinct populations of organelles move in the anterograde and retrograde directions suggests that control of directionality is associated with specific characteristics of various organelle populations. However, the mechanisms by which the neurone controls directionality of organelle transport is not yet understood. Schroer *et al.* (1985) observed the motility of individual fluorescently labelled synaptic vesicles and determined that the proportion of organelles travelling in the two directions changed following proteolysis. Similarly, Sahenk and Lasek (1988) demonstrated that treatment of axons with protease inhibitors results in the accumulation of organelles at the synaptic terminal or on the proximal side of a ligation. Since the protease inhibitors did not interfere with ongoing anterograde or retrograde translocations, the simplest interpretation of these results is that the protease inhibitors block a proteolytic event associated with conversion of anterograde to retrograde moving organelles. In order to address the possible involvement of proteolysis in synaptic vesicle targeting, we have begun to characterize transport of exogenous organelles in the squid giant synapse with and without proteolysis of the organelle surface.

Application of the giant synapse model

The giant synapse model provides a unique opportunity to analyse populations of synaptic vesicles after various treatments (Fig. 2.4). Synaptic vesicles purified from rat cerebral cortex (Huttner *et al.* 1993) were fluorescently labelled with the sulfonyl chloride derivative of Texas red, a rhodamine analogue (Schroer *et al.* 1985; Llinás *et al.* 1989). In order to ensure that the movement of Texas red-labelled synaptic vesicles (TRSVs) was not due to diffusion, the TRSV distribution was compared to the distribution of unconjugated fluorescein, a marker that is readily soluble. TRSVs and fluorescein were co-injected into the presynaptic terminal near the base of the digits [see arrow, Fig. 2.1 (b)]. Appropriate emission filters separate the Texas red signal from the fluorescein signal. Typically, TRSVs moved in the anterograde direction, with the fluorescence pattern eventually filling the digits of the terminal [Llinás *et al.* 1989; Fig. 2.4 (b)]. This movement was unidirectional in the anterograde direction while diffusion of the co-injected fluorescein was bidirectional. Accumulation of fluorescence at the distal ends of digits after extended incubations was striking in the case of TRSVs and absent in the case of fluorescein (Llinás *et al.* 1989). Moreover, inhibition of ATP synthesis in the terminal at the time of injection prevented movement of vesicles, but did not affect changes in distribution of unconjugated dye.

Translocation of organelles in the synaptic terminal differed in some respects from the well-characterized movement of organelles in the extruded axoplasm. The rate of vesicle movement in the synaptic terminal was determined to be 0.5 μm/s and was assessed by measuring the rate of movement of the leading edge of fluorescence. This rate was far below the velocity of approximately 1.9 μm/s characteristic of anterograde transport in the giant axon (Fig. 2.2). The difference between the velocities measured in the giant synapse and giant axon may reflect a difference in the organization of the cytoskeleton in each structure. However, velocity measurements of fluorescent vesicles in the giant synapse are inherently less precise than measurements in

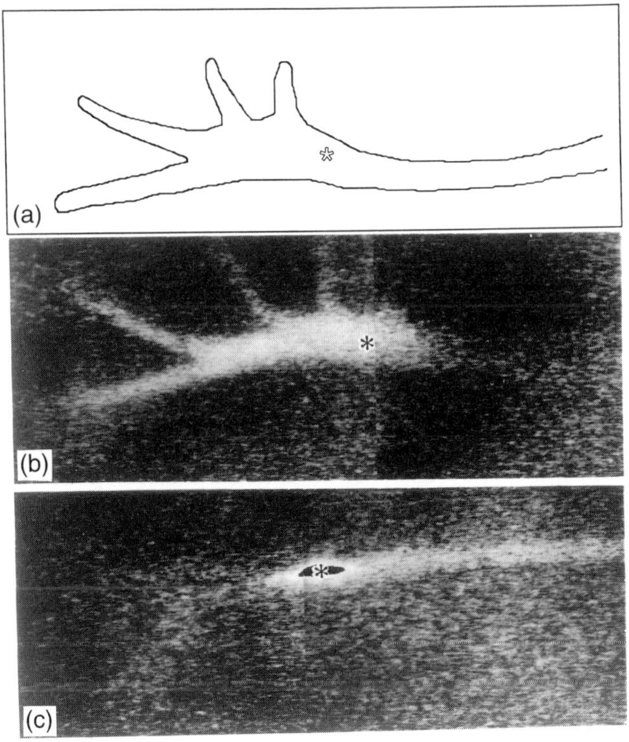

Fig. 2.4. Texas-red-labelled synaptic vesicle distribution in the squid giant synapse. (**a**) Outline of the presynaptic fibre in the stellate ganglion [see Fig. 2.1(b) for orientation]. Asterisk marks the point of injection in each panel. (**b**) Upon microinjection into the synaptic terminal of the stellate ganglion, fluorescently-labelled synaptic vesicles, prepared from rat brain, migrate in the anterograde direction. Migration was ATP-dependent but occurred at slower rates than velocities reported for organelle motility in extruded axoplasm. These results indicate that mammalian synaptic vesicles can interact with the translocation apparatus of squid. With a longer incubation, fluorescent synaptic vesicles accumulate in the distal ends of the digits (Llinás *et al.* 1989). (**c**) Proteolysis of synaptic vesicles prior to fluorescent labelling and microinjection changes the characteristics of vesicle movement in the synaptic terminal. While untreated synaptic vesicles move primarily in the anterograde direction with very little migration into the axon, chymotrypsin-treated vesicles become almost exclusively retrograde-directed. These results indicate that there are proteinaceous elements on the surface of synaptic vesicles which mediate directionality of vesicle translocation. The aggregate of vesicles located directly under the asterisk represents an area of highly concentrated vesicles which did not disperse following microinjection. Panels (b) and (c) were printed directly from a video record of the experiments using a video-printer. Each panel is a composite of two fields that cover the entire structure of the terminal. The synaptic terminals are approximately 1 mm in length.

isolated axoplasm, since axoplasm measurements are based on direct observations of individual moving organelles while the measurements in the presynaptic terminal reflect the movement of groups of fluorescent vesicles. The anterograde movement of synaptic vesicles in the giant synapse also differed from fast axonal transport with respect to the effects of calcium concentration. Low extracellular calcium blocked vesicle transport in the giant synapse while low calcium has no effect on organelle traffic in the extruded squid axoplasm (Brady *et al.* 1982, 1985; Llinás *et al.* 1989). The significance of this difference is not yet clear, but may well reflect the presence of regulatory mechanisms for organelle movement in the terminal.

Earlier experiments in the extruded axoplasm showed that proteolytic treatment of synaptic vesicles affected direction of movement (Schroer *et al.* 1985). The extruded axoplasm permits direct observation of endogenous moving vesicles, but because of the cytoarchitecture of the terminal, the giant synapse provides a much more efficient preparation for analysis of populations of microinjected synaptic vesicles. Vesicles treated with chymotrypsin prior to fluorescent labelling exhibited a nearly complete reversal in the direction of organelle translocation [Fig. 2.4(c)]. This result strongly indicates that a proteinaceous element on the vesicle surface regulates directionality, and that chymotrypsin sensitivity provides a key to the identity of that protein. Just as the identification of kinesin resulted from study of axonal transport in isolated axoplasm, these observations in the giant synapse may lead to the identification of a direction-determining protein on organelles.

The giant synapse model also has the potential to address a great number of other questions about organelle targeting in the synaptic terminal. For example, the possibility that mammalian synaptic vesicles fuse with the squid plasma membrane at the active site is currently under investigation. Preliminary evidence suggests that fusion does occur, which may lead to identification of vesicle proteins important for fusion and recycling of synaptic vesicles. The squid has already provided key insights into unravelling the physiology and biophysics of the action potential. New technologies applied to the study of the squid nervous system and new experimental models promise to facilitate continued exploration of neuronal function.

Acknowledgements

The authors would like to thank N. Joan Abbott and Roddy Williamson for organizing the Symposium on Cephalopod Neurobiology and the Wellcome Trust for providing a travel grant to P. L. L. to attend that meeting. We wish to acknowledge our collaborators, Robert Day Allen, George Bloom, Paul Greengard, John Gruner, Ray Lasek, and Teresa McGuiness, whose advice and interaction have helped to develop the use of *Loligo pealei* for motility research. We are grateful to Bob Snyder and Bruce Richards for their technical support. The authors thank the Department of Marine Resources at the Marine Biological Laboratory for providing specimens for these experiments. This work was supported by National Institutes of Health grants NS23868 and NS23320, National Science Foundation Biological Instrumentation Program grant DMB-8701164, Welch Foundation grant I-1077, and United States Air Force grant AFOSR850368.

References

Adams, R. J. (1982). Organelle movement in axons depends on ATP. *Nature*, **297**, 327–9.

Adams, R. J. and Pollard, T. D. (1986). Propulsion of organelles isolated from *Acanthamoeba* along actin filaments by myosin-I. *Nature*, **322**, 754–6.

Adelman, W. J. and Gilbert, D. L. (1990). Electrophysiology and biophysics of the squid giant axon. In *Squid as experimental animals* (ed D. L. Gilbert, W. J. Adelman, and J. M. Arnold), pp. 93–132. Plenum Press, New York.

Allen, R. D., Allen, N. S., and Travis, J. L. (1981). Video-enhanced contrast, differential interference contrast (AVEC-DIC) microscopy: A new method capable of analyzing microtubule related motility in the reticulopodial network of *Allogromia laticollaris*. *Cell Motility*, **1**, 291–302.

Allen, R. D., Metuzals, J., Tasaki, I., Brady, S. T., and Gilbert, S. P. (1982). Fast axonal transport in squid giant axon. *Science*, **218**, 1127–8.

Baker, P. F., Hodgkin, A. L., and Shaw. T. I. (1961). Replacement of the protoplasm of a giant nerve fibre with artificial solutions. *Nature*, **190**, 885–7.

Bear, R., Schmitt, F. O., and Young, J. Z. (1937). Investigations on the protein constituents of nerve axoplasm. *Proceedings of the Royal Society of London (Biology)*, **123**, 520–9.

Bisby, M. A. (1982). Retrograde axonal transport of endogenous proteins. In *Axoplasmic transport* (ed. D. G. Weiss), pp. 193–205. Springer-Verlag, New York.

Brady, S. T. (1985). A novel brain ATPase with properties expected for the fast axonal transport motor. *Nature*, **317**, 73–5.

Brady, S. T. (1991). Molecular motors in the nervous system. *Neuron*, **7**, 521–33.

Brady, S. T., Lasek, R. J., and Allen, R. D. (1982). Fast axonal transport in extruded axoplasm from squid. *Science*, **218**, 1129–31.

Brady, S. T., Lasek, R. J., and Allen, R. D. (1985). Video microscopy of fast axonal transport in extruded axoplasm: A new model for studying molecular mechanisms. *Cell Motility*, **5**, 81–101.

Brown, A. and Lasek, R. J. (1990). The cytoskeleton of the squid giant axon. In *Squid as experimental animals* (ed. D. L. Gilbert, W. J. Adelman, and J. M. Arnold), pp. 235–302. Plenum Press, New York.

Bullock, T. H. (1948). Properties of a single synapse in the stellate ganglion of squid. *Journal of Neurophysiology*, **11**, 343–64.

Cohn, S. A., Ingold, A. L., and Scholey, J. M., (1989). Quantitative analysis of sea urchin egg kinesin-driven microtubule motility. *Journal of Biological Chemistry*, **264**, 4290–7.

Cole, K. S. and Curtis, H. J. (1939). Electric impedance of the squid giant axon during activity. *Journal of General Physiology*, **22**, 649–70.

Darrow, R. A. and Colowick, S. P. (1962). Hexokinase from baker's yeast. In

Methods in enzymology, Vol. 5 (ed. S. P. Colowick and N. O. Kaplan), pp. 226–35. Academic Press, New York.

De Camilli, P., Benfenati, F., Valtorta, F., and Greengard, P. (1990). The synapsins. *Annual Reviews of Cell Biology*, **6**, 433–60.

Deffner, G. J. (1960). Chemical investigations of the giant nerve fiber of the squid. *Biochimica et Biophysica Acta*, **42**, 200–5.

Euteneuer, U., Koonce, M. P., Pfister, K. K., and Schliwa, M. (1988). An ATPase with properties expected for the organelle motor of the giant amoeba, *Reticulomyxa*. *Nature*, **322**, 176–8.

Fahim, M. A., Lasek, R. J., Brady, S. T., and Hodge, A. (1985). AVEC-DIC and electron microscope analyses of axonally transported particles in cold-blocked squid giant axons. *Journal of Neurocytology*, **14**, 698–704.

Ferro-Novick, and Jahn, R. (1994). Vesicle fusion from yeast to man. *Nature*, **370**, 191–3.

Hodgkin, A. L. and Huxley, A. F. (1952). Currents carried by sodium and potassium ions through the membrane of the giant axon of *Loligo*. *Journal of Physiology*, **116**, 449–72.

Huttner, W. B., Schiebler, W., Greengard, P., and De Camilli, P. (1983). Synapsin I (protein I), a nerve-terminal specific phosphoprotein. III. Its association with synaptic vesicles studied in a highly purified synaptic vesicle preparation. *Journal of Cell Biology*, **96**, 1374–88.

Katz, B. and Miledi, R. (1969). Tetrodotoxin-resistant electric activity in presynaptic terminals. *Journal of Physiology*, **203**, 459–87.

Lam, S. C.-T. and Packhman, M. A. (1986). Isolation and kinetic studies of nucleoside diphosphokinase from human platelets and effects of cAMP phosphodiesterase inhibitors. *Biochemical Pharmacology*, **35**, 4449–55.

Lasek, R. J. (1984). The structure of the axoplasm. *Current Topics in Membranes and Transport*, **22**, 39–53.

Lasek, R. J. and Brady, S. T. (1984). Adenylyl imidodiphosphate (AMPPNP), a non-hydrolyzable analogue of ATP, produces a stable intermediate in the motility cycle of fast axonal transport. *Biol. Bull.*, **167**, 503.

Lasek, R. J. and Brady, S. T. (1985). Attachment of transported vesicles to microtubules in axoplasm is facilitated by AMP-PNP. *Nature*, **316**, 645–7.

Leopold, P. L., Snyder, R., Bloom, G. S., and Brady, S. T. (1990). Nucleotide specificity for the bidirectional transport of membrane-bounded organelles in isolated axoplasm. *Cell Motility and the Cytoskeleton*, **15**, 210–19.

Lin, J.-W., Sugimori, M., Llinás, R. R., McGuiness, T. L., and Greengard, P. (1990). Effects of synapsin I and calcium/calmodulin-dependent protein kinase II on spontaneous neurotransmitter release in the squid giant synapse. *Proceedings of the National Academy of Sciences, USA*, **87**, 8257–61.

Llinás, R. and Nicholson, C. (1975). Calcium role in depolarization–secretion coupling: an aequorin study in squid giant synapse. *Proceedings of the National Academy of Sciences, USA*, **72**, 187–90.

Llinás, R., Blinks, R. J., and Nicholson, C. (1972). Calcium transients in presynaptic terminal of squid giant synapse: Detection with aequorin. *Science*, **176**, 1127–30.

Llinás, R., Steinberg, I. Z., and Walton, K. (1981*a*). Presynaptic calcium currents in squid giant synapse. *Biophysical Journal*, **33**, 289–322.

Llinás, R., Steinberg, I. Z., and Walton, K. (1981*b*). Relationship between presynaptic calcium current and postsynaptic potential in squid giant synapse. *Biophysical Journal*, **33**, 323–52.

Llinás, R., McGuiness, T. L., Leonard, C. S., Sugimori, M., and Greengard, P. (1985). Intraterminal injection of synapsin I of calcium/calmodulin-dependent protein kinase II alters neurotransmitter release at the squid giant synapse. *Proceedings of the National Academy of Sciences, USA*, **82**, 3035–9.

Llinás, R., Sugimori, M., Lin, J.-W., Leopold, P. L., and Brady, S. T. (1989). ATP-dependent directional movement of rat synaptic vesicles injected into the presynaptic terminal of squid giant synapse. *Proceedings of the National Academy of Science, USA*, **86**, 5656–60.

Llinás, R., Gruner, J. A., Sugimori, M., McGuiness, T. L., and Greengard, P. (1991). Regulation by synapsin I and Ca^{2+}-calmodulin-dependent protein kinase II of transmitter release in squid giant synapse. *Journal of Physiology*, **436**, 257–82.

Martin, R. and Miledi, R. (1986). The form and dimensions of the giant synapse of squids. *Philosophical Transactions of the Royal Society of London, Series B*, **312**, 355–77.

Martz, D., Lasek, R. J., Brady, S. T., and Allen, R. D. (1984). Mitochondrial motility in axons: Membranous organelles may interact with the force generating system through multiple surface binding sites. *Cell Motility*, **4**, 89–101.

Martz, D., Garner, J. A., and Lasek, R. J. (1989). Protein changes during anterograde to retrograde conversion of axonally transported vesicles. *Brain Research*, **476**, 199–203.

McGuiness, T. L., Brady, S. T., Gruner, J. A., Sugimori, M., Llinás, R., and Greengard, P. (1989). Phosphorylation-dependent inhibition by synapsin I of organelle movement in squid axoplasm. *Journal of Neuroscience*, **9**, 4138–49.

Miledi, R. (1973). Transmitter release induced by injection of calcium ions into nerve terminal. *Proceedings of the Royal Society of London (Biological Sciences)*, **183**, 421–5.

Miller, R. H. and Lasek, R. J. (1985). Crossbridges mediate anterograde and retrograde vesicle transport along microtubules in squid axoplasm. *Journal of Cell Biology*, **101**, 2181–93.

Morris, J. and Lasek, R. J. (1982). Stable polymers of the axonal cytoskeleton: The axoplasmic ghost. *Journal of Cell Biology*, **92**, 192–8.

Morris, J. and Lasek, R. J. (1984). Monomer–polymer equilibria in the axon: Direct measurement of tubulin and actin as polymer and monomer in axoplasm. *Journal of Cell Biology*, **98**, 2064–76.

Oikawa, T., Spyropoulis, C. S., Tasaki, I., and Teorell, T. (1961). Methods for perfusing the giant axon of *Loligo pealei*. *Acta Physiologica Scandinavica*, **52**, 195–6.

Paschal, B. M. and Vallee, R. B. (1987). Retrograde transport by the microtubule-associated protein MAP 1C. *Nature*, **330**, 181–3.

Pevsner, J. and Scheller, R. H. (1994). Mechanisms of vesicle docking and fusion: insights from the nervous system. *Current Opinions in Cell Biology*, **6**, 555–60.

Pollard, T. D., Doberstein, S. K., and Zot, H. G. (1991). Myosin-I. *Annual Reviews of Physiology*, **53**, 653–81.

Porter, M. E., Scholey, J. M., Stemple, D. L., Vigers, G. P. A., Vale, R. D., Sheetz, M. P., and McIntosh, J. R. (1987). Characterization of the microtubule movement produced by sea urchin egg kinesin. *Journal of Biological Chemistry*, **262**, 2794–802.

Rubinson, K. and Baker, P. (1979). The flow properties of axoplasm in a defined chemical environment: Influence of anions and calcium. *Proceedings of the Royal Society of London (Biology)*, **205**, 323–45.

Sahenk, Z. and Lasek, R. J. (1988). Inhibition of proteolysis blocks anterograde–retrograde conversion of axonally transported vesicles. *Brain Research*, **460**, 199–203

Schliwa, M., Shimizu, T., Vale, R. D., and Euteneuer, U. (1991). Nucleotide specificities of anterograde and retrograde organelle transport in *Reticulomyxa* are indistinguishable. *Journal of Cell Biology*, **112**, 1199–203.

Schroer, T. A., Brady, S. T., and Kelly, R. B. (1985). Fast axonal transport of foreign synaptic vesicles in squid axoplasm. *Journal of Cell Biology*, **101**, 568–72.

Smith, R. S. (1980). The short term accumulation of axonally transported organelles in the region of localized lesions of single myelinated axons. *Journal of Neurocytology*, **9**, 39–65.

Tsukita, S. and Ishikawa, H. (1980). The movement of membranous organelles in axons. Electron microscopic identification of anterogradely and retrogradely transported organelles. *Journal of Cell Biology*, **84**, 513–30.

Vale, R. D., Schnapp, B. J., Mitchison, T., Steuer, E., Reese, T. S., and Sheetz, M. P. (1985). Different axoplasmic proteins generated movement in opposite directions along microtubules *in vitro*. *Cell*, **43**, 623–32.

Young, J. Z. (1936). The giant nerve fibers and the epistellar body of cephalopods. *Quarterly Journal of Microscopic Science*, **78**, 367–86.

Young, J. Z. (1939). Fused neurons and synaptic contacts in the giant nerve fibers of cephalopods. *Philosophical Transactions of the Royal Society, Series B*, **229**, 465–503.

3

Cyclic nucleotide homeostasis and axonal G proteins in the squid *Loligo forbesi*

T. J. A. ALLEN and B. ROUOT

Summary

This chapter reviews studies on signal:transduction coupling in the squid giant axon and its associated Schwann cell sheath, and the role of cyclic nucleotides and G-proteins in this coupling. The cAMP content of the axon is increased by external application of serotonin (5HT), and this effect is modulated by $[Ca^{2+}]_i$ and is dependent on internal GTP; membrane G-proteins are implicated in the coupling. Lowering $[Mg^{2+}]_i$ attenuates 5HT-evoked adenylate cyclase activity. The giant axon provides too little G-protein for analysis, but it has been possible to demonstrate G-proteins in the richer source of squid optic lobes. The cGMP content of the axon is increased by carbachol, and by nitroprusside, an agent activating cGMP synthesis. The role of G-proteins in Schwann cell–axon signalling is discussed.

The study of cyclic nucleotide metabolism in nerve tends to focus on roles in synaptic transmission, with particular emphasis on cAMP and cGMP (cyclic adenosine and guanosine monophosphate, respectively). In this respect cyclic nucleotides act as messengers or modulators of neurotransmitter function, probably by use of phosphorylation–dephosphorylation reactions. Due to the complexity of mammalian nervous systems it is usually difficult to study these systems in detail and in isolation. This article aims to introduce to the reader the usefulness of the squid axon–Schwann cell preparation as a model with which to study certain aspects of signal:transduction coupling. Signal:transduction coupling involves receptor binding at an extracellular site and the eliciting of intracellular signals or messengers to evoke biochemical changes in transport or homeostasis. The physiological implications of changes in cAMP and cGMP content for the giant axon, periaxonal ionic regulation, and axon–Schwann cell signalling will be covered in detail in other chapters of this book. This chapter will try to combine simple observations on cAMP and cGMP homeostasis in squid nerve with those acquired from largely biochemical work on vertebrate systems. Evidence will also be presented on how a central component of cAMP production, G proteins, are expressed in squid, and how well conserved they are when compared to their mammalian counterparts.

Adenylate cyclase

cAMP metabolism in intact squid axons

Once dissected, squid axons from *Loligo forbesi* are 15–20 cm long and approximately 800–900 μm in diameter. After incubation in artificial sea waters of various compositions, paired axons (test versus control) can be extruded of axoplasm, and the latter can be homogenized in ethanol for later analysis for cAMP content by radioimmunoassay.

Using this simple approach it was found that serotonin (5-hydroxytryptamine, 5HT) at μM levels caused a clear increase in cAMP content (Baker and Carruthers 1982; Allen 1989). As cAMP content reflects the relative rates of production and breakdown, axons were exposed to the phosphodiesterase inhibitor, caffeine, to determine the route by which the cAMP content was increased. Caffeine alone (10 mM) caused a twofold increase in cAMP content, demonstrating the presence of a high basal rate of cAMP breakdown, but under these conditions 5HT still increased cAMP content.

Neuronal adenylate cyclase can be affected by calcium, either directly or via calmodulin activation; calmodulin activation has been utilized in purification of the enzyme (Brostrom *et al.* 1978; Minocherhomjee *et al.* 1987). Does intracellular calcium $[Ca^{2+}]_i$ affect the route that evokes a rise in cAMP in squid nerve? Free calcium in intact axons can be raised during activation of reverse Na–Ca exchange by removal of sodium from the extracellular medium, thus promoting calcium influx (Allen and Baker 1986). Under these conditions the basal cAMP content of axons was little changed; however, the effects of 5HT activation were attenuated some 60 per cent (Allen 1989); $[Ca^{2+}]_i$ can modulate 5HT-coupled signal transduction. However, as these studies were performed on intact axons, these manipulations do not tell us where $[Ca^{2+}]_i$ is exerting its effects; does it affect receptor coupling, adenylate cyclase activity, or phosphodiesterase (PDE) activity? This last consideration arises as there is good evidence that neuronal PDE is also modulated by Ca^{2+} and calmodulin (Strada *et al.* 1984).

Many responses to receptor activation are also tuned or modulated by desensitization. This mechanism normally serves to switch off or attenuate a response despite continuous presence of agonist, and can occur at various levels in a signal transduction pathway. Does desensitization occur during responses to 5HT? The change in axonal cAMP content in 5HT appears to exhibit rather slow desensitization (Fig. 3.1); the initial increase in cAMP is only reduced to 50 per cent after 3-h exposure. This may suggest that *in situ* desensitization does not play an important regulatory role. But many questions remain unanswered: is the axon normally exposed to 5HT for brief periods or could associated changes in $[Ca^{2+}]_i$ effect a faster reduction in cAMP, by activation of PDE, for example?

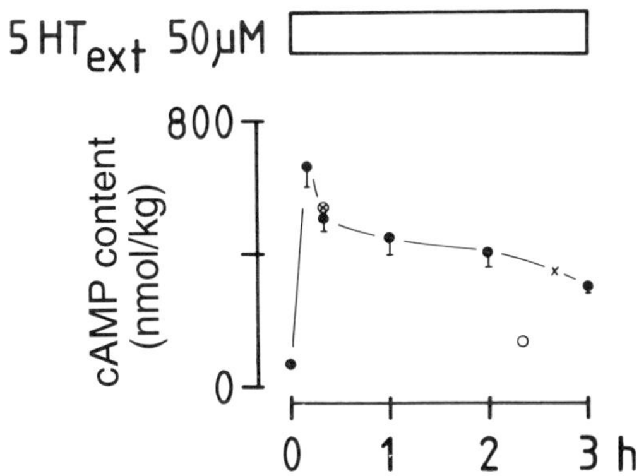

Fig. 3.1. The effects of continuous exposure to 5HT on the cAMP content of intact axons. The open box denotes the period during which intact axons were exposed to 50 μM of 5HT externally. Paired axons were used with staggered time points: t=0 min, n=2; t=10, n=8; t=20, n=3; t=60, n=10; t=120, n=4; t=180, n=3. At the end of each incubation period axoplasm was extruded, homogenized in ethanol, and the cAMP content was determined. Closed circles, mean cAMP content, with SEM bars. Open circles, an axon pair, both of which were incubated for 20 min in 5HT, but one axon was extruded of axoplasm and homogenized immediately whilst the other was washed in sea water for 2 h and then processed. Crosses, an axon pair, both of which were incubated for 20 min in 5HT, but one axon was processed immediately whilst the other was washed for 2 hours and then incubated for a further 20 min in 5HT before being processed. (Reproduced with permission from Allen (1989).)

cAMP metabolism in perfused squid axons

One of the great advantages of the squid giant axon is the possibility of cannulating it at both ends, and advancing a glass capillary along its central axis, so removing a central core of axoplasm that then serves as a conduit for intracellular perfusion. The plasmalemma remains essentially intact, so that membrane-bound adenylate cyclase activity can be simply monitored by measuring the cAMP content of the emerging perfusate. Under these conditions signal transduction coupling is retained (Fig. 3.2).

The addition of GTP (guanosine triphosphate) and 5HT in this and similar experiments demonstrates micromolar requirements for internal GTP and external 5HT. The evidence that GTP is important for 5HT activation supports a role for GTP-binding proteins as a central component in this signal transduction pathway. Indeed, in other experiments (see Fig. 3.3) non-hydrolysable GTP analogues, such as GppNHp and GTPγS, can activate cAMP production without addition of 5HT to the external medium. These analogues have been shown to activate G-protein coupling to adenylate cyclase due to their slow rate of hydrolysis (Birnbaumer *et al.* 1990); once GTP is bound to G proteins, hydrolysis must occur to form GDP and inorganic phosphate (PO_4) in order for the system to de-activate. Furthermore, NaF

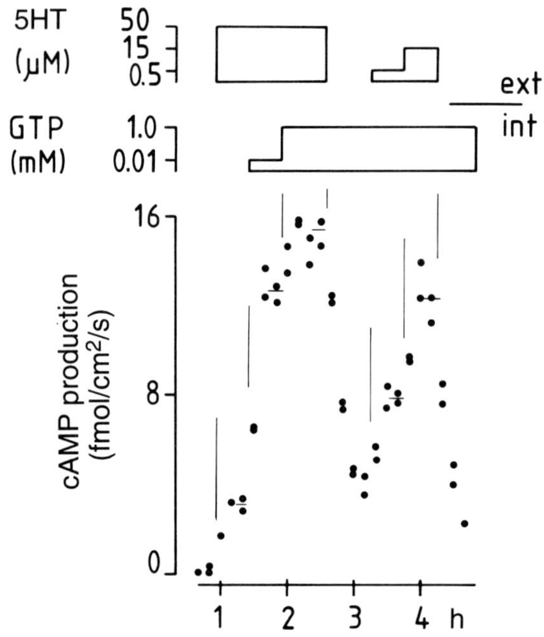

Fig. 3.2. Dependence of cAMP production on extracellular 5HT and intracellular GTP in perfused axon. Addition of 50 μM 5HT to the external solution (ext) produced a larger increase in the cAMP content of the emerging perfusate when GTP was included in the internal solution (int). Subsequent removal and readdition of 5HT led to a reduction and then stimulation of cAMP production. Perfusate used: KH_2PO_4 with 3 mM ATP, 1.5 mM $CaCl_2$, and 5 mM EGTA, pH 7.3. 10 mM of caffeine was included in all external solutions. Duplicate cAMP determinations are shown for each 10-min collection period. (Reproduced with permission from Allen (1989).)

applied intracellularly to perfused squid axons directly activates cAMP production (Allen 1989); it is thought that a contaminant (Al_3F^-) binds to guanosine diphosphate (GDP), which is usually bound to G proteins, and that this complex then mimics GTP (Birnbaumer *et al.* 1990).

The main reactions of the adenylate cyclase system can be summarized as follows.

1. The hormone-activated receptor binds to the GDP-liganded G protein and promotes dissociation of GDP. This is a Mg^{2+}-dependent reaction, as, in the absence of hormone, high levels of Mg^{2+} (50–10 mM) are required to promote GDP dissociation.
2. The hormone-receptor promotes the binding of GTP to the G protein.
3. The GTP-liganded G protein α subunit (α*) then dissociates from the βγ-subunits, and both the active α* subunit and βγ subunits dissociate from the receptor.

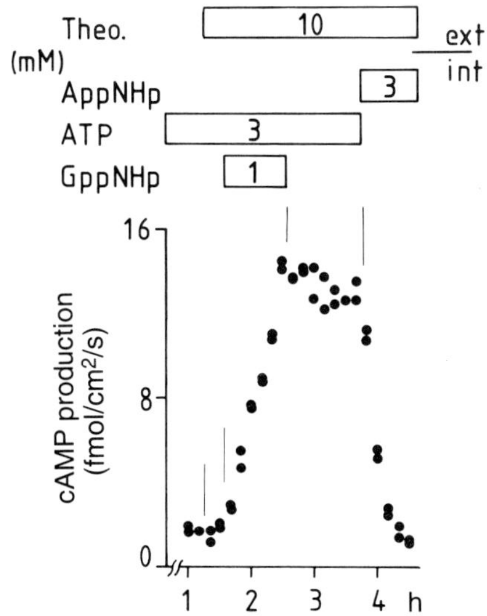

Fig. 3.3. Activation of cAMP production by GppNHp and inhibition by AppNHp. Removal of GppNHp from the internal solution did not lead to a rapid reduction in cAMP production, but subsequent substitution of ATP by AppNHp rapidly halted cAMP production. Axon perfused with KH_2PO_4-based perfusate containing 1.5 mM $CaCl_2$ and 5 mM EGTA, pH 6.8. Additions to internal (int) and external (ext) solutions are denoted by the open bars at the top of the figure. Theo, theophylline. (Reproduced with permission from Allen (1989).)

4. α* activates the effector. The life-time of the activated effector is determined by the rate of the GTPase activity of the α subunit, which is slow at 4/min (the dissociation of GDP is even slower at 0.4/min).

When known quantities of cAMP are perfused through the axon or caffeine added to external solutions, little—if any—PDE activity is found. Is PDE activity washed out during perfusion, or is an important cofactor lost? There is good evidence that some neuronal PDE activity is dependent on calmodulin, for example (Strada *et al.* 1984). This condition in the perfused axon is, however, fortuitous, allowing investigation into the ionic requirements for the activating arm of the signal transduction pathway, adenylate cyclase generation of cAMP.

The way $[Ca^{2+}]_i$ affects signal:transduction coupling has proved very interesting: $[Ca^{2+}]_i$ can stimulate or inhibit cAMP production in perfused axons depending on the route of adenylate cyclase activation (Allen 1989). With intact axons, manoeuvres that would raise $[Ca^{2+}]_i$ attenuate an increase in cAMP evoked by 5HT. Similar experiments with 5HT have not yet been done on perfused axons, but raising $[Ca^{2+}]_i$ to ~ 1 μM during activation of adenylate cyclase via G protein by the GTP analogue, GppNHp, or by NaF, leads to attenuation of cAMP production. The same manoeuvre during direct activation by forskolin stimulates cAMP production. A full Ca^{2+} activation–inhibition curve has not yet been performed, however, but the data suggest that modulation of signal:transduction coupling by $[Ca^{2+}]_i$ may depend on the prevailing physiological conditions. As $[Ca^{2+}]_i$ can also modulate PDE activity, a net change in cAMP content in intact axons is under rather complex control, which may even be further tuned by calmodulin. During continuous perfusion, calmodulin does not appear to recover the steady loss in adenylate cyclase activity, which may instead be due to loss of receptor–G protein coupling, but may affect the Michaelis constant (K_m) for $[Ca^{2+}]_i$ activation or inhibition.

Data on G-protein coupling to effectors in vesicles or reconstituted systems show that Mg^{2+} can attenuate both receptor and G-protein activation of adenylate cyclase; one of the complicating factors in these studies is to distinguish between actions on extra and intracellular faces of the membrane (Gilman 1987). In the perfused axon this problem is simply overcome; lowering $[Mg^{2+}]_i$ attenuates 5HT evoked cyclase activity. A full investigation of the dose–response relation has not yet been done. As $[Mg^{2+}]_i$ and $[Ca^{2+}]_i$ play a central role in signal transduction coupling, the squid axon should allow a detailed study of the interaction of these cations, both intra- and extracellularly, with the various steps in the coupling processes.

One disadvantage of the perfused squid axon is the physical removal of axoplasm, the cytoskeleton of which may play an important part in signal: transduction coupling (see Rasenick *et al.* 1981); indeed continuous axonal perfusion leads to a gradual loss of adenylate cyclase activity generated by 5HT, but not apparently of that activated by the GTP analogues, NaF or forskolin (Allen 1989). This loss is not prevented by inclusion of agents to limit possible proteolysis, such as dithiothreitol (DTT), bovine serum albumin (BSA), and leupeptin. The possibility remains that a cytoskeletal component may aid or influence receptor coupling to G proteins. This possibility can be addressed by using dialysed axons in which porous tubing is inserted down the central axis of the axon and the intracellular compartment controlled by dialysis.

In the experiments done so far using dialysis (Allen and Knight 1992) a new type of tubing has been utilized. This pure cellulose tubing has a molecular weight cut-off of 5 kDa and does not have to be treated with alkali, which is necessary for cellulose acetate tubing and tends to affect mechanical stability. Data from dialysed squid axons show that the 5HT activated system remains functional, can be activated by non-hydrolysable GTP analogues, and can be studied for a number of hours.

The prospect for future experiments is exciting, as the added advantage of the dialysed axon is the ability to voltage clamp by inclusion of the current-passing electrode within the dialysis tubing. This would allow an examination of how signal:transduction coupling is affected by both the chemical *and* the electrical gradients that normally exist across intact membranes.

Investigations into the characteristics and presence of GTP-binding proteins

The presence of a system in squid axons with which to study, at both membrane faces, detailed aspects of signal:transduction coupling involving GTP-binding proteins invites the following question; what types of GTP-binding protein substrates exist? In collaboration with Dr Bruno Rouot of Montpellier, we have investigated the neuronal substrates that are ADP-ribosylated by using two bacterial endotoxins, *Bordetella pertussis* and *Vibrio cholera* toxins (PTX and CTX, respectively). These toxins consist of two components, A and B, that are non-covalently linked (Gill 1977). The A component contains the active site of ADP-ribosyltransferase and has disulfide bonds that have to be reduced to obtain the active form of the enzyme, when the A component is dissociated from the B oligomer. It is the B oligomer that binds to the cell surface to allow the A promoter to enter the cell; however, in cell-free systems activation is simply achieved by a reducing agent such as DTT. The toxins catalyse the covalent attachment of ADP-ribose donated from NAD to the α subunits. By doing so, CTX attenuates GTPase activity and maintains GTP at the nucleotide binding site which leads to a permanently activated α_s subunit. In contrast, ADP ribosylation catalysed by PTX prevents the activation of G_i and G_o proteins by receptors. These toxins have proved useful in numerous molecular and recombinant DNA cloning studies identifying GTP-binding proteins (Birnbaumer *et al.* 1990); CTX has helped identify four variants of $G_{s\alpha}$, while use of PTX reveals three forms of $G_{i\alpha}$ (G_{i1}, G_{i2}, and G_{i3}) and one type of $G_{o\alpha}$. The subscripts i and s stand for inhibitory and stimulatory actions on adenylate cyclase, respectively. Two forms of transducins, involved in visual transduction, are ADP-ribosylated by both PTX and CTX (see also Chapter 29).

The primary function of α_s subunits appears to be to stimulate adenylate cyclase, but there is good evidence that they can also use Ca^{2+} channels as another effector (in cardiac and skeletal muscle: Birnbaumer *et al.* 1990; Cavalié *et al.* 1991). It is now clear that a variety of effectors exist for the PTX substrates, α_i and α_o, so that this area of subunit–effector matching emerges as rather ill defined, primarily due to limited use and availability of recombinant α subunits. The term G_i emerged due to its ability to restore inhibition of adenylate cyclase activity (Hildebrandt *et al.* 1983); it is much clearer now that the three forms of G_i can also open K^+ channels with varying potencies (Yatani *et al.* 1988) and that G_o can close Ca^{2+} channels. There is also further evidence that two isoforms of $G_{o\alpha}$ exist, with differing expression during neuronal differentiation (Rouot *et al.* 1992).

GTP-binding proteins in squid optic lobes

While single squid giant axons are useful for their size in yielding large quantities of uncontaminated axoplasm, this is a clear disadvantage for those studies requiring large quantities of membrane; it is necessary here to focus on membrane-bound GTP-binding proteins. Because of this problem squid optic lobes were used as the primary source of membrane-rich plasmalemma, due to the rich presence of many synaptic connections (Allen and Rouot, in preparation). Squid photoreceptor membranes have already proved valuable for the purification of the principal G_α protein involved in light activation of phospholipase C (Pottinger *et al.* 1991). Each optic lobe is located behind the eye of *Loligo*, can be rapidly dissected, and weighs 0.5 g. The experimental approach obviously means that this heterogeneous cellular material yields G-protein substrates of a nature that cannot be readily identified as axonal or glial, for example. However, these studies provide a useful starting point for the characterization and reconstitution of squid neuronal GTP-binding proteins.

Frozen squid optic lobes were thawed in Tris–HCl buffer and homogenized in a glass–Teflon potter. After centrifugation and further resuspension a crude membrane fraction of 6–10 mg protein/ml was made; this could be used immediately or stored at −80°C. ADP-ribosylation of membranes was achieved using DTT-preactivated PTX or CTX incubated in the presence of [^{32}P]NAD. Using crude membranes obtained by this procedure a number of CTX and PTX substrates were found (Fig. 3.4).

The figure shows that the major PTX substrate of squid optic lobes (Fig.

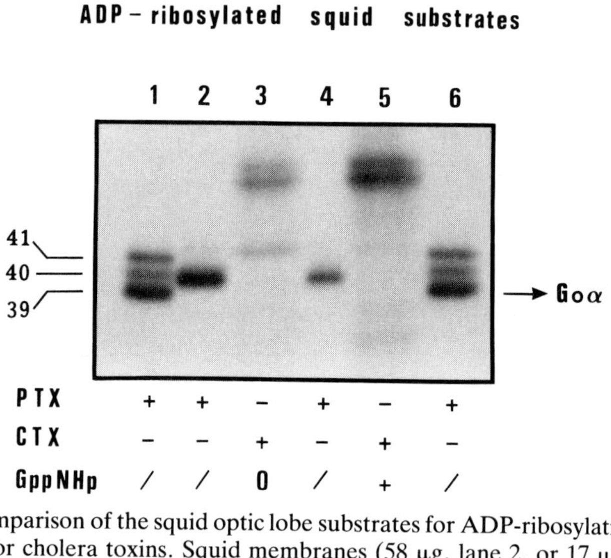

Fig. 3.4. Comparison of the squid optic lobe substrates for ADP-ribosylation catalysed by pertussis or cholera toxins. Squid membranes (58 μg, lane 2, or 17 μg, lane 4) or purified bovine brain G proteins (lanes 1 and 6) were subjected to ADP-ribosylation in the presence of pertussis toxin, PTX. 40 μg of crude membranes were also ADP-ribosylated with cholera toxin. CTX, either with GppNHp (lane 5) or with added nucleotides (lane 3). After sequential treatment with DTT and NEM samples were loaded on to an SDS gel of polyacrylamide (10 per cent) and bisacrylamide (0.13 per cent) with a separating gel longer than 20 cm (Allen and Rouot, unpublished data).

3.4, lane 4) migrates with an electrophoretic mobility comparable to that of the bovine brain α_{i2} (lanes 1 and 6, middle band); this mammalian subunit is considered to be a 40-kDa polypeptide (Katada *et al.* 1987; Brabet *et al.* 1988). By using a three-fold higher protein concentration (lane 2) another PTX substrate of slightly lower mobility (about 42 kDa) than bovine brain α_{i1} (41 kDa) could also be detected. On immunoblot, antibodies raised against the calf brain $G_{o\alpha}$ subunit (Homburger *et al.* 1987) cross-reacted with the major 40-kDa PTX substrate of squid optic lobes. It has been suggested that G_o is intimately involved in coupling receptors to neuronal Ca^{2+} channels, and G_o has also been shown to represent the major PTX-sensitive substrate in mammalian brain; this appears to be similarly the case for the squid optic lobe. The minor PTX substrate (close to 42 kDa) may represent the inhibitory G protein of the G_i family (G_{i1} or G_{i2} α subunits).

When squid optic lobe membranes were ADP-ribosylated in the presence of CTX (Fig. 3.4 lanes 3 and 5), the homologue squid α_s proteins were labelled as a doublet at about 45 kDa. The labelling was considerably lowered in the absence of GppNHp, but under these conditions another ADP-ribosylated substrate could be detected. This CTX-sensitive substrate exhibits a mobility in between the 41-kDa bovine brain α_{i1} and the minor squid PTX substrate.

It would be of great interest to establish whether the squid optic lobe PTX and CTX substrates are unique polypeptides; this is indeed possible, as exemplified by photoreceptor transducins (Birnbaumer *et al.* 1990; Bornancin and Chabre 1991). Another example has been reported with the neutrophil α_{i2} subunit (40 kDa), which can be ADP-ribosylated by PTX in the presence of the chemoattractant agonist fMLP and which stimulates phospholipase C regulation of inositol poly(phospho)lipids (Goldsmith *et al.* 1987).

If ADP-ribosylation is performed in the absence of CTX, only faint substrate labelling below 40 kDa is seen, demonstrating that the CTX substrates do not depend on the activity of an endogenous ADP-ribosyltransferase; this is shown in the autoradiogram in Fig. 3.5. However, the figure also shows that both the stimulatory (GppNHp) and the inhibitory (GDPβS) nucleotides, singularly or together, reduce the labelling of the 41–42 kDa CTX substrate. This suggests that optimal ADP ribosylation by CTX occurs when the nucleotide binding site is empty.

This is examined in more detail in Fig. 3.6. The gel shows that high nucleotide concentrations increase the overall labelling by CTX. When the lanes are compared to the central lane, which shows incubation without nucleotides ('None'), however, nucleotides at 100 μM reduced (GppNHp) or abolished (GDPβS) the labelling of the CTX substrate observed at 41–42 kDa. Reducing the guanyl-nucleotide concentrations to 1 μM still attenuated the 41–42 kDa band labelling with the order of potency,

$$\text{GTP}\gamma\text{S} > \text{GTP} > \text{GDP}\beta\text{S} > \text{GppNHp} = \text{GDP}$$

How these nucleotides affect ADP-ribosylation of the 41–42 kDa substrate

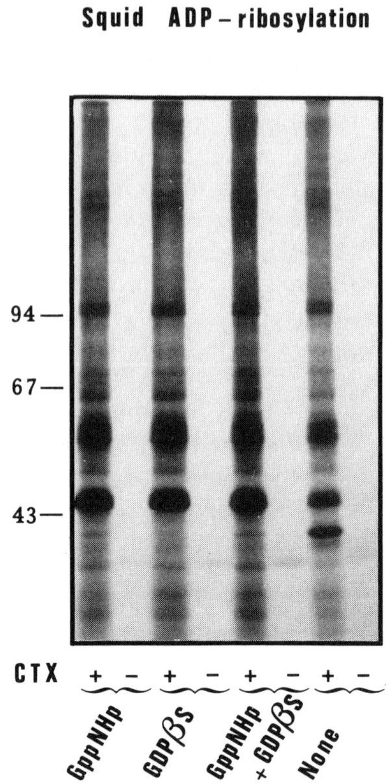

Fig. 3.5. Effects of guanyl nucleotides on the labelling of the squid cholera toxin substrate at 41–42 kDa. Squid membranes were incubated in the presence of [^{32}P]NAD with or without cholera toxin (CTX), and in the presence or absence of 100 μM guanyl nucleotides. The numbers at left indicate kDa markers (Allen and Rouot, unpublished data).

by CTX is not clear; the findings of Mattera *et al.* (1987) suggest that the substrate for PTX is not the free $G_{\alpha i}$ subunit but the GDP-liganded αβγ complex. Thus the addition of agents such as Mg^{2+} and GTP analogues that promote the activation and thus dissociation of PTX substrates greatly reduces the capability of the α_i subunit to serve as an acceptor for PTX. CTX ribosylation of G_s, on the other hand, appears to also involve another GTP-binding protein, ADP-ribosylation factor (ARF), which may be membrane-bound (Khan and Gilman 1986; Tsai *et al.* 1987). Although not yet identified, it is likely that squid membranes also possess ARF activity. As for transducin of the rod outer segment, it is most likely that ARF is involved in the ADP ribosylation of the squid 41–42 kDa polypeptide catalysed by CTX (Bornancin and Chabre 1991). This may give rise to the dual requirement for nucleotides—a GTP analogue for ARF and an absence of nucleotide for the 41–42 kDa substrate—which explains the variable level of ADP-ribose incor-

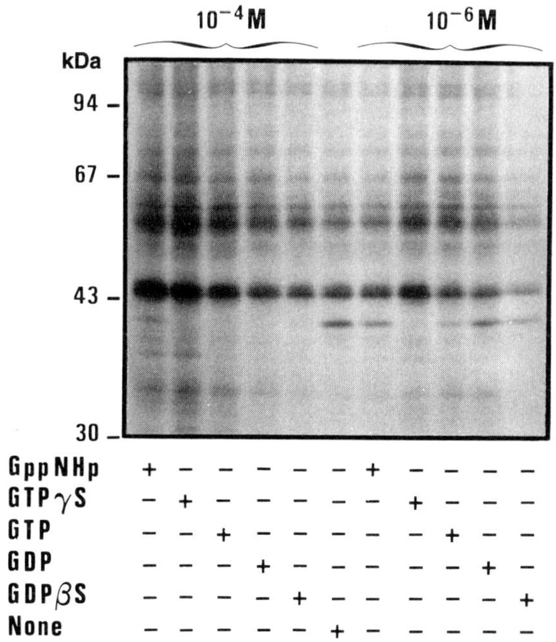

Fig 3.6. Determination of the potency of various guanyl nucleotides to inhibit ADP-ribosylation of the cholera toxin substrate at 41–42 kDa. Squid membranes were ADP-ribosylated with cholera toxin (CTX) in the presence of various nucleotides at either 10^{-4} M or 10^{-6} M (Allen and Rouot, unpublished data).

poration into the 41–42 kDa substrate observed in the presence of various nucleotides (see Figs 3.5 and 3.6).

Since squid optic lobes display good evidence for G_i and G_o-like substrates, it was of great interest to examine whether mammalian βγ subunits could affect PTX-catalysed ADP-ribosylation. In preliminary experiments (Allen and Rouot, unpublished), the addition of βγ- subunits from bovine transducin or brain, restored the ability of the purified squid α subunit at 40 kDa to be ADP-ribosylated, confirming good homology with mammalian G proteins. We have also attempted to examine βγ- interaction with 5HT activation of adenylate cyclase in perfused axons; however, these experiments have proved difficult to perform, due to the relatively large amounts of subunits needed. The use of dialysed axons should circumvent this problem, as axons can simply be injected prior to insertion of dialysis tubing.

Guanylate cyclase

Metabolism of cGMP in the nervous system has been less well elucidated when compared to that of cAMP. From extensive studies in mammalian brain, lung and smooth muscle, and sea urchin sperm, it appears that guanylate cyclase exists in two forms: soluble (haem-containing, with two subunits; cytosolic) and particulate (a single non-haem subunit; membrane-associated) (Waldman and Murad 1987). The isolation or purification of these forms to allow detailed studies has been an experimental problem; receptor activation of guanylate cyclase activity in membrane fractions or cell-free systems has proved remarkably elusive. However, despite these problems, a very interesting and novel picture is emerging which points to the use of nitric oxide (NO) as an intercellular messenger for both the soluble form (Moncada *et al.* 1991) and the particulate form of guanylate cyclase and to its receptor being one and the same (Chinkers and Garbers 1991).

It appears that endothelium-derived relaxing factor (EDRF), which is a powerful but labile vasodilator, may be similar to a factor produced by *N*-methyl-D-aspartate (NMDA) receptor activation in mammalian brain (Garthwaite *et al.* 1988); coupled with these observations, there is strong evidence that EDRF and NO are the same. In neurones and endothelial cells, NO synthesis by NO synthase requires arginine and uses reduced nicotinamide adenine dinucleotide phosphate (NADPH) as a cofactor: the coproduct formed is citrulline. NO synthase activity in the rat central nervous system is triggered by Ca^{2+} and calmodulin, with K_m for Ca^{2+} close to 0.2 μM. Since NO can, in principle, act within the cell that produces it, it is thought that the level of Ca^{2+} required to initiate NO synthesis is generally sufficient to inhibit guanylate cyclase activity, such that NO can diffuse from the cell and act as an intercellular messenger (Knowles *et al.* 1989). NO appears to bind to the prosthetic haem group of soluble cyclase to directly activate the catalytic domain. Some of the confusion that slowed the emergence of this hypothesis resulted from the experimental use of free forms of the cyclase, purified with or without its attached haem group.

Activation of the particulate form of guanylate cyclase is closely coupled to receptor activation by peptides such as diarrhoea causing heat-stable enterotoxins, natriuretic peptides, and sea urchin egg peptides that activate spermatozoa (Chinkers and Garbers 1991). Recent sequencing and expression of clones suggest that the particulate form can be functionally divided into three domains; binding (receptor), kinase (protein kinase-like), and catalytic (cyclase). This latter catalytic domain has good homology to the catalytic domain of the soluble guanylate cyclase (Thorpe and Garbers 1989).

cGMP metabolism in intact and perfused axons

The cGMP content of axoplasm extruded from axons was found to be quite variable, and lay within the range 30–200 pmol/g (Adam-Vizi *et al.* 1988). Changing $[Ca^{2+}]_i$ by electrical stimulation or prolonged potassium depolarization had little effect on resting levels of cGMP. However, Na_o substitution by lithium, which also raises $[Ca^{2+}]_i$, halved cGMP content after a 30-min incubation. Basal PDE activity appeared quite significant; caffeine evoked a threefold increase in axonal cGMP content, compared to a twofold increase seen with cAMP. Of the neurotransmitters tested on intact squid axons, carbachol (100 μM) was effective in increasing axonal cGMP content twofold over controls, while glutamate (100 μM) had little effect. Both of these agents have been demonstrated to affect the electrophysiology of squid Schwann cells (Lieberman *et al.* 1989; Evans *et al.* 1991).

As with 5HT stimulation of cAMP content, the effects of carbachol persisted after inhibition of PDE with caffeine, suggesting activation of guanylate cyclase; the form and route of cyclase coupling remains uncertain.

A potent activator of guanylate cyclase in axons is nitroprusside (Adam-Vizi *et al.* 1988). Nitroprusside and sodium azide emerged in the late 1970s as agents that could activate cGMP synthesis in intact cells and broken cell preparations (Murad *et al.* 1979). There is now good evidence that haem is the site for the formation of a nitrosyl–porphyrin complex that then activates cyclase activity. Due to the variability in axonal cGMP content among squid, the data of Fig. 3.7 were obtained from divided axons of the same squid. Part (a) shows that potent stimulation by nitroprusside occurs with sub-μM concentrations, increasing cGMP content tens of times over basal. In part

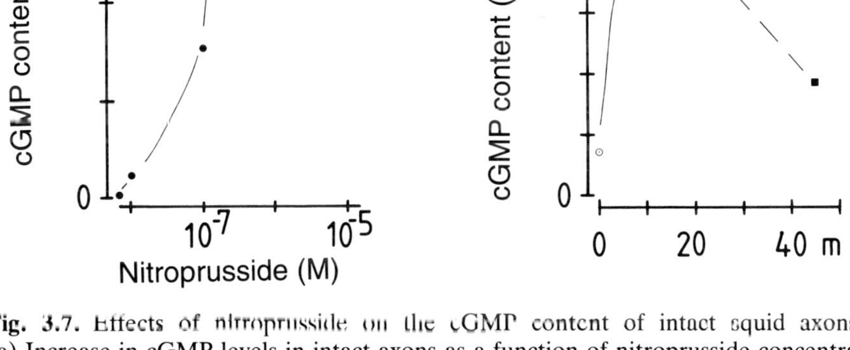

Fig. 3.7. Effects of nitroprusside on the cGMP content of intact squid axons. (a) Increase in cGMP levels in intact axons as a function of nitroprusside concentration. Incubation time 30 min. (b) Time course of the effect of 10 nM nitroprusside on cGMP levels of intact axons. Axon segments from the same squid were incubated for different times (open circles). After 15 min exposure one axon was washed a few times in the absence of nitroprusside for 30 min (closed square). (Reproduced with permission from Adam-Vizi *et al.* (1988).)

(b) the time course of nitroprusside activation is shown to be rapid and reversible; the lowering of axonal cGMP content after nitroprusside activation presumably occurs via PDE activity.

Nitroprusside stimulation of guanylate cyclase is thought primarily to occur via activation of the haem prosthetic group forming part of the soluble cyclase (Waldman and Murad 1987). One of the great advantages of the squid axon is the ease of extruding samples of axoplasm that should then facilitate studies on nitroprusside activation of the soluble guanylate cyclase moiety. However, when nitroprusside activation of cGMP production was compared between axons or isolated axoplasm injected with 0.1 μM nitroprusside, nitroprusside activation was some 10 times more effective in intact axons (Adam-Vizi *et al.* 1988). The data could suggest one of the following: pooling of soluble cyclase close to the plasma membrane (a cytoskeletal layer of 5–10 μm is left behind after axoplasmic extrusion), or location of important coupling factor(s) close to the membrane. To examine this, further experiments were performed on intracellularly perfused axons, in which axons were first cored of axoplasm and the intracellular compartment continuously washed (Fig. 3.8). Despite 1.5 h of continuous perfusion, 20-μm nitroprusside still evokes a large activation of guanylate cyclase activity. This appears transient and not greatly

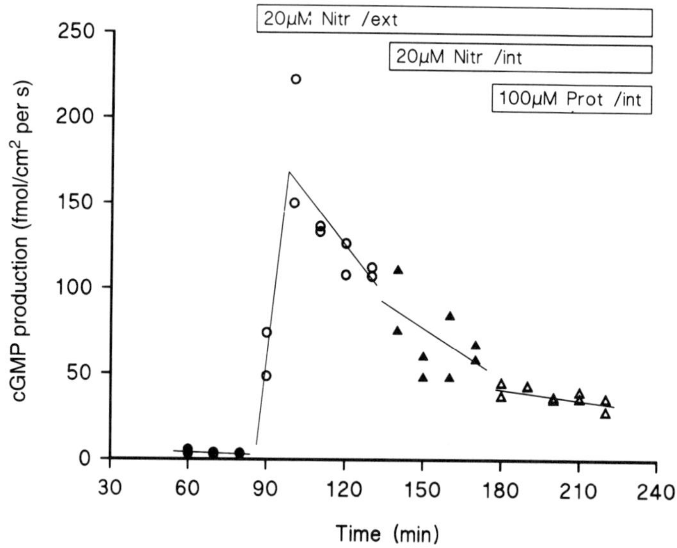

Fig. 3.8. Nitroprusside stimulation of cGMP production in the perfused squid axon. The axon was perfused with a K_2HPO_4-based perfusate containing 1 mM GTP, no ATP, 5 mM EGTA, and 1.5 mM $CaCl_2$. At the times indicated 20 μM nitroprusside was added externally (20 μM Nitr/ext) and internally (20 μM Nitr/int). Finally, 100 μM protoporphyrin IX was also added to the internal perfusate (100 μM Prot/int).

affected by further addition to the internal perfusate, suggesting that soluble cyclase may indeed be closely associated with the plasma membrane in axons. The transient nature of nitroprusside activation could be due to a number of possibilities: the slow wash-out of residual cyclases, or loss of the haem subunit needed for the formation of a nitrosyl–porphyrin complex which then activates the catalytic enzyme subunit. It has been reported that protoporphyrin IX (the demetalated precursor of haem) can activate soluble cyclase in haem- and NO-independent manner (Wolin *et al.* 1982); Fig. 3.8 shows that its addition to the perfusate failed to evoke a marked increase in cGMP production, but note that its addition may have stabilized cGMP production. It should also be noted that evidence for porphyrin regulation of particulate guanylate cyclase does exist (Waldman *et al.* 1984), but the literature remains unclear on this point (Waldman and Murad, 1987).

The perfused axon will help characterize the nature of the soluble and particulate guanylate cyclases found in squid axons, for example, in studies on ionic regulation of signal transduction coupling. Ionic regulation was mainly investigated in the 1970s (Murad *et al.* 1979), where strong activation by Mn^{2+} was noted but the effects of Ca^{2+} and Mg^{2+} were confused. This area has received little attention since then; Ca^{2+} can, albeit modestly, stimulate soluble and particulate cyclase in platelets (MacNeil 1987) but can also inhibit the NO-activated soluble cyclase in the central nervous system with K_m close to 100 nM (Knowles *et al.* 1989). Perhaps more importantly, the introduction and use of dialysis tubing that is porous to solutes up to 5 kDa, thus retaining cytosolic factors (Allen and Knight 1992), will greatly promote the experimental advantage of the giant axon preparation.

Due to the intimate association of the Schwann cell layer to the squid giant axon (Brown *et al.* 1991) it is possible that Schwann cell receptor activation could release an intercellular, NO-like substance that regulates the large reservoir of guanylate cyclase activity in the giant axon. Identification and characterization of the receptor types in Schwann cells (e.g Evans *et al.* 1991; see also Chapter 15) and the giant axon may lead the way towards a detailed study of signal:transduction coupling effected by guanylate cyclases, a research area that has been little explored.

Concluding remarks

The squid giant axon and its surrounding Schwann cells lend themselves to a number of studies involving signal:transduction coupling, particularly in the area of cyclic nucleotide homeostasis. Intact axons, perfused axons cored of axoplasm, and internally dialysed axons provide the experimenter with a means to investigate not only how cAMP or cGMP signals arise but, more importantly, how they can be modulated under conditions that retain membrane chemical and electrical asymmetry.

Acknowledgements

I am grateful to the Medical Research Council, Marine Biological Association, and the British Heart Foundation for their financial support, and to the Director of the Marine Biological Association Laboratory in Plymouth, for the use of its facilities.

References

Adam-Vizi, V., Allen, T. J. A., and Baker, P. F. (1988). The effects of nitroprusside and putative agonists on guanylate cyclase activity in squid giant axons. *Biochimica et Biophysica Acta*, **938**, 461–8.

Allen, T. J. A. (1989). Characteristics of a 5-hydroxytryptamine-sensitive adenylate cyclase in intact and intracellularly perfused squid axons. *Journal of Physiology*, **415**, 1–18.

Allen, T. J. A. and Baker, P. F. (1986). Comparison of the effects of potassium and membrane potential on the calcium-dependent sodium efflux in squid axons. *Journal of Physiology*, **378**, 53–76.

Allen, T. J. A. and Knight, D. E. (1992). The use of intracellular dialysis to study signal transduction coupling in the squid giant axon. *Journal of Neuroscience Methods*, **42**, 169–74.

Baker, P. F. and Carruthers, T. (1982). Cyclic AMP metabolism in squid giant axons and giant barnacle muscle fibres. *Journal of Physiology*, **326**, 13P.

Birnbaumer, L., Abramowitz, J., and Brown, A. M. (1990). Receptor–effector coupling by G proteins. *Biochimica et Biophysica Acta*, **1031**, 163–224.

Bornancin, F. and Chabre, M. (1991). Cholera toxin ADP-ribosylates transducin only when it is bound to photoexcited rhodopsin and depleted of its nucleotide. *Federation of European Biochemical Societies Letters*, **291**, 273–6.

Brabet, P., Pantaloni, C., Rouot, B., Toutant, M., GarciaSainz, A., Bockaert, J., Homburger, V. (1988). Multiple species and isoforms of *Bordetella pertussis* toxin substrates. *Biochemical and Biophysical Research Communications*, **152(3)**, 1185–92.

Brown, E. R., Bone, Q., Ryan, K. P., and Abbott, N. J. (1991). Morphology and electrical properties of Schwann cells around the giant axon of the squids *Loligo forbesi* and *Loligo vulgaris*. *Proceedings of the Royal Society of London*, **B 243**, 255–62.

Brostrom, C. O., Brostrom, M. A., Breckenridge, B., and Wolff, D. J. (1978). Calcium regulation of brain adenylate cyclase. *Advances in Cyclic Nucleotide Research*, **9**, 85–99.

Cavalié, A., Allen, T. J. A., and Trautwein, W. (1991). Role of the GTP-binding protein G_s in the β-adrenergic modulation of cardiac Ca channels. *Pflüger's Archiv*, **419**, 85–95.

Chinkers, M. and Garbers, D. L. (1991). Signal transduction by guanyl cyclases. *Annual Reviews of Biochemistry*, **60**, 553–75.

Evans, P. D., Reale, V., Merzon, R. M., and Villegas, J. (1991). NMDA and non-NMDA type glutamate receptors are present on squid giant axon Schwann cell. *Journal of Experimental Biology*, **157**, 593–600.

Garthwaite, J. (1991). Glutamate, NO and cell–cell signalling in the nervous system. *Trends in Neuroscience*, **14**, 60–7.

Garthwaite, J., Charles, S. L., and Chess-Williams, R. (1988). EDRF release on activation of NMDA receptors suggests a role as an intercellular messenger in brain. *Nature*, **336**, 385–8.

Gill, D. M. (1977). Mechanism of action of cholera toxin. *Advances in Cyclic Nucleotide Research*, **6**, 85–118.

Gilman, A. G. (1987). G proteins: transducers of receptor generated signals. *Annual Reviews of Biochemistry*, **56**, 615–49.

Goldsmith, P., Gierschik, P., Milligan, G., Unson, C., Malech, H. L., and Spiegel, A. M. (1987). Antibodies directed against synthetic peptides distinguish between GTP-binding proteins in neutrophil and brain. *Journal of Biological Chemistry*, **262**, 1–6.

Hildebrandt, J. D., Sekura, R. D., Codina, J., Iyengar, R., Manclark, C. R., and Birnbaumer, L. (1983). Stimulation and inhibition of adenylyl cyclases is mediated by distinct proteins. *Nature*, **302**, 706–9.

Homburger, V., Brabet, P., Audigier, Y., Pantaloni, C., Bockaert, J., and Rouot, B. (1987). Immunological localization of the GTP-binding protein G_o in different tissues of vertebrates and invertebrates. *Molecular Pharmacology*, **31**, 313–19.

Katada, T., Oinuma, M., Kusakabe, K., and Ui, M. (1987). A new GTP-binding protein in brain tissues serving as the specific substrate of islet-activating protein, pertussis toxin. *Federation of European Biochemical Societies Letters*, **213**, 353–8.

Khan, R. A. and Gilman, A. G. (1986). The protein cofactor necessary for ADP-ribosylation of Gs by cholera toxin is itself a GTP binding protein. *Journal of Biological Chemistry*, **261**, 7906–11.

Knowles, R. G., Palacios, M., Palmer, R. M., and Moncada, S. (1989). Function of NO from L-arginine in the CNS: a transduction mechanism for stimulation of the soluble guanylate cyclase. *Proceedings of the National Academy of Sciences, USA*, **86**, 5159–62.

Lieberman, E. M., Abbott, N. J., and Hassan, S. (1989). Evidence that glutamate mediates axon–Schwann cell signalling in the squid. *Glia*, **2**, 94–102.

MacNeil, S. (1987). Investigation of the role of Ca^{2+} and calmodulin in the regulation of platelet guanylate cyclase activity. *Biochemical Journal*, **242**, 607–10.

Mattera, R., Codina, J., Sekura, R. D., and Birnbaumer, L. (1987). GTP reduces ADP-ribosylation of the inhibitory guanine nucleotide regulatory protein of adenylyl cyclase by PTX without causing dissociation of the subunits of N_i. *Journal of Biological Chemistry*, **262**, 11247–51.

Minocherhomjee, A. M., Selfe, S., Flowers, N. J., and Storm, D. R. (1987). Direct interaction between the catalytic subunit of the calmodulin-sensitive adenylate cyclase from bovine brain with ^{125}I-labelled wheat germ agglutinin and ^{125}I-labelled calmodulin. *Biochemistry*, **26**, 4444–8.

Moncada, S., Palmer, R. M. J., and Higgs, E. A. (1991). Nitric oxide: physiology, pathophysiology and pharmacology. *Pharmacological Reviews*, **43**, 109–42.

Murad, F., Arnold, W. P., Mittal, C. K., and Braughler, J. M. (1979). Properties and regulation of guanylate cyclase and some proposed functions for CGMP. *Advances in Cyclic Nucleotide Research*, **11**, 175–204.

Pottinger, J. D., Ryba, N. J., Keen, J. N., and Findlay, J. B. (1991). The identification and purification of the heterotrimeric GTP-binding protein from squid (*Loligo forbesi*) photoreceptors. *Biochemical Journal*, **279**, 323–6.

Rasenick, M. M., Stein, P. J., and Bitensky, M. W. (1981). The regulatory subunit of adenylate cyclase interacts with cytoskeletal components. *Nature*, **294**, 560–2.

Rouot, B., Zumbihl, R., Charpenter, N., Chabbert, C., Carrette, J., Bockaert, J., and Homburger, V. (1992). Specific antibodies against G_o isoforms reveal the early expression of the $G_{o2\alpha}$ subunit and appearance of $G_{o1\alpha}$ during neuronal differentiation. *Molecular Pharmacology*, **41(2)**, 273–80.

Strada, S. J., Martin, M. W., and Thompson, W. J. (1984). General properties of multiple molecular forms of cyclic nucleotide phosphodiesterase in the nervous system. *Advances in Cyclic Nucleotide and Protein Phosphorylation Research*, **16**, 13–29.

Thorpe, D. S. and Garbers, D. L. (1989). The membrane form of guanylate cyclase; homology with a subunit of the cytoplasmic form of the enzyme. *Journal of Biological Chemistry*, **264**, 6545–9.

Tsai, S.-C., Noda, M., Adamik, R., Moss, J., and Vaughan, M. (1987). Enhancement of choleragen ADP-ribosyltransferase activities by guanyl nucleotides and a 19 kDa membrane protein. *Proceedings of the National Academy of Sciences, USA*, **84**, 5139–42.

Waldman, S. A. and Murad, F. (1987). Cyclic GMP synthesis and function. *Pharmacological Reviews*, **39**, 163–96.

Waldman, S. A., Sinacore, M. S., Lewicki, J. A., Chang, L. Y., and Murad, F. (1984). Selective activation of particulate guanylate cyclase by a specific class of porphyrins. *Journal of Biological Chemistry*, **259**, 4038–42.

Wolin, M. S., Wood, K. S., and Ignarro, L. J. (1982). Guanylate cyclase from bovine lung. A kinetic analysis of the regulation of the unpurified soluble enzyme by protoporphyrin IX, heme, and nitrosyl-heme. *Journal of Biological Chemistry*, **257**, 13312–20.

Yatani, A., Mattera, R., Codina, J., Graf, R., Okabe, K., Padrell, E., Iyengar, R., Brown, A. M., and Birnbaumer, L. (1988). The G-protein-gated atrial K^+ channel is stimulated by three distinct $G_{i\alpha}$ subunits. *Nature*, **336**, 680–2.

4

The detection and properties of electrogenic Na^+–K^+ transport in the squid axon membrane

U. KISHIMOTO, I. INOUE, I. TSUTSUI, and T. OHKAWA

Summary

The ouabain-sensitive current (i_p) in the squid giant axon (*Sepioteuthis lessoniana*) was extracted under varied external Na^+ and K^+ concentrations. All the i_p–V curves can be fitted with a reaction kinetic scheme that assumes a single electrogenic step in the reaction cycle. Two rate constants (k_{12} and k_{21}) in the electrogenic step, two rate constants (r_{21} and r_{12}) in the electroneutral step, and the amount of charge carried during each cycle are determined with a lumped two-state model. The charge carried in the electrogenic step (translocation of 3 Na^+ across the membrane) is +1. Not only the occlusion and deocclusion of Na^+ or K^+ but also the translocation of 2 K^+ is independent of the membrane potential. Only two rate constants in the electrogenic step, i.e., k_{12} ($=k^0_{12}$ exp($zFV/2RT$)) and k_{21} ($=k^0_{21}$ exp($-zFV/2RT$)), are assumed to be highly voltage-dependent. All the rate constants (k^0_{12}, k^0_{21} r_{21}, r_{12}) change with external Na^+ and K^+ concentrations in systematic ways. Decrease of external Na^+ concentration causes a great increase of r_{12}, a marked decrease of k^0_{21} and a slight decrease of k^0_{12}; decrease of external K^+ concentration causes a great increase of r_{21}, increase of k^0_{21}, and a slight decrease of k^0_{12}. Affinity of the enzyme for the ligand increases markedly when the ligand concentration decreases. The standard free energy for the cycle remains unchanged at − 33 kJ/mol, which corresponds to that of ATP-hydrolysis. The reversal potential (E_p) of the pump is highly dependent on the external concentrations of Na^+ and K^+, that is, E_p shifts toward hyperpolarization almost instantly with decreasing external $[Na]_o$. If $[Na]_o$ is lower than 30 mM, the sodium pump behaves as a constant current source in the physiological voltage range. On the other hand, E_p shifts toward depolarization with decreasing external $[K]_o$. If $[K]_o$ is lower than 1 mM, the pump current cannot be outward at the resting potential, but is inward.

The Na pump in the plasma membrane of animal cells mediates uphill transport of Na^+ and K^+. The energy is supplied from ATP-hydrolysis (Hodgkin and Keynes 1955). A number of investigators have sought experimental evidence for the Na/K pumping rate as a function of the membrane potential in a variety of cells and tissues (reviews by Thomas 1972; Glynn and Karlish 1975; Beaugé 1984; DeWeer *et al.* 1988; Apell 1989). Cyclic change of conformation of the enzyme, including steps of phosphorylation, dephosphorylation, occlusions, and deocclusions of ions, is proposed for its mechanism (Albers–Post scheme). It seems unequivocal that the mechanism is highly voltage-dependent and that there is a reversal potential. However, the experimental evidence for a quantitative relation between the pumping rate and the membrane potential have been lacking because of the technical difficulties involved.

Experimental procedures

Experiments were performed on giant axons isolated from the squid, *Sepioteuthis lessoniana*, at the Marine Biological Station, Tokushima University, Naruto, Japan. The axon diameter varied between 450 and 650 μm.

A standard voltage-clamp system (Cole and Moore 1960) was used. A series resistance of about 0.5 Ω cm^2 was compensated electronically. Applied voltage changes were the holding potential at the resting level, ramp hyperpolarization down to about −110 mV, holding potential for about 5 s, ramp depolarization up to about 0 mV, and back to the holding potential (Fig. 4.1). The rate of ramp was 50 mV/s both for hyperpolarization and for depolarization. This rate of ramp was so slow that the current versus voltage (*I–V*) curves obtained were the same within experimental error as those obtained under voltage clamp of different steps. These voltage changes were generated by means of a microcomputer-controlled function synthesizer (NF1731, NF, Japan). The current records together with the voltage record were sampled at a rate of 10 ms per point with a 12-bit A/D converter (S-210, Autnics, Japan) and transferred to another computer (9826, Hewlett-Packard, USA). With this method 512 data points (360 data points during the ramps for about 2.6 s) were collected. Computations such as subtraction, smoothing, averaging,

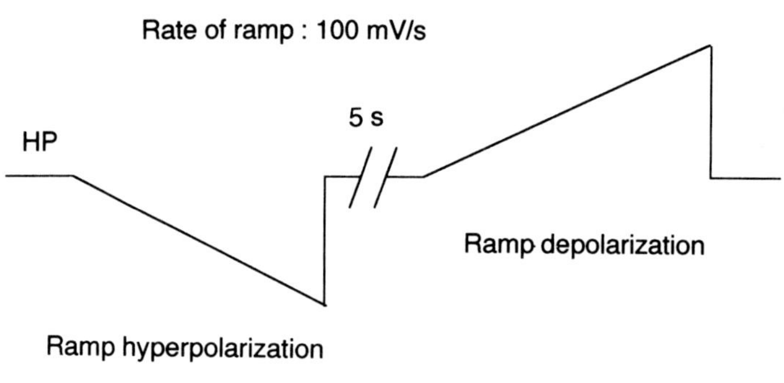

Fig. 4.1. A scheme for the ramp voltage clamp. The rate of ramp was 50 mV/s, which was slow enough to give the steady-state current–voltage curve. HP, Holding potential.

curve fitting, and regression analyses were carried out with other microcomputers (NEC 9801 VM2 and Apple Mac IIcx).

The essential point in the present experiment is to compare the steady-state *I–V* curve before addition of 100 μM ouabain (Sigma, USA) with that after its addition. Since this ouabain-sensitive current is generally one to two orders smaller than the passive currents normally carried by Na^+, K^+, and Cl^-, reduction of the latter currents is needed for a better comparison. Tetrodotoxin (300 nM TTX, Sankyo, Japan) was added to the external solution to block the Na^+ channel. Effective reduction of the outward K^+ current was achieved by injecting a small amount of 0.6 M tetraethylammonium chloride (TEA-chloride, Sigma, final concentration about 50 mM) into the axon at the centre measuring region with another micropipette that was introduced from the end of the axon. External Cl^- concentration was also reduced to 39 mM to minimize the contribution of Cl^- current (Inoue 1985). The low-Cl artificial sea water (low-Cl-ASW) contained a 429-mM mixture of tetramethylammonium methanesulfonate, Na-methanesulfonate, and K-methanesulfonate, 10 mM $CaCl_2$, and 50 mM $MgSO_4$, the pH being adjusted to 7.8 with 10 mM Tris–HCl. The concentrations of Na^+ and K^+ were changed while maintaining the total concentration of methanesulfonate at 429 mM (Table 4.1).

Table 4.1. Ionic constitution of the low-Cl ASW (artificial sea water). External concentrations of Na^+ and K^+ were changed keeping the total concentration of methanesulfonate at 429 mM. This was achieved by mixing tetramethylammonium methanesulfonate, Na-methanesulfonate, and K-methanesulfonate (all 429 mM) in the desired proportions*

	Ionic concentration (mM)	
	Axoplasm	Artificial sea water (low Cl)
K	400	Varied
Na	40	Varied
Ca		10
Mg	10	50
Tetramethylammonium		Varied
Cl	40–150	39
SO_4		50
Methanesulfonate		429
ATP	2	0
ADP	0.5	0
P_i	15	0
TEA-Cl	50	
TTX		3×10^{-4}
Ouabain		~0.1

* The pH of the axoplasm was 7.3–7.4 and that of the ASW was 7.8.

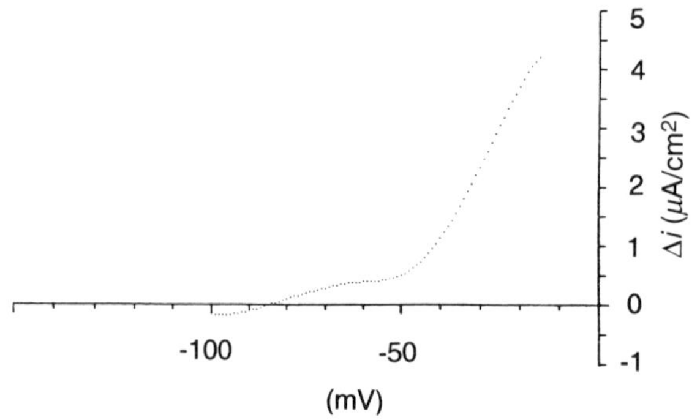

Fig. 4.2. An example of the differential current–voltage curve (Δi–V curve) which is obtained by subtracting the current–voltage curve after addition of 100 μM ouabain from that before its addition. This is the original data of 90 May 18b5–10 listed in Table 4.2, where $[Na]_o$ is 423, $[K]_o$ 20, and $[Cl]_o$ 39 mM. A potential candidate for the ouabain-sensitive current can be found generally only in a limited voltage range (i.e. in a region of $\sim\pm30$ mV around the resting potential).

The current component carried by K^+ was markedly reduced by TEA injection. However, blocking of the K^+-channel with TEA was not necessarily perfect. We found that the remaining K^+-current component continued to decrease slowly. On the other hand, hyperpolarization beyond -110 mV caused some signs of membrane breakdown, the extent of which also changed with time. A typical example is shown in Fig. 4.2.

If it were possible to block all the passive ionic channels of the living cell membrane, we could obtain a perfect I–V curve of the pumping system. So far this has proved impossible in the squid giant axon membrane. What we could do is to subtract the I–V curve after addition of ouabain from that before its addition. The extracted differential Δi–V curve would be the ouabain-sensitive current versus voltage (i_p–V) curve of the Na pump so long as the current through passive channels remained unchanged. Generally, a shift of the resting potential by a few mV occurs on addition of ouabain. This causes a slight change in the I–V curve of the passive channels, which is inevitably superimposed on the differential Δi–V curve. On the other hand, activations or inactivations of passive channels (i.e. changes of the electromotive force of the membrane) are evoked if we expand the measuring voltage range either under the ramp voltage clamp or under the step voltage clamp. These would be demonstrated by the tail current on coming back to the holding potential under the voltage clamp. Therefore, the superimposition of a part of passive ionic current components in the differential Δi–V curve is enhanced further, the farther away the voltage is from the resting potential. This situation is what occurs in the squid giant axon membrane and has made it difficult to extract a reliable i_p–V curve over a wide voltage range.

For the reason stated above, the differential Δi-V curve is reliable as a candidate for the i_p–V only in a limited voltage range such as 30–40 mV around the resting potential. It is unfortunate indeed that we could not use all the data over a wider voltage range. The curve thus extracted has its own curvature, although it is small because of the limited voltage range. However, we could collect about 30 or more data points in this limited voltage range by applying a slow ramp under the voltage clamp condition. This amount of data is enough to apply a program of non-linear optimization to find a best set of kinetic parameters characterizing the i_p–V curve. We adopted a computer program of non-linear optimization (a revised version of SIMPLX or POWELL). This is a program of successive approximation, which is repeated some hundreds of times to find the least sum of squared errors. Only if the functional shape of the adopted model is adequate, can the experimental data, which consist of about 30 data points in this voltage range, be successfully fitted and a set of parameters that characterize the shape of the i_p–V curve be determined. Our next step is to see whether or not there are systematic relations between the kinetic parameters determined and the external Na^+ and K^+ concentrations.

Kinetics with a lumped two-state model

Post and Jolly (1957) reached a conclusion from their Na and K flux measurements that the stoichiometry of the Na pump of erythrocytes is 3Na/2K/1ATP. This notion is supported by simultaneous measurements of fluxes and ouabain-sensitive currents in snail neurones (Thomas 1972), in Purkinje fibres (Eisner *et al.* 1981; Glitsch *et al.* 1982), in squid giant axons (DeWeer *et al.* 1986), in *Aplysia* neurones (Cooke *et al.* 1974), in barnacle muscle cells (Lederer and Nelson 1984), and in *Xenopus* oocytes (Eisner *et al.* 1987). Recently, Rakowski *et al.* (1989; Rakowski 1989) concluded from their simultaneous measurements of ouabain-sensitive Na^+ efflux and current in internally dialysed squid giant axons that this stoichiometry is independent of changes in voltage from −120 to +30 mV and of changes in external Na^+ concentration. In the kinetic model presented here the number of electric charges gained at the internal surface of the pump is $-4(+\mathrm{ATP})+3(+\mathrm{Na}^+)+2(-\mathrm{ADP})+2(-\mathrm{P_i})-2(-\mathrm{K}^+)= +1$, while that at the external surface is $-3(-\mathrm{Na}^+)+2(+\mathrm{K}^+)=-1$. ($P_i$, inorganic phosphate; ATP, adenosine triphosphate; ADP, adenosine diphosphate.) Therefore, it is likely that a single net positive charge is moving during each cycle of the pump.

The data presented in this report all support the idea that translocation of three Na^+ ions across the membrane is electrogenic carrying a single positive charge (z=1, Table 4.2), while translocation of 2 K^+ ions is electroneutral. Such a notion has been suggested by the measurement of strophanthidin-

Table 4.2 The results of fitting the ouabain-sensitive current–voltage curves under varied $[Na]_o$ and $[K]_o$

Date	Temper-ature	$[Na]_o$	$[K]_o$	E_p	z	sFE	k^o_{12}	k^o_{21}	r_{21}	r_{12}	A_1	A_2	A_3	A_4	$-A_4(A_1/A_2)$
	°C	mM	mM	mV		$J \cdot mol^{-1}$	sec^{-1}	sec^{-1}	$mol^{-6} sec^{-1}$	$mol^{-7} sec^{-1}$				$\mu A \cdot cm^{-2}$	$\mu A \cdot cm^{-2}$
89Jun13a8-10	15.9	399	30	-170.08	1.00	-36325	1750	14	1.39E+11	4.71E+06	0.00108	0.00804	0.0052	0.800	-0.108
90May21b10-12	12.3	423	20	-76.63	1.00	-29401	3320	80.6	1.91E+11	3.28E+07	0.0444	0.0243	0.00417	0.489	-0.893
90May18b5-10	12.6	423	20	-85.27	1.00	-30257	3070	145	3.79E+11	2.98E+07	0.0313	0.0375	0.0058	0.970	-0.811
86Jul17b13-16	13.4	423	9	-60.98	1.00	-31780	3730	298	2.03E+12	4.08E+07	0.0845	0.08	0.0058	1.050	-1.110
86Dec17c5-8	13.1	423	9	-71.52	1.00	-32770	6680	668	4.82E+12	5.05E+07	0.0551	0.1	0.0058	2.500	-1.380
90May18a9-10	12.4	423	10	-81.75	1.00	-33193	426	30.9	8.93E+11	1.04E+07	0.0361	0.0726	0.0201	0.571	-0.284
90May18a7-11	12.1	423	10	-76.17	1.02	-33512	5230	365	2.57E+12	3.65E+07	0.0421	0.0697	0.00504	0.168	-1.020
90May18a8-11	12.1	423	10	-89.72	1.00	-33936	1340	100	2.34E+12	1.91E+07	0.026	0.075	0.0151	1.500	-0.520
90May18a9-11	12.4	423	10	-68.44	1.02	-32822	2040	141	2.60E+12	5.15E+07	0.058	0.0689	0.015	1.710	-1.440
90May18a10-11	12.4	423	10	-65.67	1.00	-31643	1910	144	2.34E+12	5.09E+07	0.0693	0.075	0.0151	1.500	-1.390
89Dec01d6-7	12.1	420	9	-50.60	1.04	-32112	1410	174	6.86E+11	1.27E+07	0.117	0.123	0.00492	0.371	-0.353
90May21c9-11	12.6	423	5	-46.73	1.04	-34617	3040	146	2.50E+12	4.19E+07	0.139	0.0486	0.00506	0.415	-1.190
90May17b7-8	12.4	423	5	-46.68	1.00	-33102	2930	1100	6.63E+12	1.56E+07	0.15	0.375	0.00508	1.060	-0.425
90May17b7-11	12.4	423	5	-31.81	1.04	-32913	4440	910	5.28E+12	3.97E+07	0.262	0.205	0.00434	0.875	-1.120
90May22b4-5	13.1	423	2	-38.62	0.90	-32735	1520	425	1.58E+13	1.30E+07	0.244	0.28	0.00499	0.364	-0.318
89Jun02c14-16	12.9	400	2	-3.94	1.00	-32988	1650	2010	1.93E+13	1.50E+07	0.852	1.22	0.00508	0.493	-0.346
89Jun02c15-16	12.9	400	2	-3.94	1.00	-32988	2240	2720	2.61E+13	2.03E+07	0.852	1.22	0.00508	0.668	-0.468
89May22b9-13	15.2	210	30	-164.75	1.10	-36009	2350	2.41	7.65E+10	8.78E+07	0.00068	0.00102	0.00312	0.485	-0.322
90May19c9-14	12.4	200	10	-106.18	1.00	-30216	2260	47.4	1.36E+12	1.92E+08	0.0134	0.021	0.00629	0.868	-0.552
89May19b21-22	14.6	210	9	-131.38	1.00	-33655	1470	14.7	1.10E+12	8.55E+07	0.005	0.01	0.0058	0.570	-0.285
90May19b10-11	12.6	200	5	-93.38	1.00	-32177	1710	179	4.98E+12	6.02E+07	0.0228	0.105	0.00567	0.794	-0.173
90May20a6-8	12.9	200	2	-40.42	1.00	-31562	2560	1920	4.03E+13	9.27E+07	0.194	0.75	0.00508	1.030	-0.267
90May18d6-10	11.9	200	2	-48.73	0.98	-31537	3950	177	3.02E+13	8.56E+08	0.143	0.0449	0.0082	0.758	-2.420
89Dec03b11-13	11.9	210	1	-19.00	1.00	-33031	1520	762	2.34E+14	4.15E+08	0.461	0.5	0.0189	1.500	-1.380
90May18c12-14	11.9	200	1	-30.77	1.00	-33820	2070	776	9.33E+13	1.58E+08	0.286	0.375	0.00508	0.597	-0.455
90May15a10-14	12.4	100	2	-76.27	1.02	-30851	2660	172	3.68E+13	1.71E+09	0.0421	0.0646	0.00585	0.963	-0.628
90May15b17-19	12.4	100	2	-105.72	1.00	-32876	2620	11.2	1.29E+13	2.93E+09	0.0136	0.00426	0.00529	0.330	-1.050
90May16a7-9	12.1	100	1	-67.05	1.00	-32400	3600	1080	2.34E+14	9.07E+08	0.0653	0.3	0.00508	1.500	-0.327
90May16a8-9	12.2	100	1	-59.21	1.00	-31662	2460	738	1.50E+14	8.01E+08	0.09	0.3	0.00508	0.962	-0.289
90May16c7-10	12.4	100	0.5	-41.63	1.00	-33276	2810	1450	6.56E+14	1.04E+09	0.184	0.515	0.00508	1.050	-0.376
90May16b14-16	12.4	100	0.5	-45.06	1.02	-34219	2650	206	9.47E+14	8.40E+09	0.156	0.078	0.0172	1.540	-3.070
90May16c7-10	12.4	100	0.5	-42.83	1.05	-35120	1340	546	7.86E+14	1.39E+09	0.162	0.406	0.013	1.320	-0.524
89May22a15-17	15.6	50	9	-239.67	1.00	-33335	12000	0.0719	7.97E+11	1.24E+11	0.0000656	0.000006	0.00508	0.510	-0.558
89Dec02b22-25	11.4	10	9	-296.50	1.16	-37453	1650	0.0014	1.19E+12	1.64E+12	0.000000816	0.00000085	0.00731	0.712	-0.684
89Dec02b23-25	11.5	10	9	-402.84	0.96	-34936	1290	0.000193	6.14E+11	8.48E+11	0.000000144	0.00000015	0.00484	0.305	-0.293
89Dec03a7-8	11.5	10	9	-356.89	1.00	-33571	1520	0.000758	7.58E+11	1.05E+12	0.00000048	0.0000005	0.00508	0.393	-0.377
89Nov29b14-15	11.4	10	9	-346.48	1.00	-32567	2410	0.00193	1.24E+12	1.62E+12	0.00000073	0.0000008	0.00508	0.641	-0.585
87Jun02c5-8	12.2	10	0.1	-130.28	1.00	-33057	391	3.91	2.36E+16	2.10E+12	0.005	0.01	0.058	1.510	-0.755
87Jun02d8-9	12.5	10	0.1	-117.07	1.01	-32222	67	0.577	2.77E+15	4.71E+11	0.00821	0.00859	0.0516	0.179	-0.171
87Jun10d13-15	11.2	10	0.1	-131.03	1.00	-33057	135	1.36	8.43E+15	7.10E+11	0.00476	0.01	0.0587	0.540	-0.256
87Jun10d14-15	11.2	10	0.1	-115.85	1.01	-32113	205	1.81	8.64E+15	1.45E+12	0.00835	0.00886	0.0525	0.559	-0.527
87Jun08a20-21	12.3	10	0.01	18.26	1.00	-29662	197	147	4.69E+17	2.33E+12	2.1	0.75	0.058	0.300	-0.840
87Jun08b10-11	13.2	10	0.01	1.04	1.06	-33393	178	23	2.56E+17	3.68E+12	1.05	0.129	0.0839	0.174	-1.410
				Averages	1.01	-32904									

sensitive transient current in cardiac cells (Nakao and Gadsby 1986), by the measurements of fluxes as a function of voltage in reconstituted Na pumps in phospholipid vesicles (Karlish *et al.* 1985; Rephaeli *et al.* 1986*a*,*b*; Apell *et al.* 1987; Goldschlegger *et al.* 1987), by analyses of non-stationary pump currents after an ATP-concentration jump in purified microsomal membrane sheets bound to lipid bilayers (Apell *et al.* 1987; Borlinghaus *et al.* 1987; Fendler *et al.* 1987; Nagel *et al.* 1987; Stürmer *et al.* 1989), and also by experiments in guinea-pig ventricular myocytes (Bahinski *et al.* 1988) and in *Xenopus* oocytes (Rakowski and Paxson 1988).

We adopted a simplified scheme (Fig. 4.3) for the cyclic change of Na^+ K^+-ATPase (Post–Albers type scheme). The cycle includes binding of ATP, release of ADP, occlusion of 3 Na^+ at the inner surface, translocation of 3 Na^+ through the channel portion of the enzyme, deocclusion of 3 Na^+ at the outer surface, release of inorganic phosphate, P_i, occlusion of 2 K^+ at the external surface, and translocation of 2 K^+ and release of 2 K^+ at the inner surface. The applicability of this type of kinetic model for the pumping mechanism has already been demonstrated (Läuger 1979, 1984; Hansen *et al.* 1981; Chapman *et al.* 1983; Kishimoto *et al.* 1984, 1985; Takeuchi *et al.* 1985; Kami-ike *et al.* 1986).

Judging from the results of fitting the extracted i_p–V curve, only a step that translocates 3 Na^+ seems to be electrogenic, that is carries a single positive charge. For simplicity, we lumped together other steps into a non-electrogenic step as shown in Fig. 4.3(b). Then, the net rate of forward (clockwise) reaction in the electrogenic step is given by

$$f_{12}\ (\text{electrogenic}) = k_{12}E_1 - k_{21}E_2 \tag{4.1}$$

where E_1 and E_2 are densities of the enzyme intermediates, and k_{12} and k_{21} are rate constants.

The net rate of forward reaction in the non-electrogenic step is given by

$$f_{21}\ (\text{electroneutral}) = r_{21}[\text{ATP}][\text{Na}]_i^3[\text{K}]_o^2E_2 - r_{12}[\text{ADP}][\text{P}_i][\text{Na}]_o^3[\text{K}]_i^2E_1. \tag{4.2}$$

In the steady state of the cycle, f_{12}(electrogenic) = f_{21}(electroneutral).

$$\text{Conservation of the enzyme: } E_1 + E_2 = E_0. \tag{4.3}$$

From these two conditions E_1 and E_2 are calculated. Then, the net rate of the forward cycle can be calculated as

$$f_{12}\ (= f_{21}) = \frac{(r_{21}[\text{ATP}]\ [\text{Na}]_i^3[\text{K}]_o^2k_{12} - r_{12}[\text{ADP}]\ [\text{P}_i]\ [\text{Na}]_o^3[\text{K}]_i^2k_{21})E_0}{k_{12} + k_{21} + r_{21}\ [\text{ATP}]\ [\text{Na}]_i^3\ [\text{K}]_o^2 + r_{12}\ [\text{ADP}]\ [\text{P}_i]\ [\text{Na}]_o^3\ [\text{K}]_i^2} \tag{4.4}$$

Symmetry in the energy barrier in the electrogenic step is assumed (Läuger 1979) so that

$$k_{12} = k^0_{12}\exp(zFV/2RT); \qquad k_{21} = k^0_{21}\exp(-zFV/2RT) \tag{4.5}$$

where F is the Faraday, R the gas constant, T the absolute temperature,

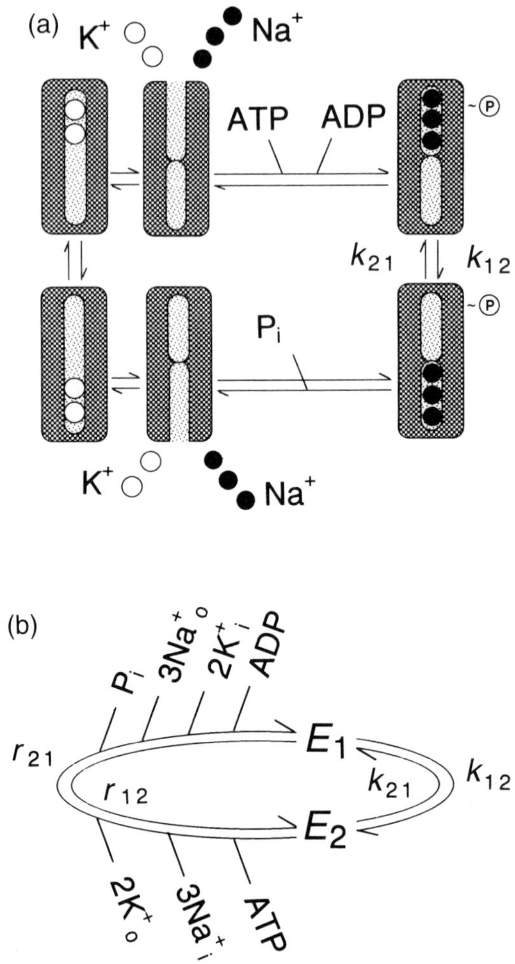

Fig. 4.3. (a) and (b) Post–Albers type kinetic schemes for the Na–K pump. Phosphorylation, dephosphorylation of the enzyme, occlusions into, and deocclusions from the enzyme of Na^+ and K^+ are supposed to be non-electrogenic. Only the translocation of 3 Na^+ in the channel portion of the enzyme is assumed to be electrogenic, while that of 2 K^+ is electroneutral. The non-electrogenic steps in (a) are lumped together as a single step in (b) having voltage-independent rate constants (r_{21} and r_{12}).

and z is the charge translocated. Introduction of k_{12} and k_{21} into the above equation and multiplication of it by zF gives the current as a function of voltage. That is,

$$i_p = \frac{\{\exp(zFV/2RT) - A_1 \exp(-zFV/2RT)\}A_4}{\exp(zFV/2RT) + A_2 \exp(-zFV/2RT) + A_3} \qquad (4.6)$$

where

$$A_1 = \frac{r_{12}\,[\mathrm{ADP}]\,[\mathrm{P_i}]\,[\mathrm{Na}]_\mathrm{o}^3\,[\mathrm{K}]_\mathrm{i}^2}{r_{21}\,[\mathrm{ATP}]\,[\mathrm{Na}]_\mathrm{i}^3\,[\mathrm{K}]_\mathrm{o}^2}\,\frac{k^0{}_{21}}{k^0{}_{12}} \tag{4.7}$$

$$A_2 = \frac{k^0{}_{21}}{k^0{}_{12}} \tag{4.8}$$

$$A_3 = \frac{r_{12}\,[\mathrm{ADP}]\,[\mathrm{P_i}]\,[\mathrm{Na}]_\mathrm{o}^3\,[\mathrm{K}]_\mathrm{i}^2 + r_{21}\,[\mathrm{ATP}]\,[\mathrm{Na}]_\mathrm{i}^3\,[\mathrm{K}]_\mathrm{o}^2}{k^0{}_{12}} \tag{4.9}$$

$$A_4 = r_{21}\,[\mathrm{ATP}]\,[\mathrm{Na}]_\mathrm{i}^3\,[\mathrm{K}]_\mathrm{o}^2\,zFE_0. \tag{4.10}$$

Evidently i_p changes with voltage sigmoidally. Furthermore, i_p has a reversal potential (E_p) where

$$E_\mathrm{p} = \frac{RT}{zF}\left\{\ln\left(\frac{k^0{}_{21}}{k^0{}_{12}}\right) + \ln\left(\frac{r_{12}}{r_{21}}\right) + \ln\left(\frac{[\mathrm{ADP}][\mathrm{P_i}]}{[\mathrm{ATP}]}\right) + 3\ln\left(\frac{[\mathrm{Na}]_\mathrm{o}}{[\mathrm{Na}]_\mathrm{i}}\right) + 2\ln\left(\frac{[\mathrm{K}]_\mathrm{i}}{[\mathrm{K}]_\mathrm{o}}\right)\right\}. \tag{4.11}$$

Note that E_p is independent of the membrane potential, but dependent on $[\mathrm{Na}]_\mathrm{o}$ and $[\mathrm{K}]_\mathrm{o}$. However, this has not been clearly demonstrated by experimental variation of the K^+ and Na^+ concentrations. Knowing E_p, the conductance (g_p) of the pump is calculated as a chord conductance,

$$g_\mathrm{p} = i_\mathrm{p}/(V - E_\mathrm{p}). \tag{4.12}$$

The shape of the i_p–V curve is characterized by a set of five parameters, A_1, A_2, A_3, A_4, and z, which can be determined by fitting the experimental i_p–V curve.

What we can expect for the voltage-dependent i_p and g_p from this kinetic model is shown in Fig. 4.4. The i_p–V curve is sigmoidal, while the g_p–V curve has a peak at a definite voltage.

Initial guesses are needed for the curve fitting. A_1 was estimated by calculation of E_p using eqn (4.11). A_2 was assumed to be of the same order as A_1, A_3 as 0.01–0.06, A_4 as 0.5–2.0 μA cm^{-2}, and z as 1–2.

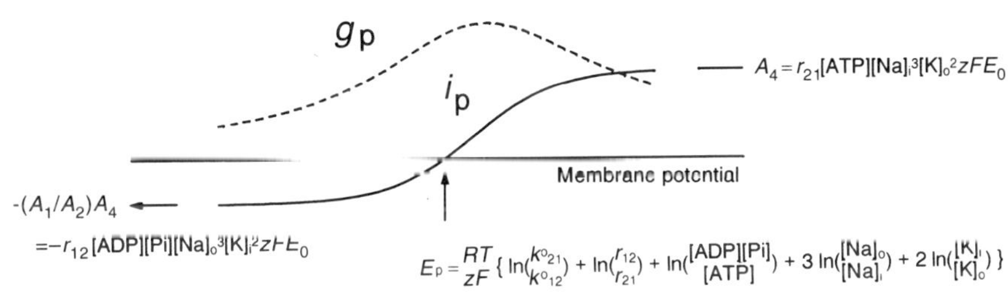

Fig. 4.4. Changes of the ouabain-sensitive current (i_p) and its chord conductance (g_p) with the membrane potential which are expected with the chosen kinetic model.

The four rate constants in the cycle can be calculated from the five parameters obtained by fitting. These are

$$r_{21} = \left(\frac{A_4}{zFE_0}\right)\frac{1}{[\mathrm{ATP}]\,[\mathrm{Na}]_i^3\,[\mathrm{K}]_o^2}, \tag{4.13}$$

$$r_{12} = \left(\frac{A_4}{zFE_0}\right)\left(\frac{A_1}{A_2}\right)\frac{1}{[\mathrm{ADP}]\,[\mathrm{P_i}]\,[\mathrm{Na}]_o^3\,[\mathrm{K}]_i^2}, \tag{4.14}$$

$$k^0{}_{12} = \left(\frac{A_4}{zFE_0}\right)\left(1 + \frac{A_1}{A_2}\right)\frac{1}{A_3}, \tag{4.15}$$

$$k^0{}_{21} = \left(\frac{A_4}{zFE_0}\right)\quad(A_1 + A_2)\frac{1}{A_3}. \tag{4.16}$$

Note that in a standard condition in which all concentrations are unity, E_p becomes

$$E_p{}^0 = \frac{RT}{zF}\left\{\ln\left(\frac{k^0{}_{21}}{k^0{}_{12}}\right) + \ln\left(\frac{r_{12}}{r_{21}}\right)\right\} \quad \text{or} \quad zFE_p{}^0 = RT\left\{\ln\left(\frac{k^0{}_{21}}{k^0{}_{12}}\right) + \ln\left(\frac{r_{12}}{r_{21}}\right)\right\}. \tag{4.17}$$

The standard free energy of the cycle can be calculated using eqn (4.17) if the four rate constants and the number of charges translocated are known.

Changes of the ouabain-sensitive current–voltage curve with $[\mathrm{K}]_o$

The normal Na-K-ASW contains 423 mM Na^+ and 9 mM K^+, the Cl^- concentration being reduced to as low as 39 mM (Table 4.1). The expected E_p is about −65 mV (eqn 4.11). The ouabain-sensitive current (i_p) under this ionic environment is shown in Fig. 4.5(b). The symbols in this figure are the experimental data, the smooth line is the theoretical i_p–V curve best fitted using eqn (4.6), and the broken line is the chord conductance calculated using eqn (4.12). For the reason stated earlier we used the i_p only in a limited voltage range. The fit of the data with a sigmoidal theoretical curve is satisfactory.

The ouabain-sensitive current–voltage curve (i_p–V curve) shifts along the voltage axis with the change in external K^+ concentration (Fig. 4.5). In these examples the external Na^+ concentration is maintained at 423 mM. If the external K^+ concentration is increased up to 20 mM, the result is not a depolarization but a hyperpolarization of E_p (Fig. 4.5(a)). This is what is expected from eqn (4.11). On the other hand, if the external K^+ concentration is reduced, not a hyperpolarization but a depolarization of E_p is expected (eqn 4.11). When the ASW contains 423 mM Na^+ and 2 mM K^+, the expected E_p is about −40 mV. The ouabain-sensitive i_p–V curve under this ionic environment is shown in Fig. 4.5(c).

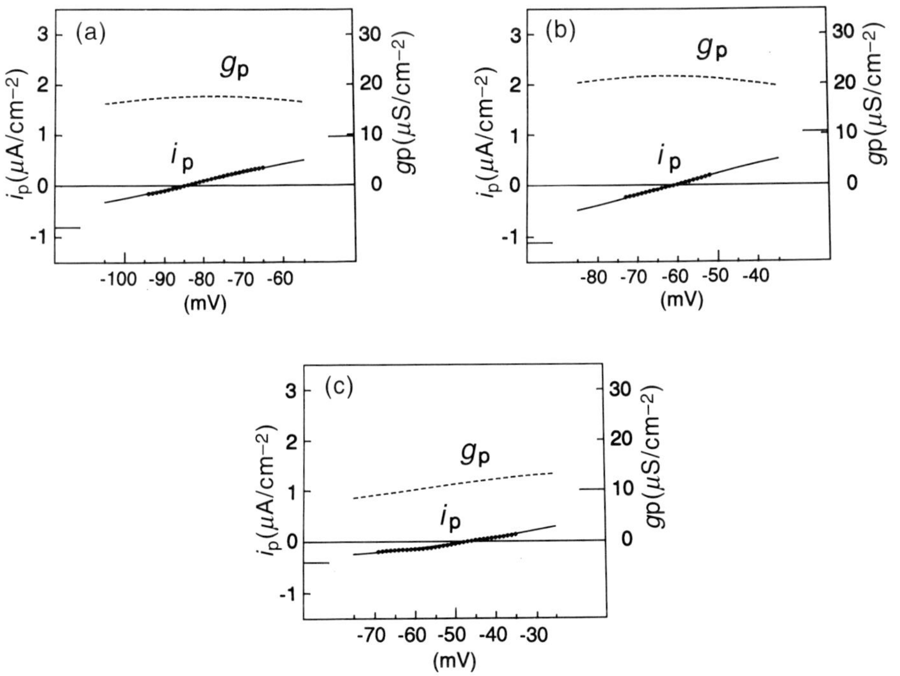

Fig. 4.5. (a) The ouabain-sensitive current–voltage (I_p–V) curve of the squid giant axon membrane in the higher K ASW ($[Na]_o$=423, $[K]_o$=20 mM). Filled circles are the experimental i_p obtained using the method of subtraction. The solid line is the result of fitting the i_p with eqn (4.6). The reversal potential of i_p in this ionic condition was −85.1 mV. The dotted line is the chord conductance g_p that was calculated using eqn (4.12). Temperature, 12.4°C. (b) A similar i_p–V curve in normal K ASW ($[Na]_o$=423, $[K]_o$=9 mM). The reversal potential of i_p in this ionic condition was −61.1 mV. The dotted line is the chord conductance g_p. Temperature, 12.4°C. (c) A similar i_p–V curve in low K ASW ($[Na]_o$=423, $[K]_o$=2 mM). The reversal potential of i_p in this ionic condition was −38.5 mV. The dotted line is the chord conductance g_p. Temperature, 13.1°C.

Changes of kinetic parameters with $[K]_o$ and $[Na]_o$

As described in the section on the kinetic model, the five parameters, that is k^0_{12}, k^0_{21}, r_{21}, r_{12}, and z that characterize the shape of the i_p–V curve can be determined by fitting the data. The results of fitting the i_p–V curves in varied external K^+ and Na^+ concentrations are listed in Table 4.2. The first point to note is that z, that is the number of charges carried in the electrogenic step, is almost exactly equal to +1 irrespective of the change in K^+ or Na^+ concentration. This indicates that the translocation of 3 Na^+ through the channel part of the pump, from its internal surface to the external surface of the membrane or vice versa, carries only a single positive charge. This may correspond to −2 charges at the internal surface of the channel portion of the pump. The translocation of 2 K^+ through it, therefore, is likely to be electroneutral.

Second, the obtained standard free energy that is calculated with eqn (4.17) is almost contant at −33 kJ/mol (Table 4.2). This is almost the same as the standard free energy of ATP-hydrolysis (−32.2 kJ/mol) reported by Rosing and Slater (1972). This result supports the notion that the kinetic cycle is supplied energy from ATP-hydrolysis.

The forward rate constant, r_{21}, in the non-electrogenic step is practically unchanged by a change of external Na^+ concentration (data not shown). However it decreases greatly with an increase of external K^+ concentration [Fig. 4.6(a)]. The plot of log (r_{21}) against log ($[K]_o$) is linear. The linear regression analysis gives

$$\log(r_{21}) = 14.43 - 1.92 \log([K]_o).$$

Note that the slope is close to −2. We cannot go into the detail of the several transition steps of enzyme intermediates with the simplified and lumped two-state model. However, these results suggest that a change of the external $[Na]_o$ has almost no influence on the deocclusion of Na^+ from the enzyme at its external surface, while a decrease of external $[K]_o$ enhances greatly the occlusion of K^+ into the enzyme at the external surface. This is the prerequisite for the enzyme to pump in K^+. The saturated i_p ($= r_{12}$ [ADP] $[P_i]$ $[Na]_o^3$ $[K]_i^2$ in Fig. 4.3) for a large hyperpolarization may remain almost unchanged for the change in $[K]_o$, since r_{12} remains almost unchanged. However, a confor-

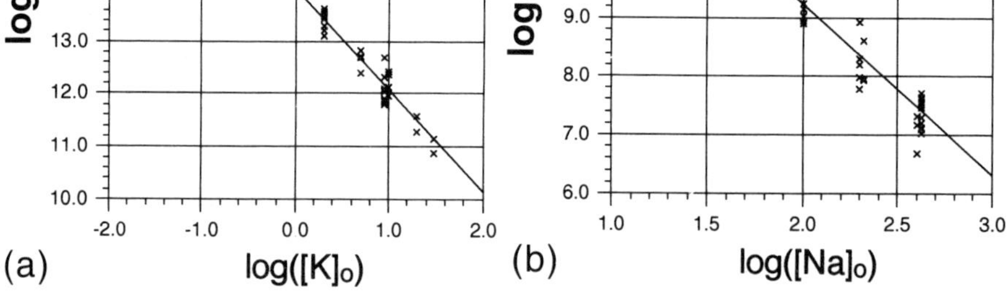

Fig. 4.6. (a) Logarithms of all the data for r_{21} (the forward rate constant in the non-electrogenic step) in Table 4.2 are plotted against the logarithm of $[K]_o$ (in mM). A linear regression analysis gives a line: $\log(r_{21}) = 14.43 - 1.92 \log([K]_o)$. r_{21} is almost independent of change in $[Na]_o$ (data not shown). (b) Logarithms of all the data for r_{12} (the backward rate constant in the non-electrogenic step) in Table 4.2 are plotted against the logarithm of $[Na]_o$ (in mM). A linear regression analysis gives a line: $\log(r_{12}) = 15.29 - 2.86 \log([Na]_o)$. r_{12} is almost independent of change in $[K]_o$.

mational change occurs in the enzyme to maintain r_{21}[ATP][Na]$_i^3$[K]$_o^2zFE_0$ or A_4 (the saturated i_p for the large depolarization in Fig. 4.3) at an almost constant level.

On the other hand, the backward rate constant r_{12} in the non-electrogenic step is practically unchanged with a change of the external K^+ concentration (data not shown). However, it increases greatly with a decrease of external Na^+ concentration [Fig. 4.6(b)]. The plot of log(r_{12}) against log([Na]$_o$) is linear. The linear regression analysis gives

$$\log(r_{12}) = 15.29 - 2.86 \log([\mathrm{Na}]_o).$$

Note that the slope is close to −3. A change of the external [K]$_o$ has almost no influence on the deocclusion of K^+ from the enzyme at its external surface, while a decrease of the external [Na]$_o$ enhances greatly the occlusion of Na^+ into the enzyme at the external surface. This is a prerequisite for the enzyme to pump in Na^+. The saturated i_p (= r_{21}[ATP][Na]$_i^3$[K]$_o^2zFE_0$ in Fig. 4.3) for a large depolarization may remain almost unchanged for the change in [Na]$_o$, since r_{21} remains almost unchanged. However, a conformational change occurs in the enzyme to maintain r_{12}[ADP][P$_i$][Na]$_o^3$[K]$_i^2$ or $-A_4(A_1/A_2)$ (the saturated i_p for the large hyperpolarization in Fig. 4.3) at an almost constant level.

The forward rate constant, $k^0{}_{12}$, in the electrogenic step increases to some extent with an increase either of external K^+ concentration [0.01–30 mM, Fig. 4.7(a)] or of Na^+ concentration [10–420 mM, Fig. 4.7(b)].

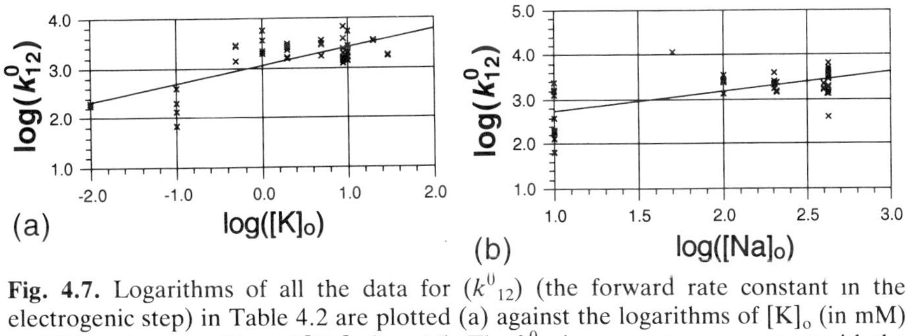

Fig. 4.7. Logarithms of all the data for ($k^0{}_{12}$) (the forward rate constant in the electrogenic step) in Table 4.2 are plotted (a) against the logarithms of [K]$_o$ (in mM) and (b) against logarithm of [Na]$_o$ (in mM). The $k^0{}_{12}$ increases to some extent with the increase of [Na]$_o$ or of [K]$_o$.

On the other hand, the backward rate constant, $k^0{}_{21}$, decreases to some extent with an increase of external K^+ concentration [Fig. 4.8(a)], while it increases greatly with an increase of external Na^+ concentration [Fig. 4.8(b)]. The linear regression analysis gives

$$\begin{aligned} \log(k^0{}_{21}) &= 3.62 - 1.39 \log([\mathrm{K}]_o) \quad &\text{for } [\mathrm{Na}]_o = 399\text{–}423\ \mathrm{mM} \\ &= 2.00 - 1.52 \log([\mathrm{K}]_o) \quad &\text{for } [\mathrm{Na}]_o = 100\ \mathrm{mM} \\ &= -0.93 - 1.69 \log([\mathrm{K}]_o) \quad &\text{for } [\mathrm{Na}]_o = 10\ \mathrm{mM} \end{aligned}$$

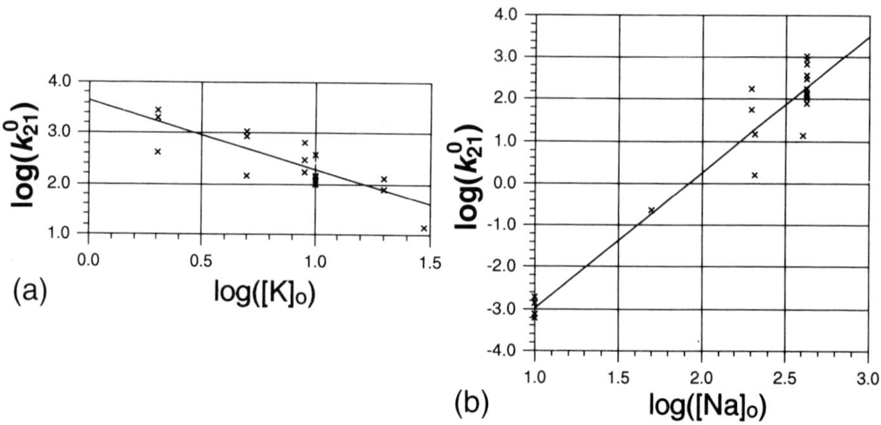

Fig. 4.8. The backward rate constant ($k^0{}_{21}$) in the electrogenic step is dependent on changes of $[Na]_o$ and of $[K]_o$. (a) Logarithms of $k^0{}_{21}$ at about $[Na]_o$=400 mM plotted against the logarithm of $[K]_o$ (in mM) give a linear relationship, i.e. $\log(k^0{}_{21})=3.62 - 1.39 \log([K]_o)$. Similar plots at different $[Na]_o$ also give linear relationships, though the slope and intersection depend on $[Na]_o$. (b) Logarithms of $k^0{}_{21}$ at $[K]_o$=10 mM plotted against the logarithm of $[Na]_o$ (in mM) give a linear relationship, i.e. $\log(k^0{}_{21})=-6.50 + 3.36 \log([Na]_o)$. Similar plots at a different $[K]_o$ also give linear relationships, though the slope and intersection depend on $[K]_o$.

and

$$\begin{aligned}\log(k^0{}_{21}) &= -\ 6.50 + 3.36 \log([Na]_o) \quad \text{for } [K]_o = 10 \text{ mM}\\ &= -\ 2.18 + 2.09 \log([Na]_o) \quad \text{for } [K]_o = 2 \text{ mM}\\ &= -\ 2.01 + 2.22 \log([Na]_o) \quad \text{for } [K]_o = 0.1\text{–}1 \text{ mM}.\end{aligned}$$

These results suggest that the energy barrier for the forward translocation of 3 Na^+ in the channel portion of the enzyme increases to some extent with a decrease of $[Na]_o$ or $[K]_o$. On the other hand, the energy barrier for the backward translocation of 3 Na^+ increases greatly with a decrease of $[Na]_o$, while it decreases to some extent with a decrease of $[K]_o$.

Reversal potential of the ouabain-sensitive current

The reversal potential of the ouabain-sensitive current is voltage-independent. However, it is highly dependent on the ligand concentration (eqn 4.11). Decrease of the external Na^+ concentration causes a great increase of r_{12}, a marked decrease of $k^0{}_{21}$ and a slight decrease of $k^0{}_{12}$, while r_{21} is almost unchanged as described above. On the other hand, decrease of the external K^+ concentration causes a great increase of r_{21}, increase of $k^0{}_{21}$ and a slight decrease of $k^0{}_{12}$, while r_{12} remains almost unchanged. However, there is a definite relation among these four rate constants. That is, $\ln(k^0{}_{21}/k^0{}_{12})$ +

ln (r_{12}/r_{21}) remains constant for the changes of external Na^+ and K^+ concentrations (eqn 4.17).

Indeed, the E_p shifts toward depolarization with a decrease in $[K]_o$. It is well known that the ouabain-sensitive efflux of Na^+ is markedly reduced, if the external $[K]_o$ is reduced (Caldwell *et al.* 1960*a*,*b*; Baker *et al.* 1969). This can be interpreted from the present kinetic analysis as the shift of E_p toward a less negative level than the resting potential. Therefore, in K^+-free medium, E_p is expected to be at a large positive level and the ouabain-sensitive current at the resting potential will not be outward, but inward. By contrast, E_p shifts to a more hyperpolarized level than the resting potential with a decrease in $[Na]_o$. In the Na^+-free medium almost a saturation of the ouabain-sensitive outward current at the resting potential is expected. Therefore, the pump behaves like a constant-current source in this ionic condition. This is clearly demonstrated by Rakowski and colleagues (Rakowski and Paxon 1988; Rakowski *et al.* 1989) in the squid giant axon in Na^+-free solution.

Concluding remarks

What seems in common is that occlusion of Na^+ at the external surface is enhanced greatly with the decrease of external Na^+ concentration and that occlusion of K^+ is also greatly enhanced with the decrease of external K^+ concentration. This seems to be a primary condition for the enzyme to pump in or pump out the ligand. (High affinity at the whence side and low affinity at the whither side.) We do not know the detailed molecular mechanism of the way in which these changes of rate constants depend on the change in ligand concentration. The enzyme seems to be so designed as to adjust its structure for the change in ligand concentration and to continue its pumping function, even though its i_p–V characteristic shifts along the voltage axis.

Changes of rate constants in the two state model with external K^+ and Na^+ concentrations are described in this report. If a similar mechanism is involved at the inner surface of the enzyme, changes of rate constants with internal K^+ and Na^+ concentrations can be anticipated. What we can expect with a decrease of internal Na^+ concentration are a great increase of r_{21}, almost no change of r_{12}, decrease of $k^0{}_{12}$ and an increase of $k^0{}_{21}$. On the other hand, what we can expect with a decrease of internal K^+ concentration are a great increase of r_{12}, almost no change of r_{21}, a decrease of $k^0{}_{12}$ and a great decrease of $k^0{}_{21}$. However, these remain to be proved.

Acknowledgements

We would like to thank Dr Nobunori Kami-ike for his help in computer programming. This work was supported by a Grant in Aid to Dr T. Ohkawa from the Ministry of Education, Science, and Culture of Japan.

References

Apell, H.-J. (1989). Electrogenic properties of the Na, K pump. *Journal of Membrane Biology*, **110**, 103–14.

Apell, H.-J., Borlinghaus, R., and Läuger, P. (1987). Fast charge-translocations associated with partial reactions of the Na, K-pump. II. Microscopic analysis of transient currents. *Journal of Membrane Biology*, **110**, 103–14.

Bahinski, A., Nakao, M., and Gadsby, D. C. (1988). Potassium translocation by the Na/K pump is voltage insensitive. *Proceedings of the National Academy of Sciences, USA*, **85**, 3412–16.

Baker, P. F., Blaustein, M. P., Keynes, R. D., Manil, J., Shaw, T. I., and Steinhardt, R. A. (1969). The ouabain-sensitive fluxes of sodium and potassium in squid giant axons. *Journal of Physiology*, **200**, 459–96.

Beaugé, L. (1984). Sodium pump in squid axons. *Current Topics in Membranes and Transport*, **22**, 131–75.

Borlinghaus, R., Apell, H.-J., and Läuger, P. (1987). Fast charge-translocations associated with partial reactions of the Na, K-pump. I. Current and voltage transients after photochemical release of ATP. *Journal of Membrane Biology*, **97**, 161–78.

Caldwell, P. C., Hodgkin, A. L., Keynes, R. D., and Shaw, T. I. (1960a). The effects of injecting 'energy-rich' phosphate compounds on the active transport of ions in the giant axons of *Loligo*. *Journal of Physiology*, **152**, 561–90.

Caldwell, P. C., Hodgkin, A. L., Keynes, R. D., and Shaw, T. I. (1960*b*). Partial inhibition of the active transport of cations in the giant axons of *Loligo*. *Journal of Physiology*, **152**, 591–600.

Chapman, J. B., Johnson, E. A., and Kootsey, J. M. (1983). Electrical and biochemical properties of an enzyme model of the sodium pump. *Journal of Membrane Biology*, **74**, 139–53.

Cole, K. S. and Moore, J. W. (1960). Ionic current measurements in the squid giant axon membrane. *Journal of General Physiology*, **44**, 123–67.

Cooke, I. M., Leblanc, G., and Tauc, L. (1974). Sodium pump stoichiometry in *Aplysia* neurons from simultaneous current and tracer measurements. *Nature*, **251**, 254–6.

DeWeer, P., Gadsby, D. C., and Rakowski, R. F. (1986). Voltage dependence of Na/K pump-mediated ^{22}Na efflux and current in squid giant axon. *Journal of Physiology*, **371**, 144P.

DeWeer, P., Rakowski, R. F, and Gadsby, D. C. (1988). Voltage dependence of the Na-K pump. *Annual Review of Physiology*, **50**, 225–41.

Eisner, D. A., Lederer, W. J., and Vaughan-Jones, R. D. (1981). The dependence of sodium pumping and tension on intracellular sodium activity in voltage-clamped sheep *Purkinje* fibres. *Journal of Physiology*, **317**, 163–87.

Eisner, D. A., Valdeolmillos, M., and Wray, S. (1987). The effects of membrane potential on active and passive Na transport in *Xenopus* oocytes. *Journal of Physiology*, **385**, 643–59.

Fendler, K., Grell, E., and Bamberg, E. (1987). Kinetics of pump currents generated by the Na^+, K^+-ATPase. *Federation of European Biochemical Societies Letters*, **224**, 84–8.

Glitsch, H. G., Pusch, H., Schumacher, T., and Verdonck, C. (1982). An identification of the K activated Na pump current in sheep *Purkinje* fibres. *Pflüger's Archiv*, **394**, 256–63.

Glynn, I. M. and Karlish, S. J. D. (1975). The sodium pump. *Annual Review of Physiology*, **37**, 13–55.

Goldshlegger, R., Karlish, S. J. D., Rephaeli, A., and Stein, W. D. (1987). The effect of membrane potential on the mammalian sodium–potassium pump reconstituted into phospholipid vesicles. *Journal of Physiology*, **387**, 331–55.

Hansen, U.-P., Gradmann, D., Sanders, D., and Slayman, C. L. (1981). Interpretation of current–voltage relationships for 'active' ion transport systems: I. Steady-state reaction-kinetic analysis of class-1 mechanisms. *Journal of Membrane Biology*, **63**, 165–90.

Hodgkin, A. L. and Keynes, R. D. (1955). Active transport of cations in giant axons from *Sepia* and *Loligo*. *Journal of Physiology*, **119**, 513–28.

Inoue, I. (1985). Voltage-dependent chloride conductance of the squid axon membrane and its blockade by some disulfonic stilbene derivatives. *Journal of General Physiology*, **85**, 519–37.

Kami-ike, N., Ohkawa, T., Kishimoto, U., and Takeuchi, Y. (1986). A kinetic analysis of the electrogenic pump of *Chara corallina*: IV. Temperature dependence of the pump activity. *Journal of Membrane Biology*, **94**, 163–71.

Karlish, S. J. D., Rephaeli, A., and Stein, W. D. (1985). Transmembrane modulation of cation transport by the Na, K-pump. In *The sodium pump* (ed. I. M. Glynn and C. Ellory), pp. 487–99. The Company of Biologists, Cambridge, UK.

Kishimoto, U., Kami-ike, N., Takeuchi, Y., and Ohkawa, T. (1984). A kinetic analysis of the electrogenic pump of *Chara corallina*: I. Inhibition of the pump by DCCD. *Journal of Membrane Biology*, **80**, 175–83.

Kishimoto, U., Takeuchi, Y., Ohkawa, T., and Kami-ike, N. (1985). A kinetic analysis of the electrogenic pump of *Chara corallina*: III. Pump activity during action potential. *Journal of Membrane Biology*, **86**, 27–36.

Läuger, P. (1979). A channel mechanism for electrogenic ion pumps. *Biochimica et Biophysica Acta*, **552**, 143–61.

Läuger, P. (1984). Thermodynamic and kinetic properties of electrogenic ion pumps. *Biochimica et Biophysica Acta*, **779**, 307–41.

Lederer, W. J. and Nelson, M. T. (1984). Sodium pump stoichiometry determined by simultaneous measurements of sodium efflux and membrane current in barnacle. *Journal of Physiology*, **348**, 665–77.

Nagel, G., Fendler, K., Grell, E., and Bamberg, E. (1987). Na^+ currents generated by the purified (Na^+ + K^+)-ATPase on planar lipid bilayers. *Biochimica et Biophysica Acta*, **901**, 239–49.

Nakao, M. and Gadsby, D. C. (1986). Voltage dependence of Na translocation by the Na/K pump. *Nature*, **323**, 628–30.

Post, R. L. and Jolly, D. C. (1957). The linkage of sodium, potassium, and ammonium active transport across the human erythrocyte membrane. *Biochimica et Biophysica Acta*, **25**, 118–28.

Rakowski, R. F. (1989). Simultaneous measurement of changes in current and radio-tracer flux in voltage-clamped squid giant axon. *Biophysical Journal*, **55**, 663–71.

Rakowski, R. F. and Paxon, C. L. (1988). Voltage dependence of Na/K pump current in *Xenopus* oocytes. *Journal of Membrane Biology*, **106**, 173–82.

Rakowski, R. F., Gadsby, D. C., and DeWeer, P. (1989). Stoichiometry and voltage dependence of the sodium pump in voltage-clamped, internally dialyzed squid giant axon. *Journal of General Physiology*, **93**, 903–41.

Rephaeli, A., Richards, D. E., and Karlish, S. J. D. (1986*a*). Conformational transitions in fluorescein-labeled (Na, K)ATPase reconstituted into phospholipid vesicles. *Journal of Biological Chemistry*, **261**, 6248–54.

Rephaeli, A., Richards, D. E., and Karlish, S. J. D. (1986*b*). Electrical potential accelerates the E1P(Na)-E2P conformational transition of (Na,K)-ATPase in reconstituted vesicles. *Journal of Biological Chemistry*, **261**, 12437–40.

Rosing, J. and Slater, E. C. (1972). The value of $\Delta G°$ for the hydrolysis of ATP. *Biochimica et Biophysica Acta*, **267**, 275–90.

Stürmer, W., Apell, H.-J., Wuddel, I., and Läuger, P. (1989). Conformational transitions and charge translocation by the Na, K pump: comparison of optical and electrical transients elicited by ATP-concentration jumps. *Journal of Membrane Biology*, **110**, 67–86.

Takeuchi, Y., Kishimoto, U., Ohkawa, T., and Kami-ike, N. (1985). A kinetic analysis of the electrogenic pump of *Chara corallina*: II. Dependence of the pump activity on external pH. *Journal of Membrane Biology*, **86**, 17–26.

Thomas, R. C. (1972). Electrogenic sodium pump in nerve and muscle cells. *Physiological Reviews*, **52**, 563–94.

PART II

Squid axon ion channels

5

Resting and active K^+ channels in the squid axon membrane

ISAO INOUE

Summary

Potassium currents (I_K) were recorded from giant axons of squid, *Sepioteuthis lessoniana* under voltage clamp. The K^+ accumulation in a space between outer membrane surface and surrounding Schwann cells layer, the so-called Frankenhaeuser–Hodgkin space (F–H space) due to K^+ efflux was completely removed by immersing the nerve fibre in a high osmolarity solution. Under such conditions I_K displays the real time course of potassium conductance (g_K) of the membrane. Even though the K^+ accumulation had been removed completely, I_K still decayed after reaching a peak. This decay is due to g_K inactivation, and is well fitted by a dual exponential function, with time constants of approximately 10 ms and of the order of 100 ms. Cytochalasin B applied intracellularly was found to block only one component of g_K (with the slow inactivation) in a concentration-dependent manner. At 50 μM cytochalasin B, g_K lost the slow inactivation component, thus the decaying g_K could be fitted by a single exponential function with time constant of approximately 10 ms. The cytochalasin B-sensitive component had quite different characteristics from the delayed rectifier g_K: 1) This component was activated along a single exponential time-course with a time constant of 2 ms at 16.5°C following depolarization and then it decayed slowly toward a steady state. 2) Prepulses of −140 to −60 mV did not alter the time course of the cytochalasin B sensitive g_K. Because of the slow activation of the drug-sensitive g_K, cytochalasin B did not affect the time course of the action potential. On the other hand, cytochalasin B greatly decreased the steady-state K^+ conductance. These results suggest that g_K of the squid axon membrane consists of at least two components, i.e. a delayed rectifier K^+ channel and a steady-state K^+ channel. The cytochalasin B resistant g_K may represent the real time- and voltage-dependent characteristics of the delayed rectifier.

The delayed rectifier of the squid axon membrane was the first K^+ channel to be studied extensively, and its time- and voltage-dependent characteristics and pharmacological properties were the first to be described (cf. Hodgkin and Huxley 1952; Hodgkin *et al.* 1952; Ehrenstein and Gilbert 1966; Armstrong 1969, 1971; Meves and Pichon 1977; Inoue 1981). This channel was considered to be involved in the generation of both spike and resting potentials

(Hodgkin *et al.* 1952). However, since the introduction of the patch-clamp technique (Neher and Sakmann 1976), many different types of K^+ channels have been identified in a range of cell types, and it is now clear that many of these channels are specialized for particular cell functions. This has led us to the idea that the delayed rectifier K^+ channel that is responsible for the highly sophisticated function of excitable cells may be separate from the K^+ channel responsible for the generation of the resting potential, a much more widespread cellular phenomenon. This idea agrees with observations of multiple types of K^+ channels in the squid axon membrane obtained with patch clamp (Llano *et al.* 1988). It is shown in this chapter that the potassium conductance g_K of the membrane of the squid *Sepioteuthis lessoniana* can be separated into two components using a pharmacological tool, cytochalasin B.

Removal of K^+ accumulation in the Frankenhaeuser–Hodgkin (F–H) space

The K^+ currents under voltage-clamp decay after reaching the peak (Fig. 5.1). This decay is attributed to two mechanisms. One is inactivation of the K^+ conductance (Inoue 1981), and the other is the successive change of the K^+ equilibrium potential (E_K) by accumulation of K^+ in the F–H space (Frankenhaeuser and Hodgkin 1957; Adelman *et al.* 1973).

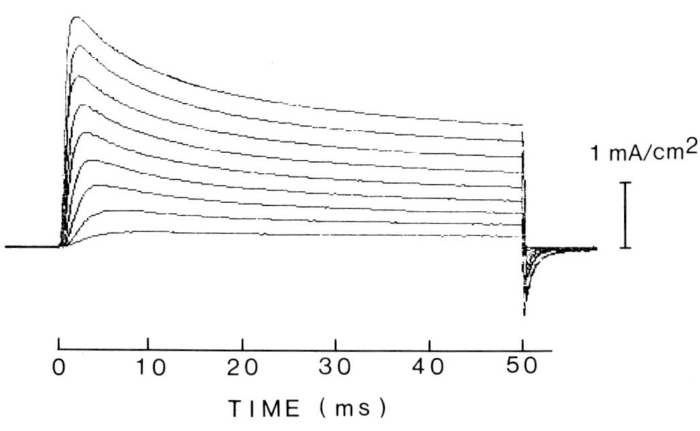

Fig. 5.1. Potassium currents (I_K) elicited from an internally perfused squid giant axon under voltage clamp. Depolarizing pulses of −40 to +40 mV at 10-mV intervals were given from a holding potential of −70 mV. The internal solution consisted of 400 mM KF, 50 mM K-phosphate (pH 7.2), and 4 per cent (by volume) glycerol. The bathing solution was artificial sea water (ASW) which consisted of 404 mM NaCl, 28 mM KCl, 10 mM $CaCl_2$, 23 mM $MgCl_2$, 25 mM $MgSO_4$, 6 mM Tris–HCl (pH 7.8). Tetrodotoxin (TTX) at 300 nM was added to the bathing solution to block currents through the Na^+ channels in all experiments. The equilibrium potential for K^+ (E_K) was nearly equal to the holding potential. Temperature, 16.5°C.

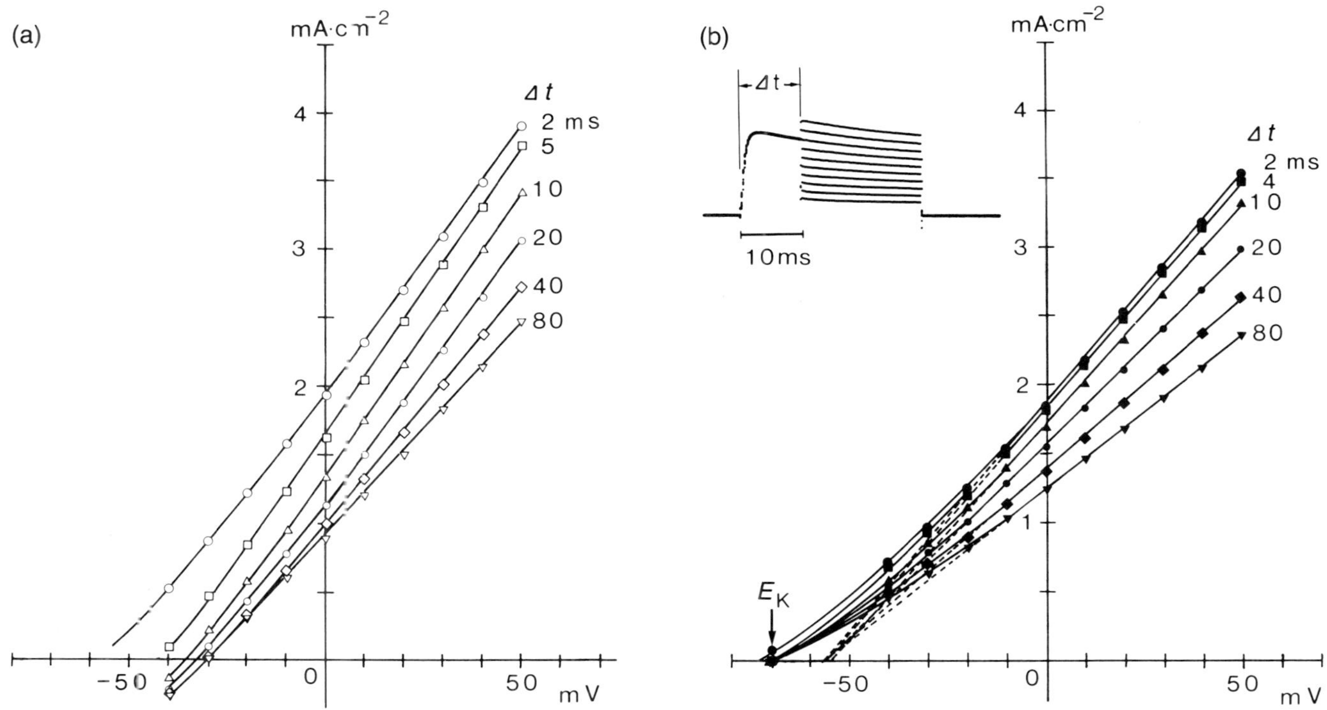

Fig. 5.2. (a) Instantaneous I–V relations at different times (Δt) after depolarization to +30 mV. Temperature, 16.0°C. (b) Instantaneous I–V relations with the same pulse protocol after elevating the external osmolarity by adding 1.2 M glucose to the ASW bath. This hypertonic ASW is referred to as glucose-ASW.

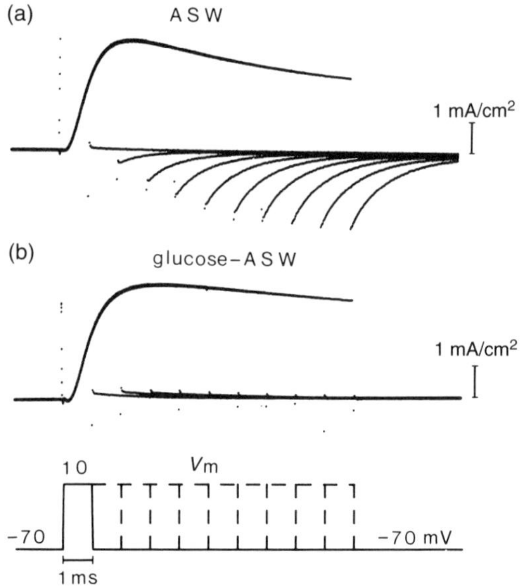

Fig. 5.3. The disappearance of the I_K tail current associated with repolarization after switching from the ASW bath to a glucose-ASW bath demonstrates that the K^+ accumulation in the F–H space is completely removed. The pulse protocol is given at the bottom; V_m is the membrane voltage under voltage clamp. Ten traces are superposed in each set of records. Temperature, 15.3°C.

Instantaneous current–voltage (I–V) relations at different times after depolarization demonstrate the two mechanisms [Fig. 5.2(a)]. The slope of each I–V relation indicates the K^+ conductance, and the voltage obtained by extrapolation of the I–V curve to the voltage axis gives E_K of the axon membrane. The change in E_K is due to K^+ accumulation in the F–H space. In the experiment of Fig. 5.2(a) the K^+ concentration in the F–H space changed from 28 to 110 mM during the 10 ms after depolarization to +30 mV from a holding potential of −70 mV. To examine only the K^+ conductance change, it is necessary to remove the K^+ accumulation.

One method of removal of K^+ accumulation in the F–H space is an elevation of the osmolarity of the external solution that produces outward water flow across the axolemma and clears away K^+ from the F–H space (Kukita and Yamagishi 1985). When 1.2 M glucose was added to the artificial sea water (ASW) both, the shift of E_K after depolarization decreased and eventually converged to one point, indicating that the K^+ accumulation was completely removed [Fig. 5.2(b)].

When the K^+ accumulation in the F–H space has been completely removed, the K^+ current must display the real time course of the K^+ conductance of the squid axon membrane. The absence of K^+ accumulation is characterized by the absence of a tail current associated with repolarization to the original holding level as can be seen in Fig. 5.3(b). In all the experiments described here, g_K was studied in ASW containing 1.2 M glucose (glucose-ASW).

The effect of cytochalasin B

Cytochalasin B is one of a family of cytochalasins that are blockers of actin polymerization. Among cytochalasins A, B, and C, only cytochalasin B was found to have a clear effect on g_K when applied intracellularly. Therefore, the effect of cytochalasin B described below is probably not due to cytoskeletal conformational change, but due to direct effects on the K^+ conductance system.

Cytochalasin B enhanced the decay of g_K in a concentration-dependent manner (Fig. 5.4). There was no use- or voltage-dependent inhibition. The effect was reversible unless the drug concentration exceeded 100 μM. The K^+ currents remaining at 50 μM cytochalasin B lost their slowly decaying phase, and the inactivation was fitted by a single exponential function (Fig. 5.5). The time constant was voltage-dependent, and was in the range 5–20 ms. The cytochalasin B-sensitive potassium current (I_K) component that was obtained by subtracting I_K at 50 μM from that at 10 μM at each corresponding voltage has only slowly activating and inactivating phases [Fig. 5.4(b) and (c)].

To further study the physiological properties of the two g_K components, large negative prepulses were given before depolarization. This revealed the Cole–Moore effect: that hyperpolarizing prepulses increase the amplitude of the delayed rectifier I_K and cause a delay of activation (Cole and Moore 1971). As shown in Fig. 5.6(b), the cytochalasin-resistant current clearly exhibits the Cole–Moore effect. In contrast, the cytochalasin B-sensitive current amplitude was totally insensitive to hyperpolarizing prepulses [Fig. 5.6(a), (b)]. This strongly suggests that the cytochalasin B-sensitive g_K is not involved in the delayed rectifier, and thus that this is a separate K^+ conductance system from the delayed rectifier K^+ channel.

Figure 5.7 shows that the activating phase of the cytochalasin B-sensitive g_K can be fitted by a single exponential function. The time constant of the activation kinetics is voltage-sensitive, and is greater than 2 ms. This activation is much slower than the process of the action potential. Therefore, cytochalasin B, even at 100 μM, had no effect on the time course of the action potential (Fig. 5.8). In contrast, cytochalasin B remarkably decreased the steady-state g_K of the membrane when it had been depolarized by elevating external K^+ concentration (Fig. 5.9).

Discussion

The results presented above suggest that the g_K of the squid axon membrane consists of at least two components that can be pharmacologically separated.

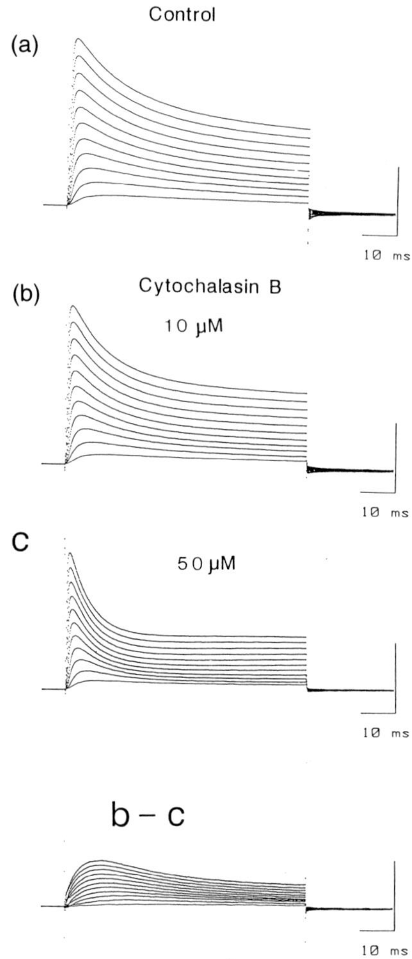

Fig. 5.4. Effect of internally applied cytochalasin B on I_K. The cytochalasin B concentrations tested are given. (a)–(c) Pulses of −50 to +50 mV at 10 mV intervals were given from the holding potential of −70 mV. The vertical bars indicate 1 mA/cm^2. The traces labelled (b) − (c) were obtained by subtracting the current trace in (c) from that in (b) at each test voltage, and display the time course of the cytochalasin B-sensitive g_K. (To extract the pure cytochalasin B-sensitive g_K component, we used the current traces at 10 μM instead of the control traces.) Bath, glucose-ASW; temperature, 16.5°C.

One is the delayed rectifier K^+ channel and the other is a K^+ channel responsible for the steady-state g_K generation. This view is quite consistent with that of Hendry and Haydon (1991) who used anaesthetics for the pharmacological separation of g_K of the squid axon membrane.

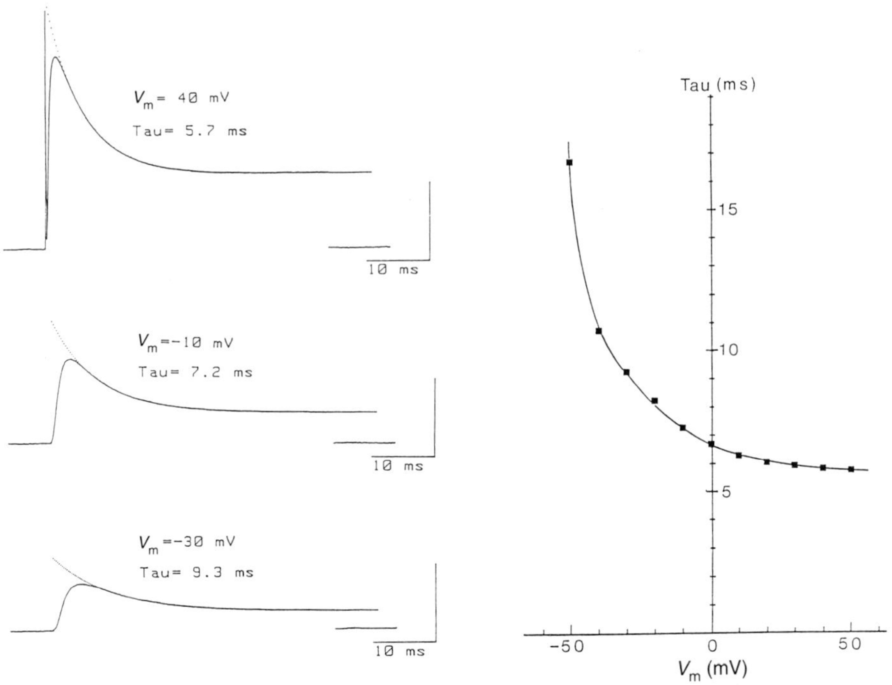

Fig. 5.5. Left, curve fitting to the inactivating phase of the cytochalasin B-resistant g_K of Fig. 5.4(c) with a single exponential function. The vertical bars indicate 1 mA/cm^2. Right, voltage dependence of the time constant (tau) of the decaying g_K. V_m is the membrane voltage under voltage clamp.

The delayed rectifier g_K has a large amount of inactivation in its nature. Therefore, g_K can be expressed in terms of an inactivation term, k, as

$$g_K = \bar{g}_K\, n^4 k \tag{5.1}$$

where $\bar{g}_K$ is the maximum value of the delayed rectifier K^+ conductance and n is $(1-\exp(-t/\tau_n))$. The inactivation can be fitted by a single exponential function. Therefore eqn (5.1) can be arranged as

$$g_K = \bar{g}_K\,(1 - e^{-t/\tau_n})^4\,[k_\infty - (k_\infty - k_o)e^{-t/\tau_k}]. \tag{5.2}$$

where τ_n and τ_k are the activation and inactivation time constants of the delayed rectifier K^+ conductance (g_K) respectively

This formula resembles that of g_{Na} of the squid giant axon membrane under certain conditions, for example internally perfused with NaF solution (Adelman 1971). In fact the delayed rectifier g_K can generate an all-or-none type spike if the equilibrium potential across the K^+ channel is set to be inside positive

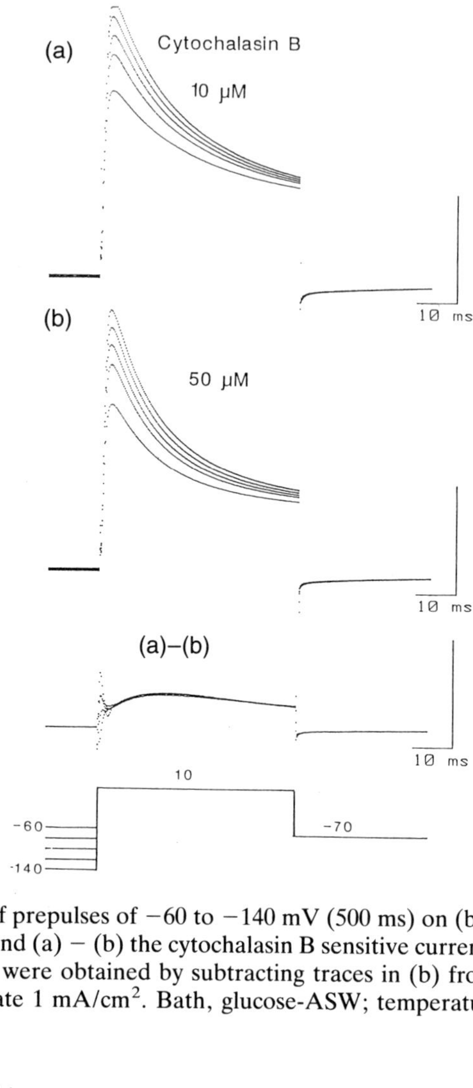

Fig. 5.6. Effects of prepulses of -60 to -140 mV (500 ms) on (b) the cytochalasin B-resistant current, and (a) $-$ (b) the cytochalasin B sensitive current (a $-$ b). The traces labelled (a) $-$ (b) were obtained by subtracting traces in (b) from traces in (a). The vertical bars indicate 1 mA/cm^2. Bath, glucose-ASW; temperature, 15.5°C.

(Inoue 1980). This suggests that there is a homology in the voltage-sensing (gating) mechanism between the Na^+ and the delayed rectifier K^+ channels.

The other g_K, which is sensitive to cytochalasin B, slowly activates with a single exponential, and slowly inactivates toward a steady level. This K^+ conductance (g'_K) can be formulated as

$$\begin{aligned} g'_K &= \bar{g}'_K\, sv \\ &= \bar{g}'_K(1 - e^{-t/\tau_s})\, [v_\infty - (v_\infty - v_o)\, e^{-t/\tau_v}]. \end{aligned} \tag{5.4}$$

Here s and v represent the activation and inactivation of the cytochalasin B sensitive K^+ conductance (g'_K), and τ_s and τ_v are the activation and

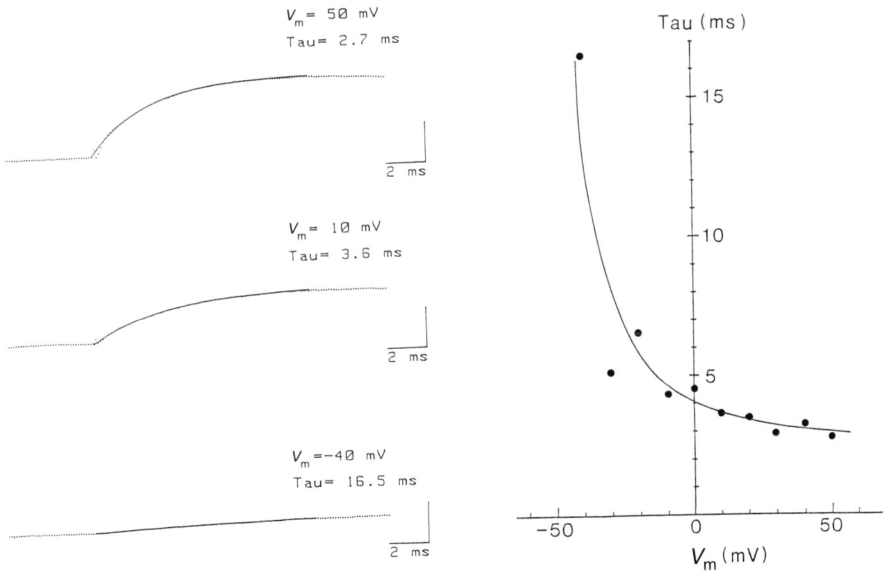

Fig. 5.7. Left, curve fitting to the activating phase of the cytochalasin B-sensitive currents in Fig. 5.4 (b)–(c) with a single exponential function. The vertical bars indicate 0.25 mA/cm^2. Right, voltage-dependence of the time-constant (tau) of the activation of the cytochalasin B-sensitive g_K; V_m is the membrane voltage under voltage clamp.

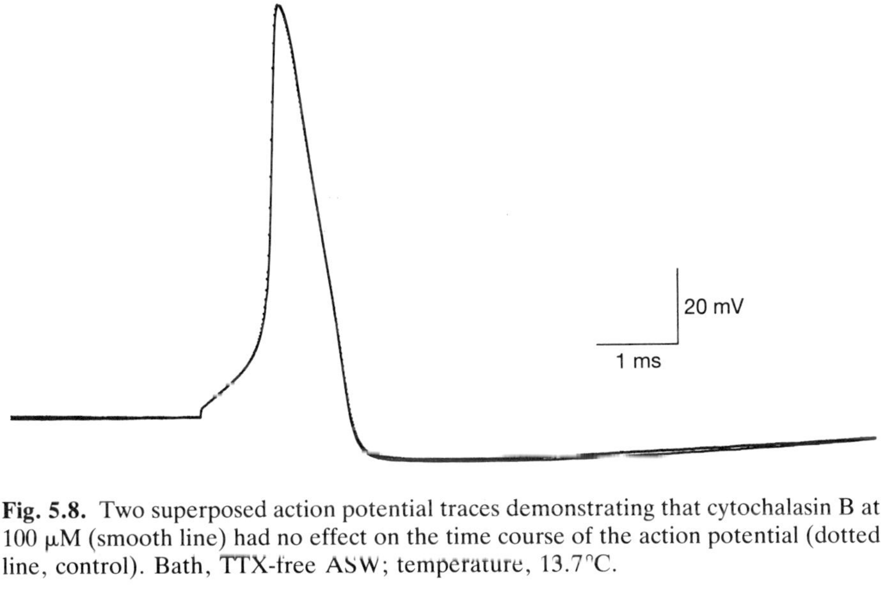

Fig. 5.8. Two superposed action potential traces demonstrating that cytochalasin B at 100 μM (smooth line) had no effect on the time course of the action potential (dotted line, control). Bath, TTX-free ASW; temperature, 13.7°C.

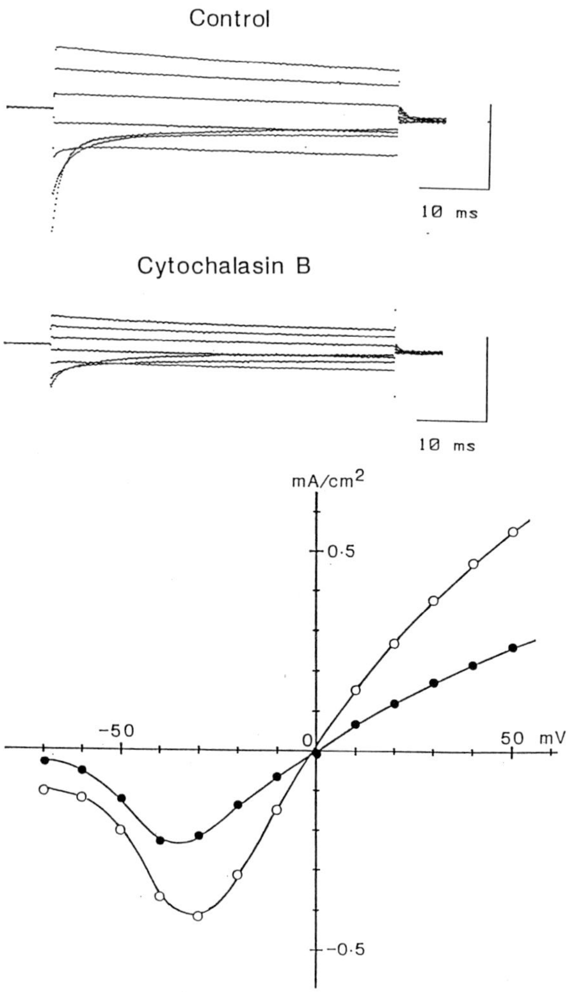

Fig. 5.9. Effect of cytochalasin B (50 μM) on the steady-state membrane conductance. The membrane was depolarized by elevating the K^+ concentration of glucose-ASW to 450 mM. Temperature, 15.5°C. Above, current traces under voltage clamp. Below, *I–V* relations plotted from the current traces in control (open circles) and after cytochalasin B (closed circles). The vertical bars indicate 1 mA/cm^2.

inactivation time constants, of g'_K respectively. The activation time-constant is much slower than that of the delayed rectifier g_K. Therefore, this conductance has no effect on action potential generation. This or a similar type of K^+ channel may exist in living cells more generally, and may play a role in generation of the steady-state membrane potential and conductance.

References

Adelman, Jr, W. J. (1971). Electrical studies of internally perfused squid axons. In *Biophysics and physiology of excitable membranes* (ed. W. J. Adelman Jr), pp. 274–319. W. J. Van Nostrand Reinhold Co, New York.
Adelman Jr, W. J., Palti, Y., and Senft, J. P., (1973). Potassium ion accumulation in a periaxonal space and its effect on the measurements of membrane potassium ion conductance. *Journal of Membrane Biology*, **23**, 387–410.
Armstrong, C. M. (1969). Inactivation of the potassium conductance and related phenomena caused by quaternary ammonium ion injected in squid axons. *Journal of General Physiology*, **54**, 553–75.
Armstrong, C. M. (1971). Interaction of tetraethylammonium ion derivatives with potassium channels of giant axons. *Journal of General Physiology*, **58**, 513–37.
Cole, K. S. and Moore, J. W. (1971). Potassium ion current in the squid giant axon: Dynamic characteristics. *Biophysical Journal*, **1**, 1–14.
Ehrenstein, G. and Gilbert D. L. (1966). Slow changes of potassium permeability in the squid giant axon. *Biophysical Journal*, **6**, 553–66.
Frankenhaeuser, B. and Hodgkin, A. L. (1957). The after effect of impulses in the giant fibres of *Loligo*. *Journal of Physiology*, **131**, 341–76.
Hendry, B. M. and Haydon, D. A. (1991). Alterations in resting K conductance in nerve by convulsants and anaesthetics. *Annals of the New York Academy of Sciences, USA*, **625**, 355–64.
Hodgkin, A. L. and Huxley, A. F. (1952). A quantitative description of membrane current and its application to conduction and excitation in nerve. *Journal of Physiology*, **117**, 500–44.
Hodgkin, A. L., Huxley, A. F., and Katz, B. (1952). Measurements of current–voltage relations in the membrane of giant axon of *Loligo*. *Journal of Physiology*, **116**, 424–48.
Inoue, I. (1980). Separation of the action potential into a Na-channel spike and a K-channel spike by tetrodotoxin and by tetraethylammonium ion in squid giant axons internally perfused with dilute Na-salt solutions. *Journal of General Physiology*, **76**, 337–54.
Inoue, I. (1981). Activation–inactivation of potassium channels and development of potassium channel spike in internally perfused squid giant axons. *Journal of General Physiology*, **78**, 43–61
Kukita, F. and Yamagishi, S. (1985). Effect of K^+ on K-channel in a squid giant axon. *Journal of the Physiological Society of Japan*, **47**, 399.
Llano, I., Webb, C. K., and Bezanilla, F. (1988). Potassium conductance of the squid giant axon. Single channel studies. *Journal of General Physiology*, **92**, 179–96.
Meves, H. and Pichon, Y. (1977). The effect of internal and external 4-aminopyridine on the potassium currents in internally perfused squid giant axons. *Journal of Physiology*, **268**, 511–32.
Neher, E. and Sakmann, B. (1976). Single-channel currents recorded from membrane of denervated frog muscle fibres. *Nature*, **260**, 799–802.

6

Studies of the kinetics of the ionic and gating currents in the axons of *Loligo forbesi* as a guide to modelling of the sodium channel

RICHARD D. KEYNES

Summary

My purpose here is to show how I have arrived at a new model of the voltage-gated sodium channel that for the first time successfully reconciles almost all the data obtained in recent years from a variety of experimental approaches. I have achieved this by exploiting to the full the superlative temporal resolution that can be achieved in recording the ionic and gating currents in the squid giant axon, and by relating the results to Numa's structural evidence (Noda *et al.* 1984) that each channel is controlled by four voltage-sensors arranged in parallel. For an updated account of these studies and a further revised model of the sodium channel, the reader is referred to a review by Keynes (1994).

The effect of tetramethylammonium (TMA) on the two open states of the sodium channel

Let me deal first with some basic properties of the macroscopic sodium current, I_{Na}. Observations on I_{Na} by Bekkers *et al.* (1987) revealed marked differences between the behaviour of the initial fast-inactivating open state and that of the non-inactivating open state (Chandler and Meves 1970), when dialysing with TMA^+ ions as compared with Cs^+. As may be seen in Fig. 6.1 (and see also Keynes 1991, fig. 3), the effect of TMA was to reduce the initial peak of $I_{Na,fast}$ but at the same time to increase the final steady-state current $I_{Na,non}$. Thus, for a test pulse to −20 mV, the ratio of the peak of the fast-inactivating current when dialysing with TMA to that with internal caesium was 0.83, whereas for the non-inactivating current it was 2.2. For a test pulse to +80 mV, the ratios were, respectively, 0.72 and 1.6, while for a pulse to +160 mV they had fallen further to 0.52 and 1.1 (Keynes, Meves, and Hof 1992). Analysis of the original data of Bekkers *et al.* (1987) was complicated by the fact that all the measurements had been made with low $[Na^+]$ bathing solutions where four-fifths of the sodium was replaced by Tris; Tris has since

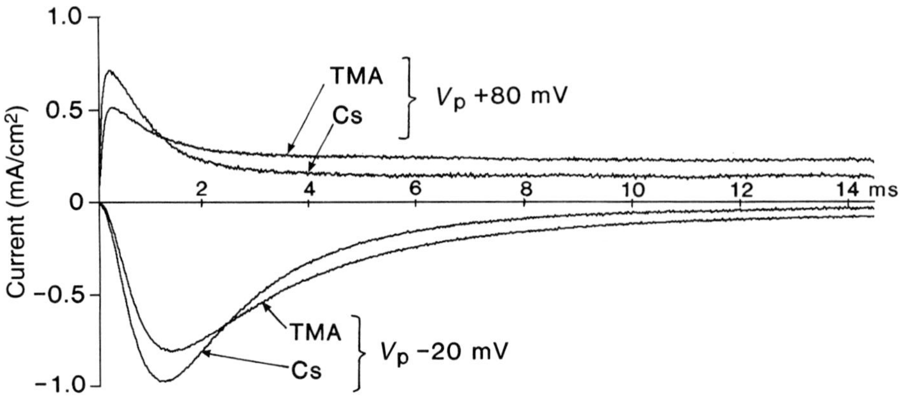

Fig. 6.1. Traces showing the macroscopic outward sodium current on application of pulses with V_p of −20 and +80 mV for a squid axon bathed in a solution containing 103 mM Na + 411 mM choline sea water, and dialysed first with 20 mM NaF + 330 mM CsF and then with 20 mM NaF + 330 mM TMAF. Holding potential −80 mV; temperature, 5°C. (After Keynes, Meves, and Hof (1992).)

been shown by Keynes *et al.* (1991*b*) to cut down the single-channel conductance in a voltage-dependent fashion that appears to differ somewhat for the two open states. Account also needs to be taken of possible interactions between the external binding of Tris and the internal binding of TMA and other cations. However, there is no doubt that the blocking of I_{Na} by TMA in the initial open state of the channels is invariably reversed to a substantial extent when they pass to the final steady state (see Keynes (1994)).

The mode of action of TMA

An alteration in total ionic current may arise from a change either in the conductance of the open channels or in the number of channels that are open at any moment. The block caused by TMA is normally thought to arise from a slowing down of the passage of sodium ions through open channels because of a binding at blocking sites located in their walls that increases with pulse potential as was found by Horn *et al.* (1981) for single sodium channels in rat myotubes. In a similar fashion, the single-channel conductance of squid axons was observed in the fluctuation analyses of Bekkers *et al.* (1986) to be halved by 50 mM TEA (tetraethylammonium). If in the steady-state TMA elicits a rise of I_{Na} instead of a fall, this must therefore be attributed to an effect on the number of open channels that overrides the simultaneous drop in their conductance.

There are two ways in which TMA might affect the number of channels that are opened by a given test pulse. One would be to alter the fraction of fully

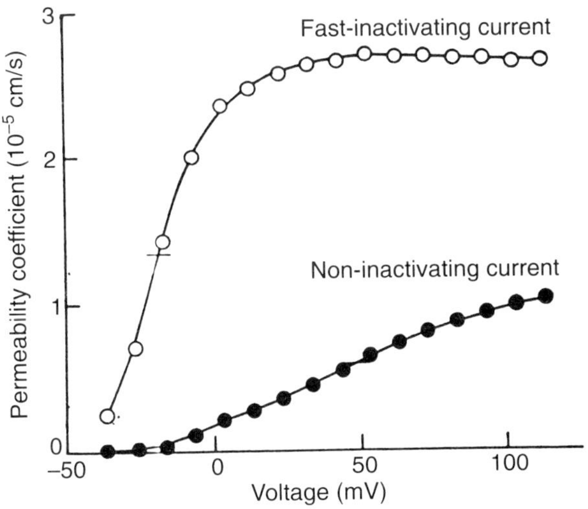

Fig. 6.2. Voltage dependence of the permeability coefficient for sodium in the fast-inactivating and non-inactivating open states, calculated by applying the constant-field equation to values of I_{Na} for an axon bathed in full [Na] sea water + 16 nM TTX, and dialysed with 20 mM NaF + 330 mM TMAF. Temperature, 10°C. The values of P_{Na} should be increased by a factor of 6 to allow for the blocking effect of TTX. (Reproduced with permission from Keynes (1991).)

activated channels, and the other would be to alter their mean open time. From the measurements of gating charge when dialysing with Cs and TMA shown in Fig. 6.4, it may be concluded that the total charge ΣQ was wholly insensitive to the internal ionic environment. A direct effect of TMA on the displacement of the voltage-sensors by the electric field can therefore be ruled out, and consideration must be given to the idea that TMA may act in a hitherto unrecognized fashion on the probability of opening of the channels that have arrived at the steady state.

It may be seen in Fig. 6.2 that the macroscopic permeability coefficient P_{Na} in the normal open state reaches a plateau at V_p = 50 mV, and that the potential V_{mid} for half-maximal activation lies at about −20 mV, as it does in an intact axon (Hodgkin and Huxley 1952). This behaviour fits well with the predictions of the model incorporating four voltage-sensors operating in parallel that I shall describe later. As had been noted by Chandler and Meves (1970), the steady-state current also rose with potential, though much less steeply, and it is apparent in Fig. 6.2 that V_{mid} was shifted to about +50 mV. It has been shown in recordings made by Bezanilla and Correa (1991) of single-channel currents in squid axons that $I_{Na,non}$ arises from repeated late openings of the channels, and it may be suggested that, in the final steady state, either the duration or the frequency of these openings is somehow governed not only by V_p but also by the presence of internal TMA. These findings on the nature of the steady state have provided a valuable lead in thinking about a revised mechanism of coupling between activation and inactivation for incorporation in my new model of the sodium channel.

The sodium gating current

The other main line of approach for which squid axons provide the best available material is the measurement of the sodium gating current I_g (Keynes *et al.* 1990). In the low-noise and high-resolution records of I_g illustrated in Fig. 6.3, five components could be distinguished on kinetic grounds. First, there is a very fast component I_{g1} that relaxes with a time constant of the order of 10 μs and forms the initial small blip. It is relatively small, and its precise role in gating is still undecided (Bekkers *et al.* 1989; see also Chapter 7), so that for my present purpose it need not be further discussed. The main part of I_g consists of the rapidly relaxing fraction designated I_{g2}, but there are in addition two fairly large components that relax at an intermediate rate. I_{g3a} is present over the whole potential range, and like I_{g1} and I_{g2} has a fast rise. I_{g3b} comes in only on the positive side of −40 mV, and has a relatively slow rise, thus being responsible for the shoulder in the records that follows the initial blip. Lastly, it is possible under favourable conditions to record a small component I_{g4} that relaxes very slowly.

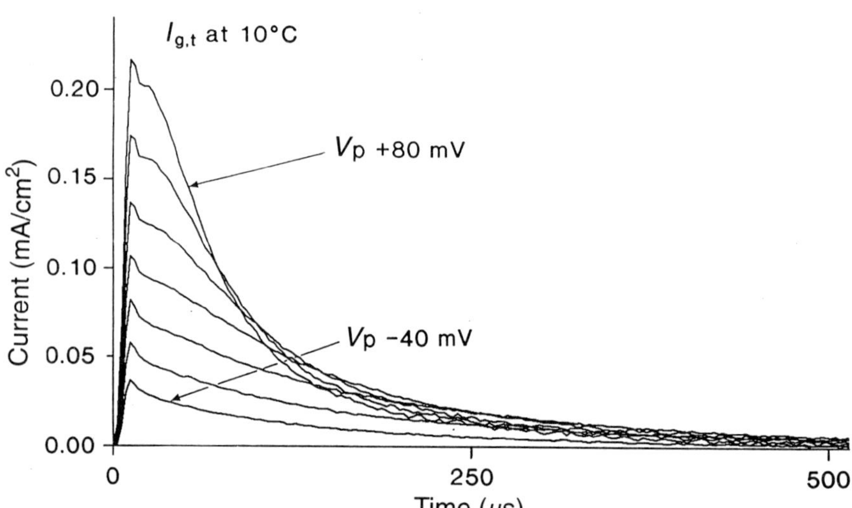

Fig. 6.3. Superimposed low-noise records of the sodium gating current for an axon bathed in Tris sea water + 1 μM TTX, and dialysed with 350 mM TMAF. Test potential $V_p = -40, -20, 0, 20, 40, 60$, and 80 mV; holding potential, −80 mV; back reference pulse −150 to −180 mV; sampling period 3 μs per point; number of sweeps averaged, 64; temperature, 10°C. (After Keynes *et al.* (1990).)

The slowly rising component I_{g3b} has proved to be a side issue as far as modelling is concerned, because it seems probable that it does not arise as was initially assumed from movement of the charged voltage-sensors in the electric field, but from unit events of a different kind. Thus I_{g3b} appears only for pulses that take the potential above the threshold at which the channels begin to open, beyond which it grows quickly in size, while it is knocked out completely by an inactivating prepulse. Its association with the opening of the individual channels rather than the displacement of the individual voltage-sensors agrees with the effect of negative prepulses on the time course of its relaxation. It was shown in Keynes *et al.* (1990, fig. 2) that, when the starting potential was made more negative, the tail of I_{g3b} was progressively delayed, in parallel with the delay in the rise of I_{Na} first reported by Keynes and Rojas (1976). In the experiment illustrated in Fig. 6.4, the extra delay in the relaxation of I_{g3b} when the starting potential was lowered from −60 to −160 mV was about 60 μs, which was roughly matched by a corresponding delay of 45 μs in the rise of I_{Na}.

It has been pointed out by Edmonds (1990) that ionizable groups that occupy fixed sites in the wall of the channel will, because of the influence on their p*K* of the dielectric environment, be liable to undergo a change in their state of ionization when the aqueous lumen expands. Any charge created in this fashion at the moment of hydration of the channel, which corresponds to the transfer from state 4A to state O(I) of the model shown in Fig. 6.6, will be registered as a unit event of asymmetrical displacement charge that will not be

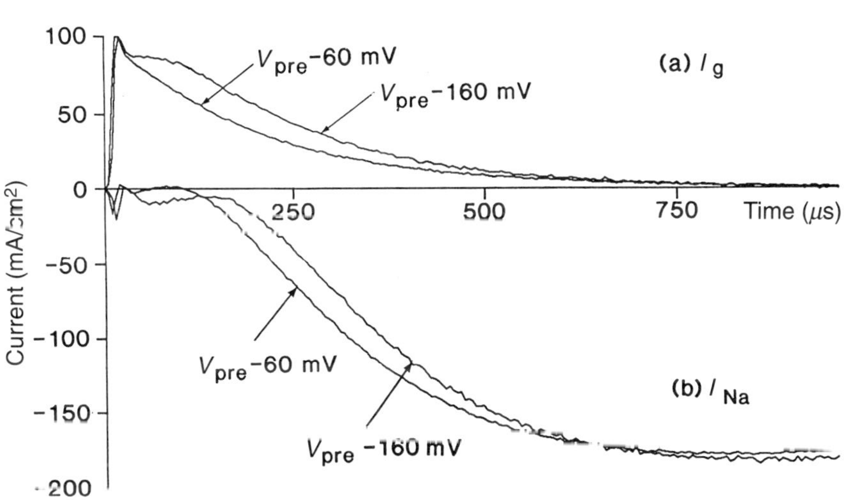

Fig. 6.4. The effect of a negative prepulse V_{pre} on the delays of (a) the relaxation of the gating current I_g, and (b) the rise of I_{Na}. Time marks, 250 μs apart; holding potential, −80 mV; number of sweeps averaged 32; temperature 5°C. (Reproduced with permission from Keynes *et al.* (1991*a*).)

reversed until the channel is emptied of water at the end of the pulse. Hence there are good grounds for identifying I_{g3b} as a fixed-site gating charge of the type proposed by Edmonds (1990).

There was, on the other hand, good reason to suppose that I_{g3a} was related to a movement of the voltage-sensors, because it had a fast rise and was present across the whole range of potentials between −40 and −140 mV, where no channels would have opened whether the test pulses were positive or negative. There was an instructive relationship between I_{g2} and I_{g3a} that emerged on fitting double exponentials to the records, and plotting the values of the charges Q_2 and Q_3 and of the relaxation time constants τ_2 and τ_3 against V_p. As shown in Fig. 6.5, on the negative side of −50 mV, Q_2 and Q_3 were roughly equal whatever the value of V_p, and τ_3 was uniformly about five times larger than τ_2, so that both curves peaked at slightly under 0 mV. This behaviour suggests that I_{g2} and I_{g3a} may have a common origin, and would fit very neatly with the screw helical hypothesis for the operation of the voltage-

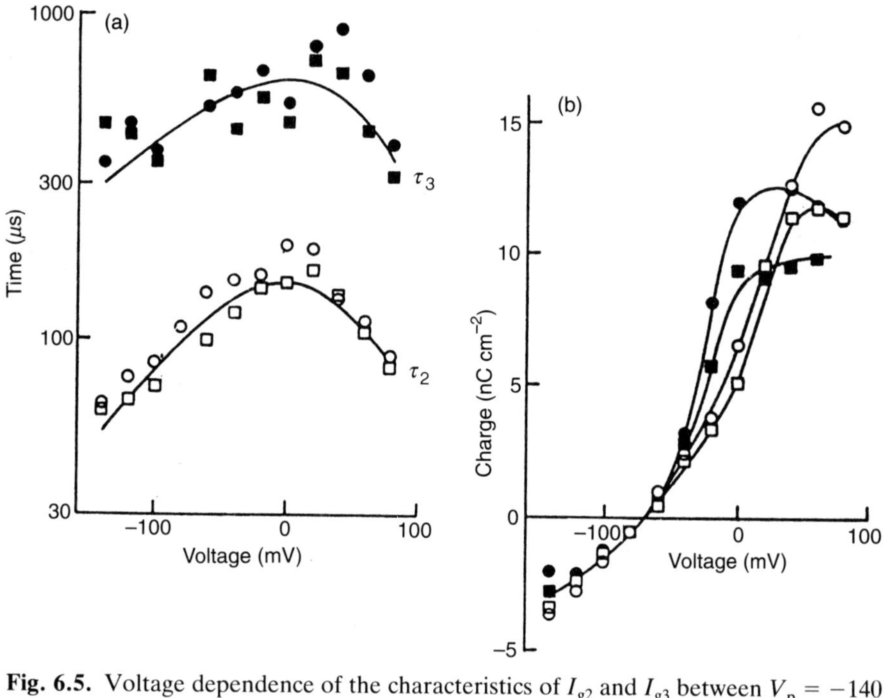

Fig. 6.5. Voltage dependence of the characteristics of I_{g2} and I_{g3} between $V_p = -140$ mV and $V_p = +80$ mV. Temperature 5°C; holding potential −70 mV. (a) Average values of τ_2 (open symbols) and τ_3 (solid symbols) from four runs for two axons bathed in Tris sea water + 1 μM TTX and dialysed with 350 mM CsF (squares) and three runs for the same axons dialysed with 350 mM TMAF (circles). (b) Corresponding averages of the sizes of the charges Q_2 and Q_3, plotted with the same symbols. (Reproduced with permission from Keynes *et al.* (1990).)

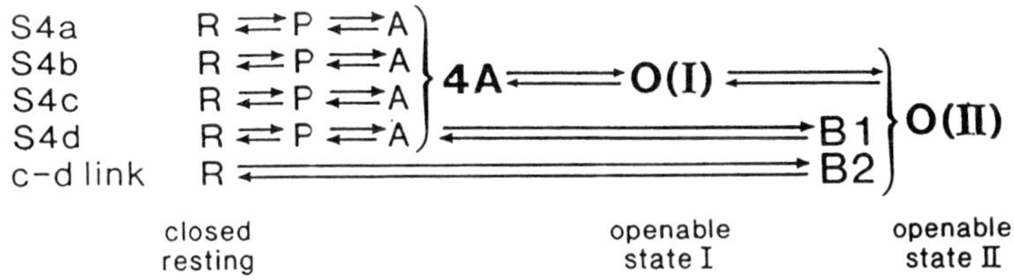

Fig. 6.6. State diagram of a revised series–parallel model of the kinetics of the sodium channel. The transitions between states R, P, A, and B1 of the S4 units carry a charge of the order of 1.3 *e*, but the units are not necessarily identical. The step between states 4A and O(I) of the channel is solvent-sensitive but not voltage-sensitive. The energy wells that determine the probability of re-opening in state O(II) of the channel are governed in parallel by the step from state A to state B1 of voltage-sensor S4d, and by the short link connecting homology domains c and d, which behaves as if it carries a charge of 0.75 *e* and is sensitive to internal TMA. (Reproduced with permission from Keynes (1992).)

sensors (Catterall 1986; Guy and Seetharamulu 1986) if I_{g2} was generated by the first 60° twist of the α-helix and I_{g3a} was generated by a second 60° twist of the same α-helix. Activation of the channel would then involve two steps for each of the four voltage-sensors, from R to P and then from P to A, as indicated in Fig. 6.6.

The proposition that the opening of a given channel can take place only after all four of its voltage-sensors have first been fully activated, that is when it has arrived at state 4A, follows the line adopted by Hodgkin and Huxley (1952) to account for the delay in the rise of I_{Na}, and explains why the threshold of the system lies at about −50 mV while that of the individual voltage-sensors is appreciably more negative. It also accounts satisfactorily for the effect of negative prepulses.

The coupling between activation and inactivation

I must next consider the mechanism of coupling between activation and inactivation. The case has been presented elsewhere (Keynes 1991) for rejecting the idea (Armstrong and Gilly 1979; Cota and Armstrong 1989) that the transition between the open and inactivated states involves little or no transfer of charge. My principal reason was that omission of a charged step would not explain the relatively steep voltage-dependence of recovery from inactivation, which is also observed in other species. I did not, however, provide a convincing explanation for the tendency of the inactivation time constant τ_h to reach a lower limit for large test pulses, and this deficiency in my arguments must now be put right.

My proposal for introducing an appropriate degree of voltage dependence into the processes of inactivation and recovery, which has been shown by computer simulation [see Keynes 1991, fig. 2(d)] to fit the experimental data rather well, is that it involves a third 60° twist of voltage-sensor S4d taking it from state A to state B1. It has been shown in studies employing site-directed mutagenesis (Stühmer *et al.* 1989) and specific antibodies (Vassilev *et al.* 1989) that the short link between homology domains c and d of the sodium channel is closely involved in inactivation, and S4d is the obvious voltage-sensor to choose because it carries more positively charged arginine groups and is longer than S4a, b or c. A charge transfer of the order of 1.3 *e* to accompany the A–B1 step of S4d agrees with Greeff and Forster's (1991) estimate of quantal inactivation charge, with the dependence on potential of τ_h between −20 and +30 mV and of $\tau_{recovery}$ between −100 and −140 mV, and with the size of the small component of gating current I_{g4} whose relaxation time constant τ_4 is close to τ_h (Keynes *et al.* 1990).

As was recognized by Hodgkin and Huxley (1952), the rate of reduction of τ_h with an increase of pulse potential does not correspond accurately to the behaviour of a Boltzmann system for V_ps beyond about 30 mV. The departure might arise from an asymmetry of the electric field seen by S4d (Keynes 1991), but this could not readily account for the continued flattening off of the τ_h curve until above 100 mV its slope corresponds to a charge of no more than 0.2 *e*. It therefore appears that there must be an additional rate-limiting step in the path to the open and inactivated states that prevents τ_h from falling any further.

Evidence for the inclusion in models of voltage-gated ion channels of a solvent-sensitive but voltage-insensitive transition in order to account for the action of heavy water and hyperosmolar media on the rate of opening and closing of the channels without affecting the gating current, has been presented for other species and types of channel by Schauf and Bullock (1979), Alicata *et al.* (1990), and Chen and Hess (1990). That it also applies to squid has been shown by Conti and Palmieri (1968), Meves (1974), and Conti *et al.* (1984). Such a hydration process has therefore been incorporated into the model of Fig. 6.6 as the step from state 4A of the channel to state 0(I). Although further data is certainly needed on its kinetics and precise role in the system, it provides a promising way of accounting for the flattening of the τ_h curve at large pulse potentials.

The mechanism of inactivation and the nature of the steady state

This brings me finally to a consideration of the nature of the non-inactivating open state. It is clear from the single-channel recordings of Bezanilla and Correa (1991) that, whereas in the initial state O(I) the channels have a high probability of opening with a brief latency, in the final steady state O(II) they have a low probability of opening repeatedly with a much longer latency.

Once in this state, their mean open time increases both with V_p and with the presence of internal TMA, suggesting that a modulation takes place of the energy wells that determine the Markovian probability of opening and closing. For the same reasons as before, the part of the channel responsible for this modulation seems most likely to be the c–d link, to which TMA might be bound at some point, and which would be influenced by the electric field at the inner edge of the membrane, though it should not be thought to function as a voltage-sensor in exactly the same way as the S4 units.

The conclusion that a modulation of the energy wells that determine the probability of opening and closing of the channels is involved in the steady state has implications for the model as a whole, and leads me to put forward a new but admittedly speculative concept of the nature of inactivation. The idea of Hodgkin and Huxley (1952) that, after the channels have been opened by the activating *m* particles, they are closed again by the movement into position of the blocking *h* particle, has not previously been questioned. The blocking group has been likened to a ball and chain by Armstrong and Bezanilla (1977), but nothing is known about its precise structure apart from the recent evidence of Zagotta *et al.* (1990) that inactivation can be restored to non-inactivating mutants of *Shaker* potassium channels from *Drosophila* by the insertion of a particular sequence of 20 amino acid residues. However, an alternative to the closure of the channels by a specific blocking group would be that it depends in the normal fashion on the magnitude of the energy wells in the initial state O(I) and that, rather than swinging a blocking group into position, the subsequent A–B1 transition of S4d has the effect of altering the conformation of the channel more subtly by a modulation of the energy wells that greatly lowers the probability of its reopening. On arrival at the inactivated steady state, the energy wells would be subject to further modulation by the c–d link in the manner already discussed.

By thus making it unnecessary to postulate the existence of a specific blocking group, the mechanism of inactivation would be appreciably simplified, for a less drastic modification of the internal structure of the channel would be involved. Another advantage over the conventional mechanism would be that an explanation might be provided for the observations of Aldrich and Stevens (1987) on the failure of the initial open time in single sodium channels from mammalian neuroblastoma cells to vary as predicted for a voltage-dependent process. As far as the sodium channel is concerned, the findings on the abolition of inactivation by perfusion with pronase and other agents (Bezanilla 1985) might be explained by supposing that any disruption of the structure of the c–d link affects the energy wells. The evidence of Zagotta *et al.* (1990) and others on the mechanism of inactivation of potassium channels cannot necessarily be dismissed in a similar way, but the closeness of the parallel between the properties of channels of other types and from other species still remains to be fully established.

The ultimate test of any model of the sodium channel must be to insert parameters based strictly on measurements of the gating current, and to demonstrate that this yields correct predictions of the macroscopic ionic current, that is to say for the behaviour of the two open states. The preliminary exercise of this kind that was reported by Keynes (1991) was based on a version of the model that did not include all the features shown in Fig. 6.6, and on a simplified computer program, which allowed for an inadequate number of states and left out of account the possible effect of co-operativity in the activation of the system (Keynes 1990). It therefore provided no more than a qualitative indication that the approach which I have described here is on sound lines. A properly rigorous examination of the model has yet to be conducted.

Acknowledgements

I am indebted to Drs J. M. Bekkers, N. G. Greeff, and I. C. Forster for assistance with the experiments, which were conducted at the Station Biologique de Roscoff. The work was supported by a series of Project Grants from the Medical Research Council.

References

Aldrich, R. W. and Stevens, C. F. (1987). Voltage-dependent gating of single sodium channels from mammalian neuroblastoma cells. *Journal of Neuroscience*, **7**, 418–31.

Alicata, D. A., Rayner, M. D, and Starkus, J. G. (1990). Sodium channel activation mechanisms: insights from deuterium oxide substitution. *Biophysical Journal*, **57**, 745–58.

Armstrong, C. M. and Bezanilla, F. (1977). Inactivation of the sodium channel. II. Gating current experiments. *Journal of General Physiology*, **70**, 567–90.

Armstrong, C. M. and Gilly, W. F. (1979). Fast and slow steps in the activation of sodium channels. *Journal of General Physiology*, **74**, 691–711.

Bekkers, J. M., Greeff, N. G., and Keynes, R. D. (1986). The conductance and density of sodium channels in the cut-open squid giant axon. *Journal of Physiology*, **377**, 463–86.

Bekkers, J. M., Forster, I. C., Greeff, N. G., and Keynes, R. D. (1987). Non-inactivating sodium channels in the squid giant axon. *Journal of Physiology*, **396**, 83P.

Bekkers, J. M., Forster, I. C., and Greeff, N. G. (1989). High resolution recording of Na gating currents from squid reveals a fast initial component. *Biophysical Journal*, **55**, 316a.

Bezanilla, F. (1985). Gating of sodium and potassium channels. *Journal of Membrane Biology*, **88**, 97–111.

Bezanilla, F. and Correa, A. M. (1991). Single sodium channels in high internal sodium in the squid giant axon. *Biophysical Journal*, **59**, 12a.

Catterall, W. A. (1986). Voltage-dependent gating of sodium channels: correlating structure and function. *Trends in Neurosciences*, **9**, 7–10.

Chandler, W. K. and Meves, H. (1970). Evidence for two types of sodium conductance in axons perfused with fluoride solutions. *Journal of Physiology*, **211**, 653–78.
Chen, C. and Hess, P. (1990). Mechanism of gating of T-type calcium channels. *Journal of General Physiology*, **96**, 603–30.
Conti, F. and Palmieri, G. (1968). Nerve fiber behaviour in heavy water under voltage-clamp. *Biophysik*, **5**, 71–7.
Conti, F., Inoue, I., Kukita, F., and Stühmer, W. (1984). Pressure dependence of sodium gating currents in the squid giant axon. *European Biophysics Journal*, **11**, 137–47.
Cota, G. and Armstrong, C. M. (1989). Sodium channel gating in clonal pituitary cells. The inactivation step is not voltage dependent. *Journal of General Physiology*, **94**, 213–32.
Edmonds, D. T. (1990). Gating charge transfer due to fixed ionizable sites. *European Biophysics Journal*, **18**, 135–7.
Greeff, N. G. and Forster, I. C. (1991). The quantal gating charge of sodium channel inactivation. *European Biophysics Journal*, **20**, 165–76.
Guy, H. R. and Seetharamulu, P. (1986). Molecular model of the action potential sodium channel. *Proceedings of the National Academy of Sciences, USA*, **83**, 508–12.
Hodgkin, A. L. and Huxley, A. F. (1952). A quantitive description of membrane current and its application to conduction and excitation in nerve. *Journal of Physiology*, **117**, 500–44.
Horn, R., Patlak, J., and Stevens, C. L. (1981). The effect of tetramethylammonium on single sodium channel currents. *Biophysical Journal*, **36**, 321–7.
Keynes, R. D. (1990). A series–parallel model of the voltage-gated sodium channel. *Proceedings of the Royal Society of London, Series B*, **240**, 425–32.
Keynes, R. D. (1991). On the voltage dependence of inactivation in the sodium channel of the squid giant axon. *Proceedings of the Royal Society of London, Series B*, **243**, 47–53.
Keynes, R. D. (1992). Biological transistors: the voltage-gating of ion currents in neurons. *Proceedings of the American Philosophical Society*, **136**, 51–60.
Keynes, R. D. (1994). The kinetics of voltage-gated ion channels. *Quarterly Reviews of Biophysics* (in press).
Keynes, R. D. and Rojas, E. (1976). The temporal and steady-state relationships between activation of the sodium conductance and movement of the gating particles in the squid giant axon. *Journal of Physiology*, **255**, 157–89.
Keynes, R. D., Greeff, N. G., and Forster, I. C. (1990). Kinetic analysis of the sodium gating current in the squid giant axon. *Proceedings of the Royal Society of London, Series B*, **240**, 411–23.
Keynes, R. D., Greeff, N. G., Forster, I. C., and Bekkers, J. M. (1991*a*). The effect of tetrodotoxin on the sodium gating current in the squid giant axon. *Proceedings of the Royal Society, London*, **B246**, 135–40.
Keynes, R. D., Meves, H. and Hof, D. (1992). The dual effect of internal tetramethylammonium ions on the two open states of the sodium channel in the squid giant axon. *Proceedings of the Royal Society of London, Series B*, **249**, 101–6.
Keynes, R. D., Rojas, E., and Ceña, V. (1991*b*). Blocking of the sodium channel by external Tris in the squid giant axon. *Proceedings of the Royal Society of London, Series B*, **246**, 129–33.
Meves, H. (1974). The effect of holding potential on the asymmetry currents in squid giant axons. *Journal of Physiology*, **243**, 847–67.
Noda, S., Shimizu, S., Tanabe, T., Takai, T., Kayano, T., Ikeda, T., Takahashi, H.,

Nakayama, H., Kanaoka, Y., Minamino, N., Kangawa, K., Matsuo, H., Raftery, M. A., Hirose, T., Inayama, S., Hayashida, H., Miyata, T., and Numa, S. (1984). Primary structure of *Electrophorus electricus* sodium channel deduced from cDNA sequence. *Nature*, **312**, 121–7.

Schauf, C. L. and Bullock, J. O. (1979). Modification of sodium channel gating in *Myxicola* giant axons by deuterium oxide, temperature, and internal cations. *Biophysical Journal*, **27**, 193–208.

Stühmer, W., Conti, F., Suzuki, H., Wang, X., Noda, M., Yahagi, I., Kubo, H., and Numa, S. (1989). Structural parts involved in activation and inactivation of the sodium channel. *Nature*, **339**, 597–603.

Vassilev, P., Scheuer, T., and Catterall, W. A. (1989). Inhibition of single sodium channels by a site-directed antibody. *Proceedings of the National Academy of Sciences, USA*, **86**, 8147–51.

Zaggotta, W. N., Hoshi, T., and Aldrich, R. W. (1990). Restoration of inactivation in mutants of *Shaker* potassium channels by a peptide derived from ShB. *Science*, **250**, 568–71.

7

An improved voltage clamp for gating current recording from the squid giant axon

IAN C. FORSTER and NIKOLAUS G. GREEFF

Summary

The performance of the conventional cylindrical axon voltage clamp has been improved for high-resolution recording of gating currents from the squid giant axon. First, the requirement for fast, smooth settling of the membrane potential was met by applying a set of design equations derived from an analysis of an electrical model of the clamp. In practice, a membrane settling time down to 3 μs with up to 100 per cent compensation of the membrane series resistance was possible. Second, the overall instrumentation rms noise was reduced more than 10-fold by identifying the extrinsic noise sources and balancing their contributions against the intrinsic noise from the membrane. Third, instrumentation linearity was improved to reduce distortion of the early time course of gating currents such that this accounted for <1 per cent of the total gating charge recorded. Gating current components with relaxations faster than 15 μs and slow components during sodium channel inactivation which were hitherto buried in the background noise have been detected with this system.

The study of gating currents generated by charge movements accompanying molecular conformational changes in voltage-dependent ion channel proteins, provides important information for understanding the structure–function relationships of these channels (see Chapters 6 and 8). Since gating currents were first recorded from the squid giant axon by Armstrong and Bezanilla (1974) and Keynes and Rojas (1974), the squid still remains one of the best preparations for studying these charge movements in native sodium or potassium channels. This is because of the large channel density and ease with which the internal and external membrane environment can be controlled. Nevertheless, the usefulness of the preparation is dependent on the fidelity and precision of the instrumentation used to record and process the gating currents, in particular, the voltage-clamp hardware. The theoretical principles for the design of cylindrical axon voltage clamps have been presented by a number of authors (e.g. Moore and Cole 1963; Katz and Schwartz 1974; Levis 1981; Bezanilla *et al.* 1982, Fishman 1982). However, some of these designs have not been addressed specifically to the unique requirements of gating current recording and do not take advantage of the current state-of-the-art electronic devices.

In this chapter we deal with three aspects of cylindrical voltage-clamp design that we have found important in achieving a predictable clamp performance for high-resolution gating current recordings: dynamic response; noise performance; and system linearity.

Clamp configuration

Figure 7.1(a) depicts the cylindrical axon recording chamber, showing the arrangement of the axon and electrodes. The dialysed axon (AX) rests in two slotted Perspex partitions (PP) that isolate the central recording section from the surrounding guard sections. A special composite internal electrode (IE) is inserted in the dialysis tubing (DT) and runs along the length of the recording chamber. Figure 7.1(b) gives a simplified electrical equivalent of the axon and electrodes together with the voltage-clamp circuitry, which is similar to previously described designs (e.g. Bezanilla *et al.* 1982). A differential amplifier (DA), having a fixed gain A_d and very high input impedance, senses and amplifies the voltage V_m' appearing across the series-shunt combination of membrane capacitance C_m, membrane leak R_l, and the series access resistance R_s. Resistor R_s accounts for the resistance of the internal and external solutions and access pathway through the Schwann cell layer surrounding the axon. The output of the DA and the command input V_c are summed as currents at the summing node input of the high-gain loop summing amplifier (SA). The SA then drives a current I_m through the R_s–C_m/R_l pathway to maintain the negative feedback conditions. The current-to-voltage converter (IVC) stage, having a transresistance R_i, then converts this current to the output signal I_mR_i. Resistor R_{ll} accounts for the leakage between the guard and central sections of the chamber. To control the true membrane voltage V_m, we employ the usual positive feedback loop from the IVC output via a compensation amplifier (CA) having a gain factor α, to effect R_s compensation in the manner of Hodgkin and Huxley (1952). Full details of the realization of our voltage-clamp system have been published elsewhere (Bekkers *et al.* 1986; Forster and Greeff 1990).

Dynamic performance

The clamp dynamic performance affects both the speed and accuracy of membrane potential control. Also, because the membrane potential may be treated as an electrical driving function that initiates the membrane-bound gating charge movements, the clamp dynamic characteristics influence the accuracy and intrinsic bandwidth of the recorded gating currents. A further

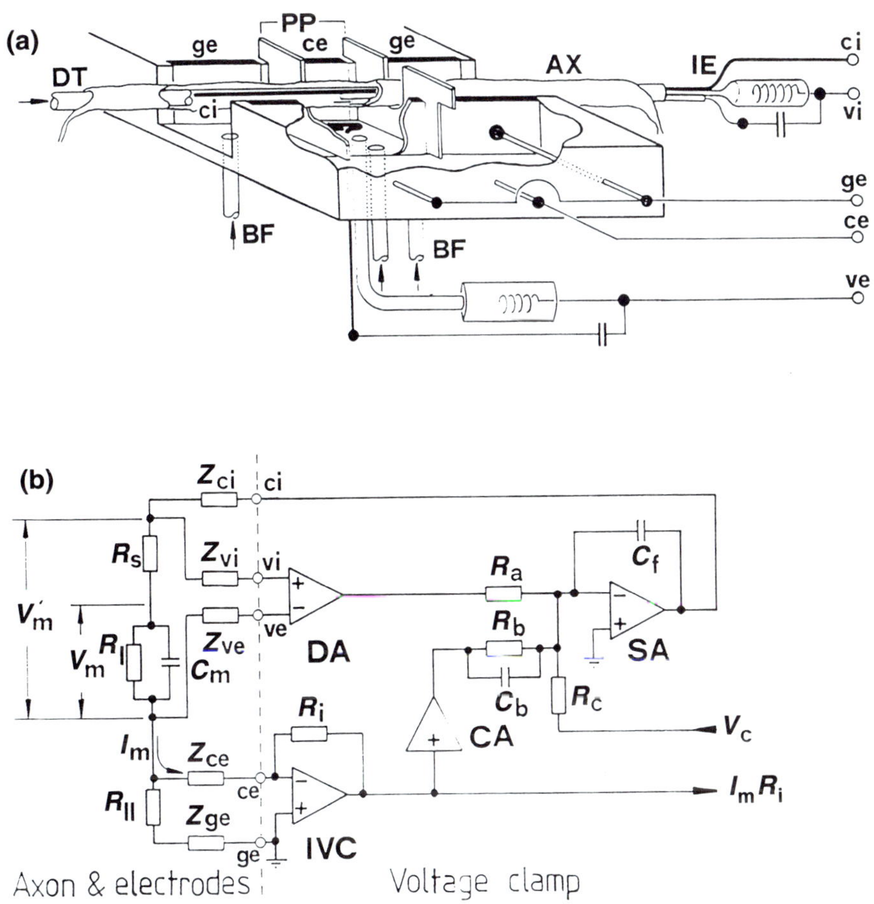

Fig. 7.1. The cylindrical-axon voltage-clamp configuration. (**a**) Recording chamber with cutaway view of central current recording section and axon. AX, axon; BF, bath solution feed inlet; ce, ci, external and internal current electrodes; DT, dialysis tubing; ge, guard electrodes; IE, internal composite electrode; PP, Perspex partitions; ve, vi, external and internal voltage electrodes comprising KCl filled capillary and tubing, respectively, together with capacitively coupled platinized Pt–Ir wire. For electrode vi, wire is exposed only at tip. Plate electrodes are platinized Pt. Chamber opening cross-section is 4 mm × 4 mm; guard plate (ge) length, 6.5 mm; central plate (ce) length, 5 mm. (**b**) Simplified equivalent circuit of axon, electrodes, and clamp, omitting the chamber guard sections. R_s models the series access resistance and C_m/R_l model the membrane. Electrodes are modelled by impedances Z where subscripts refer to electrodes in (a). R_{ll} represents the leak between central recording section and guard sections in chamber. DA, differential amplifier; IVC, current-to-voltage converter; CA, compensation amplifier; SA, summing amplifier. V_c is the command input; I_m is the current through the membrane in the central chamber section and I_mR_i is IVC output for $R_{ll} \gg Z_{00}$; V_m' and V_m are the voltages sensed by the DA and the true membrane voltage, respectively. (Modified from Forster and Greeff (1990).)

practical requirement was to achieve the desired performance through the adjustment of as few external parameters as possible.

Real active circuit elements have finite bandwidths that will affect the stability of the clamp. Capacitor C_f and resistor R_a create a dominant pole around the SA stage in the main feedback loop so that, in the frequency domain, the transfer function of interest (V_m/V_c), approximates that of a second-order network. An additional capacitor C_b in shunt with resistor R_b allows independent setting of the natural frequency (related to the risetime of V_m) and damping factor (related to the smoothness of settling of V_m) with 100 per cent of R_s being compensated (Levis 1981). Despite the usefulness of this simplification, we found it necessary to take into account the finite bandwidths of the DA, IVC, CA, and SA. For the simplest case, in which all stages have single pole roll-offs in the frequency domain and the CA and IVC are lumped together, a fifth-order network results. Its complexity makes practical realization difficult because of the large number of circuit parameters that must be specified and adjusted. However, by choosing the SA to have a bandwidth much larger than the other stages, the clamp network can be reduced to a third-order system (Forster and Greeff 1990). Now, to avoid distortion of the initial time course of the gating currents, it is necessary to have V_m settle smoothly in response to a step input for V_c (Forster and Greeff 1992). This will be guaranteed if the coefficients of the third-order characteristic equation are adjusted to coincide with those of a Bessel polynomial of the same order (Levis 1981). If the DA time constant τ_d ($= 1/2\pi f_d$, where f_d is the DA 3-dB bandwidth) and membrane time constant τ_m ($= R_sC_m$) are specified, then for a desired 10–90 per cent true membrane risetime τ_r, we can derive three design equations for the unknown time constants: τ_f ($= R_aC_f$); τ_b ($= R_bC_b$); and τ_i ($= 1/2\pi f_i$, where f_i is the 3-dB bandwidth of the IVC and CA) (Forster and Greeff 1990), that is

$$\tau_f = 0.5\,A_d\{[0.798\tau_r + \tau_d] + \surd[\tau_r^2((1.02\tau_d - 0.156\tau_r)/\tau_m + 0.637) + \tau_d(5\tau_d - 1.596\tau_r)]\}, \tag{7.1}$$

$$\tau_i = 0.798\tau_r - \tau_f/A_d, \tag{7.2}$$

$$\tau_b = \tau_r\,(0.798 - 0.255\,\tau_r/\tau_m) - \tau_d, \tag{7.3}$$

and the transfer function V_m/V_c in the steady state $=R_a/(R_cA_d)$. It is assumed that the CA gain α has been set for 100 per cent R_s compensation (i.e. $\alpha = A_dR_sR_b/(R_aR_i)$) and the membrane leak resistance (R_l) has been neglected.

If eqns 7.1–7.3 are used to design the clamp for a specified membrane risetime, very good agreement between the expected and measured performance is obtained (Forster and Greeff 1990). Typical true membrane voltage risetimes of 3 μs can be achieved using standard wideband operational amplifiers for the gain stages in Fig. 7.1(b) when clamping a dummy model membrane. When clamping an axon, access to the true membrane potential is not possible and we therefore measured this indirectly using the capacitive current transient at the IVC output and adjusted the R_s compensation circuitry for the fastest settling of this signal without overshoot, termed critical compensation (Greeff *et al.* 1982). This condition is routinely monitored during the course of an experiment and readjustments made to ensure that any changes in R_s resulting, for example, from temperature or solution changes and membrane access resistance inhomogeneities in the Schwann cell sheath (Zwahlen *et al.* 1988) do not cause over- or under compensation. Such variations can result in significant distortion of the early time course of the gating current (Forster and Greeff 1992).

With this design, the 3-dB bandwidth of the gating currents is given by $1.1/\pi\tau_r$ Hz so that for $\tau_r = 3$ μs, the intrinsic bandwidth is about 120 kHz. In the time domain such band limiting causes a rising phase and rounding of the exponentially decaying gating currents resulting from a step input for V_c. Usually the gating current record contains multiple exponential decays (e.g. Bekkers *et al.* 1990; Keynes *et al.* 1990) with relaxation time constants ranging from 10 μs to > 1 ms. Here accurate separation of these components using exponential curve fitting procedures requires rapid charging of the membrane to minimize distortion of the time course of the fastest relaxing components in addition to that mentioned above (Forster and Greeff 1992).

Noise performance

Random noise, superimposed on the gating current signal at the IVC output, arises from two sources: intrinsic—from the preparation itself, specifically R_s; and extrinsic—from the voltage clamp, recording chamber, and electrodes. For a given membrane recording area, temperature, and bath/perfusion solutions, R_s ideally remains constant. The noise optimization procedure that we have adopted involves reducing the extrinsic noise to a level close to that attributable to R_s alone. A useful design equation for the equivalent mean square noise current $\overline{i_{neq}}^2$ at the IVC output can be derived from analysis of the circuit of Fig. 7.1(b) (Forster and Greeff 1990) whereby all circuit elements are modified to include the equivalent voltage and current noise generators (e.g. Motchenbacher and Fitchen 1973); such that

$$\overline{i_{neq}}^2 = F(\omega)\,(4kT/R_s)\,(1 + \Sigma R_{extr}/R_s) \qquad (7.4)$$

where k is Boltzmann's constant and T the absolute temperature. The factor $F(\omega)$ is a frequency-dependent term that depends on the R_s compensation level and clamp dynamic response. For no R_s compensation $F(\omega) \approx 1$. The term ΣR_{extr} is the sum of the equivalent noise resistances of all the extrinsic components in Fig. 7.1(b) referred to the input of the DA. The second factor in parentheses then represents the increase in noise over the intrinsic component due to R_s alone ($\times\ 4kT/R_s$) and shows the importance of having ΣR_{extr} as small as possible compared with R_s. Based on equivalent noise resistances,

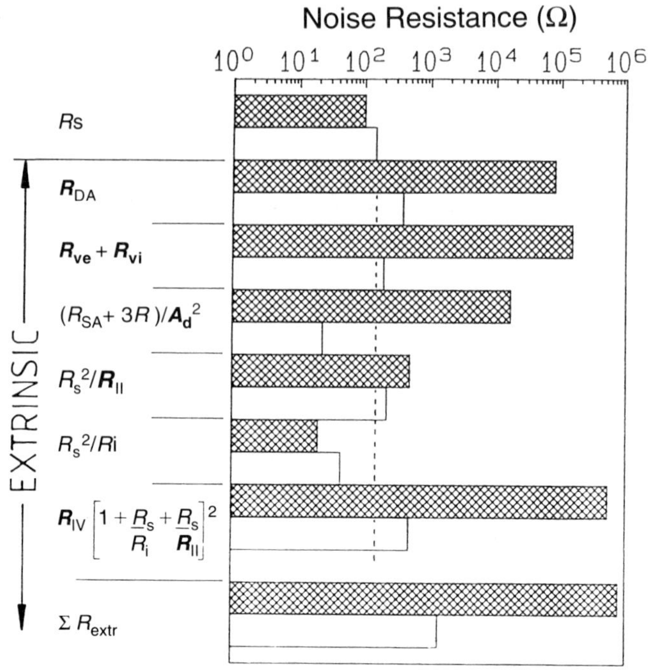

Fig. 7.2. Clamp noise performance. Equivalent noise resistances for specific elements of Fig. 7.1(b) are compared on a logarithmic scale for a conventional (hatched bars) and the improved clamp design (open bars). The improved design brings all extrinsic noise terms close to R_s (indicated by the broken line) by modifying the bold-faced parameter in the expression for each term. All resistors at the SA input have the same value R. (Modified from Forster and Greeff (1990).)

an evaluation of the clamp noise performance can then be made to identify which components contribute significantly to the overall noise. This is shown in Fig. 7.2, where we compare the noise resistances of the elements in Fig. 7.1(b) from an earlier clamp design using conventional components (e.g. Bezanilla *et al.* 1982) with our improved design.

To improve the noise performance we paid special attention to the contributions from the DA and IVC stages, voltage-sensing electrodes, and leak resistance R_{ll}. For the DA and IVC, we used specially designed low-noise wide-band circuitry to achieve a 10-fold reduction in noise resistance over previous designs (Forster and Greeff 1990) so that the effective noise resistances were now comparable with R_s (i.e $\approx 100\ \Omega$). Furthermore, by having the DA gain $A_d > 1$, the contributions from the SA were concomitantly reduced. The internal voltage-sensing electrode had, in addition to the usual fluid-filled capillary, an insulated platinum wire exposed to the internal solution and platinized only at the tip in the vicinity of the capillary end [see Fig. 7.1(a)]. This wire was capacitively coupled to the capillary (Levis and Bezanilla 1983;

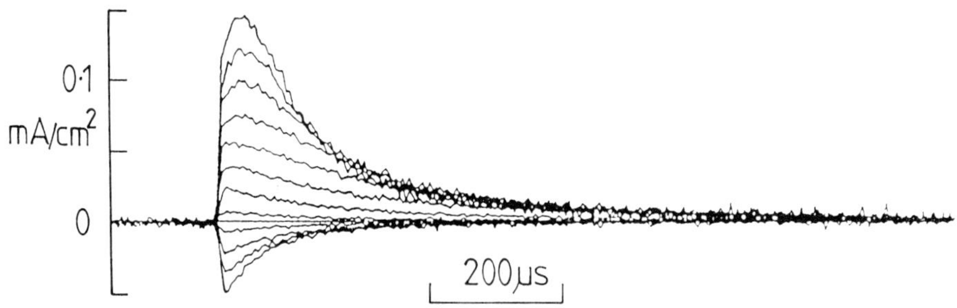

Fig. 7.3. Typical gating current recording for a series of voltage steps from −80 mV to +140 mV recorded at 5°C. Single sweep records are shown with no signal averaging to demonstrate the noise performance of the clamp. Records filtered at 100 kHz. (From an unpublished experiment by J. M. Bekkers, I. C. Forster, N. G. Greeff, and R. D. Keynes.)

Bekkers *et al.* 1986) so that the noise resistance of this component (typically > 100 kΩ) was now close to R_s. A similar principle was also used for the bath voltage-sensing electrode (Forster and Greeff 1990). Finally, by supporting the axon in slotted insulated partitions as shown in Fig. 7.1(a), the leak resistance R_{ll} was increased 10-fold to about 100 Ω which reduced significantly the effective noise resistance of the IVC (see Fig. 7.2). The overall improvement in noise performance is found by comparing the factor $\sqrt{R_{extr}}$ for each clamp since this term is proportional to the extrinsic rms noise, assuming the same dynamic characteristics ($F(\omega)$ in eqn (4)). In the present case this amounted to a 24-fold improvement.

For gating current measurements in which steady-state errors in V_m caused by R_s are not relevant, we obtained a further improvement in the noise performance at high frequencies by reducing the percentage of R_s compensation (typically 80 per cent) and re-adjusting C_b and C_f to maintain the same membrane risetime (Forster and Greeff 1990). This gave a typical equivalent input noise current of 86 nA_{rms} for the voltage clamp alone with a 100-kHz recording bandwidth, that is a peak-to-peak noise amplitude that was typically less than 5 per cent of the peak gating current amplitude. In practice, such a noise reduction meant fewer signal averages were required for a given signal-to-noise ratio and the effects of preparation rundown could be minimized. Figure 7.3 shows a typical gating current recording obtained using only 100-kHz filtering and no signal averaging.

System linearity

Linearity of the complete signal pathway between electrodes and data acquisition hardware is necessary for accurate registration of gating currents,

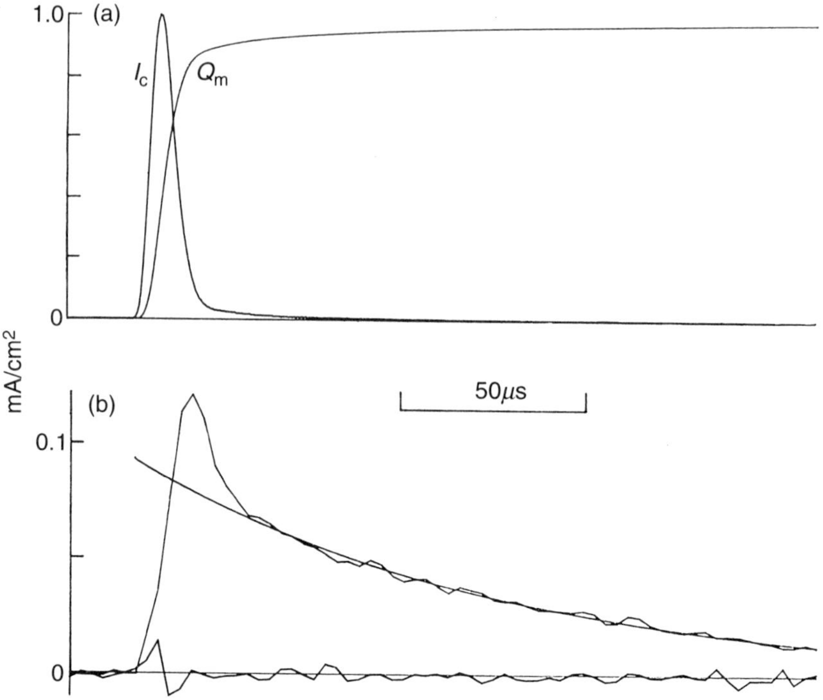

Fig. 7.4. Instrumentation asymmetry time course and magnitude. (**a**) The membrane capacitive transient (I_c) and membrane charging time course (Q_m) found by integrating I_c for a voltage step from −80 mV to −110 mV where contamination from gating currents is small. Apart from the slow tail caused by inhomogeneities in the membrane access resistance, the bulk of the membrane has been charged within 10 μs when taking account of the additional 100 kHz filtering of the I_c signal. Sampling at 0.5 μs/point. (**b**) Voltage-clamp system residual asymmetries (lower trace) compared with typical gating current recording (upper trace) for a step to +40 mV from a −100 mV holding potential plotted on same time scale as (a). The test pulse was preceded by an inactivating prepulse for 20 ms at 0 mV to reveal a fast component when the main relaxation is fitted to a single exponential. All traces have been filtered at 100 kHz for comparison purposes. (Modified from Bekkers *et al.* (1990).)

particularly during their early time course. Since the currents of interest result from the subtraction of two signals that can be up to 1000 times larger (i.e. the linear charging transient of the membrane capacitance C_m), any instrumentation non-linearities will not be readily distinguished from the gating currents themselves. Furthermore, this subtraction procedure usually involves first scaling up one signal assumed to contain no gating currents [e.g the P/4 procedure of Armstrong and Bezanilla (1977) or the prescaled back-reference procedure seen in Bekkers *et al.* (1984, fig. 1)] so that non-linearities in the back reference signal will be further enhanced. Precautions were taken to avoid non-linear operation of the electronic hardware (e.g. slew rate limiting) and ensuring, where possible, that risetimes were dependent on passive rather than active components. Particular attention was paid to the digital-to-analogue converter specifications and the design of the capacitive transient suppression circuitry. For large voltage steps, we found it necessary to trade off available bandwidth against size of transient asymmetries by keeping the amplitude of the membrane charging transient within conservative limits. This meant compromising on the speed of membrane charging. The complete system non-linearity was measured using a dummy membrane/electrode network and the same voltage step protocols to be used with the axon preparation. This was typically ≤ 0.1 per cent over the ± 200 mV membrane voltage swing used.

Figure 7.4 compares the instrumentation asymmetry for the clamp system with the membrane capacitance charging time course and a typical gating current record in which a fast component has been revealed through exponential curve fitting to the main relaxation. The residual asymmetry accounted for less than 1 per cent of the total gating charge (or only 0.2 nC/cm^2) and is negligible after the bulk of the membrane has been charged following the command step onset.

Conclusions

The results of the dynamic and noise analyses have facilitated the realization of a cylindrical axon voltage clamp having a predictable performance compatible with the requirements for high-resolution gating current measurements. These improvements to the clamp design and realization have enabled us to make high-resolution recordings of fast (Bekkers *et al.* 1990; Forster and Greeff 1992), intermediate, and slow (see Greeff and Forster 1991; also Chapters 6 and 8), charge movements associated with voltage-dependent processes in the squid giant axon.

Acknowledgements

We wish to acknowledge the technical expertise provided by S. B. Cross (Cambridge) and G. Mäder (Zürich) for the chamber construction. R. D. Keynes and J. M. Bekkers provided valuable assistance in constructing the internal electrodes. This work was funded by Swiss NF grant 3.143–0–85 to NGG.

References

Armstrong, C. M. and Bezanilla, F. (1974). Charge movements associated with the opening and closing of the activation gates of the Na channel. *Journal of General Physiology*, **63**, 533–52.

Armstrong, C. M. and Bezanilla, F. (1977). Inactivation of the sodium channel II. Gating current experiments. *Journal of General Physiology*, **70**, 567–90.

Bekkers, J. M., Greeff, N. G., Keynes, R. D., and Neumcke, B. (1984). The effect of local anaesthetics on the asymmetry current in the squid giant axon. *Journal of Physiology*, **352**, 653–68.

Bekkers, J. M., Forster, I. C., Greeff, N. G., and Keynes, R. D. (1986). Improvements in the recording of gating currents in the squid giant axon. *Journal of Physiology*, **381**, 9P.

Bekkers, J. M., Forster, I. C., and Greeff, N. G. (1990). Gating current associated with inactivated states of the squid giant axon sodium channel *Proceedings of the National Academy of Sciences, USA*, **87**, 8311–15.

Bezanilla, F., Vergera, J., and Taylor, R. E. (1982). Voltage clamping of excitable membranes. *Methods of Experimental Physics*, **20**, 445–511.

Fishman, H. M. (1982). Current and voltage clamp techniques. In *Techniques in the life sciences—Physiology* (ed. P. F. Baker), pp. 1–11. Elsevier Biomedical, New York.

Forster, I. C. and Greeff. N. G. (1990). Technical aspects of voltage clamping the squid giant axon. *Journal of Neuroscience Methods*, **33**, 185–205.

Forster, I. C. and Greeff, N. G. (1992). The early phase of sodium channel gating currents in the squid giant axon. *European Biophysics Journal*, **21**, 99–116.

Greeff, N. G. and Forster, I. C. (1991). The quantal gating charge of sodium channel inactivation. *European Biophysics Journal*, **20**, 165–76.

Greeff, N. G., Keynes, R. D., and Van Helden D. F. (1982). Fractionation of the asymmetry current in the squid giant axon into inactivating and non-inactivating components. *Proceedings of the Royal Society of London, Series B*, **215**, 375–89.

Hodgkin, A. L. and Huxley, A. F. (1952). Measurement of current–voltage relations in the membrane of the giant axon of *Loligo*. *Journal of Physiology*, **116**, 424–48.

Katz, G. M. and Schwartz, T. L. (1974). Temporal control of voltage-clamped membranes: an examination of principles. *Journal of Membrane Biology*, **17**, 275–91.

Keynes, R. D. and Rojas, E. (1974). Kinetics and steady state properties of the charged system controlling sodium conductance in the squid giant axon. *Journal of Physiology*, **239**, 393–434.

Keynes, R. D., Greeff, N. G., and Forster, I. C. (1990). Kinetic analysis of the sodium gating current in the squid giant axon. *Proceedings of the Royal Society of London, Series B*, **240**, 411–23.

Levis, R. A. (1981). Patch and axial wire voltage clamp techniques and impedance measurements of cardiac purkinje fibers. Unpublished PhD thesis, University Microfilms 8120995, University of California at Los Angeles.

Levis, R. A. and Bezanilla, F. (1983). Noise performance of a giant axon voltage clamp. *Biophysical Journal*, **41**, 53a.

Moore, J. W. and Cole, K. C. (1963). Voltage clamp techniques. In *Physical techniques in biological research*, Vol. 6B (ed. W. L. Nastuk), pp. 263–321. Academic Press, New York.

Motchenbacher, C. D. and Fitchen, F. C. (1973). *Low noise amplifier design*. Wiley, New York.

Zwahlen, M. J., Sandri, C., and Greeff, N. G. (1988). Transglial pathway of diffusion in the Schwann cell sheath of the squid giant axon. *Journal of Neurocytology*, **17**, 145–59.

8

Voltage dependence of sodium channel inactivation in the squid giant axon

NIKOLAUS G. GREEFF and IAN C. FORSTER

Summary

The current views about voltage dependence of sodium channel inactivation are rather divergent. In this study we show that high resolution recording of gating current from the squid giant axon detects a slow component during the inactivation phase of sodium ionic current. From the combined data of ionic and gating current we obtained with a new approach (isochronic analysis) an estimate of about 1.2 electron charges for the voltage dependence of microscopic sodium inactivation, i.e. the quantal gating charge of the open-to-inactivated transition. This method also reveals that conventional rate analysis of the inactivation of sodium ionic currents will estimate the voltage dependence correctly between −10 and +40 mV but not at lower voltages. Taking this constraint into account would resolve most of the previous controversies.

The initiation of a nerve action potential depends on the activation of voltage-gated sodium channels, which leads to their opening and the inward flow of sodium ions. The termination of the action potential is then ensured by the subsequent inactivation process. The conformational state of the inactivated channel is a closed one that differs from the resting closed state to which the channels return upon repolarization. Both processes, activation and inactivation, occur faster at more depolarized voltages indicating that they are driven by the electric field across the membrane.

There is general agreement that activation is voltage-driven. In contrast, the views about the voltage dependence of the inactivation process of sodium channels have recently become rather controversial. The views have changed not so much because old data have been shown to be wrong but rather because new techniques (gating currents, patch clamp of single channels, molecular biology, and others) have allowed new insights and different interpretations.

Here we will discuss how ionic and gating currents on the microscopic (single-channel) and macroscopic level (channel population) are related and what the term inactivation rate signifies in these cases. By means of a high-resolution voltage-clamp system we have recently succeeded in recording gating currents from the squid giant axon during the macroscopic inactivation phase that were hitherto buried in the baseline noise. In addition, we have developed a new analysis for the gating current and provide evidence that the controversy about voltage dependence of inactivation may be mainly due to experimental constraints with respect to the voltage range chosen. A full treatment is given in Greeff and Forster (1991).

Sodium channel inactivation: signals and concepts

Figure 8.1 shows schematically the main features of the seemingly contradictory hypotheses about sodium channel inactivation and the signals to be

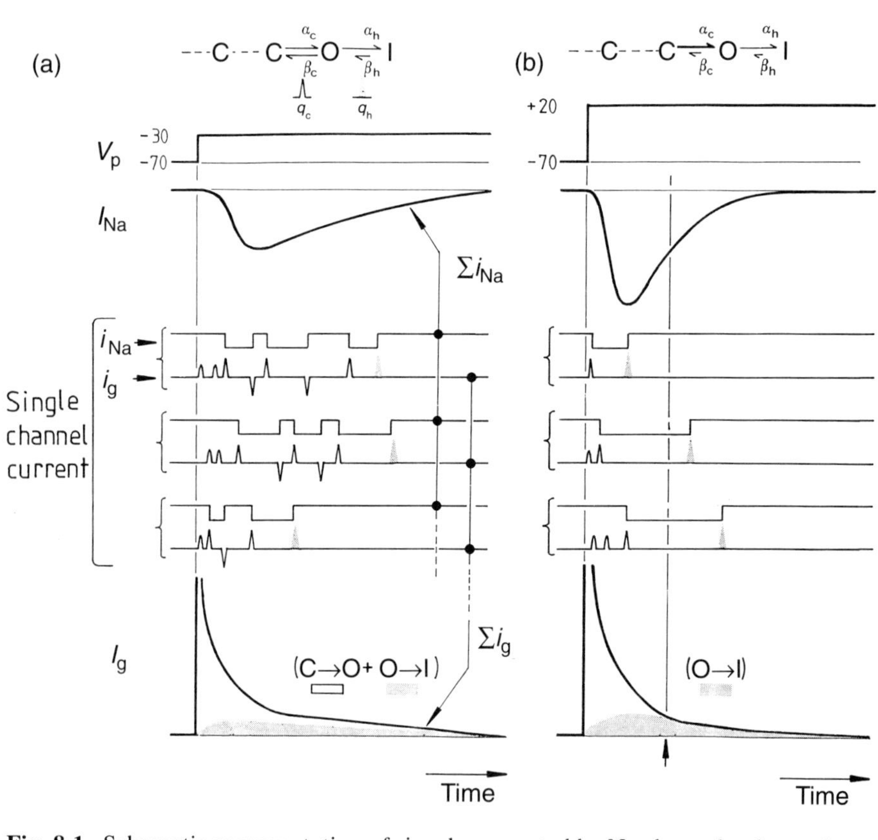

Fig. 8.1. Schematic representation of signals generated by Na channels when voltage-clamped to two different voltage steps (V_p) showing for $V_p = -30$ mV (a) and $V_p = +20$ mV (b) two hypothetical cases with respect to the processes during the inactivation phase. See text for interpretation. (Reproduced with permission from Greeff and Forster (1991).)

expected. The more traditional view, dating back to Hodgkin and Huxley (1952), is shown in Fig. 8.1(b). According to this, single channels (represented by the current i_{Na}, which could not be seen at that time) would open during the activation phase of the recorded macroscopic current (I_{Na}). During the following inactivation phase of I_{Na}, only inactivation transitions of single channels would occur. The simplified scheme at the top reflects the transitions from closed (C) to open (O) and open to inactivated (I) states and the rate constants around the open state. If a conformational transition C–O or O–I is accompanied by an intramolecular charge displacement q_c or q_h respectively, this causes a shot of displacement current to flow (i_g trace). These shots coincide with the time and the direction of the transition seen in the corresponding single-channel ionic record. The size of these quantal shots determines the intrinsic voltage dependence of a transition. Therefore, the size of q_h (shaded in Fig. 8.1) determines the mean open time T^o of single channels as well as the inactivation time constant τ_h of I_{Na}. The specific question of this study concerns the estimation of q_h, that is the microscopic voltage dependence of the O–I transition.

These quantal charges are on the order of one electron charge e^-, and are far too small to be recorded directly in contrast to the several hundred times larger single ionic currents. The size of q_h has been traditionally estimated from the voltage dependence of τ_h or the h_∞ curve (steady-state inactivation; Hodgkin and Huxley 1952). It was found to be around 3 e^-, which would suggest a substantial voltage dependence.

Just as I_{Na} is the sum of many single ionic currents, one can record a macroscopic gating current I_g as the sum of charge shots from a population of channels (Armstrong and Bezanilla 1974; Keynes and Rojas 1974). Here the squid giant axon preparation, giving access to typically more than 10^9 channels, provides a relatively large signal. However, since gating current arises from the transitions between states, it is proportional to the time derivative of the state occupancy. Therefore, I_g will be larger during the activation phase where transitions occur at high frequency but much smaller during the slow inactivation of I_{Na} as seen further below.

Figure 8.1(a) shows another view about inactivation as suggested from patch-clamp data from single sodium channels. Already in the first recordings by Sigworth and Neher (1980) it was found that channels may open late and sometimes repeatedly during the macroscopic inactivation phase. This was studied more systematically and reported in two influential papers by Aldrich *et al.* (1983) and Aldrich and Stevens (1987). They confirmed the existence of late openings and furthermore found that the mean open time of single channels is constant between −60 and −30 mV. In this voltage range the macroscopic time constant τ_h varies substantially, corresponding to the known q_h of 3 e^-. The authors then proposed the following reinterpretation of sodium inactivation: the intrinsic voltage dependence of the O–I step would be very small, q_h only about 0.3 e^-; that is, 1/10 of the previous estimate. The apparent τ_h in I_{Na} would gain its voltage dependence indirectly from C–O transitions that occur earlier at higher voltages.

It should be noted here that single sodium channels were usually recorded below 0 mV where, far from the sodium reversal potential, the amplitudes are larger, and this together with the slower kinetics facilitated to resolve the openings. Thus, the constant T^o was actually seen in a rather limited voltage region only. With respect to gating current this concept would predict a slow component of mixed origin, that is quanta from activation and inactivation transitions as shown in Fig. 8.1(b).

Gating current recording during the inactivation phase

In their detailed study on inactivation, Armstrong and Bezanilla (1977) could hardly detect a gating current component in the baseline noise during macroscopic inactivation in squid. That which was seen had a smaller time constant than τ_h and a voltage-independent O–I transition was therefore suggested. Some slow components in crayfish (Swenson 1983) and the node of Ranvier (Meves and Pohl 1990) were reported without being conclusive about the relation to sodium inactivation.

Since we had developed a low-noise voltage clamp in squid (Forster and Greeff 1990, and Chapter 7, this volume) we decided to reinvestigate I_g and succeeded in recording a clearly visible slow component with a similar time course to the inactivation of I_{Na}. An example for a voltage pulse (V_p) to 0 mV is given in Fig. 8.3(a). Special precautions were made to suppress low-frequency contaminations. Perfusion with tetramethylammonium ions (TMA^+) instead of Cs^+ decreased the leak through potassium channels almost completely. The complete absence of K^+ and Cs^+ on either side of the membrane was assumed to destroy these channels and minimize slow contaminations by them. Protocols were used that minimized 50-Hz pick-up and droop in the baseline (for references and discussion of details see Greeff and Forster 1991).

The next step was to identify the origin of this slow I_g component, that is to clarify whether it arises from C–O or O–I or a mixture. Quantitative fitting still proved to be difficult (Keynes *et al.* 1990), probably because one does not know the overlap of different transitions in an arbitrarily chosen time window used for fitting. Fits to slow processes are also more susceptible to baseline distortions. However, it may be added here that we have examined single exponential fits to I_g in the time window and voltage range where, according to the isochronic analysis given below, the τ_h of inactivation of the sodium current would be controlled by O–I only. Here we found the τs of gating current and τ_h to agree rather well (unpublished). With respect to the fits given in Keynes *et al.* (1990), this slow gating current would most likely correspond to $I_{g,4}$ as suggested by the 15°C data. But it always has to be kept in mind that a component extracted by fitting does not necessarily reflect charge from one state transition only.

Isochronic plot analysis demonstrated with simulated gating currents

For the identification of the slow gating current we have developed a new analytical tool which is different from rate analysis (Greeff and Forster 1991). Here we shall demonstrate its application using simulated currents from the three-state C–O–I model in Fig. 8.2. At first, we chose values for the rate constants and the quantal charges q_c and q_h equal to 2 e^- and 0.5 e^-, respectively, and curves for the sodium conductance g_{Na} and gating current I_g were computed at different voltages. The task was then to recover q_h from these curves.

The principal idea can be grasped by comparing the two cases of inactivation shown in Fig. 8.1(a) and (b). If O–I transitions dominate, as seen in (b) after the vertical line, it becomes evident that I_g is exactly proportional to the rate at which I_{Na} decreases. Each O–I transition contributes one quantal charge unit q_h to I_g and reduces the I_{Na} amplitude by one unit current i_{Na}. This can be tested by plotting I_g versus the time derivative of I_{Na} or, rather, in order to compare different voltages, the derivative of conductance $-\mathrm{d}g_{Na}/\mathrm{d}t$, where $g_{Na} = I_{Na}/(V_p - V_{Na})$. The expected constant proportion between these two quantities would then show up as a straight line through zero which has the slope q_h/γ where γ is the single channel conductance equal to $i_{Na}/(V_p - V_{Na})$. In this so-called isochronic plot shown in Fig. 8.2(c) it can be seen that this is indeed so in the modelled case for voltages above 20 mV. At lower voltages steeper slopes are obtained which indicates that additional charge quanta q_c contribute to I_g, corresponding to the case in Fig. 8.1(a). Note that these additional slow I_g may well have the same time course (same eigenvalues). Therefore straight lines still appear but the proportional relationship to $\mathrm{d}g_{Na}/\mathrm{d}t$ varies with voltage.

Thus, only if the final straight lines at different voltages show the same constant limiting slope S_l in the isochronic plot does this indicate that O–I transitions only are involved. We have tested this for other values of q_h including zero and other state diagrams (e.g. having a parallel C–I pathway) and could always recover q_h correctly from the limiting slope S_l as the product of S_l and γ. This analytical tool appears to be very robust with respect to model type.

The isochronic method also shows clearly the constraints for a conventional Eyring rate analysis of the inactivation transition or time constants. In Fig. 8.2(d) the inactivation time constants of the simulated ionic currents are plotted logarithmically against voltage. Only above 20 mV where, as we now know, the O–I transition dominates does the rate analysis give q_h correctly (but see below). From the τ_hs below 20 mV one would overestimate the true q_h.

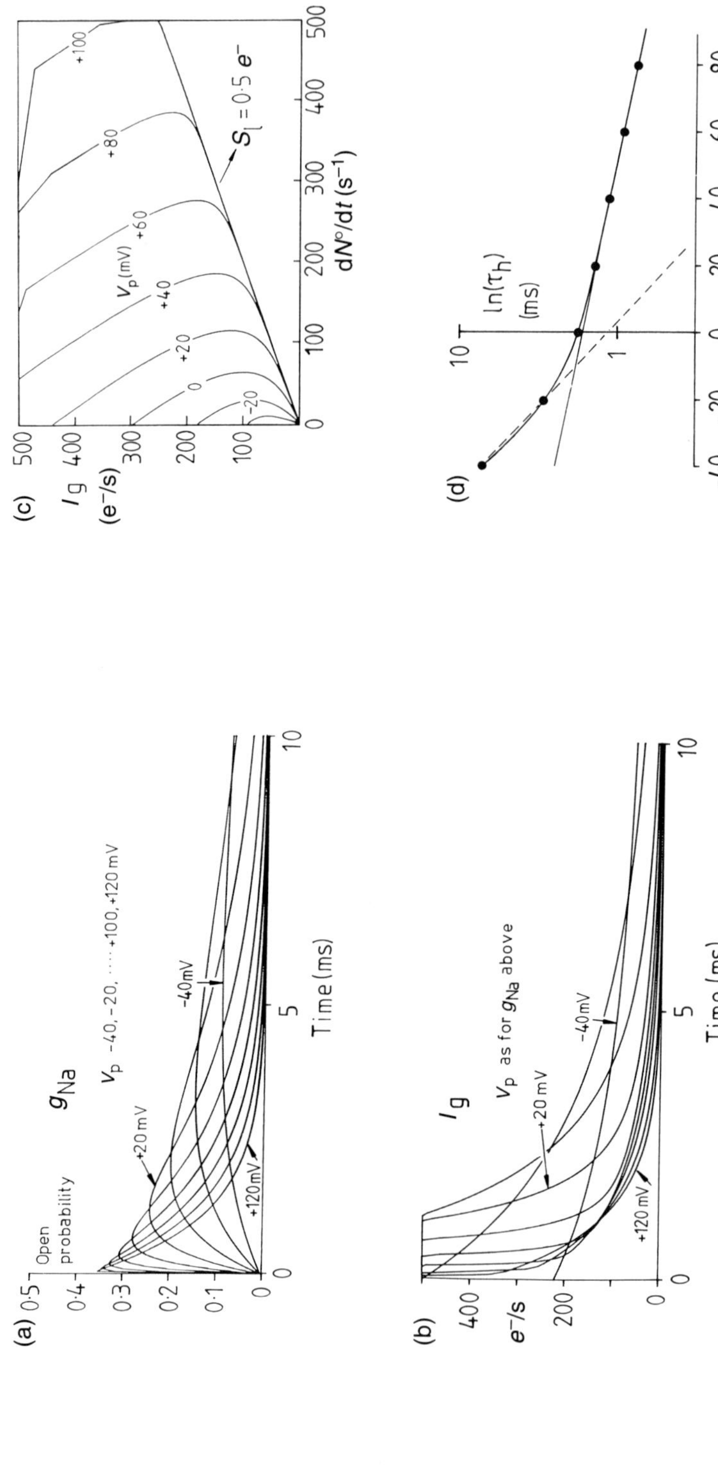
(a)
Open probability
g_{Na}
V_p −40, −20, ·····+100, +120 mV
+20 mV
−40 mV
+120 mV
Time (ms)
(b)
I_g
e^-/s
V_p as for g_{Na} above
+20 mV
−40 mV
+120 mV
Time (ms)
(c)
I_g (e^-/s)
V_p (mV)
$S_l = 0.5\, e^-$
dN^o/dt (s^{-1})
(d)
ln(τ_h) (ms)
V_p (mV)

Fig. 8.2. Simulation of isochronic analysis using a three-state model (C–O–I). (a) Curves for ionic conductance g_{Na} at indicated voltages V_p. Quantal charges were $q_c = 2\,e^-$ for C–O and $q_h = 0.5\,e^-$ for O–I. Rate constants at 0 mV (in s^{-1}) were $\alpha_c = 1500$, $\beta_c = 60$, $\alpha_h = 600$, and $\beta_h = 0.4$; for other voltages they were computed assuming a symmetrical energy barrier position ($\delta = 0.5$). Single channel conductance was taken as unity. (b) Gating currents I_g on an enlarged vertical scale to allow visualization of the slow component. (c) Isochronic representation of $I_g(t)$ traces against corresponding $-dg_{Na}/dt$ given here as $-dN^o/dt$ since one channel has unity conductance; N^o is the fraction of open channels. The limiting slope S_l can be read off directly to be $0.5\,e^-$ equal to q_h. (d) τ_h from the g_{Na} curves plotted logarithmically against V_p at 20-mV intervals. The continuous line gives computed τ_hs from the simulation. Between -20 and -40 mV an *e*-fold decay per 22.5 mV is found corresponding to 2.3 e^- (broken line) and only above 20 mV is an *e*-fold decay of 105 mV giving the correct q_h of 0.5 e^- for the voltage dependence found. $\delta = 0.5$ is taken into account; $kT = 26$ mV in the simulation. (Modified from Greeff and Forster (1991).)

Isochronic plot analysis applied to real gating currents

Figure 8.3(a) shows I_{Na} and a high-resolution recording of I_g on two scales for a pulse to 0 mV from a holding potential of -80 mV at 5°C. Since we found the isochronic analysis to be model-independent, we could apply it to the experimentally recorded I_g even without knowing the gating mechanism of sodium channels. Figure 8.3(b) shows I_g on a larger vertical scale and $-dg_{Na}/dt$ as obtained from I_{Na}. It can be seen that these two signals decay in equal proportion after the time marked by the vertical line. This can be tested better in the isochronic representation in Fig. 8.3(c). Here it is seen how the labelled trace (0 mV) joins the common final straight line of the isochronic curves for voltages of -10 to $+40$ mV. The isochronic curves for -20 mV and lower voltages are steeper which can be better seen in Fig. 8.3(d) for another experiment, which shows the isochronic curves for voltages of -30, -20, -10, and 0 mV with traces separated from each other.

Thus a behaviour is observed that is similar to that in the simulation and we can conclude that, for voltages between -10 and $+40$ mV, the case of only O–I transitions occurring holds true in the later phase of the macroscopic inactivation. The exact beginning of this phase corresponds to the arrow and vertical line in Fig. 8.3(c) and (b), i.e. about half-way down the inactivation phase. These are now the constraints for a rate analysis of τ_h or mean open time T^o. At lower voltages, or earlier inactivation phase, C–O transitions are mixed in.

From the limiting slope S_l of 0.247 μC/S or 1.54 e^-/pS we obtained an estimate for the quantal inactivation charge q_h of 1.21 e^- by multiplication with the single-channel conductance γ (Greeff and Forster 1991).

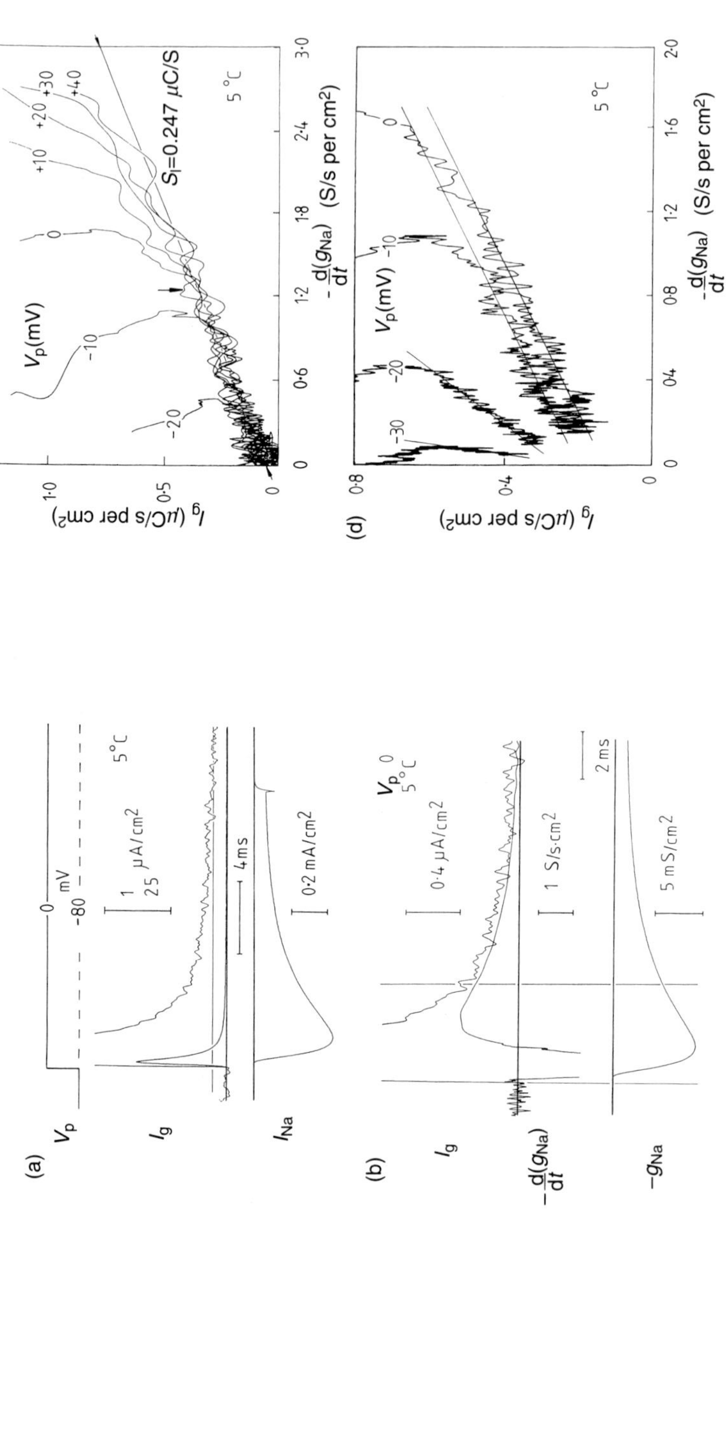
(a)
V_p
0
mV
-80
5°C
I_g
1
25 μA/cm²
4 ms
I_{Na}
0·2 mA/cm²
(b)
V_p 0
5°C
I_g
0·4 μA/cm²
$-\frac{d(g_{Na})}{dt}$
1 S/s·cm²
2 ms
$-g_{Na}$
5 mS/cm²
(c)
I_g (μC/s per cm²)
V_p(mV)
-20
-10
0
+10
+20
+30
+40
S_i=0.247 μC/S
5 °C
$-\frac{d(g_{Na})}{dt}$ (S/s per cm²)
(d)
I_g (μC/s per cm²)
V_p(mV)
-30
-20
-10
0
5 °C
$-\frac{d(g_{Na})}{dt}$ (S/s per cm²)

Fig. 8.3. Experimental gating current I_g and sodium current I_{Na} and isochronic analysis. (a) $V_p = 0$ mV from holding potential of -80 mV. I_g is shown on two scales; its trace lasts longer than shown to verify stable baseline. (b) The same data enlarged. g_{Na} is obtained from $I_{Na}/(V_p-V_{Na})$; in addition, the time derivative $-dg_{Na}/dt$ is shown and scaled for comparison with the decay in the I_g trace. The vertical line indicates the start of the common decay rate. (c) Isochronic plots of I_g versus $-dg_{Na}/dt$ for -20 to $+40$ mV. The arrow indicates, for the 0 mV trace, the start of the limiting slope and corresponds to the time of the vertical line in (b). (Note the characteristic noise hump in the I_g trace.) (d) Isochronic curves for lower voltages -30 to -10 mV from a different experiment. The traces are separated to show a final straight line with steeper slopes at -30 and -20 but a common slope at -10 mV and above. Temperature, 5°C. (Modified from Greeff and Forster (1991); experiments K14O, K5N1, K5N2.)

Impact on rate analysis of inactivation time constant and mean open time

By taking into account the constraints obtained from the isochronic plot analysis we now fitted single exponential time constants to the later part of the inactivation phase of the sodium ionic currents. They are plotted logarithmically against voltage in Fig. 8.4(a). It can be seen that indeed, from -10 to $+40$ mV, the τ_hs lie on a straight line with an e-fold change per 48 mV. This indicates a voltage dependence of 0.50 e^-. It is important to note here that rate constants depend not on the full quantal charge q_h displaced between stable conformational states (O and I) but only on the charge $q_h\delta$, δ being the fractional distance from O to the energy barrier (see inset). This is detailed in Greeff and Forster (1991) where we also show how to obtain the fraction q_h $(1-\delta)$ from recovery experiments. We then found a good agreement with the estimate for the full q_h resulting from the isochronic plot analysis. At $+50$ mV close to the sodium reversal potential and above we have not done the isochronic analysis. Here the τ_h (V) curve flattens as seen in Fig. 8.4(a). This may indicate a further rate-limiting transition.

Below -20 mV τ_h shows a steeper voltage dependence much as found in the simulation of Fig. 8.2(d). Here the isochronic analysis suggests mixed C–O and O–I transitions. This is also the region where single channels showed constant mean open times T^o (Aldrich and Stevens 1987). We now think that this and other divergent views about the voltage dependence of inactivation are complementary and the controversies can thus be resolved as shown in Fig. 8.4(b) with a simulation of τ_h, and T^o for a three-state model. The mean open time T^o (circles on dashed line) is given by the rates leaving the open state as $1/(\beta_c + \alpha_h)$. The curve has a flat peak and the filled circles in the voltage region of the study may well be taken as nearly constant T^os as done by these authors. τ_h has a steep voltage dependence in this voltage region but above $-$ 10 mV it joins the T^o curve. It is here that both show the

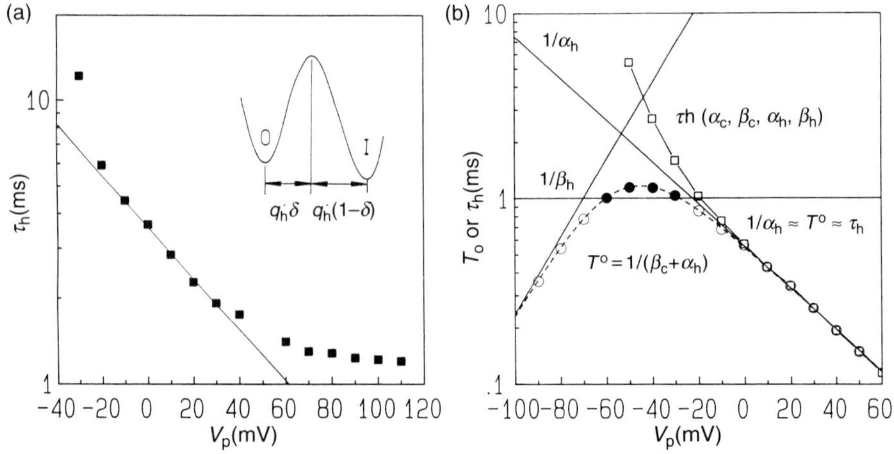

Fig. 8.4. (a) Experimental inactivation time constants τ_h plotted logarithmically against V_p. A line fitted by eye shows *e*-fold decay per 48 mV of τ_h between -20 and $+$ 40 mV. See text for discussion. Experiment K18O; temperature, 5°C. Inset shows free energy barrier between open (O) and inactivated (I) states. τ_h would depend on $q_h{}^*\delta$ (δ is the fractional distance from O to barrier) and not on the full q_h. (b) Comparison of mean open time T^o (circles) and τ_h (squares) for a simulated C–O–I scheme. q_c = 2.3 e^- and q_h = 1.25 e^-; rate constants in s^{-1} at 0 mV where α_c = 5000, β_c = 35, α_h = 1800, and β_h = 1; δ is taken as 0.5. Filled circles in voltage region where single channel analysis showed constant T^os (Aldrich and Stevens 1987).

voltage dependence of the O–I transition, although only the fraction $q_h{}^*\delta$ is directly obtained from the slope that is about half of 1.2 e^-. This figure is in good agreement with earlier estimates of 0.6 e^- in squid (Stimers *et al.*, 1985) or in GH3 cells (Vandenberg and Horn 1984). Furthermore, recent results extending T^o measurements from single cardiac channels up to + 20 mV support our interpretation (e.g. Yue *et al.* 1989).

Conclusions

We have shown that the isochronic analysis of high-resolution gating currents allows identification of O–I gating charge and provides the constraints, most importantly the voltage range, where the apparent time constants are predominantly influenced by this transition alone. Only here will rate analysis estimate the voltage dependence of this transition correctly, giving the fraction δ of the total charge displaced. This is about 0.5–0.6 e^- between $-$ 10 and +40 mV and the full q_h is about 1.2 e^-. Below -10 mV, q_h will be underestimated from the $T^{o}(V)$ relation but overestimated from the $\tau_h(V)$. The knowledge of the quantal inactivation gating charge leads to the question about the molecular structural element in the channel protein that represents the voltage sensor of the inactivation gate. This, as well as the relation to a postulated cytoplasmic loop involved in inactivation, is discussed elsewhere (Greeff and Forster, 1991.

Acknowledgement

This work was supported by the Swiss National Science Foundation grant No. 3.143–0.85.

References

Aldrich, R. W. and Stevens, C. F. (1987). Voltage-dependent gating of single sodium channels from mammalian neuroblastoma cells. *Journal of Neuroscience*, **7**, 418–31.

Aldrich, R. W., Corey, D. P., and Stevens, C. F. (1983). A reinterpretation of mammalian sodium channel gating based on single channel recording. *Nature*, **306**, 436–41.

Armstrong, C. M. and Bezanilla, F. (1974). Charge movements associated with the opening and closing of the activation gates of the Na channel. *Journal of General Physiology*, **63**, 533–52.

Armstrong, C. M. and Bezanilla, F. (1977). Inactivation of the sodium channel. II. Gating current experiments. *Journal of General Physiology*, **70**, 567–90.

Forster, I. C. and Greeff, N. G. (1990). High resolution recording of asymmetry currents from the squid giant axon—technical aspects of voltage clamp design. *Journal of Neuroscience Methods*, **33**, 185–205.

Greeff, N. G. and Forster, I. C. (1991). The quantal gating charge of sodium channel inactivation. *European Biophysical Journal*, **20**, 165–76.

Hodgkin, A. L. and Huxley, A. F. (1952). A quantitative description of membrane current and its application to conductance and excitation in nerve. *Journal of Physiology*, **117**, 500–44.

Keynes, R. D. and Rojas, E. (1974). Kinetic and steady state properties of the charged system controlling sodium conductance in the squid giant axon. *Journal of Physiology*, **239**, 393–434.

Keynes, R. D., Greeff, N. G., and Forster, I. C. (1990). Kinetic analysis of the sodium gating current in the squid giant axon. *Proceedings of the Royal Society of London, Series B*, **240**, 411–23.

Meves, H. and Pohl, J. A. (1990). A slow component in the gating current of the frog node of Ranvier. *Pflüger's Archiv*, **416**, 162–9.

Sigworth, F. J. and Neher, E. (1980). Single Na^+ channel currents observed in cultured rat muscle cells. *Nature*, **287**, 447–9.

Stimers, J. R., Bezanilla, F., and Taylor, R. E. (1985). Sodium channel activation in the squid giant axon. Steady state properties. *Journal of General Physiology*, **85**, 65–82.

Swenson, R. P. (1983). A slow component of gating current in crayfish giant axons resembles inactivation charge movement. *Biophysical Journal*, **41**, 245–9.
Vandenberg, C. A. and Horn, R. (1984). Inactivation viewed through single sodium channels. *Journal of General Physiology*, **84**, 535–64.
Yue, D. T., Lawrence, J. H., and Marban, E. (1989). Two molecular transitions influence cardiac sodium channel gating. *Science*, **244**, 349–52.

9

Tetrodotoxin affects sodium gating current in squid giant axon

MICHINORI ICHIKAWA and GEN MATSUMOTO

Summary

Tetrodotoxin (TTX), a cationic toxin which specifically blocks the sodium current, has long been believed to plug the ion-selective filter without any significant effect on the gate current. The guanidinium group of TTX is thought to bind to the charge of the filter in particular. In the present paper we provide experimental evidence that, instead, TTX binds to the gating site of the sodium channel in the squid giant axon to inhibit the transition primarily between the final closed state and the open state. This has been concluded as a result of the following experimental findings: (1) The gating current contains a TTX-sensitive component, a part of the gating current which can be specifically blocked by TTX in a concentration-dependent manner. (2) The variation with time of the running integral of the TTX-sensitive component is consistent with that of the rising phase of the sodium current. (3) The TTX-sensitive gating component is greatly affected by the presence and absence of sodium inactivation.

Molecular rearrangements of the sodium channels associated with membrane depolarization can be electrophysiologically studied by the analysis of the single-channel records (see review by Hille 1984) or, more directly, by measurements of the gating current (Armstrong and Bezanilla 1973, 1974; Keynes and Rojas 1973, 1974; Meves 1974). Since the gating charge movement reflects the transition of the channel between protein conformations that include all the closed and the open state(s) (Hodgkin and Huxley 1952; Stimers *et al.* 1987), the movement provides information that cannot be obtained as readily from macroscopic ionic currents or single-channel records. This is especially true for states that are increasingly removed from the open state (Stimers *et al.* 1987). The squid giant axon is suitable for the measurement of the sodium and potassium gating currents because of its enormously large diameter (see reviews by Armstrong 1981; Matsumoto *et al.* 1984*a*; Matsumoto and Ichikawa 1985; White and Bezanilla 1985; Stimers *et al.* 1985, 1987; Bekkers *et al.* 1990; Keynes 1990; Ichikawa *et al.* 1991).

Tetrodotoxin (TTX), a cationic toxin that specifically blocks the sodium currents, has long been believed to plug the ion-selective filter without any significant effect on the gate current (Narahashi *et al.* 1964; Moore and

Narahashi 1967). Therefore, TTX or saxitoxin (STX) have conveniently been used to block sodium ionic currents specifically, and have been reported essentially not to affect gating currents in squid giant axons (Armstrong and Bezanilla 1974; Keynes and Rojas 1974; Gilly and Armstrong 1982). Based on these observations, gating experimentalists have long used TTX or STX to separate the gating current from the sodium ionic current. On the other hand, several experiments have demonstrated that TTX (or STX) does affect sodium channel gating. For example, Heggeness and Starkus (1986) found in crayfish axons that the binding of TTX (or STX) displaced a divalent cation from a negatively charged site near the external opening of the sodium channel, thereby producing a voltage offset sensed by the channel gating apparatus. In the present chapter, we have carefully compared sodium gating currents of giant axons of squid (*Loligo budo* and *Doryteuthis bleekeri*) in the presence and absence of external TTX. In contrast to the lack of effect previously reported for squid giant axons, we note significant effects of TTX on sodium gating currents. A preliminary report of the present work was presented in abstract form at the second bilateral symposium between Japan and China on biophysics (Ichikawa and Matsumoto 1988), and 19th Annual Meeting of Society for Neuroscience (Ichikawa and Matsumoto 1989).

Experimental procedures

Materials

Experiments were performed on isolated giant axons of *Loligo budo* or *Doryteuthis bleekeri*, captured in Sagami Bay and maintained in a special aquarium tank in our laboratory (Matsumoto 1976; Matsumoto and Shimada 1980; Ichikawa *et al.* 1991).

Intracellular perfusion, solutions, and voltage-clamp

We adopted an improved method of intracellular perfusion and used the hyperosmotic external solutions according to Stimers *et al.* (1987) to reduce the slow component of the linear capacitative transient. In the improved perfusion method, one extreme side portion of the central axoplasm was extruded with a tiny roller. This was done just to preserve a steady outlet of the perfusion fluid while maintaining excitability of the extruded portion. In the central portion of the axon, where the internal current and potential electrodes were placed, the axoplasm was gently removed by orally drawing in through an inlet glass pipette. In order to maintain a steady flow, pronase was used at the initial stage of perfusion (Tasaki 1968). As an internal perfusion solution, we used 200 TMA consisting of 200 mM tetramethylammonium (TMA, a potassium current blocker; see Hille 1984), 150 mM glutamate, and 50 mM F (pH was adjusted to 7.25 by 20 mM Tris–HCl, and osmolarity was adjusted to 980 mOsm/kg by sucrose), as long as any specific alternative was not used. Two (or three) kinds of external solutions, H0 NaSW and H55 NaSW (or H110 NaSW), were used; H0 NaSW contained 50 mM Ca, ~675 mM Cl, and ~575 mM Tris–HCl, and H55 NaSW (or H110 NaSW) contained 55 (or 110) mM Na, 50 mM Ca, ~675 mM Cl, and ~520 (or 465) mM Tris–HCl. In both solutions, Trizma −7.5 (Sigma Chemical Co) was used to adjust the osmolarity to 1100 mOsm/kg. Both the improved perfusion and the use of hypertonic external solutions enabled us to perform the axon-only experiments in which the capacity current decayed within 20–30 μs to 1 per cent of its maximum amplitude. The voltage-clamp system with compensated feedback (Bezanilla and Armstrong 1977) was improved as compared with our previous system (Matsumoto *et al.* 1984*a*,*b*).

Results

The TTX-sensitive component of the gating current

The gating current obtained in the absence of TTX exhibits a shoulder in its falling phase [solid curve in Fig. 9.1(a); Matsumoto *et al.* 1984*a*; Matsumoto and Ichikawa 1985; Ichikawa *et al.* 1991], but the one obtained in the presence of TTX lacked the shoulder [dotted curve in Fig. 9.1(a)], although both currents exhibited instantaneous rises on application of a voltage step (Stimers *et al.* 1987). In other words, the gating current contained a TTX-sensitive component [(Fig. 9.1(b)], which was obtained by subtracting the dotted curve from the solid one in record A.

We have concluded that the TTX-sensitive component is characterized as an intrinsic part of the gating current for the following three reasons.

1. One naturally doubts that the shoulder may originate artefactually from an insufficiently compensated series resistance as has been argued by Stimers *et al.* (1987). In fact, the rising phase previously observed in the gating current (Armstrong and Gilly 1979) was caused by the uncompensated series resistance in the Frankenhaeuser–Hodgkin space (Frankenhaeuser and Hodgkin 1956; Adelman *et al.* 1973; Stimers *et al.* 1987; Abbott *et al.* 1988; Astion *et al.* 1988). However, this was not the cause of the shoulder occurring in the falling phase in the TTX-untreated axon, because the shoulder was lost in the TTX-treated phase, despite the similar decay of both capacity currents with time constants of 5 μs and without any prominent slow component (see Fig. 9.2). The shoulder was dependent on test and holding potentials; it was not obvious for small test potentials, and was more obvious for holding potentials of larger hyperpolarization.
2. It was further confirmed that the shoulder of the gating current in the TTX-untreated axon was temperature-dependent, while the time courses of the capacity currents were almost temperature-independent (Fig. 9.2). This shoulder exhibited greater delay and was more evident at lower temperatures.

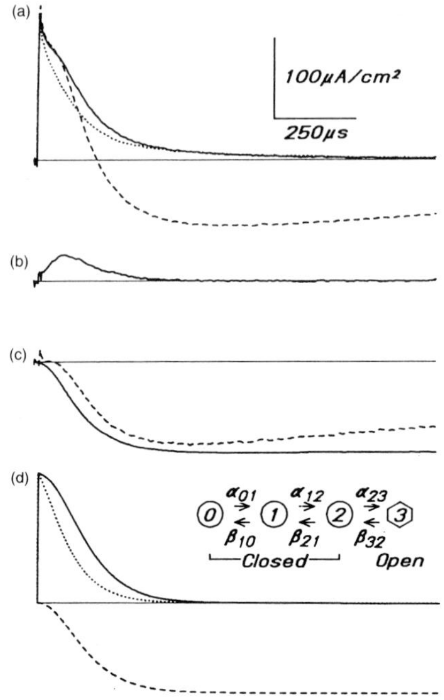

3. The existence of the shoulder contradicts results from previous reports on gating currents (Stimers *et al.* 1985, 1987). However, if we carefully review recent literature (Armstrong and Gilly 1979; Taylor and Bezanilla 1983; Stimers *et al.* 1985) on simultaneous measurements on the sodium and gating currents, our conclusion receives support when compared to those studies performed in the absence of TTX.

Correlation between the TTX-sensitive component of the gating current and the sodium conductance

The TTX-sensitive component of the gating current was well correlated with the sodium conductance.

1. Amounts of the gating charge carried by the TTX-sensitive component and peak sodium current were both dependent on the TTX concentration. These amounts were linearly and reversibly correlated (Fig. 9.3). It is seen in the figure that the effect of TTX on both the sodium and gating currents became saturated for a TTX concentration over 160 nM.
2. The variation with time of the running integral of the TTX-sensitive

Fig. 9.1. (**a**) Sodium gating currents in the absence (solid curve) and presence (dotted curve) of external TTX. The gating current was obtained from a pair of current records without using TTX; one record was taken when the voltage-clamped axon, internally perfused with 200 TMA, was bathed in H0 NaSW, and the other was taken when the same axon under the same clamping condition was immersed in H55 NaSW (see Experimental procedures). The former current record mainly contained the gating current but also contained a small amount of the sodium ionic current even if the clamp voltage was placed close to the reversal potential. (**c**) Using the latter current record (dashed curve in (a)) where the sodium and gating currents are comparable in size, we could easily determine the variation with time of the sodium current (broken curve in (c)) by subtracting the former current record (data not shown) from the latter one (dashed curve in (a)). This, in turn, allowed us to separate the gating current (solid curve in (a)) from the sodium current (dashed curve in (c)), without using TTX. The gating current in the presence of TTX was then obtained by bathing the same axon in H0 NaSW containing 300 nM TTX (dotted curve in (a)). Both the gating currents in the absence (solid curve in (a) and presence (dotted curve in (a)) of TTX exhibited the same instantaneous rise time but decayed differently. (**b**) The component ('TTX-sensitive' gating current) of the difference between the two gating currents could be obtained by subtracting the dotted curve in (a) from the solid curve in (a). (**c**) The solid curve is a running integral Q_g of the 'TTX-sensitive' component of the gating current. Q_g was numerically calculated using the 'TTX-sensitive' component I_g through an equation, $Q_g(t) = \int_0^t I_g(t')dt'$, where time t was measured from the start of the test potential step. The temperature was 8.4°C. (**d**) The results of computer simulation based on the four-state model of the gate (Hodgkin and Huxley 1952; Armstrong 1981) with three successive closed states and one open state (see inset). The solid and dotted curves were obtained to simulate the experimentally determined gating currents in the absence and presence of TTX, corresponding to the solid and dotted curves in (a). In the simulation, the values of transition velocities α_{01}, α_{12} and α_{23} in the absence of TTX, were chosen to be 17.6, 15.5, and 18.6 ms^{-1}, respectively, while α_{01}, α_{12}, and α_{23} in the presence of TTX were 17.6, 15.5, and 0 ms^{-1}, respectively. The dashed curve is the result of calculating the fraction of the open state as a function of time on a potential step application to 50 mV. The test potential height was 120 mV (corresponding to the membrane potential of 50 mV): Holding potential, −60 mV; preholding potential, −70 mV; and subtracting holding potential, −170 mV.

---

component was in good agreement with that of the rising phase of the sodium current, except for the parallel shift along the time axis [compare the solid curve in Fig. 9.1(c) with the dashed curve]. The agreement was confirmed for different potential steps. A typical experiment is illustrated in records of Fig. 9.4(a)–(c) where the axon was voltage-clamped to −25, 5, and 35 mV, respectively. Note that the running integral curves for the TTX-sensitive component always preceded the corresponding sodium current records. We found that time differences for the precedence were temperature-dependent but membrane-potential-independent. They became larger at lower temperatures; i.e. 100 ± 20, 50 ± 10, and 20 ± 5 μs at 4.8, 8.4, and 16°C, respectively.

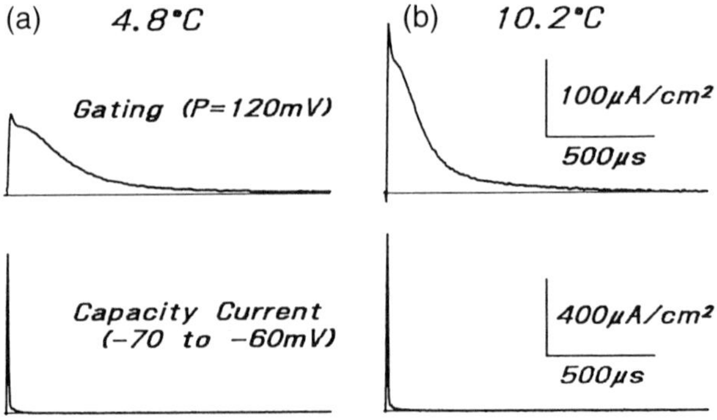

Fig. 9.2. Temperature dependence of the gating currents (upper records) and the capacitative transients (lower records). The gating currents were measured in the absence of external TTX at (a) 4.8°C and (b) 10.2°C ($P=V_p$).

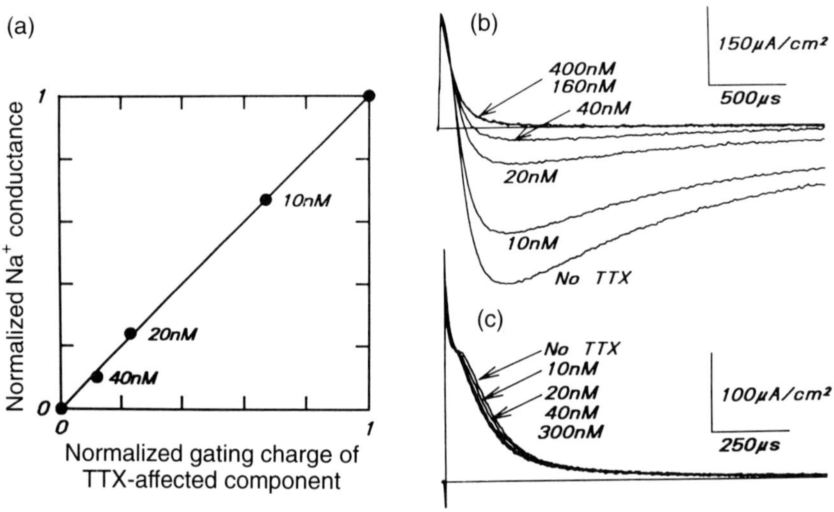

Fig. 9.3. TTX concentration dependence of the TTX effect on sodium and gating currents. The concentrations used are designated in the figure. (a) Correlation diagram between the gating charge carried by the TTX-affected component and the peak sodium current. The data used here were obtained from (b) and (c). (b) TTX concentration dependence on the sodium current. The sodium and the gating currents were observed for the axon internally perfused with 200 TMA and externally bathed in H110 NaSW on a potential step to 50 mV at 10.8°C. (c) TTX concentration dependence on the gating currents. The gating currents were obtained as a function of the TTX concentration for the axon internally perfused with 200 TMA and externally bathed in H0 NaSW on a potential step to 50 mV at 10.2°C.

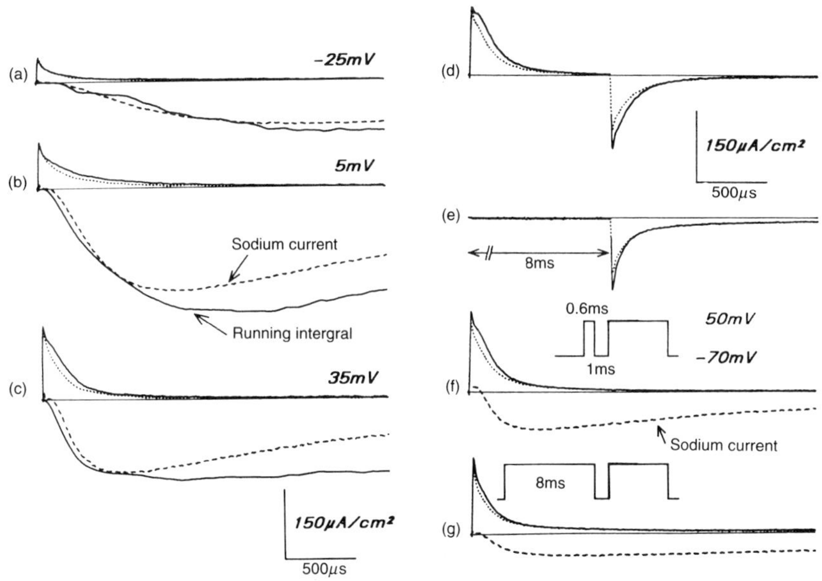

Fig. 9.4. Membrane-potential dependence of the TTX effect on the sodium and gating currents, and (d)–(g) effects of inactivation on the TTX suppression of the sodium gating current. In (a)–(c) the experiments were performed for the same axon used in Fig. 9.1 under the same experimental conditions, except for the application of potential steps to (a)-25, (b)5, and (c)35 mV. The solid and dotted curves above the baselines represent the gating currents in the absence and presence of TTX, respectively. The solid curves below the baselines represent the running integrals of the TTX-sensitive gating components, and the broken curves represent the sodium currents.

3. The TTX-sensitive gating component was greatly affected by the sodium inactivation while the TTX-resistant component was weakly affected. One of the typical experiments is shown in Fig. 9.4(d)–(g), where the TTX-sensitive component was reduced as the axon was inactivated by either an 8-ms depolarization to 50 mV (e) or a prepulse of 50 mV for 8 ms (g). In the 8-ms depolarization to 50 mV, the TTX-sensitive component of the gating OFF current was observed to decrease substantially [see Fig. 9.4(e) as compared with a control, Fig. 9.4(d)]. The effect of the inactivation on the gating ON currents in the presence and absence of TTX was also tested by applying an 8-ms prepulse to 50 mV, and the TTX-sensitive component was again observed to decrease significantly [compare Fig. 9.4(g) with a control, Fig. 9.4(f)]. When either a depolarization or a prepulse was applied in order to exclude the inactivation, slowly decaying components of the OFF and ON currents were greatly enhanced by the same amount as shown in both TTX-untreated [solid curves in Fig. 9.4(e) and (g)] and TTX-treated [dotted curves in Fig. 9.4(e) and (g)] axon. Further, when

the inactivation was completely removed by an internal application of pronase, the gating currents in the presence and absence of TTX were totally unaffected by either the depolarization or prepulse application (data not shown). These findings suggest that the TTX-sensitive gating component is closely related to the sodium conductance.

Discussion

Our present experiments have led us to propose that TTX can bind to the gating site of the sodium channel to inhibit the transition between the final closed state of the gate and the open state. This is, however, in sharp contrast with the previously accepted view that TTX binds to the site of the ion-permeable pore (see review by Hille 1984) without exhibiting any detectable effect on the ON or OFF gate current (Armstrong and Bezanilla 1974; Keynes and Rojas 1974; Armstrong 1981; Gilly and Armstrong 1982). The following points indicate the advantageous aspects of our theory, although further investigations are required to confirm these points:

The fact that the running integral of the TTX-sensitive gate component always preceded the sodium current [see Fig. 9.1(c) and Fig. 9.4(a)–(c)] is of advantage to our theory, although the molecular origin of the time delay between the gate and the channel was not apparent.

We performed some tests, as described in 'Results', to ascertain whether the TTX-sensitive component could be characterized as a counterpart of the gating current or whether it simply originated from some artefacts. With reference to the other possible artefacts, it was considered whether the binding of TTX to the sodium channel could neutralize part of the gating charge resulting in a reduction of the gating current, since TTX is a cationic toxin with one guanidium group per molecule (Narahashi *et al.* 1964; Hille 1984). However, this is unlikely to be the case because the effects of TTX were all accurately simulated through the use of STX, a cationic toxin with two guanidium groups per molecule (Kao and Nishiyama 1965; Catterall 1980; Hille 1984), instead of TTX. We also tested the possibility that an ion environment or other type of cell environment surrounding the sodium channel could induce the TTX effect by interacting with TTX. The following modifications did not cause any changes in the effect (data not shown): substitution of D_2O for H_2O in the external medium; alternation of internal TMA with Cs; external and internal pH changes; dilution of external Na with Tris or choline; external application of glycine in the presence of Na; use of HEPES buffer instead of Tris; change of the ratio of internal to external concentration of Cl; change of the internal anion ratio of F to that of glutamate; or internal application of TTX and use of TTX without a citric acid stabilizer. These modifications indicate that the TTX-sensitive component is an intrinsic part of the gating current.

Our theory that TTX inhibits the gate transition between the final closed state and the open state could be successfully simulated by assuming the four-state model (Armstrong 1981; Stimers *et al.* 1985), as shown in Fig. 9.1(d). In this figure the solid and dotted curves represent the gating current in the absence and presence of TTX, respectively, corresponding to the experimental records of the solid and dotted curves in Fig. 9.1(a). The dashed curve in Fig. 9.1(d) represents the fraction of the open state as a function of time on a potential step to 50 mV, corresponding to the experimentally determined running integral in Fig. 9.1(c). The calculated records bore an excellent correlation to the experimental results. It is noted, however, that the values of the transition velocities adopted here were more or less the same (α_{01} = 17.6, α_{12} = 15.5, and α_{23} = 18.6, ms^{-1}, while those previously used were generally fast for the first two steps and slow for the last step (Armstrong 1981; Stimers *et al.* 1987). Further, it was also noted that the charge movements for each step were equally distributed in our calculations, while those previously used in each of the fast transitions were only about half the value of the last one (Armstrong 1981; Stimers *et al.* 1987).

Effects of external TTX and STX on the sodium gating current in the crayfish axon were extensively studied by Heggeness and Starkus (1986). They found that these toxins suppressed the gating current without any visible change in its kinetics, and that the degree of suppression was sensitive only to the holding potential. They concluded that the binding of TTX or STX displaced a divalent cation from a negatively charged site near the external opening of the sodium channel, thereby producing a voltage offset sensed by the channel gating apparatus. Their findings are in sharp contrast to ours in which the degree of suppression was test-potential-dependent and the gating kinetics were modified. Further, we examined the holding-potential-dependence of the TTX effect on the gating current in the squid giant axon when the holding potential was changed from −50 through −120 mV, and found that the effect was weak, if it occurred at all (data not shown). Finally, we note the assertion by Keynes and colleagues (Keynes 1990; Keynes *et al.* 1990) and Bekkers *et al.* (1990) that the gating current has a peculiar shoulder when it is measured even in the presence of 1 μM TTX. This is in sharp contrast with all the data of the gating current obtained with TTX to date. We have repeated their experimental conditions to measure the gating current, but failed; that is, we have confirmed our results that the gating current has a shoulder in the absence of TTX that it loses in the presence of TTX. At present we have no answer to the question as to the source of this discrepancy

In conclusion, we propose that TTX (or STX) binds to the gating site of the sodium channel to inhibit the transition primarily between the final closed state and the open state. This alters the previously accepted view of the TTX-binding site (Narahashi *et al.* 1964; Kao and Nishiyama 1965) but is consistent with the physiological data of the gating current concerning TTX obtained to date (Hille 1984).

References

Abbott, N. J., Lieberman, E. M., Pichon, Y., Hassan, S., and Latmet, Y. (1988). Periaxonal K^+ regulation in the small squid *Alloteuthis*. Studies on isolated and in situ axons. *Biophysics Journal*, **53**, 275–9.

Adelman, W. J., Palti, Y., and Senft, J. P. (1973). Potassium ion accumulation in a periaxonal space and its effects on the measurement of membrane potassium ion conductance. *Journal of Membrane Biology*, **13**, 387–410.

Armstrong, C. M. (1981). Sodium channels and gating currents. *Physiological Reviews*, **61**, 644–83.

Armstrong, C. M. and Bezanilla, F. (1973). Currents related to movement of the gating particles of the sodium channels. *Nature*, **242**, 459–61.

Armstrong, C. M. and Bezanilla, F. (1974). Charge movement associated with the opening and closing of the activation gates of the Na channels. *Journal of General Physiology*, **63**, 533–52.

Armstrong, C. M. and Gilly, W. F. (1979). Fast and slow steps in the activation of sodium channels. *Journal of General Physiology*, **74**, 691–711.

Astion, M. L., Coles, J. A., Orkand, R. K., and Abbott, N. J. (1988). K^+ accumulation in the space between giant axon and Schwann cell in the squid *Alloteuthis*. Effects of changes in osmolarity. *Biophysics Journal*, **53**, 281–5.

Bekkers, J. M., Forster, I. C., and Greeff, N. G. (1990). Gating current associated with inactivated states of the squid axon sodium channel. *Proceedings of the National Academy of Sciences*, **87**, 8311–15.

Bezanilla, F. and Armstrong, C. M. (1977). Inactivation of the sodium channel. I. Sodium current experiments. *Journal of General Physiology*, **70**, 549–66.

Catterall, W. A. (1980). Neurotoxins that act on voltage-sensitive sodium channels in excitable membranes. *Annual Review of Pharmacology and Toxicology*, **20**, 15–43.

Frankenhaeuser, B. and Hodgkin, A. L. (1956). The after-effects of impulses in the giant nerve of fibers of *Loligo*. *Journal of Physiology*, **131**, 341–76.

Gilly, W. F. and Armstrong, C. M. (1982). Slowing of sodium channel opening kinetics in squid axon by extracellular zinc. *Journal of General Physiology*, **79**, 935–64.

Heggeness, S. T. and Starkus, J. G. (1986). Saxitoxin and tetrodotoxin. Electrostatic effects on sodium channel gating current in crayfish axons. *Biophysics Journal*, **49**, 629–43.

Hille, B. (1984). *Ionic channels of excitable membranes*. Sinauer Associates Inc, Sunderland, Massachusetts.

Hodgkin, A. L. and Huxley, A. F. (1952). A quantitative description of membrane current and its application to conduction and excitation in nerve. *Journal of Physiology*, **117**, 500–44.

Ichikawa, M. and Matsumoto, G. (1988). Sodium channel gating currents. A tetrodotoxin sensitive component. *Proceedings of the Second Japan–China Bilateral Symposium on Biophysics*, pp. 22–3. Japan Biophysical Society.

Ichikawa, M. and Matsumoto, G. (1989). Effect of TTX and amiloride on the sodium channel gating. *19th Annual Meeting of the Society for Neuroscience*, **15**, 536 (Abstract 219.1).

Ichikawa, M., Urayama, M., and Matsumoto, G. (1991). Anticalmodulin drugs block the sodium gating current of squid giant axons. *Journal of Membrane Biology*, **120**, 211–22.

Kao, C. Y. and Nishiyama, A. (1965). Action of saxitoxin on peripheral neuromuscular systems. *Journal of Physiology*, **180**, 50–66.

Keynes, R. D. (1990). A series–parallel model of the voltage-gated sodium channel. *Proceedings of the Royal Society of London, Series B*, **240**, 425–532.

Keynes, R. D. and Rojas, E. (1973). Characteristics of the sodium gating current in the squid giant axon. *Journal of Physiology*, **233**, 28P–30P.

Keynes, R. D. and Rojas, E. (1974). Kinetic and steady-state properties of the charged system controlling sodium conductance in the squid giant axon. *Journal of Physiology*, **239**, 393–434.

Keynes, R. D., Greeff, N. G., and Forster, I. C. (1990). Kinetic analysis of the sodium gating current in the squid giant axon. *Proceedings of the Royal Society of London, Series B*, **240**, 411–23.

Matsumoto, G. (1976). Transportation and maintenance of adult squid *(Doryteuthis bleekeri)* for physiological studies. *Biological Bulletin*, **150**, 279–85.

Matsumoto, G. and Ichikawa, M. (1985). Kinetics of sodium activation in giant axons of squid *(Doryteuthis bleekeri)*. *Neuroscience*, **14**, 327–34.

Matsumoto, G. and Shimada, J. (1980). Further improvement upon maintenance of adult squid (*Doryteuthis bleekeri*) in a small circular and closed-system aquarium tank. *Biological Bulletin*, **159**, 319–24.

Matsumoto, G., Ichikawa, M., and Tasaki, A. (1984*a*). Axonal microtubules necessary for generation of sodium current in squid giant axons. II. Effect of colchicine upon asymmetrical displacement current. *Journal of Membrane Biology*, **27**, 93–9.

Matsumoto, G., Ichikawa, M., Tasaki, A., Murofushi, H., and Sakai, H. (1984*b*). Axonal microtubules necessary for generation of sodium current in squid giant axons. I. Pharmacological study on sodium current and restoration of sodium current by microtubule proteins and 260K protein. *Journal of Membrane Biology*, **77**, 77–91.

Meves, H. (1974). The effect of holding potential on the asymmetrical displacement current in giant axons of *Loligo forbesi*. *Journal of Physiology*, **267**, 377–93.

Moore, J. W. and Narahashi, T. (1967). Tetrodotoxin's highly selective blockage of an ionic channel. *Federation Proceedings*, **26**, 1655–63.

Narahashi, T., Moore, J. W., and Scott, W. R. (1964). Tetrodotoxin blockage of sodium conductance increase in lobster giant axons. *Journal of General Physiology*, **47**, 965–74.

Stimers, J. R., Bezanilla, F., and Taylor, R. E. (1985). Sodium channel activation in the squid giant axon. Steady state properties. *Journal of General Physiology*, **85**, 65–82.

Stimers, J. R., Bezanilla, F., and Taylor, R. E. (1987). Sodium channel gating currents. Origin of the rising phase. *Journal of General Physiology*, **89**, 521–40.

Tasaki, I. (1968). *Nerve excitation. A macromolecular approach*. Charles C. Thomas, Springfield, Illinois.

Taylor, R. E. and Bezanilla, F. (1983). Sodium and gating current time shifts resulting from changes in initial conditions. *Journal of General Physiology*, **81**, 773–84.

White, M. M. and Bezanilla, F. (1985). Activation of squid axon K^+ channels. Ionic and gating current studies. *Journal of General Physiology* **85**, 539–54.

10

Single-channel properties and gating of Na^+ and K^+ channels in the squid giant axon

FRANCISCO BEZANILLA and ANA M. CORREA

Summary

This chapter describes patch clamp recordings made from the inner membrane surface of the cut-open squid giant axon, allowing examination of macroscopic ionic currents, gating currents, and single channel fluctuations. Sodium currents could be studied in the absence of divalent cations, and showed a ~ 15 mV shift to more negative potentials of the voltage-dependence of all parameters of activation and inactivation, suggesting that divalent cations neutralize external negative charge near the voltage sensor of the channel. Batrachotoxin (BTX) eliminated fast and slow inactivation of a fraction of the Na channels and reduced single channel conductance from 33 pS to 11 pS. The conductance of the Na channel increased with $[Na^+]$, saturating at about 1M. The results fitted a single-ion pore, 3 barrier 2 site model, with fixed charge at the pore vestibules. Modelling of the open–closed states of the Na channels requires several inactivated states to account for charge immobilization, and several closed states to account for the lag of the ionic current and absence of rising phase in the gating current. At positive potentials the currents show a second open state with the same conductance value, and this second state is responsible for the incomplete inactivation observed at positive potentials. Use of BTX permits analysis of the effect of voltage and temperature on Na channel kinetics, providing information about the underlying molecular events, and refinement of the channel model. Several types of K channels can also be detected: a 20 pS (delayed rectifier) channel, and less abundant 10 pS and 40 pS channels. The delayed rectifier is modulated by internal ATP, and phosphorylation increased the probability of opening; the results could be modelled by shifts in the voltage-dependence of activation and inactivation as a result of addition of negative charges.

The squid giant axon has provided a wealth of information on the initiation and propagation of the nervous impulse. The impedance measurements done in the giant axon by Cole and Curtis (1939) showed that during the action potential there is a large transient conductance increase with only a small capacitance change. Hodgkin and Katz (1949) demonstrated that during the action potential the giant axon membrane became transiently selective to Na^+ ions, explaining the overshoot of the spike. These results led Hodgkin *et al.*

(1952) to use the voltage clamp, previously introduced by Cole (1949), to describe the ionic currents underlying the main conductance changes during the action potential. Hodgkin and Huxley (1952) made a quantitative description of the ionic conductances and predicted the time course and general properties of the axon membrane during excitation and propagation of the impulse. Their outstanding work has formed the basis for the understanding of properties of voltage-dependent channels as we know them today. These authors predicted the existence of ionic channels and gating currents which were confirmed many years later when techniques were dramatically improved. In their view, there were two basic pathways for ions to flow through the membrane, one selective to Na^+ and the other selective to K^+. The conductances of these pathways were voltage-dependent and Hodgkin and Huxley proposed that the passage of ions was through a few specialized regions whose probability of being open was regulated by voltage, via rearrangement of charges within the membrane. This view predicted the existence of gating currents, which was confirmed by Armstrong and Bezanilla (1974) and by Keynes and Rojas (1974) in the squid axon, and the existence of single channels, which was first reported by Conti and Neher (1980) for K^+ channels and by Bezanilla (1987) for Na^+ channels in the squid giant axon.

The high density of channels on the squid axon surface makes it ideal for recording gating currents with a large signal-to-noise ratio, and these measurements have provided insights into the possible molecular mechanisms underlying the opening and closing of the ionic channels (Armstrong 1981; Bezanilla 1985; Keynes 1986). To constrain the number of possible models of the gating process it is necessary to gather information on macroscopic ionic and gating currents and measurements of single-channel fluctuations. In addition, these three types of measurements must be done in the same biological preparation, because the characteristics of ionic channels differ in different species and tissues. The squid axon has proven ideal for macroscopic ionic and gating currents but the presence of the Schwann cell layer prevents the access from the outside of regular patch pipettes (Hamill *et al.* 1981) for single-channel recording. In this chapter we describe the cut-open axon method that allows the access of patch pipettes from the interior of the axon. We also describe the properties of the Na^+ channel in the squid giant axon as seen with single-channel recording. At the end, the characteristics of potassium channels of the squid axon will be briefly summarized.

Methods

Giant axons from the squid *Loligo pealei* were used in this study. A summary of the technique follows; details have been published previously (Bezanilla 1987; Llano *et al.* 1988; Bezanilla and Vandenberg 1990). A segment of axon of about 5 mm long was carefully cleaned of adjacent tissue and other fibres and tied at both ends with surgical thread (5–0). The segment was removed and transferred to a chamber filled with artificial sea water, ASW (440 mM NaCl, 50 mM $MgCl_2$, 10 mM $CaCl_2$, 10 mM HEPES). This chamber was mounted on an inverted microscope; its bottom was made of a thin coverslip covered with a thin layer of Sylgard. Under view through a dissection microscope, the segment was pinned to the Sylgard bed and slit open longitudinally with microscissors. After a period of 3–5 min, the chamber was flushed with ASW to remove the axoplasm, followed by a change to the experimental external solution, such as 540 mM NaCl, 10 mM HEPES. Under these conditions the axon sheet showed grooves when inspected with the inverted microscope and it was essentially free of debris and axoplasm. At this point, it was possible to approach the axon with patch pipettes made of Corning 7052 glass and filled with internal solution. When the pipette touched the membrane, the release of the positive pressure would usually result in a gigaseal formation; on a few occasions, it was necessary to apply slight suction.

Results and discussion

The cut-open axon technique allows the measurement of macroscopic ionic currents, gating currents, and single-channel fluctuations. The number of channels in a small patch pipette ($\leq 1\ \mu m$ in diameter) varied considerably among patches; frequently about 10 channels were found, although very often one or none were visible or, alternatively, large macroscopic currents could be recorded. We think that this variability reflects the non-homogeneous distribution of channels in the axon membrane. Single-channel studies were done with patches showing one active channel such as that shown in Fig. 10.1. Macroscopic currents were studied from patches with 50 or more active channels (see Fig. 10.2).

The effects of Ca^{2+} on the Na^+ conductance

The cut-open axon preparation has provided a unique opportunity to study the Na^+ conductance in the virtual absence of external divalent cations, in contrast to the voltage-clamped whole axon, either perfused or dialysed, which does not survive under these conditions. In the whole axon preparation, low Ca^{2+} concentrations (less than 3 mM) have deleterious effects observed in the form of increased leakage conductance. This increase in leakage forces the axial wire inside the axon to pass more current, with the consequent increase of its surface resistance. This resistance increase decreases the effectiveness of the axial wire and the space clamp fails; this failure in turn produces anomalous currents with faulty kinetics. These problems make the study of sodium current kinetics in the whole axon at low

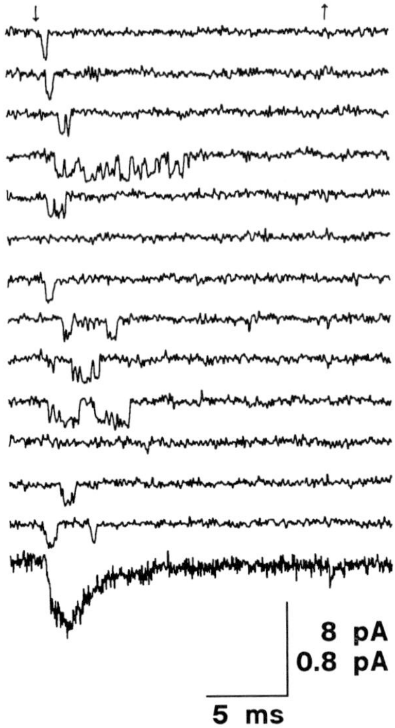

Fig. 10.1. Traces showing single-channel activity in a patch with only one channel. The top 13 traces are examples of single-channel activity selected from a set of 145 records under the same conditions. The trace at the bottom is an average of the data in the 145 records. The recordings were made in 1 M NaCl external and 50 mM NaCl internal (pipette) at 4°C. The calibration bar corresponds to 8 pA for the first 13 traces and 0.8 pA for the average shown in the bottom trace. The 18-ms depolarizing pulses (onset and offset shown by the arrows) to −10 mV were applied from a holding potential of −80 mV. Data filtered at 4.5 kHz for illustration.

calcium concentrations unreliable and extrapolation to zero Ca^{2+} concentration is not warranted.

Patches can be formed easily on the cut-open axon bathed with zero added divalents (probably a total Ca^{2+} concentration of less than 10 μM). These patches are of high resistance (more than 10 GΩ) and can last as much as several hours; in a few cases EGTA was added to the external solution with the same results (Vandenberg and Bezanilla 1991*a*). We do not have an explanation for the difference in behaviour between the whole axon and the cut-open axon under low external calcium concentration. It might be speculated that low Ca^{2+} may induce changes in the interaction between Schwann cells and the axon membrane of the whole axon producing tear and consequent leakage increase.

It is well known that external Ca^{2+} reduces macroscopic Na^+ currents in the

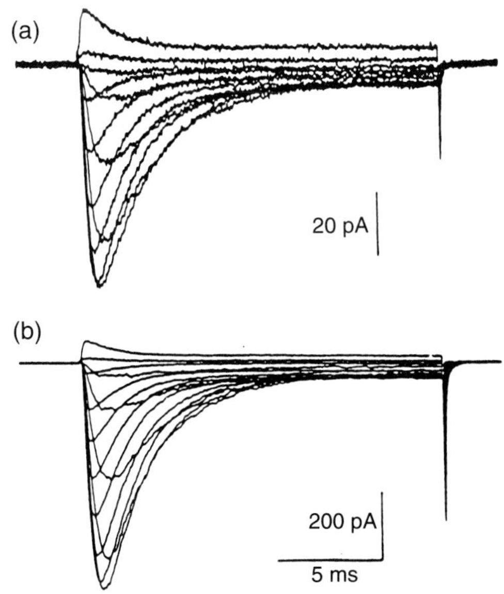

Fig. 10.2. Patch recordings of macroscopic currents with and without external divalents. (a) Family of currents obtained in 50 mM Mg^{2+}, 10 mM Ca^{2+} artificial sea water. Pulses (18 ms long) to potentials from −30 to +80 mV were applied from a holding potential of −90 mV, with a 50-ms prepulse to −100 mV. (b) Currents recorded from a patch in 546 mM Na external solution without added divalents. Pulses from −40 to +90 mV every 10 mV from a holding of −90 mV with a 50-ms prepulse to −100 mV. Currents are averages of 16 cycles in (a) and 8 cycles in (b). Filter was 7 kHz. (Taken with permission from Vandenberg and Bezanilla (1991*a*).)

squid axon (Taylor *et al.* 1976) as well as in other preparations (cf. Yamamoto *et al.* 1984). As explained below, external Ca^{2+} decreases the amplitude of the single-channel event and for this reason all recordings were made in the absence of external divalents. In fact, a fit of the data of instantaneous current–voltage (I–V) curves, recorded from the patch with and without Ca^{2+}, to a single site blocking model revealed a dissociation constant (K_d) of 150 mM at 0 mV (Vandenberg and Bezanilla 1991*a*).

To make valid comparisons with the currents recorded in the whole axon, a characterization of the effect of Ca^{2+} on the kinetics of the Na^{2+} conductance was required (Vandenberg and Bezanilla 1991*a*). The currents recorded without divalents have very similar characteristics to the currents in normal calcium and magnesium (see Fig. 10.2) and no anomalous kinetics, such as very slow tail currents, are observed in the absence of divalents. A detailed analysis of the kinetics revealed a shift of 15 mV to more negative potentials of the voltage dependence of all the parameters of activation and inactivation. This effect is observed for ionic currents as well as gating currents and it could be due to neutralization by divalent cations of an external negative surface charge located near the voltage sensor of the channel.

The single Na^+ channel conductance

Patches revealed a single amplitude for Na^+ channels and, as we only rarely observed intermediate conductance values, we analysed our data as if the Na^+ channel has open states of one conductance value. We will see below that at large positive potentials there are two open states but both show the same conductance value.

As mentioned above, all recordings were done in the absence of divalent cations. In an initial study it was found that the usual calcium and magnesium concentrations present in artificial sea water decreased the open-channel Na^+ conductance to levels that made event detection marginal (Vandenberg and Bezanilla 1991*a*). This decreased conductance can be explained by Ca^{2+} blocking the Na^+ channel from the outside (see above) as had been seen in macroscopic currents in the squid axon (Taylor *et al.* 1976; Tanguy and Yeh 1988) and in other preparations (cf. Woodhull 1973; Yamamoto *et al.* 1984).

A detailed study of the open Na^+ channel conductance was done by varying the concentration of Na^+ on both sides of the membrane. To avoid any blocking effects, no other ions were used to replace Na^+ when working at low concentrations; instead a low osmolarity was preferred and it did not affect patch stability. The range of concentrations studied was from 15 mM to 1 M symmetrical Na^+ and up to 4 M in the asymmetrical case (with internal 108 mM Na^+; Correa *et al.* 1991).

The conductance properties of Na^+ channels at the single-channel level have been studied mainly in bilayers in channels modified with batrachotoxin (BTX), a potent specific activator of Na^+ channels (French *et al.* 1984; Moczydlowski *et al.* 1984; French *et al.* 1986; Garber and Miller 1987; Green *et al.* 1987; Garber 1988; Behrens *et al.* 1989; Naranjo and Latorre 1993). We took advantage of the fact that in the cut-open axon both normal and BTX-treated channels could be studied simultaneously to compare their conductance properties. The single-channel events of treated and untreated channels were measured and a comparison was done for all sodium concentrations used. BTX was used at a concentration of 200 nM, most of the time added to the patch pipette. In patches with several channels, BTX transformed only a fraction of the channels and very frequently only one, allowing a detailed study of both conductance and kinetics.

In the presence of BTX, fast and slow inactivation are apparently eliminated, the open times of the channels are greatly prolonged, and the activation curve is shifted towards more hyperpolarized potentials (Fig. 10.3). Hence, just by changing the holding potential the activity and conductance properties of the channels can be studied for prolonged times within a wide range of potentials. The transient nature of the activity of the unmodified channel, that

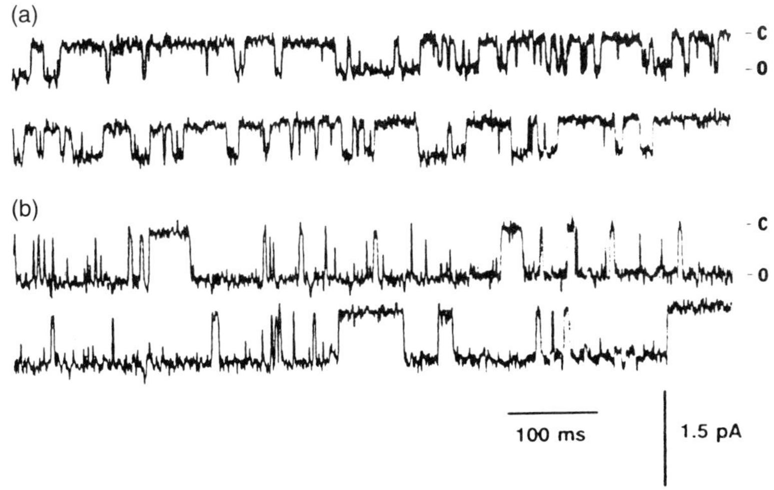

Fig. 10.3. Activity of a BTX-modified squid Na^+ channel. The traces show single BTX-modified channels recorded in symmetrical solutions at two different sodium concentrations: (a) 15 mM Na^+; and (b) 540 mM Na^+. The closed and open levels are indicated by C and O to the right of the first trace in each set. Patches were held at the same potential, −60 mV. This figure illustrates the different gating kinetics of the BTX-modified channel: channels have a high probability of opening at potentials where they are normally mainly closed, open times are dramatically increased, and inactivation is absent. Data filtered at 0.5 kHz for illustration. (Reproduced with permission from Correa *et al.* (1991).)

is the channel activating upon depolarization and inactivating shortly after, required a different protocol. The measurement of the normal, untreated channel conductance was done following a protocol in which channels were first primed by holding at negative potentials, very often −100 mV, and then changing the holding potential to the desired potential, and measuring at short times after the depolarization when fast and slow inactivation had not yet eliminated the activity completely. In patches with many channels we waited some time after the establishment of the depolarization until single-channel events could be detected, that is after the majority of the channels became inactivated, and overlapping activity was greatly reduced.

The conductance of the squid Na^+ channel increases with Na^+ concentration saturating at about 1 M but it does not follow a single-site simple Michaelis–Menten type relationship [Fig. 10.4(a)]. Instead, as seen in the figure, the conductance seems to reach a low limiting value at about 15 mM. This effect could be explained by the presence of fixed charges of about −0.1 e/nm^2 (internal) and −0.8 e/nm^2 (external) near the conduction pore. (The surface charge density obtained when the analysis was done not taking into account the conductance data obtained in asymmetrical conditions was

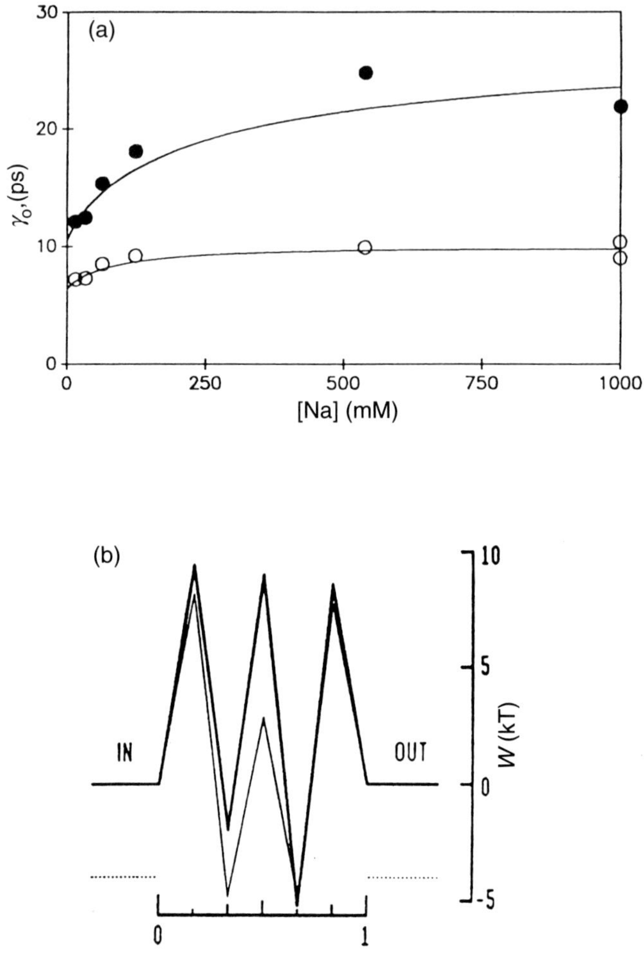

$-0.25\ e/\mathrm{nm}^2$ on both sides). An internal fixed surface charge density of $-0.4\ e/\mathrm{nm}^2$ had been proposed by Smith-Maxwell and Begenisich (1987) to explain the conduction properties of the whole axon of the squid. In addition, our results confirmed the observations, made by Quandt and Narahashi (1982) and Huang *et al.* (1984) in neuroblastoma cells, that BTX induced a decrease in the single-channel conductance. We found that the conductance of the BTX-treated channel was 1.7–2.5 times smaller than that of the normal channel [Fig. 10.4(a)]. The maximum conductance of the non-treated channels was 33 pS and 11 pS in BTX. The results for both the normal and the BTX-treated channel were fitted to a single-ion pore, three-barrier two-site (3B2S) model [Fig. 10.4(b)] with fixed charge at the pore vestibules. This fitting gave the values of energy wells and barriers that accounted for the

Fig. 10.4. (a) Conductance (γ_o) versus concentration ([Na]) relationship of normal and BTX-modified squid sodium channels. Filled circles correspond to the normal channel and open circles to the BTX-treated one. Each point is the fitted value of the current versus voltage relation at 0 mV obtained from all the data from several patches at each concentration denoted. The curves are predicted by the three-barrier, two-site model (3B2S model) shown in part (b) of this figure, with the proviso that no more than one ion can be in the channel at any given time, and with surface charge densities of -0.8 and -0.1 e^-/nm^2 in the internal and external vestibules, respectively. The parameters of the model were obtained by fitting the complete set of *I–V* relations in both symmetrical and asymmetrical conditions. (b) 3B2S energy profile of a one-ion pore representing the steps of ion movement through a normal (thin line) and a BTX-modified (thick line) Na^+ channel at 0 mV. The profiles shown do not include surface charge effects. *W* is the energy level (*kT*). The reference level of 0 *kT* corresponds to a molar fraction of one, i.e. at a salt concentration of 55.5 M. The dotted lines located at -4.02 *kT* at each side of the profile represent the reference level for a 1 M salt solution. The major difference between the two profiles is in the height of the centre peak. Both this peak and the inner well were significantly raised by BTX modification. The abscissa is given in fractional electrical distance. (Taken with permission from Correa *et al.* (1991).)

---

conduction in symmetrical and asymmetrical Na^+ conditions and they are illustrated in Fig. 10.4(b). The main conclusion is that BTX reduced the conductance by causing a substantial increase in the central energy barrier. A less dramatic increase in the energy level of the internal well was also seen. Both effects would make the transit of an ion less likely and thus decrease the conductance. A model including double occupancy would have given a smaller increase by BTX in the energy level of the central barrier due to ion repulsion itself. These results, however, were well fitted assuming single occupancy. BTX did not affect the binding affinity for external sodium (K_ds of 314 mM in BTX and 458 mM without it) nor did it modify the value of the fixed charge density (Correa *et al.* 1991).

The results at the single-channel level are consistent with those obtained at a macroscopic level in the same preparation. The conductance properties of the squid giant axon had also been previously modelled following a three-barrier two-site model by Begenisich and Cahalan (1980*a,b*). The K_ds for Na^+ ions were also similar: 0.4–1 M in the whole axon (Begenisich and Cahalan 1980*b*; Yamamoto *et al.* 1985) and 458 mM at the single-channel level. The results obtained in BTX are also comparable to those obtained in bilayers for Na^+ channels from other membrane preparations, but differences among preparations are frequently observed. Apparently, Na^+ channels from different sources have different conduction characteristics, not only in the value of the maximum conductance achieved but also in the presence of fixed surface charges close to the conduction pathway. In general, high-affinity channels, that is those showing low K_ds, have conductance–concentration relationships that follow simple Michaelis–Menten type curves. Conversely, channels that show saturation of the conductance at very low sodium concentrations, indicating the possible existence of surface charge at the mouth of the channel, normally exhibit low affinity for sodium. A particular case of interest is that of Na^+ channels from the squid retinal nerve, obtained from *Sepiotheutis sepioidea*, incorporated into neutral lipid bilayers, and modified by BTX. Behrens *et al.* (1989) found that the conductance–concentration curves followed a simple Michaelis–Menten relationship. No need was found to incorporate surface charge effects to fit the data. The difference could be species-related, but sodium channels from different tissues may also be different.

Sodium-channel kinetics

Model for Na^+-channel gating

In a kinetic scheme, Na^+ channels are envisioned to go through several closed (non-conducting) states before they finally open and let ions flow through when the membrane is depolarized; then the channels would become inactivated, again non-conducting. Repolarization of the membrane would restore the equilibrium towards the resting, closed state(s). This basic kinetic scheme has been pieced together using evidence obtained at different levels by making use of various techniques, starting with the macroscopic currents measurements by Hodgkin and Huxley (1952). In their classic work, these authors proposed a mathematical model (HH model) for the Na^+ and K^+ conductances of the squid axon; this model can be interpreted in terms of molecular events in the channel. In the case of the Na^+ conductance, their formulation indicates the presence of two sets of charged groups (or dipoles) that reorient themselves in the membrane field. The first group consist of three identical and independent units that each have two stable positions, and the channel is activated only when the three units are in the same active position. The other group comprises only one charged unit with two stable positions, which can be in the inactivated or non-inactivated position. The channel conducts only when the first group is in the active position and the other in the non-inactivated position. Consequently, this model generates eight states with only one conducting and seven non-conducting states.

The kinetics of the charge movement predicted by the HH model could only be tested when gating currents measurements became possible. The first discrepancy occurred when Armstrong and Bezanilla (1977) found that the tails of the ionic currents upon a repolarization were not three times as fast as the gating current tails, a necessary consequence of the model with three identical and independent particles. This result implies that channel activation is the result of the movement of several non-identical and possibly interacting subunits. The second interesting result provided by gating currents was that the inactivation process was concomitant with activation-charge immobilization (Armstrong and Bezanilla 1977). This result implied that the same charge responsible for the activation process is immobilized by the inactivation process which in turn has only a small charge of its own. In this view, the voltage dependence of inactivation is a consequence of the coupling to the activation process, which is not in agreement with the independence of activation and inactivation of the HH model. Gating currents studied with prepulses to more hyperpolarized potentials provided evidence for more than three closed non-inactivated states (Taylor and Bezanilla 1983). The combination of all these measurements of macroscopic and gating currents made possible the proposal of more elaborated kinetic models for the Na^+ channels (e.g. Armstrong and Gilly 1979; Stimers *et al.* 1985; Keynes 1986; see also Chapter 6) but details of the transition rates between states have not been determined with any degree of certainty.

Single-channel recordings (Fig. 10.1) provide direct information on the events occurring close to the open state(s); therefore this type of experiment combined with the results on macroscopic currents and gating currents should constrain the models of channel gating. As stated before, with the cut-open axon technique it is possible to measure not only aspects of the opening and closing of single channels, but also gating currents and macroscopic currents in patches with high channel density. Vandenberg and Bezanilla (1991*a,b*) recorded single-channel events in the range −50 to −20 mV and, analysing the data using maximum-likelihood estimation of model parameters (Horn and Lange 1983; Horn and Vandenberg 1984), found a minimum model that accounts very well for all the single-channel properties. This model required three closed states, one inactivated state, and one open state. The inactivated state could be reached from a closed state (not contiguous with the open state) as well as from the open state. An important new result from the single-channel studies was that in this potential range (more negative than 0 mV) there is only one open state that is connected to a closed state and the inactivated state. It is interesting to note that this model was unable to predict gating currents correctly both in its kinetics and in the features of charge immobilization, even though the single-channel and macroscopic currents were well predicted. The reason for these discrepancies is that single-channel recordings are very insensitive to transitions occurring far from the open state. With the small number of closed states that satisfy the single-channel results, to account for the lag in channel opening, the rate constants fitted to the model were slower between the initial closed states as compared to the later transitions. This relation between rate constants predicts a rising phase in the gating current that is not observed experimentally (Stimers *et al.* 1987); therefore the channel must have more closed states than the number required by single-channel results. The other features not predicted by the single-channel model were the details of charge immobilization; this is due to the fact that the model best fitted by the single-channel data contained, for simplicity, only one inactivated state.

Vandenberg and Bezanilla (1991*b*) gathered all the results from gating currents, macroscopic currents, and single-channel recordings from patch experiments in the cut-open axon and fitted them to a more complex model that considered several inactivated states to account for charge immobilization, and several closed states to account for the lag of the ionic current and the absence of rising phase in the gating current. All the rate constants were assumed to be exponential functions of voltage, and the fitted parameters were the voltage dependence and the charge moved in each step of the gating sequence. The kinetic model with its rate constants is presented in Fig. 10.5(a). The first three transitions are fitted quite well to the same rate constants (y and z) and each moves a total of 1.5 e^{-}. With these rates and

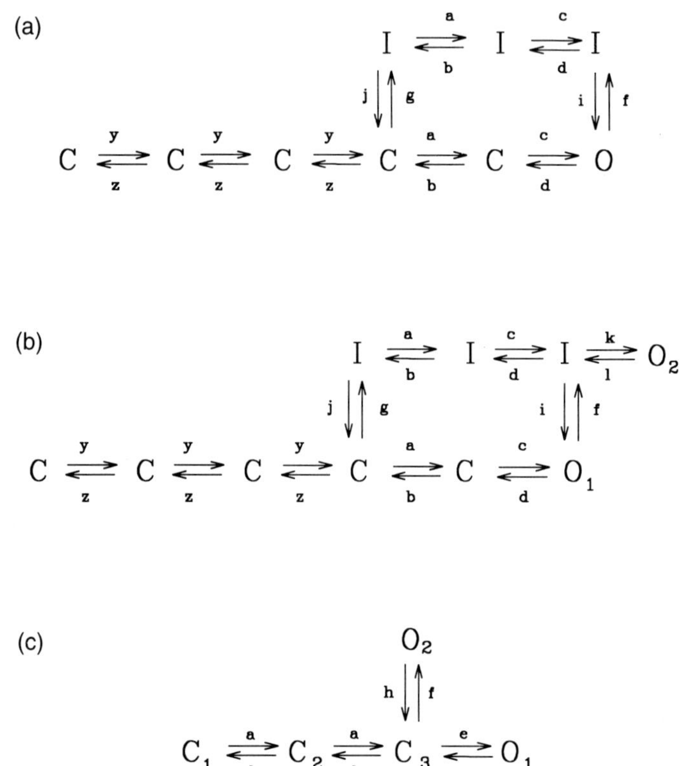

Fig. 10.5. Kinetic models for voltage-activated sodium channels. (a) Kinetic model that describes the gating behaviour of the squid sodium channel in the negative voltage range. This model fits well the data obtained measuring macroscopic, gating, and single-channel current experiments. The main features are described in the text. (b) Kinetic model describing the gating kinetics of sodium channels at positive potentials. The data used to construct the model comes from recordings made with reverse gradients and gating currents at large depolarizations. (c) Kinetic model describing the kinetics of BTX-modified sodium channels. At least three closed states and two open states were determined from the distributions of open and closed times.

charges, the model predicts gating currents without a rising phase but with a flat top, as recorded experimentally. The other two transitions with rates a, b and c, d move a total of 2.33 e^-. These last two transitions are responsible for the charge immobilization that occurs when the channel is populating the three I states which moves 2.33/(1.5 + 1.5 + 1.5 + 2.33) = 0.34 of the total charge in agreement with the results of charge immobilization reported before (Armstrong and Bezanilla 1977). The rate constants f, i (and g, j) carry a total charge of 0.91 e^- but most of the voltage dependence is in the return paths (i or j) to account for the strong voltage dependence of the recovery from inactivation (Bezanilla and Armstrong 1977). The return from the open state when the membrane is repolarized after short pulses is limited by rates d and b which are much slower than rate z, and this produces a similar rate of decay for gating and ionic current as described by Armstrong and Bezanilla (1977).

The fact that the three first transitions are the same in the best fit to the data has important consequences in the correlation of the kinetic model with possible molecular rearrangements of the molecule. If one assumes that three of the four domains of the channel have identical voltage sensors, then the equality of the rate constants implies that there is positive co-operativity among the domains as explained in the appendix of Vandenberg and Bezanilla (1991*b*). Another possibility is that they are not identical but similar and also interact. In a recent publication by Sato and Matsumoto (1992), the sequence of a putative squid Na^+ channel was deduced from cDNA, and it has a similarity to other Na^+ channels. The S4 regions of the first three domains contain less charge (3, 4, and 5 positive charges) than the last domain (8 positive residues). Under the assumption that the gating charges are proportional to the S4 charges, it is tempting to speculate that the first three transitions correspond to the movement of the first three domains, while the movement of the last domain produces the last two transitions which carry more charge. Although this is highly hypothetical, it provides a number of predictions that could be tested by mutagenesis of the channel protein.

Kinetics of gating at positive potentials

All the information used to build the model presented in the previous section comes from experiments done in the negative voltage range. Chandler and Meves (1970) had previously reported that in the squid axon Na^+ currents at positive potentials and under inverse gradient conditions exhibited incomplete inactivation. They suggested a second conductance pathway to explain the non-inactivating component of the current. This phenomenon is dramatic at potentials of about +80 mV where the peak current is about twice the value of the steady current. Incomplete inactivation is not due to the reverse gradient because it is also observed in a normal gradient (Bezanilla and Armstrong 1977; Correa and Bezanilla 1994*a*).

Single-channel studies with reverse gradient allowed a detailed study of conductance and kinetics at positive potentials (Correa and Bezanilla 1994*b*). It was found that, in a single-channel patch, openings occurred throughout an 18-ms pulse, and that the ensemble average of these single-channel currents showed incomplete inactivation with the same characteristics observed in the whole axon. Logarithmic plots of open times revealed two open states, although the conductance of both open states was the same. As these studies were done on patches at potentials where the open probability is about 0.7, and in several hundred trials, no second levels were observed, they demonstrate that there is only one type of channel with two open states, as opposed to two types of Na^+ conductance. A more detailed study of the open times revealed that the longer open times occurred at the beginning of the depolarization and the shorter open times more towards the end of an 18-ms pulse. Furthermore, when a depolarizing prepulse preceded the test pulse, mainly the short-lived openings were observed and the ensemble current showed no inactivation (Correa and Bezanilla 1994*b*). These results indicate that the second open state is populated after the channel goes into inactivation after populating the first open state. These studies were completed with measurements of gating currents at positive potentials to assess the extent of charge immobilization (Correa and Bezanilla 1994*a*). If the second open state observed at positive potentials occurs because the inactivation is less complete, then it would be expected that the charge immobilization should be less than that at less depolarized potentials such as 0 mV. Charge immobilization measured at different test potentials showed that about 0.6 of the total charge is immobilized at 100 mV, which is the same as it is at 0 mV although it occurs faster. The conclusion of these experiments is that the second open state is connected to the inactivated state. According to this view, a depolarization would first populate the first open state; then the inactivating particle, would inactivate the channel and, finally, if the depolarization was very large, the channel would again become conducting by entering into the second open state [see Fig. 10.5(b)]

In conclusion, we believe that the sodium channel of the squid has at least two open states of the same conductance value. At low depolarizations, only one of the open states get populated, but at large depolarizations, the second open state gets populated from the inactivated state and this second state is responsible for the incomplete inactivation observed at positive potentials.

Gating kinetics of batrachotoxin-modified channels

As mentioned above, one of the most dramatic effects brought about by BTX is the prolongation of the open times of the channels and a total absence of inactivation. In a typical experiment the range of potentials can easily cover from −50 to 100 mV. The resolution of the events is limited in many cases by the size of the events, which will vary with the driving force. These characteristics make the BTX-modified channel an ideal way to study effects of voltage and temperature on channel kinetics. The thermodynamic information provided by this study gives better insight into the nature of the changes occurring during channel opening and closing.

We have studied BTX-modified channel activity at temperatures between 0 and 15°C in the same patch and for the range of voltages covering the open probability curves (Correa *et al.* 1992). Figure 10.6 shows the open probability (P_o) of a BTX-modified channel as a function of potential (V) at various temperatures. P_o was measured as the total time spent in the open state over the total time of the recording at a given potential. The relation is sigmoidal: the probability approaches 1 at potentials more positive than −20 mV and 0 at potentials more negative than −100 mV. The closest fit to the data is achieved with several Boltzmann distributions as expected from processes with multiple voltage-dependent steps. In fact, we have found that the BTX-modified channel could be modelled with at least three closed and two open states (Correa *et al.* 1992). But single Boltzmann distributions described the data well; the simplification allows comparison between sets of data with variable numbers of experimental points. The P_o versus V data at 0°C shown in Fig. 10.6 (filled circles) could be fitted with an equivalent number of gating particles of 3.7 (on average: 3.58) and a potential for half activation (V_o) of −67.5 mV. The value for V_o depended on the Na^+ concentration and, most importantly, on temperature. An increase in temperature of 14°C shifted the

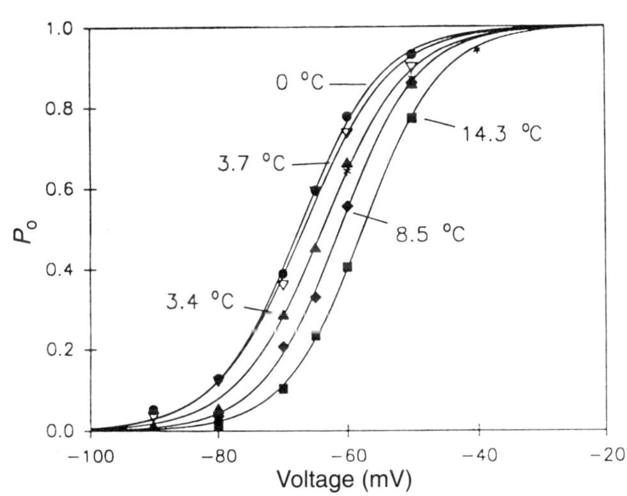

Fig. 10.6. Opening probability (P_o) as a function of potential of single BTX-modified channels at different temperatures. The control curve was obtained at 3.4°C (filled triangles). The variations in the bath temperature followed the sequence: 8.5°C (filled diamonds), 14.3°C (filled squares), 3.9°C (asterisks), 0°C (filled circles), and 3.7°C (open inverted triangles). The curves are fits to single Boltzmann distributions. Recordings made in external 540 mM Na^+ solution and with internal 50 mM Na^+. (Taken with permission from Correa *et al.* (1992).)

whole P_o versus V curve towards more depolarized potentials by about 10 mV (Fig. 10.6). The shift to the right of the P_o versus V relation with an increase in temperature has important implications because it indicates a decrease in entropy with channel opening (see below) as described by the expression

$$V_o(T_2) - V_o(T_1) = -\Delta S(T_2 - T_1)/zF$$

where T_1 and T_2 are two different temperatures, ΔS is the entropy change, z is the valence and F is the Faraday constant (Correa *et al.* 1992).

Approximately the same shift was obtained when the mean open and mean closed times, were analysed. Logarithmic histogram distributions of open and closed times revealed several exponential components, the weight of each of which depended on the potential in such a way that the main component of the open times shifted towards smaller values with hyperpolarization. Conversely, the closed times became longer with hyperpolarization as would be expected from the P_o observed. Also, with an increase in temperature the main component of the open times shifted towards shorter values. The same tendency was observed in the distribution of closed times.

A model with three sequential closed states and two open states branching from the last closed state fitted the data very well [Fig. 10.5(c)]. The values for the rate constants between states were well predicted and showed a monotonic voltage dependence. Predicted open time distributions also overlapped with the actual data. The transitions between closed states involved about half the gating charge; the rest was accounted for in the step from last closed to open state. The main voltage-dependent step was the return from open to closed. Raising the temperature increased all rates. The shift in P_o towards more positive potentials would be the result of an increase with temperature of the closing rates predominant over the opening rates.

Finally, this study provided information regarding the energy profile accompanying each transition between states in the model (Fig. 10.7). The most interesting finding relates to the energetic changes that occur upon channel opening. We found that there was a negative change of entropy in the transition from the last closed to the main open state (the second open state is rarely populated), the activation energy of this transition being purely entropic. The activation energy involved in the closing of the channel had large enthalpic and positive entropic components. This agreed well with the finding that the closed states were favoured over the open configuration (more ordered) with an increase in temperature. The energy changes involved in the transitions between closed states were mainly enthalpic with a small, net entropic increase in the pathway between the first (C_1) and last closed (C_3) states (Fig. 10.7). On somewhat speculative grounds the findings related to the energy profile suggested a series of chemical reactions taking place in the transit between the successive closed states which would correspond to an initial rearrangement of the structure of the channel. In the final step, a major reorganization and ordering of the molecule, in fact a reduction in the degrees

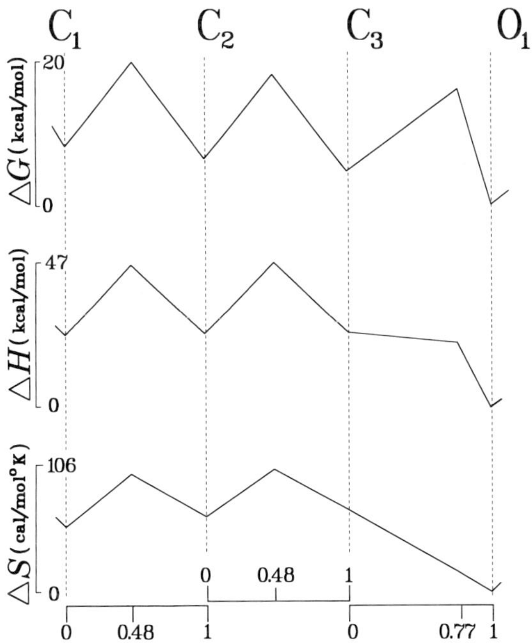

Fig. 10.7. The energy profile of the model of the BTX-modified channel shown in Fig. 10.5(c). In this case, only one open state (the most populated state) is being considered. The fit to the data gave values for the effective gating charge for each transition (forward and backward). In the diagram, the fraction of the field where the peak of the barrier for each transition was located is indicated at the bottom (scale bars from 0 to 1). The fraction of the field times the total valence is the effective charge. Also, the fits to the rate constants for each transition gave four values (one per temperature) for ΔG where G is the Gibbs function. These ΔG values were fitted to: $\Delta H - T\Delta S$, to obtain the entropy (ΔS) and enthalpy (ΔH) profiles represented in the figure. (Modified with permission from Correa *et al.* (1992).)

of freedom (as expected from the decrease in the entropy) occurs when the channel finally opens.

K^+ channels as seen with single-channel recording

Hodgkin and Huxley (1952) described the K^+ conductance or delayed rectifier as a voltage-dependent conductance selective to K^+ ions. Conti and Neher (1980) recorded single-channel events with a bent pipette inside a perfused axon with a sub-gigaohm seal and their properties agreed with the characteristics of the delayed rectifier. Llano *et al.* (1988), using gigaseal patches with the cut-open technique, were able to record several types of K^+ channels from the giant axon. The predominant type is a 20-pS channel that has all the properties of the delayed rectifier and probably is the major contributor to the macroscopic K^+ current. In addition, a 40-pS and a 10-pS channel were found but they were less abundant. Recently, Caputo *et al.* (unpublished) have found that, in the squid *Sepiotheutis sepioidea*, still another type of K^+ channel is found. This channel has the characteristics of A-type channels and shows marked inactivation. Inoue (Chapter 5) has reported that macroscopic K^+ currents from *Sepiotheutis lessoniana* show inactivation suggesting that this A-type current may be characteristic of this genus.

The delayed rectifier of the squid is modulated by internal ATP (Bezanilla *et al.* 1986; Augustine and Bezanilla 1989; Perozo *et al.* 1989). The modulation is a phosphorylation reaction that shifts the activation and the inactivation curves to more depolarized potentials (Perozo *et al.* 1989) producing changes in the kinetics and amplitude of the K^+ currents. Single-channel studies have been extremely useful in understanding the effects of phosphorylation on the K^+ conductance. Perozo *et al.* (1991*a*) found that the probability of opening of the 40-pS channel in the axon is increased by phosphorylation and this increase occurs as a consequence of a decrease of the closed times; the single-channel conductance was unaffected. This channel is slower in its activation than the regular delayed rectifier and most probably does not play a significant role in the macroscopic current. In a further study Perozo *et al.* (1991*b*) found that the 20-pS delayed rectifier channel was modulated by phosphorylation. Consistent with the finding of macroscopic currents, the 20-pS channel slow inactivation was shifted to more positive potentials by phosphorylation as revealed by a decrease in the number of blanks in consecutive trials in single-channel records. In addition, the first latency was affected in a manner consistent with the change of kinetics of the macroscopic currents. The single-channel conductance was not affected by phosphorylation; therefore the increase in macroscopic conductance observed was the result of the removal of slow inactivation which produces an effective increase in the open probability. Most of the effects of phosphorylation can be explained by shifts produced in the voltage dependence of activation and inactivation as a result of the addition of negative charges, presumably supplied by the transferred phosphate near the voltage sensor of the channel. This explanation was tested quantitatively by fitting the ionic currents to a sequential activation model with voltage-dependent rate constants. This model was able to fit the currents in the phosphorylated state by only adjusting one parameter, the voltage shift of the voltage-dependent parameters (Perozo and Bezanilla 1990).

Acknowledgement

This work was supported by USPHS grant GM30376.

References

Armstrong, C. M. (1981). Sodium channels and gating currents. *Physiological Reviews*, **61**, 644–83.
Armstrong, C. M. and Bezanilla, F. (1974). Charge movement associated with the opening and closing of the activation gates of the Na channels. *Journal of General Physiology*, **63**, 533–52.
Armstrong, C. M. and Bezanilla, F. (1977). Inactivation of the sodium channel. II. Gating current experiments. *Journal of General Physiology*, **70**, 567–90.
Armstrong, C. M. and Gilly, W. F. (1979). Fast and slow steps in the activation of sodium channels. *Journal of General Physiology*, **74**, 691–711.
Augustine, C. K. and Bezanilla, F. (1989). Phosphorylation modulates potassium conductance and gating current of perfused giant axons of squid. *Journal of General Physiology*, **95**, 242–72.
Begenisich, T. and Cahalan, M. D. (1980*a*). Sodium channel permeation in squid axons. I. Reversal potential experiments. *Journal of Physiology*, **307**, 217–42.
Begenisich, T. and Cahalan, M. D. (1980*b*). Sodium channel permeation in squid axons. II. Non-independence and current–voltage relations. *Journal of Physiology*, **30**, 243–57.
Behrens, M. I., Oberhauser, A., Bezanilla, F., and Latorre, R. (1989). Batrachotoxin-modified sodium channels from squid optic nerve in planar bilayers. Ion conduction and gating properties. *Journal of General Physiology*, **93**, 23–41.
Bezanilla, F. (1985). Gating of sodium and potassium channels. *Journal of Membrane Biology*, **88**, 97–111.
Bezanilla, F. (1987). Single sodium channels from the squid giant axon. *Biophysical Journal*, **52**, 1087–90.
Bezanilla, F. and Armstrong, C. M. (1977). Inactivation of the sodium channel. I. Sodium current experiments. *Journal of General Physiology*, **70**, 549–66.
Bezanilla, F. and Vandenberg, C. (1990). The cut-open axon technique. In *Squid as experimental animals* (ed. D. L. Gilbert, W. J. Adelman Jr, and J. M. Arnold), pp. 153–9. Plenum Press, New York.
Bezanilla, F., Caputo, C., DiPolo, R., and Rojas, H. (1986). Potassium conductance of the squid giant axon is modulated by ATP. *Proceedings of the National Academy of Sciences, USA*, **83**, 2743–5.
Chandler, W. K. and Meves, H. (1970). Evidence for two types of sodium conductances in axons perfused with sodium fluoride solutions. *Journal of Physiology*, **211**, 653–78.
Cole, K. S. (1949). Dynamic electrical characteristics of the squid axon membrane. *Archives des Sciences Physiologiques, Paris*, **3**, 253–8.
Cole, K. S. and Curtis, B. (1939). Electrical impedance of the squid giant axon during activity. *Journal of General Physiology*, **22**, 649–70.
Conti, F. and Neher, E. (1980). Single channel recording of K^+ currents in squid axons. *Nature*, **285**, 140–3.
Correa, A. M. and Bezanilla, F. (1994*a*). Gating kinetics of the squid axon sodium channel at positive potentials. I. Studies of macroscopic ionic and gating currents. *Biophysical Journal*, **66**, 1853–63.
Correa, A. M. and Bezanilla, F. (1994*b*). Gating kinetics of the squid axon sodium channel at positive potentials. II. Single channel studies reveal two open states. *Biophysical Journal*, **66**, 1854–78.
Correa, A. M., Latorre, R., and Bezanilla, F. (1991). Ion permeation in normal and

Batrachotoxin-modified Na^+ channels in the squid giant axon. *Journal of General Physiology*, **97**, 605–25.
Correa, A. M., Bezanilla, F., and Latorre, R. (1992). Gating kinetics of Batrachotoxin-modified Na^+ channels in the squid giant axon. Voltage and temperature effects. *Biophysical Journal*, **61**, 1332–52.
French, R. J., Worley III, J. F., and Krueger, B. K. (1984). Voltage-dependent block by saxitoxin of sodium channels incorporated into planar lipid bilayers. *Biophysical Journal*, **45**, 301–10.
French, R. J., Worley III, J. F., and Krueger, B. K. (1986). From brain to bilayer: sodium channels from rat neurons incorporated into planar lipid membranes. In *Ionic channels in cells and other model systems* (ed. R. Latorre), pp. 273–90. Plenum Publishing Corporation, New York.
Garber, S. (1988). Symmetry and asymmetry of permeation through toxin-modified Na^+ channels. *Biophysical Journal*, **54**, 767–76.
Garber, S. and Miller, C. (1987). Single Na^+ channels activated by Veratridine and Batrachotoxin. *Journal of General Physiology*, **89**, 459–80.
Green, W. N., Weiss, L. B., and Andersen, O. S. (1987). Batrachotoxin-modified sodium channels in lipid bilayers. Ion permeation and block. *Journal of General Physiology*, **89**, 841–72.
Hamill, O. P., Marty, A., Neher, E., Sakmann, B., and Sigworth, F. J. (1981). Improved patch-clamp techniques for high-resolution current recording from cells and cell-free membrane patches, *Pflügers Archiv*, **391**, 85–100.
Hodgkin, A. L. and Huxley, A. F. (1952). A quantitative description of the membrane current and its application to conduction and excitation in nerve. *Journal of Physiology*, **117**, 500–44.
Hodgkin, A. L. and Katz, B. (1949). The effect of sodium ions on the electrical activity of the giant axon of the squid. *Journal of Physiology*, **128**, 61–88.
Hodgkin, A. L., Huxley, A. F., and Katz, B. (1952). Measurement of current–voltage relation in the membrane of the giant axon of *Loligo*. *Journal of Physiology*, **116**, 424–48.
Horn, R. and Lange, K. (1983). Estimating kinetic constants from single channel data. *Biophysical Journal*, **43**, 207–23.
Horn, R. and Vandenberg, C. A. (1984). Statistical properties of single sodium channels. *Journal of General Physiology*, **84**, 505–34.
Huang, L.-Y., Moran, N., and Ehrenstein, G. (1984). Gating of Batrachotoxin-modified sodium channels in neuroblastoma cells determined from single-channel measurements. *Biophysical Journal*, **45**, 313–22.
Keynes, R. D. (1986). Modeling the sodium channel. In *Ion channels in neural membranes* (ed. J. Ritchie, R. D. Keynes, and L. Bolis), pp. 86–101. Alan R. Liss, New York.
Keynes, R. D. and Rojas, E. (1974). Kinetics and steady-state properties of the charged system controlling sodium conductance in the squid giant axon. *Journal of Physiology*, **239**, 393–434.
Llano, I., Webb, C. K., and Bezanilla, F. (1988). Potassium conductance of the squid giant axon. Single channel studies. *Journal of General Physiology*, **92**, 179–96.
Moczydlowski, E., Garber, S., and Miller, C. (1984). Batrachotoxin-activated Na^+ channels in planar bilayers. Competition of Tetrodotoxin block by Na^+. *Journal of General Physiology*, **84**, 665–86.
Naranjo, D. and Latorre, R. (1993). Ion conduction in substates of the batrachotoxin-modified Na^+ channel from toad skeletal muscle. *Biophysical Journal*, **64**, 1038–50.

Perozo, E. and Bezanilla F. (1990). Phosphorylation affects voltage gating of the delayed rectifier K channel by electrostatic interactions. *Neuron*, **5**, 685–90.

Perozo, E., Bezanilla, F., and DiPolo, R. (1989). Modulation of K channels in dialyzed squid axons: ATP-mediated phosphorylation. *Journal of General Physiology*, **93**, 1195–218.

Perozo, E., Vandenberg, C. A., Jong, D. S., and Bezanilla, F. (1991*a*). Single channel studies of the phosphorylation of K channels in the squid giant axon. I. Steady-state conditions. *Journal of General Physiology*, **98**, 1–18.

Perozo, E., Jong, D. S., and Bezanilla, F. (1991*b*). Single channel studies of the phosphorylation of K channels in the squid giant axon. II. Nonstationary conditions. *Journal of General Physiology*, **98**, 19–34.

Quandt, F. N. and Narahashi, T. (1982). Modification of single Na^+ channels by Batrachotoxin. *Proceedings of the National Academy of Sciences*, **79**, 6732–6.

Sato, C. and Matsumoto, G. (1992). Primary structure of squid sodium channel deduced from the complementary DNA sequence. *Biochemical and Biophysical Research Communications*, **86**, 61–8.

Smith-Maxwell, C. and Begenisich, T. (1987). Guanidinium analogues as probes of the squid axon sodium pore. Evidence for internal surface charges. *Journal of General Physiology*, **90**, 361–74.

Stimers, J. R., Bezanilla, F., and Taylor, R. E. (1985). Sodium channel activation in the squid giant axon. Steady-state properties. *Journal of General Physiology*, **85**, 65–82.

Stimers, J. R., Bezanilla, F., and Taylor, R. E. (1987). Sodium channel gating currents. Origin of the rising phase. *Journal of General Physiology*, **89**, 521–40.

Tanguy, J. and Yeh, J. Z. (1988). Divalent cation block of normal and BTX-modified sodium channels in squid giant axons. *Biophysical Journal*, **53**, 229a.

Taylor, R. E. and Bezanilla, F. (1983). Sodium and gating current time shift resulting from changes in initial conditions. *Journal of General Physiology*, **81**, 773–84.

Taylor, R. E., Armstrong, C. M., and Bezanilla, F. (1976). Block of sodium channels by external calcium ions. *Biophysical Journal* **16**, 27a.

Vandenberg, C. A. and Bezanilla, F. (1991*a*). Single-channel, macroscopic, and gating currents from sodium channels in the squid giant axon. *Biophysical Journal*, **60**, 1499–510.

Vandenberg, C. A. and Bezanilla, F. (1991*b*). A sodium channel gating model based on single-channel, macroscopic ionic and gating currents in the squid giant axon. *Biophysical Journal*, **60**, 1511–33.

Woodhull, A. M. (1973). Ionic blockage of sodium channels in nerve. *Journal of General Physiology*, **61**, 687–708.

Yamamoto, D., Yeh, J. Z., and Narahashi, T. (1984). Voltage-dependent calcium block of normal and Tetramethrin-modified single sodium channels. *Biophysical Journal*, **45**, 337–44.

Yamamoto, D., Yeh, J. Z., and Narahashi, T. (1985). Interactions of permeant cations with sodium channels of the squid axon membranes. *Biophysical Journal*, **48**, 361–8.

11

The effects of internal Ca^{2+} and Mg^{2+} on ion channels in the squid giant axon

SHUNICHI YAMAGISHI, KISHIO FURUYA, and FUMIO KUKITA

Summary

During intracellular perfusion with a 25 mM K^+ solution containing Ca^{2+} in the range 0.1 to 10 mM, the action potential of the squid giant axon gradually decreased in amplitude, while the duration was transiently prolonged. The threshold potential for the initiation of the action potential was markedly lowered towards the resting potential level. A standard voltage clamp method was applied for the measurement of Na^+ (I_{Na}) and K^+ currents (I_K). The amplitude of I_{Na} and I_K were reduced by internal addition of 1–30 mM Ca^{2+} or Mg^{2+} to a 100 mM K^+ solution. At the same time, a shift of the I–V curve in the negative voltage direction was observed. Up to 4 mM, the values for the voltage shift of I_{Na} for an *e*-fold change in Ca^{2+} and Mg^{2+} concentration were 7 mV and 8 mV respectively. I_{Na} and I_K were gradually reduced by the increase in internal Ca^{2+} or Mg^{2+} concentration. The time constants for I_{Na} reduction by Ca^{2+} and Mg^{2+} were 2.0 min and 1.8 min, respectively. The apparent dissociation constants (K_D) of I_{Na} reduction by Ca^{2+} and Mg^{2+} were 2.3 mM and 7.0 mM, respectively. I_K was reduced more quickly but less effectively than I_{Na} by Ca^{2+} and Mg^{2+} application.

Low concentrations of Ca^{2+} and Mg^{2+} exist in the cytoplasm of animal cells. Changes in concentration of Ca^{2+} are closely related to several functions, including membrane processes such as excitability and synaptic transmission, but less is known about the role of Mg^{2+}. In the case of squid axons, the free Ca^{2+} concentration in the axoplasm has been estimated to be in the range 50–150 nM (Dipolo *et al.* 1976; Requena *et al.* 1986; Baker and Umbach 1987), while the free Mg concentration has been reported as 2–4 mM (Baker and Crawford 1972; Brinley and Scarpa 1975; DeWeer 1976).

This chapter deals with the regulatory effects of internally added Ca^{2+} and Mg^{2+} on the Na^+ and K^+ channel activities of an excitable membrane. Using squid giant axons, the action of internal Ca^{2+} on Na^+ channel activity was reported as causing deterioration (Tasaki *et al.* 1965), or little effect (Begenisich and Lynch, 1974). We have previously shown that internally added Ca^{2+} produced a Na^+ action potential without any electrical stimulation in squid axons (Yamagishi and Furuya 1981).

Concerning the action of internal Ca^{2+} on K^+ channel activity, the existence of Ca^{2+} activated K^+ channels has been well established, starting with studies on *Aplysia* nerve cells (Meech and Strumwasser 1970) and cat motor neurones (Krnjevic and Lisiewicz 1972). Concerning Mg^{2+} effects on K^+ channels, Matsuda *et al.* (1987) reported the blockage of outward K^+ current in heart muscle cells. However, our understanding of the effect of divalent cations on the kinetics of Na^+ and K^+ channels is still very incomplete.

In this report, we show that internally applied Ca^{2+} and Mg^{2+} commonly have two types of effect on ion channel activities. First, they cause a lowering of the threshold voltage level for Na^+ and K^+ channel activation. Second, they reduce Na^+ and K^+ currents by a concentration-dependent amount that differs for internally applied Ca^{2+} and Mg^{2+}, respectively. The similarities and differences between Ca^{2+} and Mg^{2+} effects on ion channel activities are described.

Effects of internal Ca^{2+} and Mg^{2+} on the action potential

The first intracellular recording of the action potential was made by A. L. Hodgkin (1939), using a giant axon of squid *Loligo forbesi*. He showed a beautiful action potential of an amplitude of about 90 mV, including a 40-mV overshoot. It was a great discovery and a starting point for the elucidation of the membrane phenomena of excitable cells. In the early 1960s, the groups of Hodgkin (Baker *et al.* 1961, 1962) and Tasaki (Oikawa *et al.* 1961) independently introduced successful intracellular perfusion of the squid giant axon.

This study discusses the regulatory effects of internally applied Ca^{2+} and Mg^{2+} ions, using intracellularly perfused squid axons. A giant axon of the squid *Doryteuthis bleekeri* was mounted in a Lucite chamber filled with artificial sea water (ASW; 450 mM NaCl, 10 mM KCl, 10 mM $CaCl_2$, 20 mM $MgCl_2$, 10 mM Tris–Cl, pH 8.0). As a standard internal solution, 25–100 mM KCl, K-glutamate, or KF solution was used, and the pH was adjusted to 7.4 with 4 mM Tris–HEPES. The tonicity was adjusted to that of the external solution by adding glycerol or glucose.

Figure 11.1 illustrates the typical change in action potential caused by the addition of 1 mM $CaCl_2$ to the internal 25 mM KCl solution. The resting potential and action potential amplitudes were −35 and 140 mV, respectively. On adding Ca^{2+} internally, the resting potential shifted slightly to the negative direction (hyperpolarization), while the action potential amplitude decreased to 110 mV over 2.2 min. At that time, the duration of the action potential at the level of half amplitude was prolonged to 12 ms, which was four times longer than the control. The most prominent feature was a lowering of the threshold potential at which the action potential initiates. The threshold value during resting conditions was −19 mV, which was 16 mV depolarized from the resting potential. On adding Ca^{2+}, the threshold quickly dropped to

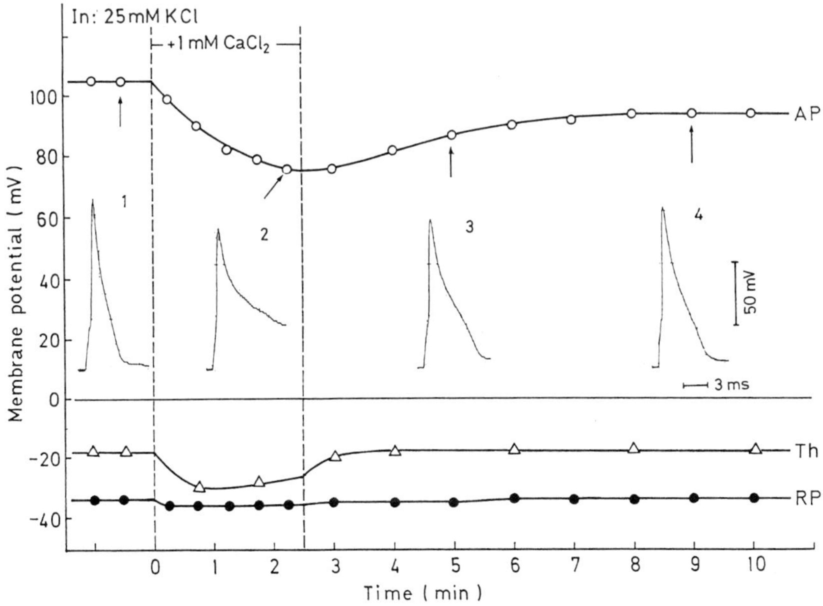

Fig. 11.1. Effect of internally (In) added 1 mM $CaCl_2$ on resting (RP), threshold (Th), and action potentials (AP) of a squid axon. The axon immersed in artificial sea water (ASW) was internally perfused with a 25 mM KCl solution as a control.

−30 mV. On washing out Ca^{2+}, the threshold value returned to the control level and the action potential recovered to 93 per cent of the initial value. When Mg^{2+} was added internally instead of Ca^{2+}, the change of threshold potential was almost the same as with Ca^{2+}, while the action potential amplitude was less reduced.

This large lowering of threshold potential caused by Ca^{2+} and Mg^{2+} is a notable and important effect, and suggests that the membrane becomes more excitable in the presence of increased levels of intracellular divalent cations.

Shift of Na^+ channel activation in a negative voltage direction

A standard voltage-clamp method was applied for the measurement of membrane currents. The axon immersed in ASW was perfused internally with a 100-mM K-glutamate and 10-mM tetraethylammonium-Cl solution for the measurement of Na^+ current (I_{Na}). Tetrodotoxin at a concentration of 0.3 μM was added to the ASW to block Na^+ current and to allow measurement of K^+ current (I_k).

Figure 11.2 shows the effect of internally applied Ca^{2+}. As shown in record 2 of the figure, the amplitude of the Na^+ current was reduced by a 4-mM

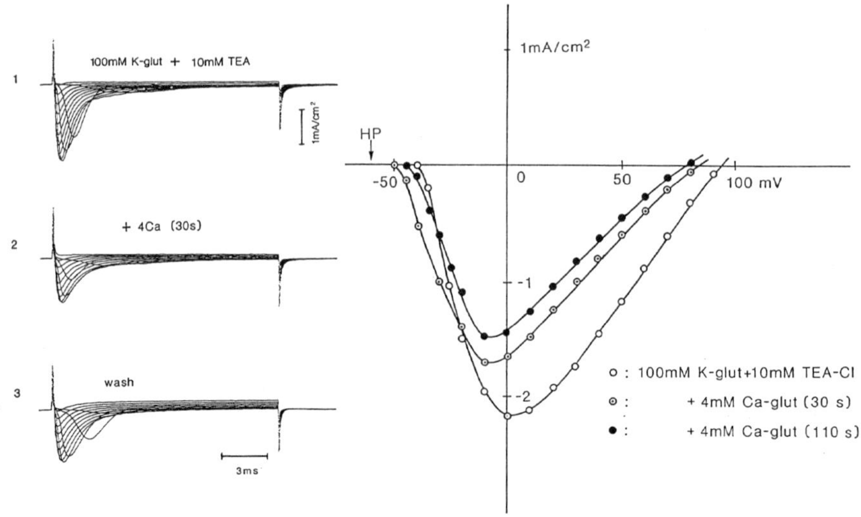

Fig. 11.2. Effect of internally added 4 mM Ca-glutamate on the inward Na^+ current (I_{Na}). The external solution was ASW, and the internal standard solution was 100 mM K-glutamate and 10 mM TEA–Cl solution. The holding potential (HP) was −60 mV. Thirty seconds after internal Ca^{2+} addition, the current–voltage (*I–V*) curve had shifted to the left and the peak Na^+ current (I_{Na}) was reduced.

addition of Ca^{2+}, but Na^+ activation and inactivation kinetics did not change. The current–voltage (*I–V*) curve of the figure shows that I_{Na} was shifted in a negative voltage direction by 4 mM Ca^{2+}, and the peak I_{Na} was reduced.

Figure 11.3 shows the values of the voltage shift caused by a range of concentrations of Ca^{2+} and Mg^{2+} (0.3–30 mM). The voltage shifts were plotted as the voltage difference at the level of 1/2 peak I_{Na} before and after the Ca^{2+} or Mg^{2+} application. For an *e*-fold increase of Ca^{2+} and Mg^{2+} concentration, the voltage shifted by 7 and 8 mV, respectively, within the concentration range 0.3–4 mM. Compared with the effect of externally applied Ca^{2+} (Frankenhaeuser and Hodgkin 1957), the direction of the voltage shift of the Na^+ activation curve was opposite, but the value of the shift was similar.

In the present experiments, it is proposed that negatively charged sites at the inner mouth of the Na^+ channel are neutralized by the divalent cations, whereas the experiment of Frankenhaeuser and Hodgkin neutralized charged sites at the outer mouth of the Na^+ channel. It would be predicted that charge neutralization at the two sites would shift the *I–V* curve in opposite directions, as observed. Shift of the *I–V* curve to the right (in the positive voltage direction) is associated with membrane stabilization, while shift to the left (negative voltage direction) increases excitability. The effects of elevated

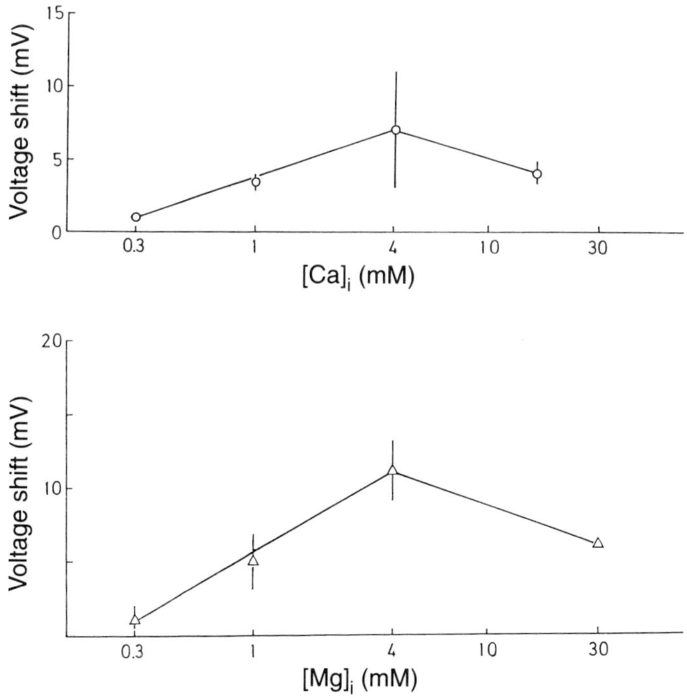

Fig. 11.3. The shifts of Na-channel activation to the negative voltage direction caused by internal addition of Ca^{2+} and Mg^{2+}. The voltage shifts were estimated from the difference of 1/2 peak I_{Na} values before and after the addition of Ca^{2+} or Mg^{2+}.

internal calcium on the threshold for action potential generation (Fig. 11.1) can be understood from analysis of the shift of the *I–V* plot.

Reduction of Na^+ and K^+ currents

Figure 11.4 shows the time course of reduction of I_{Na} and I_K by different concentrations of internal Ca^{2+} and Mg^{2+}. I_{Na} and I_K were reduced more strongly at higher Ca^{2+} or Mg^{2+} concentrations. The degrees and time course of I_{Na} and I_K reduction were different. I_{Na} was reduced more slowly but more strongly compared with I_K. The time constants of I_{Na} and I_K reduction were 2.0 and 0.5 min, respectively, while the apparent dissociation constants (K_D) with 100-mM internal K^+ were 2.3 and 5.8 mM for I_{Na} and I_K reduction, respectively.

I_{Na} and I_K were also reduced by Mg^{2+}. The patterns of reduction caused by Mg^{2+} were quite similar to those caused by Ca^{2+}; the time constants of

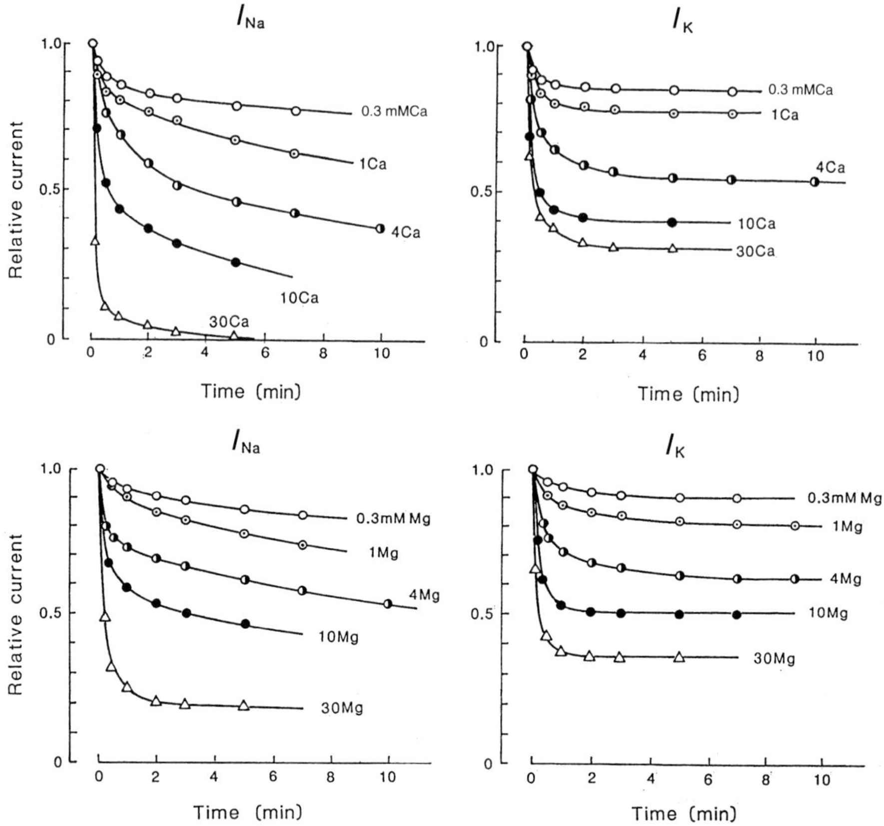

Fig. 11.4. Time course of reduction of I_{Na} and I_K by different concentrations of internal Ca^{2+} and Mg^{2+}. The experimental curves through the points are used to calculate the time constants (see text).

reduction were 1.8 min for I_{Na} and 0.5 min for I_K. The only difference between the Ca^{2+} and Mg^{2+} effects was potency; Mg^{2+} was less effective than Ca^{2+} at each concentration.

Figure 11.5 shows the dose–response curves for the effect of Ca^{2+} and Mg^{2+} on I_{Na} and I_K with 100 mM internal K^+. The K_D for I_{Na} reduction was 2.3 mM for Ca^{2+} and 7 mM for Mg^{2+}. The Ca^{2+} effect was three times greater than the Mg^{2+} effect. The K_D for I_K reduction was 5.8 mM for Ca^{2+} and 10.2 mM for Mg^{2+}.

We have mainly used an internal K^+ concentration in the range up to 100 mM, in order to avoid the reduction of membrane excitability caused by high KCl or K^+-glutamate solution. In the axoplasm, the normal internal K^+ concentration is about 400 mM (Hodgkin 1964). In the present experiments, the Ca^{2+} and Mg^{2+} effects were dependent on the internal K^+ concentration $[K^+]_i$, and were roughly proportional to the $[Ca^{2+}]_i/[K^+]_i$ or $[Mg^{2+}]_i/[K^+]_i$

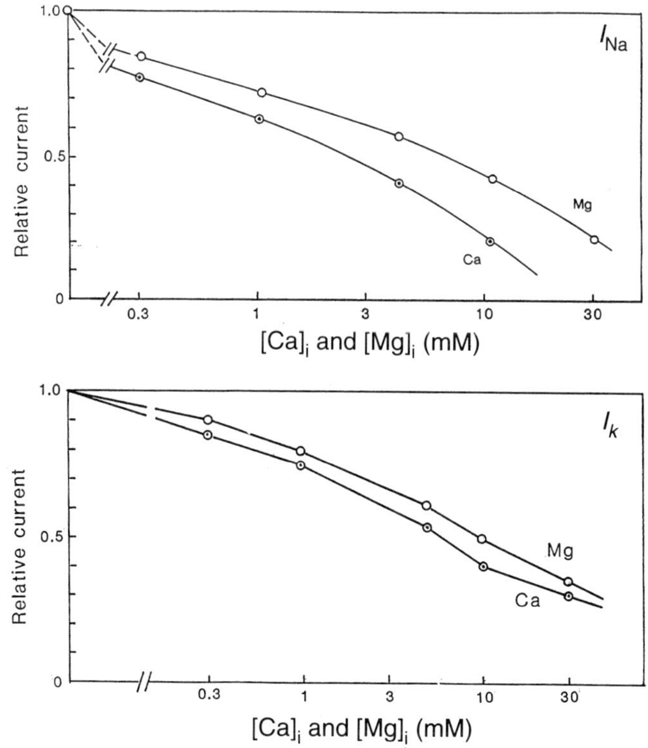

Fig. 11.5. Dose–response curves for the reduction of I_{Na} and I_K by Ca^{2+} and Mg^{2+}.

ratio. For example, the effect of 1 mM Ca^{2+} or Mg^{2+} in 100 mM K^+ internal solution was equal to that of 4 mM Ca^{2+} or Mg^{2+} in 400 mM K^+ solution. Therefore, the above described Ca^{2+} and Mg^{2+} concentration effects in the 100 mM K^+ solution are likely to be around one-quarter of those present under physiological conditions.

Conclusion

Intracellularly applied Ca^{2+} and Mg^{2+} exhibited dual effects on ion channel activities of squid axons. First, increased internal Ca^{2+} and Mg^{2+} caused the membrane to become more easily excitable, by charge neutralization at the inner mouth of the Na^+ channels. Second, they reduced the I_{Na} and I_K amplitudes in a concentration-dependent manner. The Ca^{2+} concentration used was higher than that present in axoplasm, whereas the applied Mg^{2+} concentration was of the same order as that in axoplasm (2–4 mM). Thus, internal Mg^{2+} may play a role in physiological regulation of membrane excitability.

Acknowledgements

The authors wish to thank Drs J. N. Abbott, I. Inoue, and I. Tsutsui for their helpful suggestions and comments on the manuscript.

References

Baker, P. F., Hodgkin, A. L., and Shaw, T. I. (1961). Replacement of the protoplasm of a giant nerve fibre with artificial solutions. *Nature*, **190**, 885–7.

Baker, P. F. and Crawford, A. C. (1972). Mobility and transport of magnesium in squid giant axons. *Journal of Physiology*, **227**, 855–74.

Baker, P. F. and Umbach, J. A. (1987). Calcium buffering in axons and axoplasm of *Loligo*. *Journal of Physiology*, **383**, 369–94.

Baker, P. F., Hodgkin, A. L., and Shaw, T. I. (1962). Replacement of axoplasm of giant nerve fibres with artificial solutions. *Journal of Physiology*, **164**, 330–54.

Begenisich, T. and Lynch, C. (1974). Effects of divalent cations on voltage clamped squid axons. *Journal of General Physiology*, **63**, 675–89.

Brinley, F. J. and Scarpa, A. (1975). Ionized magnesium concentration in axoplasm of dialysed squid axons. *Federation of European Biochemical Societies Letters*, **50**, 82–5.

DeWeer, P. (1976). Axoplasmic free magnesium levels and magnesium extraction from squid giant axons. *Journal of General Physiology*, **68**, 159–78.

Dipolo, R., Requena, J., Brinley, Jr, F. J., Mullins, L. J., Scarpa, A., and Tiffert, Y. (1976). Ionized Ca concentration in squid axons. *Journal of General Physiology*, **67**, 433–67.

Frankenhaeuser, B. and Hodgkin, A. L. (1957). The action of calcium on the electrical properties of squid axons. *Journal of Physiology*, **137**, 218–44.

Hodgkin, A. L. (1939). Action potentials recorded from inside a nerve fibre. *Nature*, **144**, 710–11.

Hodgkin, A. L. (1964). *The conduction of nerve impulses*. The Sherrington Lecture VIII. Liverpool University Press.

Krnjevic, K. and Lisiewicz, A. (1972). Injection of calcium ions into spinal motoneurones. *Journal of Physiology*, **225**, 363–90.

Matsuda, H., Saigusa, A., and Irisawa, H. (1987). Ohmic conductance through the inwardly rectifying K channel and blocking by internal Mg^{2+}. *Nature*, **325**, 156–9.

Meech, R. W. and Strumwasser, F. (1970). Intracellular calcium injection activates potassium conductance in *Aplysia* nerve cells. *Federation Proceedings*, **29**, 834.

Oikawa, T., Spyropoulos, C. S., Tasaki, I., and Teorell, T. (1961). Methods for perfusing the giant axon of *Loligo pealii*. *Acta Physiologica Scandinavica*, **52**, 195–6.

Requena, J., Mullins, L. J., Whittembury, J., and Brinley, Jr, F. J. (1986). Dependence of ionized and total Ca in squid axons on Na_o-free or high K_o conditions. *Journal of General Physiology*, **87**, 143–59.

Tasaki, I., Singer, I., and Takenaka, T. (1965). Effects of internal and external ionic environment on excitability of squid giant axon: A macromolecular approach. *Journal of General Physiology*, **48**, 1095–123.

Yamagishi, S. and Furuya, K. (1981). Initiation of the spike by intracellularly perfused calcium in squid giant axons. *Proceedings of the Japan Academy*, **57B**, 54–8.

12

Anaesthetics, convulsants, and the squid axon membrane

B. M. HENDRY

Summary

The use of the squid axon as a model system for the study of anaesthetic mechanisms is described. Effects are divided into actions on the voltage-gated Na and K channels and actions on a voltage-independent K conductance system of the nerve membrane. Inhibition of axonal function by anaesthetics arises largely from effects on the Na channel. The precise nature of these effects varies with the physico-chemical properties of the anaesthetic considered, but certain general principles emerge. For example, small non-polar anaesthetics tend to increase the resting fraction of Na channels in the inactivated state. The actions of certain fluorinated anaesthetics and convulsants on the squid axon are described in the context of the close relationship between anaesthetic and convulsant actions. At low concentrations these molecules all appear capable of increasing axonal excitability and reducing the threshold for action potential generation. This action is related to the inhibition of a resting K permeability system in the axon membrane. The anaesthetics and convulsants can be distinguished in this axonal model on the basis of their relative potencies for inhibition of resting K permeability and Na channels. Convulsants appear to be characterized by an ability to inhibit the resting K permeability, but are inactive, even at high concentrations, on the Na channel. It is suggested that these properties may parallel the wider biological actions of anaesthetics and convulsants including actions on the central nervous system.

The squid axon has proved to be a productive and stimulating model for the study of anaesthetic actions on excitability (Haydon *et al.* 1984). The central problem confronting research in anaesthesia is that the relevant experiments on the central nervous system (CNS) in man are difficult to perform and difficult to interpret. Accordingly, model systems have been sought which reproduce some of the properties of the human CNS in an accessible form. Early work on bulk lipid solutions was clearly too simplistic. Peripheral nerve work using the cephalopod giant axon has distinct advantages and benefits. General anaesthetics do block nerve conduction with a correlation between potency and hydrophobicity which mirrors their CNS actions. Concentrations required for nerve block appear to be some 3–10 times greater than the minimum alveolar concentration for anaesthesia, but there are actions in the squid nerve at concentrations similar to the minimum alveolar concentration. These low concentration actions include alterations in excitation threshold and may be of great significance (Haydon *et al.* 1984, 1988; Haydon and Simon 1988).

The excitable properties of the squid axon can be directly attributed to three electrical properties of the axon membrane, each of which has its origins in membrane protein ion-transport systems. These are: (1) the voltage-activated Na^+ conductance arising from the sodium channel; (2) the delayed rectifier K^+ conductance arising from the potassium channel; and (3) the resting or background K^+ conductance of the membrane which arises from a poorly characterized transport system. While some authorities believe that property (3) arises from a residual group of open K^+ channels, it is likely that (2) and (3) are separate entities (Chang 1986; Clay 1988; Haydon *et al.* 1988). Accordingly, the actions of anaesthetics (and convulsants) on the axon membrane can be described by considering effects on these three systems in turn.

Anaesthetic actions on the Na^+ channel

All general anaesthetics will inhibit and block nerve conduction; the mechanism of this action is an inhibition of the voltage-gated Na^+ channel. The precise effects on the Na^+ channel current vary with the physical properties of the anaesthetic concerned. The description of these distinct actions is aided by a model for Na^+ channel function and use of the Hodgkin–Huxley formalism has proved useful here. However, the actions of anaesthetics on activation and inactivation, on steady-state properties, and on time constants are not model-dependent and can be distinguished in other ways. In Hodgkin–Huxley terms the actions can be divided into effects on the steady-state voltage-dependence of activation and inactivation (m_∞ and h_∞, respectively), on the time constants of activation and inactivation (τ_m and τ_h, respectively), and the maximum total Na^+-channel conductance ($\bar{g}_{Na}$).

Steady-state parameters

Non-polar hydrophobic anaesthetic molecules (e.g. *n*-pentane) cause a hyperpolarizing shift in the voltage-dependence of h_∞. The origins of this cannot be precisely defined but it appears to be related to the ability of these substances to thicken the hydrocarbon leaflet of the lipid bilayer portion of the cell membrane (Haydon and Kimura 1981). This thickening can be detected by electrical measurements of membrane capacity in both artificial membranes and in the squid axon (Elliott *et al.* 1985*a*). The shifts of h_∞ can be quantitatively accounted for by the reduction in intramembrane electric field which arises from the thickness change (Haydon *et al.* 1984). More polar substances (e.g. the *n*-alkanols) do not thicken the bilayer and (in the squid) do not shift h_∞ (Haydon and Urban 1983). These ideas of the importance of thickness are further supported by the effects of short-chain phospholipids on the steady-state inactivation in squid (Hendry *et al.* 1985). These short-chain phospholipids cause thinning in the hydrocarbon leaflet of the axon membrane and cause a depolarizing shift in h_∞. Despite the success of these simple physico-chemical explanations in the cephalopod, it has been difficult to describe anaesthetic actions on mammalian nerves in such terms. For example, in cultured rat spinal neurones the *n*-alkanols do cause shifts in h_∞ (Elliott and Elliott 1991).

Anaesthetics may shift m_∞ in both depolarizing and hyperpolarizing directions. Polar amphipathic anaesthetics (including the clinically important fluoro compounds) shift m_∞ in the depolarizing direction and this effect accounts in a large part for their inhibition of excitability. Shifts of m_∞ are particularly effective in causing alterations in threshold behaviour. The depolarizing m_∞ shifts produced by polar molecules cannot be fully explained at the molecular level. It has been proposed that they may arise from asymmetric alterations in dipole potential (Haydon *et al.* 1984). Certainly, the surface dipole potentials present are very large and small fractional changes (for example, by reorientation of the glycerol ester groups of phospholipids) could have large electrical consequences. Monolayer studies do suggest that anaesthetics reduce dipole potentials but proof that such mechanisms operate in biological bilayers is lacking. The non-polar anaesthetics shift m_∞ in the hyperpolarizing direction. This action appears to be related to bilayer thickness in a similar manner to that of the h_∞ shift.

Time constants

In general all anaesthetics decrease the time constants (τ_m and τ_h) and speed up the gating of the Na^+ channel. Certain local anaesthetics that interact directly with the ion channel can reduce observed rates of gating in macroscopic current analyses, but these actions are exceptional (Elliott *et al.* 1985*b*; Elliott and Hendry 1987). The faster gating may contribute to the mechanism of nerve block. For example, reductions in τ_h will reduce peak inward currents through the Na^+ channel. The mechanisms of these time-constant effects are unclear. Anaesthetics do tend to increase the mobility of lipid chains in the bilayer and this action may be responsible for some of the observed increases in gating rate.

The maximum Na^+ channel conductance $\bar{g}_{Na}$

The parameter $\bar{g}_{Na}$ is problematical as actions on it are often irreversible so that anaesthetic effects can be difficult to distinguish from an overall decline in axon viability with time. The non-polar anaesthetics reduce $\bar{g}_{Na}$ in a manner that semiquantitatively can be accounted for by alterations in membrane thickness. The functional ion channel must link the two aqueous phases across the bilayer leaflet and this may be increasingly difficult as the bilayer is made thicker. As a result there may be desorption of ion channels or some other conformational change to an inactive state in the presence of these anaesthetics. This mechanism of action is able to provide a quantitative explanation for the effects of *n*-alkane anaesthetics on the model ion channel formed by dimers of gramicidin (Hendry *et al.* 1978).

Overall effects on the Na^+ current

The integrated result of these actions is to reduce the Na^+ current and, at sufficient concentrations of anaesthetic, to suppress nerve excitability altogether. The most important specific effect varies from molecule to molecule. For example, the clinical anaesthetics, halothane and enflurane, suppress Na^+ current largely through a depolarizing shift in m_∞ linked to a reduction in τ_h. Conversely, carbon tetrachloride suppresses Na^+ current essentially through reductions in $\bar{g}_{Na}$ and hyperpolarizing shifts in h_∞. Detailed knowledge of the separate mechanisms is sufficient to provide a fairly accurate description of the overall properties of each class of anaesthetic.

Effects of low concentrations of anaesthetic are of particular interest as they correspond to the clinical range of 1–2 times the minimum alveolar concentration. The fluorinated anaesthetics do not suppress nerve conduction at this concentration range but they do have significant effects on the threshold for Na^+ current activation (Haydon and Simon 1988; Haydon *et al.* 1988). This is of potential relevance to general anaesthesia as threshold changes may alter the firing rate of CNS neurones in the presence of a particular level of depolarizing current entering the cell body (Haydon *et al.* 1984).

Anaesthetic actions on the K^+ channel

For the most part anaesthetics inhibit K^+ conductance via the K^+ channel at a similar concentration range to that required for effects on Na^+ current. Most anaesthetics tend to exhibit a slight selectivity for Na^+ channels (Haydon *et al.* 1984; Haydon and Urban 1986). However, certain fluorinated anaesthetics are more active as inhibitors of the K^+ channel (Urban and Haydon 1987). Effects on the K^+ channel are similar to the Na^+ channel actions. For example, the non-polar anaesthetics reduce the maximum K^+ channel conductance $\bar{g}_K$ and shift the steady state parameter for K^+ channel activation (n_∞) in the hyperpolarizing direction (Haydon *et al.* 1984; Haydon and Urban 1986). Polar anaesthetics tend to shift n_∞ in the depolarizing direction. In the absence of anaesthetics the K^+ channel current shows little or no inactivation.

Many anaesthetics induce a form of inactivation in the K^+ channel current; molecules with this property include the fluorinated ethers and *n*-alkanols (Haydon and Urban 1986; Urban and Haydon 1987). This inactivation in the measured K^+ currents is not due to altered accumulation of K^+ in the Frankenhaeuser–Hodgkin space and appears to be a true effect on the molecular properties of the K^+ channel. The origins of the effect are unknown.

As discussed below, certain anaesthetics can produce a paradoxical hyperexcitability in the squid nerve. This action is noted particularly in the case of the fluorinated ethers, most notably methoxyflurane, enflurane, and isoflurane. Hyperexcitability can be produced by inhibition of K^+ channels. However, it appears that these effects occur at anaesthetic concentrations below those that are associated with significant K^+-channel modulation (Haydon and Simon 1988; Haydon *et al.* 1988; Hendry and Haydon 1991). This observation that the K^+ channel itself cannot account for these pro-excitable actions prompted an examination of the effects of anaesthetics on the resting K^+ permeability systems of the nerve membrane.

Convulsants, anaesthetics, and resting K^+ conductance

The anaesthetic–convulsant paradox

There is an intimate relationship between anaesthetic and convulsant actions. Induction of anaesthesia by volatile agents is associated with transient hyperexcitability and occasional convulsions in animal studies and in humans. This appears to be associated with the transient presence of low concentrations of the anaesthetic before the higher concentrations required for anaesthesia are reached. Another anaesthetic–convulsant connection can be illustrated by examining the structural formulae of a number of anaesthetics and convulsants. It is clear that there are very closely related molecules, particularly in the fluorinated ether group, some of which are anaesthetics and some of which are convulsants. Selected examples are shown in Table 12.1.

Work on the squid axon has shed some light on these relationships. Low concentrations of fluorinated anaesthetics produce axonal hyperexcitability and may cause spontaneous generation of action potentials (Haydon and

Table 12.1 Chemical formulae of a number of anaesthetic and convulsant substances

Anaesthetics	Convulsants
CCl_3H (chloroform)	CCl_3F (trichlorofluoromethane)
$CF_3{\cdot}CBrClH$ (halothane)	$CF_3{\cdot}CH_2{\cdot}O{\cdot}CH_2{\cdot}CF_3$ (fluorothyl, indoklon)
$CF_2H{\cdot}O{\cdot}CF_2{\cdot}CClFH$ (enflurane)	$CCl_2F{\cdot}O{\cdot}CF_2{\cdot}CClF_2$ ('466')
$CF_2H{\cdot}O{\cdot}CClH{\cdot}CF_3$ (isoflurane)	$CCl_3{\cdot}O{\cdot}CF_2{\cdot}CCl_2F$ ('314')
$CH_3{\cdot}O{\cdot}CF_2{\cdot}CCl_2H$ (methoxyflurane)	$CF_3{\cdot}CH_2{\cdot}O{\cdot}CF_2CFH_2$

Simon 1988). These properties are very close to the phenomena seen in the CNS. Certain experiments on the membrane actions of fluorinated anaesthetics and convulsants are of particular interest and will be described here. Chloroform (CCl_3H), halothane ($CF_3CBrClH$), enflurane (CF_2HOCF_2CClFH), and isoflurane ($CF_2HOCClHCF_3$) are anaesthetics. Conversely, the similar substances trichlorofluoromethane (CCl_3F) and flurothyl ($CF_3CH_2OCH_2CF_3$) are convulsants. Effects on the threshold for excitation of the squid axon can be tested by measuring the threshold current (I_C) required to elicit an action potential when applied intra-axonally for 100 μs. An example of the effects of enflurane on this parameter is shown in Fig. 12.1. A concentration of 2.0 mM enflurane causes a transient reduction in I_C indicating that the axon was rendered hyperexcitable. Following this I_C increased until excitability was completely lost. The removal of enflurane and replacement of a bathing solution of pure artificial sea water (ASW) allowed return of normal action potential generation. In contrast, Fig. 12.2 shows the effect of a lower concentration of enflurane (0.8 mM) on I_C. Here I_C is reversibly reduced to a new steady-state value in the presence of the enflurane. These results show that the effects of enflurane on excitability are concentration-dependent and that a low concentration can cause sustained hyperexcitability.

The steady-state inhibition of excitability caused by 2.0 mM enflurane can be explained by its actions on the Na^+ channel. The hyperexcitability caused by 0.8 mM enflurane can be related to actions on the K^+ permeability of the

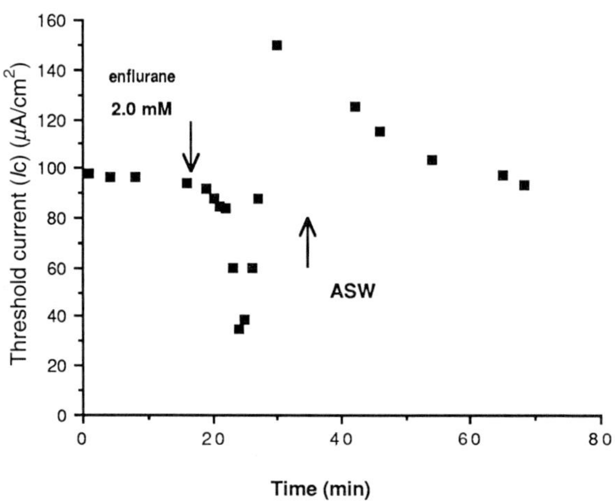

Fig. 12.1. The effects of application of 2.0 mM enflurane in artificial sea water (ASW) on the threshold 100-μs current I_C required to elicit an action potential in a squid axon. Arrows indicate the time of enflurane application and return to pure ASW. No action potentials could be elicited for the 2 min prior to reversal. The temperature was 6°C. The ASW was of composition: 430 mM NaCl; 10 mM KCl; 50 mM $MgCl_2$; 10 mM $CaCl_2$; 10 mM Trizma buffer; pH 7.6.

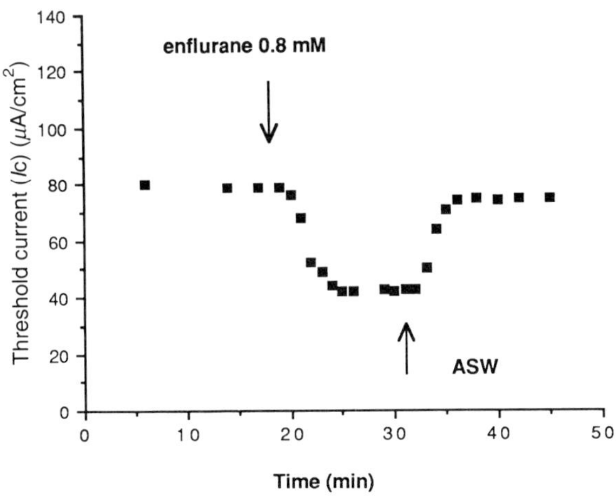

Fig. 12.2. The effects of application of 0.8 mM enflurane on the threshold current (I_c) for action potential generation in a squid axon. Arrows indicate the timing of application of enflurane and reversal to pure artificial sea water. Solutions and conditions as for Fig. 12.1.

resting membrane. This induction of excitability by low anaesthetic concentrations can be seen with isoflurane, methoxyflurane, and halothane as well as with enflurane (Haydon *et al.* 1988; Hendry and Haydon 1991). The convulsants trichlorofluoromethane and flurothyl at 0.5 mM also cause a sustained reduction in I_C (Hendry and Haydon 1991). These actions on I_C are associated with reversible resting potential depolarizations of the order of 2–5 mV (Haydon *et al.* 1988; Haydon and Simon 1988; Hendry and Haydon 1991). Membrane conductance measurements have been made with a.c. bridge techniques in the presence of tetrodotoxin and 3,4-diaminopyridine to examine these effects in more detail (Haydon *et al.* 1988; Hendry and Haydon 1991). The hyperexcitable actions are clearly associated with parallel reductions in membrane conductance. These effects are particularly clear if the measurements are made in high-K ASW (external [K] 440 mM). All these results point to the conclusion that the increased axonal excitability is due to a decrease in the permeability of the resting membrane to K^+ ions. The actions of four fluorinated substances on the threshold current (I_C) and on the membrane conductance in high-K ASW (g_K) are shown in Table 12.2.

The origins of the resting conductance in squid nerve

The resting potential of the squid axon *in vivo* is probably in the region of −70 mV. This arises through a selective permeability of the resting membrane to K^+. Ion substitution experiments suggest that the contribution of anion conductance systems (such as Cl) is dominated by the K^+ conductance. However,

Table 12.2 Summary of the effects of four fluorinated substances on some electrical properties of the resting squid nerve membrane (adapted from Hendry and Haydon 1991)

Concentration (mM)	ΔI_C (%)†	Δg_K (mS/cm)^2‡	$\Delta g'_K$ (%)§	Number of experiments
Enflurane				
0.8	−41	−1.40	−79	3
Isoflurane				
0.4	−28	0.78	−40	2
Flurothyl				
0.33	−15	–	–	1
0.5	−29	−1.25	−76	4
1.0	−26	−1.45	–	1
CCl_3F				
0.25	−11	−1.13	–	1
0.51	−25	−1.25	−64	3

† ΔI_C is the mean percentage change in the 100-μs stimulus current (I_C) required to elicit an action potential.
‡ Δg_K is the mean change in the electrical conductance of the resting membrane in high-K ASW measured with the a.c. bridge technique in the presence of tetrodotoxin and 3,4-diaminopyridine.
§ $\Delta g'_K$ is the mean percentage change in the conductance of the resting membrane in high-K ASW (with tetrodotoxin and 3,4-diaminopyridine present) measured using voltage-clamp pulses with a −90 mV prepulse.

the exact basis of the resting K^+ permeability is controversial. Clay (1988) has argued that a small fraction of resting K^+ channels are involved. Haydon and colleagues (1988; Hendry and Haydon 1991) have provided evidence for a separate, voltage-independent K^+ permeability system in the resting axon membrane. One difficulty arises from the behaviour of the available pharmacological blockers of the K^+ channel. These substances (e.g. tetraethylammonium ions, 3,4-diaminopyridine) are not highly selective and also exhibit a complex voltage-dependent block.

The evidence for a separate voltage-independent K^+ permeability has been obtained by combining pharmacological and electrophysiological methods for the inhibition of the K^+ channel. Haydon *et al.* (1988) examined axons in the presence of 1.0 mM 3,4-diaminopyridine to establish an initial substantial block of K^+ channels. Electrical measurements were made in the presence of 440 mM extracellular K (high-K ASW) in order to focus on the K^+ permeability of the membrane. In these conditions the membrane resting potential is close to 0 mV and the steady-state value of n_∞ is close to 1.0, indicating that K^+ channel gates will be open. Final removal of the residual K^+ channel current was accomplished in a voltage-clamp protocol. The axon was clamped at its resting potential and then a 50-ms duration hyperpolarizing pulse of −90 mV was applied to close the K^+ channels. Following this conditioning a test pulse was applied to membrane potentials in the range −40 to +60 mV. The membrane current elicited by these test pulses was measured. The current–voltage relationship during the first millisecond of this pulse was then used to calculate the resting conductance of the membrane without K^+ channel contributions (Haydon *et al.* 1988; Hendry and Haydon 1991). An example of the current–voltage relationship measured in these conditions is shown in Fig. 12.3. In addition, Fig. 12.3 shows the effect of 0.5 mM flurothyl on the current–voltage relationship. There is a clear reduction in membrane conductance caused by the convulsant. It is now clear that the resting voltage-independent K^+ permeability of the axon is peculiarly sensitive to low concentrations of these fluorinated substances (Hendry and Haydon 1991). The slope of the current–voltage relationship measured in this manner can be described as g'_K, the resting voltage-independent K^+ conductance. The effects of four fluorinated substances on g'_K are illustrated in Table 12.1. These substantial reductions in g'_K are far greater than any effects on the voltage-gated K^+ channel reported for these substances at equivalent concentrations.

Effects on the K^+ channel at membrane potentials close to the physiological

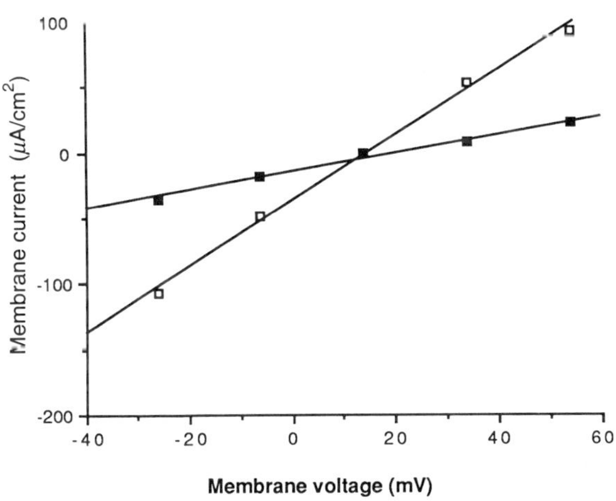

Fig. 12.3. The effect of 0.5 mM flurothyl on the current–voltage (*I–V*) relationship in high-K ASW containing tetrodotoxin (300 nM) and 3,4-diaminopyridine (1.0 mM). The axon was voltage-clamped at its resting potential of 14 mV. Control data were obtained following a prepulse of −90 mV for 50 ms to close residual Hodgkin–Huxley K^+ channels (open squares). Results were then obtained in the presence of flurothyl with an identical prepulse (closed squares). The slopes of the linear regression lines are an estimate of the voltage-independent membrane conductance (g'_K) and their values (in mS/cm^2) are: control, 2.50; flurothyl, 0.72. The effect of flurothyl appears to be to strongly inhibit g'_K. Eighty per cent reversal occurred on removal of flurothyl.

resting potential are difficult to investigate. One of the probable actions of the fluorinated ethers on the K^+ channel is a depolarizing shift in the voltage dependence of n_∞. This shift could have significant actions on the resting membrane, particularly if the K^+ channel does contribute significantly to the resting membrane K^+ conductance (Clay 1988). However, this explanation for the pro-excitant actions of the fluorinated ethers appears unlikely as the measurements of g_K and of g'_K described above were obtained at membrane voltages where n_∞ would be very close to unity.

The relationship between anaesthetic and convulsant activities

The results obtained on the squid giant axon and described above are of interest with respect to the underlying differences between anaesthetic and convulsant actions. All the fluorinated substances tested appear to be capable of producing hyperexcitability in the axon at certain ranges of concentration. For the anaesthetic compounds these pro-excitable actions occur at concentrations below those associated with nerve block. These properties are similar to those exhibited by these substances in the mammalian CNS. In the axon model it is possible to ascribe these pro-excitant actions with some confidence to an inhibition of a membrane K^+ permeability system other than the voltage-gated K^+ channel. This immediately suggests that resting membrane K^+ permeability changes in the CNS could be responsible for the chemical convulsant properties of these fluorinated molecules. Experiments on mammalian neurones should allow this idea to be tested.

Effects on the resting K^+ permeability of the axon appear to be common to all the fluorinated substances and cannot therefore account for the gross biological differences between the convulsants and the anaesthetics. Although the anaesthetic ethers may cause convulsions on induction, their equilibrium action in the doses used clinically is to induce anaesthesia. Flurothyl appears to be quite different and behaves as a pure convulsant in mammals (Orth and Mackey 1963). An explanation for such differences may be found in a comparison of the relative potencies for inhibition of g'_K and for inhibition of the Na^+ channel current (I_{Na}). This is illustrated in Table 12.3 where the mean concentrations for 50 per cent inhibition of g'_K and I_{Na} are displayed. The characteristic feature of the convulsant molecules is that they are extremely inactive as inhibitors of I_{Na}. Accordingly, the ratios of potencies for the two actions ($ED50_{I_{Na}}/ED50_{g'_K}$), which are shown in the right-hand column of Table 12.3, show values of less than 10 for the anaesthetics and $\geqslant 30$ for the convulsants. It appears unlikely that inhibition of the Na^+ channel is the mode by which anaesthetics affect the CNS. Nevertheless, these data do suggest that differential actions on ion-conductance systems might account for the varied biological actions of these structurally similar molecules.

Table 12.3 Inhibition of Na^+ current and K^+ conductance by fluorinated substances † (data from Urban and Haydon 1987 and Hendry and Haydon 1991)

Substance	$ED50_{I_{Na}}$ (mM)	$ED50_{g'_K}$ (mM)	$ED50_{I_{Na}}/ED50_{g'_K}$
Anaesthetics			
Methoxyflurane	1.8	‡	–
Halothane	2.7	‡	–
Isoflurane	3.7	0.5	7.4
Enflurane	3.9	0.5	7.8
Convulsants			
Flurothyl	$\geqslant$11.0	0.3	$\geqslant$37
CCl_3	$\geqslant$5	0.4	$\geqslant$13

† A comparison of the concentrations of fluorinated substances required for 50% inhibition of the I_{Na} current ($ED50_{I_{Na}}$) and for 50% inhibition of g'_K ($ED50_{g'_K}$). The ED50 ratios are shown in the right-hand column.

‡ The values of $ED50_{g'_K}$ have not been directly measured but examination of the effects of these substances on I_C and the resting potential suggests that their $ED50_{g'_K}$ values are in the range 0.5–2.0 mM; these estimates would correspond to ED50 ratios of <10.

References

Chang, D. C. (1986). Is the K permeability of the resting membrane controlled by the excitable K channel? *Biophysical Journal*, **50**, 1095–100.

Clay, J. R. (1988). On the relationship between resting potential and the delayed rectifier in squid axons. *Biophysical Journal*, **54**, 969–70.

Elliott, J. R. and Elliott, A. A. (1991). Actions of alcohols on sodium channels in dorsal root ganglia. *Annals of the New York Academy of Sciences*, **625**, 344–54.

Elliott, J. R. and Hendry, B. M. (1987). The effects of internal and external application of disopyramide on the ionic currents of the squid giant axon. *British Journal of Pharmacology*, **92**, 77–86.

Elliott, J. R., Haydon, D. A., Hendry, B. M., and Needham, D. (1985*a*). The inactivation of the sodium current in squid giant axons by hydrocarbons. *Biophysical Journal*, **48**, 617–22.

Elliott, J. R., Haydon, D. A., and Hendry, B. M. (1985*b*). Dual actions of internal n-alkyltrimethylammonium ions on the sodium current of the squid axon. *Journal of Physiology*, **361**, 47–64.

Haydon, D. A. and Kimura, J. E. (1981). Some effects of n-pentane on the sodium and potassium currents of the squid giant axon. *Journal of Physiology*, **312**, 57–70.

Haydon, D. A. and Simon, A. J. B. (1988). Excitation of the squid giant axon by general anaesthetics. *Journal of Physiology*, **402**, 375–89.

Haydon, D. A. and Urban, B. W. (1983). The action of alcohols and other non-ionic surface active substances on the sodium current of the squid giant axon. *Journal of Physiology*, **341**, 411–27.

Haydon, D. A. and Urban, B. W. (1986). The action of hydrophobic, polar and some inhalation general anaesthetic substances on the potassium currents of the squid giant axon. *Journal of Physiology*, **373**, 311–27.

Haydon, D. A., Elliott, J. R., and Hendry, B. M. (1984). Effects of anesthetics on the squid giant axon. *Current Topics in Membranes and Transport*, **22**, 445–82.
Haydon, D. A., Requena, J., and Simon, A. J. B. (1988). The potassium conductance of the resting squid axon and its blockage by clinical concentrations of general anaesthetics. *Journal of Physiology*, **402**, 363–74.
Hendry, B. M. and Haydon, D. A. (1991). Alterations in resting K conductance in nerve by convulsants and anesthetics. *Annals of the New York Academy of Sciences*, **625**, 355–64.
Hendry, B. M., Urban, B. W., and Haydon, D. A. (1978). The blockage of electrical conductance in a pore-containing membrane by the n-alkanes. *Biochimica et Biophysica Acta*, **513**, 106–16.
Hendry, B. M., Elliott, J. R., and Haydon, D. A. (1985). Further evidence that membrane thickness influences voltage-gated sodium channels. *Biophysical Journal*, **47**, 841–5.
Orth, O. S. and Mackey, J. P. (1963). Laboratory evaluation of Indoklon. *Journal of Neuropsychiatry*, **4**, 164–73.
Urban, B. W. and Haydon, D. A. (1987). The actions of halogenated ethers on the ionic currents of the squid giant axon. *Proceedings of the Royal Society of London, Series B*, 231, 13–26.

13

Control of the spatial distribution of sodium channels in the squid giant axon and its cell bodies

W. F. GILLY, M. T. LUCERO, M. PERRI, and J. ROSENTHAL

Summary

This chapter summarizes our progress in developing a model system for studying the control of neuronal Na channel distribution based on the squid giant axon and its cell bodies located in the giant fibre lobe (GFL) of the stellate ganglion. We have employed patch clamp methods to test the functional integrity of Na channels in GFL neurones maintained in primary culture and to map the spatial distribution in cell bodies and axons. GFL neurones *in vitro* establish and maintain a strongly polarized Na channel distribution similar to that displayed by the system *in vivo*. We have identified several manipulations that disrupt this cellular polarity, including a novel effect of the glycosylation inhibitor tunicamycin. This drug appears to selectively inhibit high-level expression of Na channels in axonal membrane.

Specificity of neuronal function at the cellular level is largely dictated by the precise spatial distribution of membrane receptors and channels. For example, the high density of Na^+ channels at the axon hillock and nodes of Ranvier is essential for impulse conduction in myelinated axons, and densely packed acetylcholine receptors at the vertebrate neuromuscular junction enable powerful synaptic excitation. In general, the functional properties of many channels and receptors have been well studied, and in some cases their spatial distributions have been carefully mapped. Although this information is vital to understanding nerve cell function, we also need to learn much more about the cell biological dynamics that control both the properties and the spatial distributions of these important membrane proteins.

All neurones are functionally and morphologically polarized, and the number of cellular control elements is large (Simons and Wandinger-Ness 1990). First, posttranslational processing events in the Golgi network (Griffiths and Simons 1986) mediate sorting and packaging of membrane (Wandinger-Ness *et al.* 1990) and secretory (Sossin *et al.* 1990) proteins into specific vesicles destined for delivery to certain domains, including axons and dendrites of neurones (Dotti and Simons 1990). Second, intracellular transport of these proteins to distant target sites prior to insertion occurs via 'axoplasmic' transport of vesicles along polarized microtubule arrays (Pfeffer and Rothman 1987; Baas *et al.* 1988; Breitfeld *et al.* 1990), but lateral diffusion in the cell membrane also occurs. Third, anchoring of membrane proteins to cytoskeletal elements such as ankyrin plays a role in restricting lateral mobility and achieving high local concentrations of channels, receptors, and pumps (Baines 1990). Finally, differentially regulated turn-over of membrane proteins may occur at various peripheral sites.

With such an array of complex processes involved, it is not surprising that we presently do not understand the detailed mechanisms by which the spatial distribution of any endogenous membrane protein is controlled in a neurone. Progress in this area is imperative if we hope to understand the molecular and cellular bases of long-term modification of neuronal function. With these ideas in mind, we turn to the squid giant axon.

The giant fibre lobe neurone–giant axon system

In lolliginid squid, the giant motor axons are formed by fusion of the initial axonal segments of several hundred giant fibre lobe (GFL) neurones. Each stellate ganglion gives rise to eight or more giant axons, and a GFL thus contains several thousand neurones of a single type—and no others. This constitutes a major experimental advantage when working with primary cultures. After exiting the GFL, each fused axon enters the neuropil of the stellate ganglion proper where finger-like dendrites form the postsynaptic sites of the giant synapse. The giant axon then runs out to innervate the mantle musculature.

This multicell system thus comprises a single functional motor neurone. The extremely large size of the giant axon and postsynaptic element not only facilitates voltage-clamp experiments in either domain, but also permits selective isolation of axoplasmic components of the protein trafficking machinery by simple extrusion procedures (Gilbert *et al.* 1990). Recently, voltage-dependent Na^+ and K^+ channels, presumably destined for incorporation into the axonal membrane, were found in vesicles isolated from extruded axoplasm (Wonderlin and French 1991).

An equally important feature of the GFL neurone–giant axon system is its well defined functional polarization. Although the giant axons have a high density of Na^+ channels and propagate action potentials, these impulses do not invade the inexcitable cell bodies (Miledi 1967). This result suggests that the unfused axonal segments as well as the somata are greatly lacking in Na^+ channels. In support of this idea, voltage-clamp experiments using two intracellular microelectrodes and intact ganglia of *Loligo opalescens* revealed outward K^+ currents but no sign of inward Na^+ current (I_{Na}) or of 'escaping'

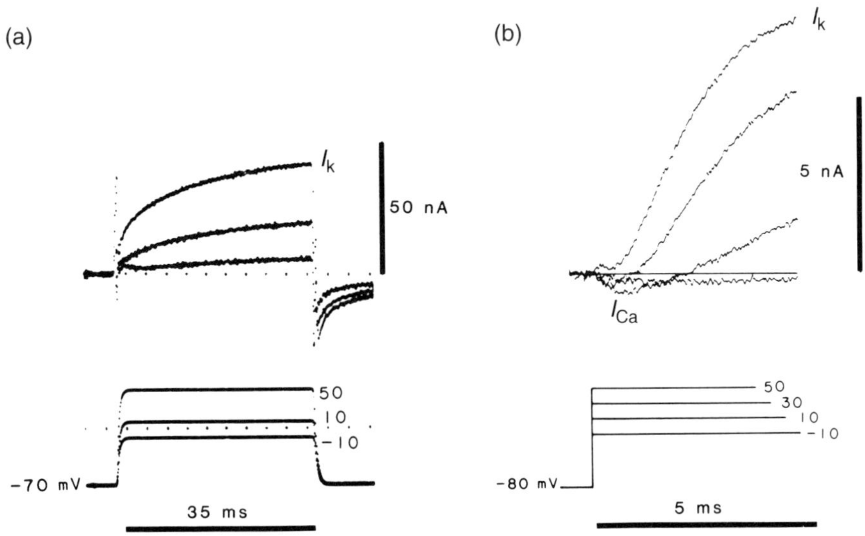

Fig. 13.1. Ionic currents in squid GFL neurones studied (a) *in situ* and (b) *in vitro*. (a) Voltage-clamp records were obtained with a conventional two-microelectrode method at the voltages illustrated. Only outward K^+ currents (I_K) are seen. The manually desheathed stellate ganglion, with GFL attached and all giant axons ligated, was bathed in sea water at 10°C. (b) Whole-cell patch-clamp recordings were made shortly after cell isolation (day 0 *in vitro*) at the indicated voltages. Large outward I_K and small inward Ca^{2+} current (I_{Ca}) are present; I_{Na} is not detectable. External solution (10°C) was ASW containing 470 mM NaCl, 10 mM KCl, 10 $CaCl_2$, and 50 $MgCl_2$ buffered to pH 7.8 with 10 mM HEPES. The internal (pipette) solution contained 100 mM Na glutamate, 100 mM K glutamate, 200 mM tetramethylammonium (TMA) glutamate, 50 mM NaF, 50 mM NaCl, and 10 mM Na_2-EGTA buffered to pH 7.8 with 10 mM HEPES.

axonal spikes [Fig. 13.1(a)]. Higher-resolution recordings made with whole-cell patch clamp using acutely dissociated GFL cell bodies also failed to detect inward I_{Na}, but these studies did reveal inward Ca^{2+} currents [Fig. 13.1(b); see also below]. Thus, the sharp regional difference in excitability of GFL neurones reflects a highly polarized distribution of Na^+ channels.

Appearance of functional Na^+ channels in cultured GFL neurones

If GFL neurones are dissociated by a combination of enzymatic (non-specific protease) and mechanical (vortexing) treatments and maintained in primary culture for several days, functional Na^+ channels can be detected in whole-cell patch clamp experiments. This was shown first by Llano and Bookman (1986) in *Loligo pealei*. These authors described a tetrodotoxin (TTX)-

sensitive I_{Na} that was similar to that recorded in giant axons but of much smaller amplitude—the maximum somatic current density was less than 1 per cent of that in the axon.

In order to reduce mechanical trauma during dissociation of GFL neurones, we modified the procedures used by Llano and Bookman (Brismar and Gilly 1987). After enzymatic treatment (0.5–1 per cent Sigma Type XIV protease in sea water for 1 h at 18 °C) we simply cultured the entire lobe on a small piece of Nitex screen in Leibovitz's L-15 supplemented with 2 mM L-glutamine and the following salts: 9.0 mM $CaCl_2$, 4.6 mM KCl, 49.5 mM $MgCl_2$, and 263 mM NaCl. Enzyme-treated lobes stick avidly to the Nitex rafts, and cells are harvested for study by dipping the raft and adherent lobe into 10–20 μl of medium in the recording chamber. Cells sloughed off by this procedure are allowed a few minutes to settle before the chamber is filled with external solution.

Figure 13.2 shows whole-cell recordings from GFL neurones of *Loligo*

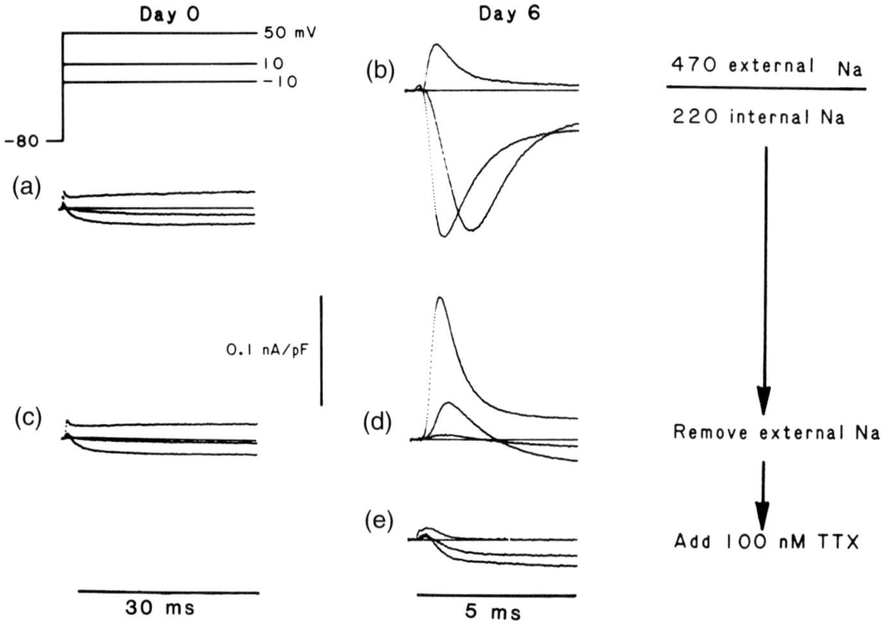

Fig. 13.2. Identification of I_{Na} in cultured GFL neurones. Illustrated voltage steps apply to all panels, but the time scales for days 0 (a), (c) and 6 (b), (d), (e) differ. The external solution was artificial sea water (ASW); the internal solution was described in the legend to Fig. 13.1, but K glutamate was replaced by 100 mM TMA glutamate and 25 mM tetraethylammonium (TEA) chloride. Large inward and outward transient currents are present (a) on day 6 but not (b) on day 0. All Na concentrations are mM. (d) Replacement of external Na with Tris abolishes inward current and augments outward current on day 6; (c) this procedure has little effect on day 0. (e) 100 nM TTX blocks transient currents on day 6 but has no effect on day 0 (not illustrated).

opalescens on the day of GFL excision and enzyme treatment (day 0) and after 6 days in culture at 15 °C. Except for the age of the cells, the recording conditions were identical (Brismar and Gilly 1987; Gilly and Brismar 1989). K^+ currents are absent, because the internal solution was K^+-free and contained 25 mM tetraethylammonium (TEA). With high Na^+ present internally and externally, large inward and outward transient currents are present on day 6 [Fig. 13.2(b)], but not on day 0 [Fig. 13.2(a)]. Removing external Na^+ abolishes the transient inward current on day 6 and augments outward current [Fig. 13.2(d)] but has no effect on day 0 [Fig. 13.2(c)]. Similarly, day 6 transient currents are blocked by TTX [Fig. 13.2(e)], whereas inward currents on day 0 are not (not illustrated).

These results firmly identify the large transient currents that appear *in vitro* as I_{Na} with conventional properties. TTX-insensitive inward currents that persist in Na^+-free external medium on days 0 and 6 flow through Ca^{2+} channels that are blocked by mM levels of Zn or Cd in the external solution (Brismar and Gilly 1987; Gilly and Brismar 1989). The amount of I_{Ca} does not change much over time in culture; this is also true for I_K.

Characterizing the properties of I_{Na} and its development in culture requires measuring this current in isolation of all others when using whole-cell recording techniques. Figure 13.3 shows the two ways in which we do this. TTX-sensitive I_{Na} [Fig. 13.3(a)] is obtained by subtracting control records from corresponding ones obtained in the presence of at least 100 nM TTX. Prepulse-sensitive I_{Na} [Fig. 13.3(b)] is the result of subtracting records with and without a brief depolarizing prepulse. The two procedures yield comparable I_{Na} records.

Figure 13.4 illustrates the time course over which I_{Na} (defined by TTX- and/or prepulse-sensitivity) appears in cultured GFL neurones. For these plots, measurement of I_{Na} has been converted to peak Na^+ conductance density (g_{Na}) achieved at positive voltages (see legend to Fig. 13.4). g_{Na} is proportional to the number of functional Na channels per unit area of membrane.

Functional Na^+ channels start appearing shortly after placing GFL neurones from *Loligo opalescens* in culture [Fig. 13.4(a)]. g_{Na} is small but routinely detectable on days 1 and 2 and becomes much larger over subsequent time in culture. Data represented by closed symbols were obtained with cells from lobes that had been proteased on day 0 in the usual manner. Open symbols are from cells cultured in lobes that were excised on day 0 but not proteased until the day of recording (day 4). Protease treatment cannot be responsible for the absence of g_{Na} on day 0 or for its development *in vitro*.

Figure 13.4(b) shows analogous results from another squid, *Lolliguncula brevis*, from the Gulf of Mexico. In this case, the time course of g_{Na} development is distinctly slower than that in *Loligo*. This may be a result of our culturing at 15 °C, a temperature close to the local environment of *Loligo*, but well below that for *Lolliguncula*. We have not studied the temperature dependence of g_{Na} development in either species.

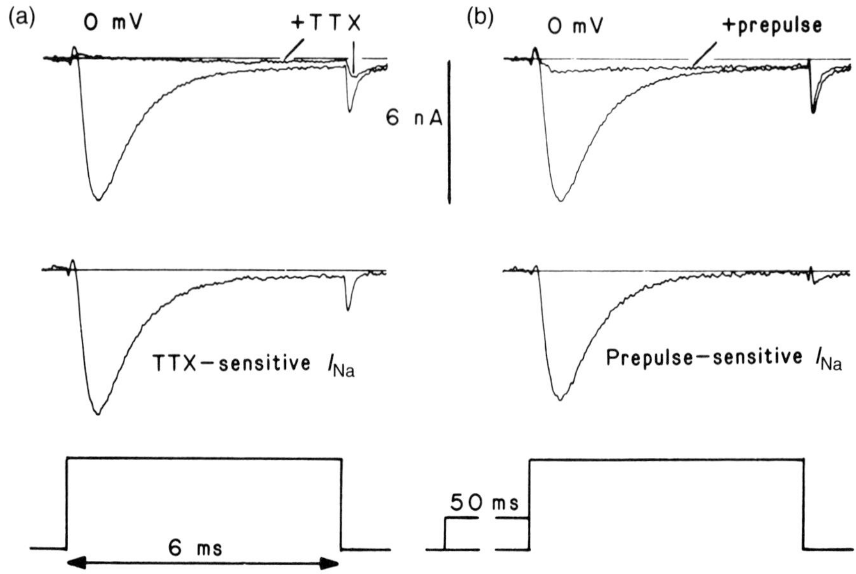

Fig. 13.3. Two methods for separating I_{Na} from other currents in an 8 day cultured GFL neurone. Recording conditions are described in the legend to Fig. 13.2. (a) Upper pair of traces shows currents recorded at 0 mV before and after applying 500 nM TTX. The TTX-sensitive I_{Na} generated by the difference of these records is illustrated below. (b) Upper pair of traces shows currents recorded at 0 mV with and without a depolarizing prepulse (50 ms to −38 mV) that inactivates I_{Na}. Lower panel illustrates prepulse-sensitive I_{Na} generated by subtraction.

Inexcitable somata with a very small g_{Na} *in vivo* or on day 0 *in vitro*, and development of much larger g_{Na} over time in culture are characteristics of stellate ganglion neurones of other cephalopods as well, and these properties appear to be widespread. Results similar to those in Fig. 13.4 have been obtained in *Sepioteuthis lessoniana*, another warm-water squid, *Sepia officinalis*, the common cuttlefish, and even *Octopus rubescens* (unpublished data). Neither *Sepioteuthis* nor *Sepia* have a well-defined GFL, so the identity of cell type is not guaranteed. *Octopus*, of course, has no giant axon system at all.

Properties of Na^+ channels in cultured GFL neurones

Na^+ channels that appear in cultured GFL neurones are present in the cell body, a spatially inappropriate region for Na^+ channel expression. Are these Na^+ channels the same as those normally found in the axon? At present we must rely on voltage-clamp data to answer this question. Dependence of the

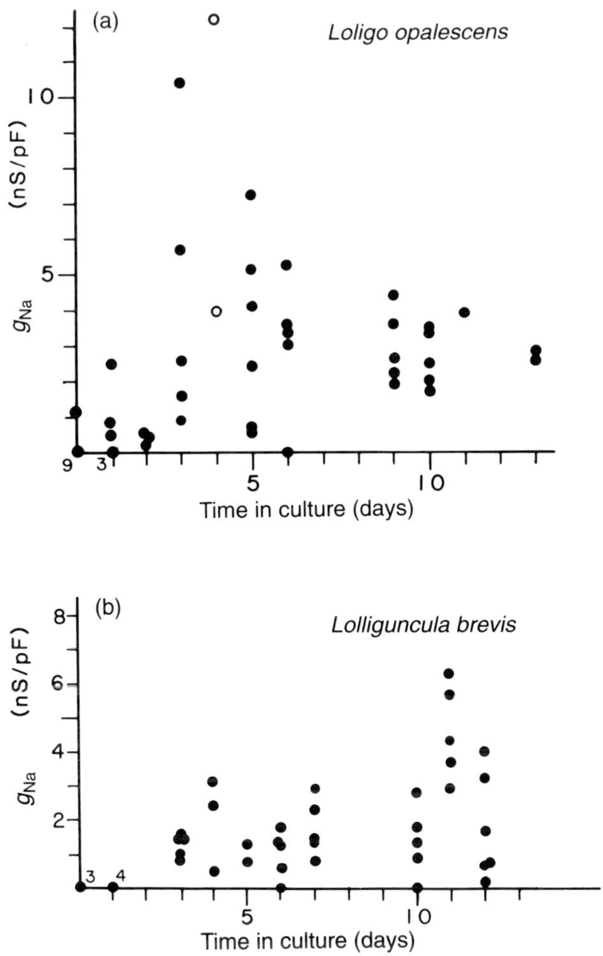

Fig. 13.4. Time course of g_{Na} development in somata of cultured squid GFL neurones. TTX- or prepulse-sensitive I_{Na} was measured over the voltage range of -50 to $+60$ mV. Na^+ conductance was estimated as $I_{Na}/(V-V_{Na})$, where V is the pipette voltage and V_{Na} is the reversal potential. Specific g_{Na} equals Na^+ conductance divided by input capacitance. Each symbol denotes a determination from one cell, except for the zero-value points on days 0 and 1, for which the number of cells studied is given. (a) Results from *Loligo opalescens*, the local species in Monterey Bay. (b) Results from *Lolliguncula brevis*, a species from the Gulf of Mexico.

functional properties of neuronal Na^+ channels on the site of expression is not well documented, and this question is of general interest.

Three types of Na^+ channels, all TTX-sensitive, have been described for giant axons of *Loligo pealei*. The predominant channel is the rapidly inactivating one that has been extensively studied and gives rise to practically all

I_{Na} at voltages below zero. Only this channel has been studied at the single-channel level (Vandenberg and Bezanilla 1991). At positive voltages, a slowly activating I_{Na} appears that does not inactivate. This current has been ascribed both to a second type of 'sleepy' Na^+ channel (Matteson and Armstrong 1982) and to normal inactivating channels that have entered a second, flickery open state (Bezanilla and Correa 1991). The third type of Na^+ channel accounts for only a few per cent of total g_{Na} and displays unusual voltage dependence and kinetics that cause preferential activation at voltages near action potential threshold (Gilly and Armstrong 1984).

How many of these channel subtypes are functional in cultured GFL neurones, and how closely do they correspond to their counterparts in the giant axon? Because we can carry out the same measurements on the cell bodies with whole-cell patch clamp and on internally perfused axons with axial wire methods, careful comparison of the functional properties of Na^+ channels expressed in the inappropriate and appropriate portions of this giant neurone is possible (Gilly and Brismar 1989).

As in the giant axon, all I_{Na} in GFL neurones is TTX- and saxitoxin (STX)-sensitive with dissociation constant (K_d) values of less than 10 nM. I_{Na} with properties associated with axonal threshold channels cannot be detected in GFL somata. Most importantly, the bulk of I_{Na} (85 per cent on average) in GFL cells is basically identical to normal axonal I_{Na} with regard to the voltage dependence and kinetics of both activation and inactivation. Thus, the isolated GFL cell body inappropriately expresses the same inactivating Na^+ channels that are normally found only in the axon.

Non-inactivating I_{Na} also occurs in cultured GFL cells. We routinely observe a prepulse-resistant component of I_{Na} (5–50 per cent of total peak I_{Na}) that cannot be differentiated from the normal, inactivating current by criteria of TTX- or STX-sensitivity, activation or inactivation voltage dependence, or activation kinetics (Fig. 13.5) but can be distinguished by its unusual sensitivity to block by Cd ions (Gilly and Brismar 1989). Procedures that 'awaken' sleepy axonal channels (Matteson and Armstrong 1982) do not potentiate this non-inactivating I_{Na} (unpublished data). Based on these features, the prepulse-resistant I_{Na} in GFL cell bodies appears to be distinct from axonal non-inactivating I_{Na}, and in many ways resembles I_{Na} recorded from the giant axon after pronase treatment to remove inactivation. What sort of channels give rise to this current in cultured GFL somata, and what their relationship is to the ordinary channels, are intriguing problems.

Controlling development of g_{Na} *in vitro*

Although the appearance of functional Na^+ channels in cultured GFL neurones is a reliable phenomenon, it is influenced by many factors. We have identified several ways of preventing inappropriate expression of Na^+ channels that

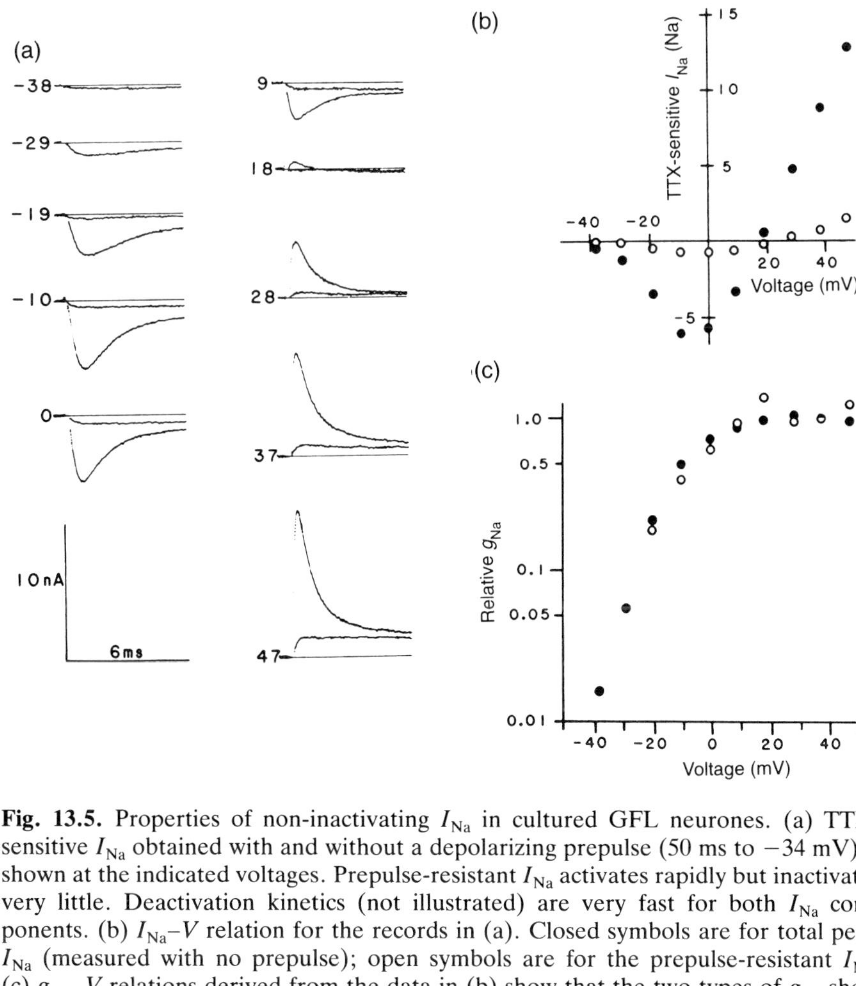

Fig. 13.5. Properties of non-inactivating I_{Na} in cultured GFL neurones. (a) TTX-sensitive I_{Na} obtained with and without a depolarizing prepulse (50 ms to −34 mV) is shown at the indicated voltages. Prepulse-resistant I_{Na} activates rapidly but inactivates very little. Deactivation kinetics (not illustrated) are very fast for both I_{Na} components. (b) I_{Na}–V relation for the records in (a). Closed symbols are for total peak I_{Na} (measured with no prepulse); open symbols are for the prepulse-resistant I_{Na}. (c) g_{Na}–V relations derived from the data in (b) show that the two types of g_{Na} show similar voltage dependence.

have important implications about the cell biological processes that control Na^+ channel distribution.

First, protein synthesis is necessary for 'proper' growth of somatic g_{Na} as defined in conjunction with Fig. 13.4. This can be shown by brief application of emetine, an irreversible translation inhibitor, on day 0 of culture (Gilly *et al.* 1990; see also below). These results suggest that establishing a significant level of g_{Na} in GFL cell bodies requires synthesis of new Na^+ channels.

Two other ways of preventing Na^+ channel expression are pertinent to intracellular trafficking of channels. As indicated in Fig. 13.6, development of g_{Na} is inhibited by demecolcine (*N*-deactyl-*N*-methylcolchicine), a drug that

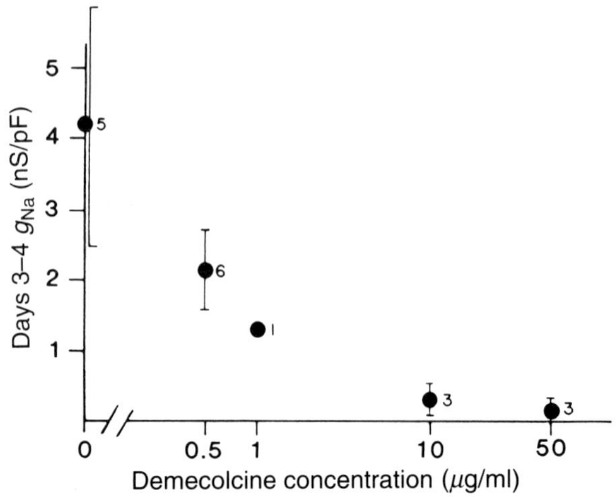

Fig. 13.6. Dose–response curve for the inhibitory action of demecolcine, a colchicine derivative, on g_{Na} development in somata of cultured GFL neurones. Demecolcine was applied on day 0 and remained in the medium until recordings were made on days 3–4. Means ± 1 SEM are plotted; numbers of cells studied are indicated.

blocks microtubule-based transport in the giant axon at similar concentrations (Brady *et al.* 1985). Transport of newly synthesized Na^+ channels to the cell surface via microtubular pathways thus appears to be necessary even for spatially inappropriate expression in the cell body.

Well-defined tracts of microtubules are prominent cytoskeletal features of both the giant axon (Schnapp and Reese 1986) and the unfused axons of GFL neurones in *vivo* (Martin and Miledi 1986) and *in vitro* (unpublished data). Presumably these elements provide the major thoroughfare for Na^+ channel traffic from the site of synthesis and processing in the cell body to the proper target membrane in the giant axon. This idea is supported by the finding that Na^+ channels do not appear inappropriately in GFL cell bodies when the whole stellate ganglion is cultured for 4 days with all of its giant axons ligated outside the ganglion but still attached proximally to the GFL cells (Fig. 13.7).

Presumably Na^+ channels are being synthesized in this case just as they are when cells are cultured in axotomized lobes, but, if the normal target membrane and the axonal pathway to it remain intact, Na^+ channels are preferentially exported from the somata and are not expressed there. This is the normal *in vivo* picture, and the cellular processes that mediate Na^+ channel sorting and trafficking thus continue to operate in culture.

Maintenance of polarized Na^+ channel distribution

In the studies described thus far, it is important to remember that Na^+ channel expression *in vitro* occurs in the soma and that this is inappropriate.

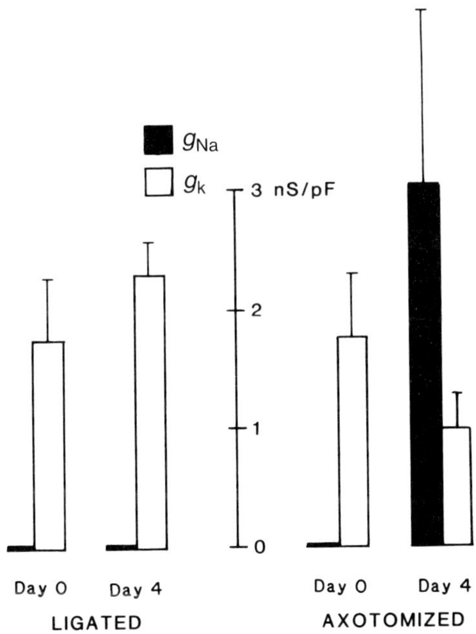

Fig. 13.7. GFL neurones cultured in 'intact' stellate ganglia with the giant axons ligated and still attached do not develop inappropriate g_{Na} in the cell body. Histograms show mean levels (± 1 SEM) of g_{Na} (filled bars) and g_K (open bars) recorded on days 0 and 4 from cells cultured in intact (ligated) ganglia and from cells cultured in excised (axotomized) lobes. g_{Na} develops *in vitro* only when the GFL neurones are severed from the giant axons. g_K was estimated in the same cells from the outward current during a 100-mV pulse recorded within 2 s after breaking into the cell for whole-cell recording, because the internal solution blocked I_K. All ganglia and lobes were proteased for 5 min (10 mg/ml Sigma Type XIV at room temperature) on day 0 and for 30 min on day 4 before isolating cells for study.

Although the functional properties of the inactivating channels are normal, channel density as judged by g_{Na} (2.6 ± 0.34 nS/pF after 10 days *in vitro*) is definitely lower than that in the giant axon (10–20 nS/pF; Gilly and Armstrong 1984 and unpublished data). Why doesn't somatic g_{Na} build up to the axonal level? What is special about the axonal environment that permits high-level Na^+ channel expression? How do Na^+ channels get to and recognize the axonal target site?

In order to address such questions, we have developed a method of culturing GFL neurones that allows us to study high-level Na^+ channel expression in axonal membrane. After protease treatment, GFLs are rinsed in filtered sea water and then neurones are gently teased out with a glass micropipette into the usual medium in a plastic culture dish precoated with the concanavalin A (see Gilly *et al.* 1990). Many cells thus isolated, often more than 100 per lobe,

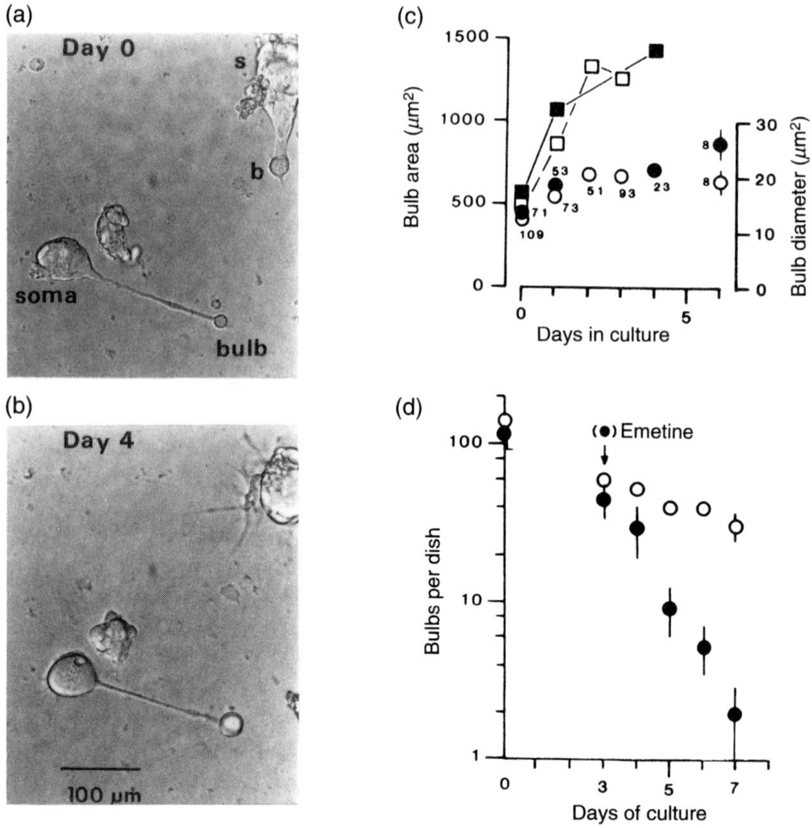

Fig. 13.8. Survival and growth of axonal bulbs in cultured GFL neurones plated out on concanavalin A-coated coverslips and plastic dishes. (**a**) Photomicrograph of cells on day 0 *in vitro*. Soma (s) and axonal bulb (b) are labelled for two cells. (**b**) The same cells on day 4 *in vitro* show two patterns of morphological change. In one case (lower left), the axon has remained stable and the axonal bulb is enlarged. Axon and bulb resorption can also occur (upper right). (**c**) Growth in optically measured diameter (circles) and calculated spherical area (squares) of axonal bulbs over time *in vitro*. Open symbols are from control cells; closed symbols are for tunicamycin-treated (2 µg/ml) cultures. Tunicamycin does not affect bulb growth (see text). (**d**) Survival of axonal bulbs *in vitro* is indicated for control cultures (open circles). Half of the control bulbs survive for 4 days or longer. Closed circles show data from cultures in which protein synthesis was blocked by emetine (50 µM for 1 h on day 3). Bulb loss is greatly accelerated (see text). Means ± 1 SEM are plotted, and the numbers of bulbs measured are indicated in (c) and (d).

retain a length of unfused axon, the distal end of which seals over and forms a small bulb [lower right of Fig. 13.8(a)]. Over time in culture the axonal bulb grows in diameter, approximately doubling in 4 days [Fig. 13.8(b) and (c)]. Nearly half of the axons and bulbs persist for at least this amount of time

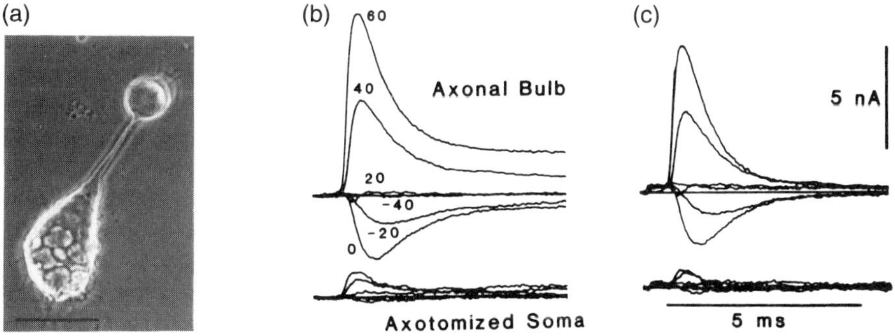

Fig. 13.9. I_{Na} measured in an axonal bulb and acutely axotomized soma of a cultured GFL neurone (day 6 *in vitro*) indicates that functional Na^+ channels are expressed at high level only in the axonal pole. (a) Photomicrograph of a GFL neurone cultured for 6 days; the axonal bulb lies at the upper right. Scale bar, 50 μm. (b) I_{Na} was measured first in the axonal bulb (22-pF input capacitance) after cutting the axon and then in the acutely axotomized soma (60 pF). Bulbar I_{Na} is much larger, even though the surface area is smaller than that of the cell body. (c) Prepulse-sensitive I_{Na} corresponding to the voltages in (b) is plotted for the same bulb and soma. The external solution was the culture medium itself (salts are adjusted to match artificial sea water); the internal solution is described in Fig. 13.2.

[Fig. 13.8(d)]. Axons that are not stable are resorbed (upper right of Fig. 13.8(a) and (b)]. Such cells thus become 'axonless' *in vitro* and appear to be similar to cells cultured in axotomized lobes on rafts.

Growth of axonal bulbs *in vitro* undoubtedly occurs because newly assembled membrane is incorporated into this region. Along with this new membrane comes a large number of Na^+ channels. I_{Na} can be recorded from both the soma and axonal bulb of a cell like that in Fig. 13.9(a) after cutting the axon with an electrolytically sharpened tungsten needle. Voltage control is not adequate if the axon remains intact. Figure 13.9(b) illustrates I_{Na} recorded from the axonal bulb and from the axotomized soma of the same cell. The corresponding inactivating fractions of I_{Na} are shown in Fig. 13.9(c). I_{Na} in the bulb is very large, whereas that in the soma is barely detectable. Functional differences between inactivating I_{Na} in bulbs and somata are not evident (Gilly *et al.* 1990).

Trafficking of Na^+ channels to the axonal domain evidently continues in such cells, although a low level of somatic expression also occurs. Figure 13.10 plots the time course of g_{Na} development in axonal bulbs and in acutely axotomized somata from experiments like that just described. Bulbar g_{Na} is detectable within hours after plating out the cells, continues to grow for 5–6 days, and then levels off at 10 nS/pF, a value similar to that found in the giant axon. Somatic g_{Na} never develops appreciably. Spatially appropriate, high-level expression of Na^+ channels thus occurs in the axonal pole just as it does *in vivo*.

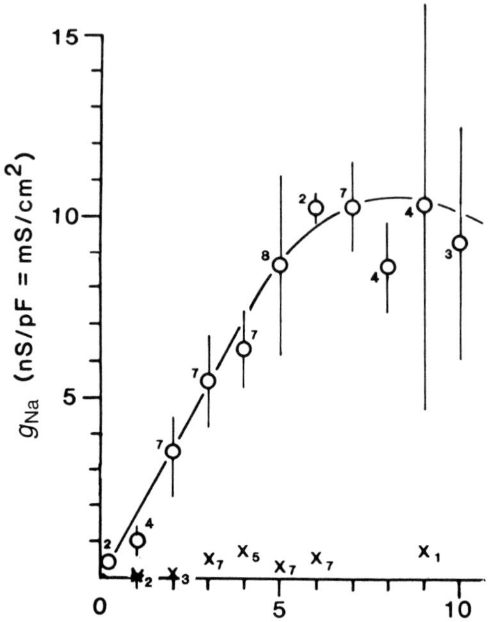

Fig. 13.10. Time course of g_{Na} development *in vitro* for axonal bulbs (circles) and acutely axotomized somata (crosses) of GFL neurones. Inactivating I_{Na} was measured, and the corresponding g_{Na} was derived from the I_{Na}–V relation. In this analysis, V has been corrected for series resistance errors (see Gilly *et al.* 1990). Mean ± 1 SEM and the number of 'cells' are given for each point. The smooth curve is drawn by eye. Solutions are described in Fig. 13.9.

Axonless cells studied in these cultures also develop g_{Na} along the same time course as that in bulbs (not illustrated) but level off at only 2.5 nS/pF, a level similar to that for neurones cultured in lobes. Although g_{Na} is low in axonless cells, the total *amount* of Na^+ conductance produced (g_{Na} × input capacitance = 333 ± 89 nS, n = 30) is similar to that computed for pairs of matching somata and bulbs of comparably aged cells with axons (288 ± 63 nS, n = 11). Thus, total Na^+ channel production is independent of the morphological status of the axon, but high-density expression requires the presence of an axonal bulb.

Differential turnover time cannot account for polarized Na^+ channel distribution

One possible way of controlling Na^+ channel density is by selective degradation in certain regions. Thus, if somatic Na^+ channels are degraded more rapidly than those in axonal bulbs, bulbar g_{Na} could build up to a higher level. Two

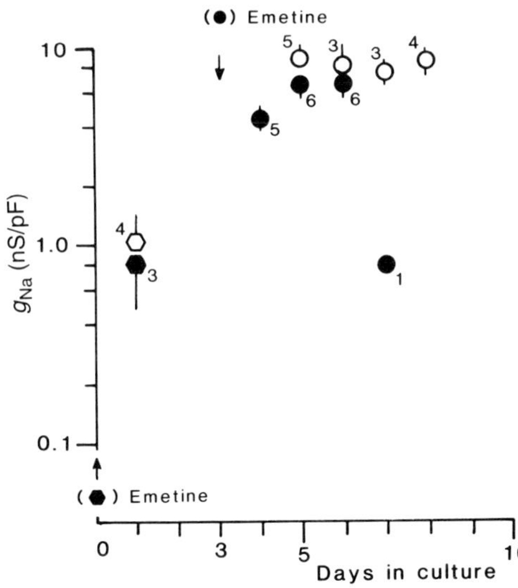

Fig. 13.11. Persistence of g_{Na} in axonal bulbs in the absence of protein synthesis. Emetine (50 μM) was applied for 1 h on either day 0 or day 3 as indicated by arrows. Prepulse-sensitive g_{Na} was then determined over time until cells were no longer viable. Control data is shown by open symbols; closed symbols represent data from emetine-treated cells. g_{Na} remains at normal levels for up to 3 days after drug treatment on day 3. Means ±1 SEM and numbers of bulbs studied are indicated.

approaches were taken to estimate the apparent lifetime of Na^+ channels in axonal versus somatic membrane.

First, protein translation was irreversibly inhibited by emetine on either day 0 or day 3 *in vitro*, and g_{Na} was then determined in axonal bulbs and axonless cells over time for a maximum of 4 more days. Application of emetine on day 0 had no effect on g_{Na} recorded 1 day later in either bulbs (Fig. 13.11) or cell bodies where this level persisted for 3 days (not illustrated). Emetine-treated bulbs could not be identified after day 1, because the axons had broken down. If emetine was applied on day 3, when a large g_{Na} was already present, a similar result was obtained. g_{Na} remained at the normal level in both bulbs (Fig. 13.11) and cell bodies (not illustrated) for 3 days. By the fourth day after treatment only one bulb was identifiable [see also Fig. 13.8(d)]. At this time the cells lacked any cytoskeletal organization and resembled oil droplets, but seven axonless cells still showed a large mean g_{Na} of 2 nS/pF (Gilly *et al.* 1990).

A second method of estimating degradation time for bulbar Na^+ channels avoided pharmacological intervention. Instead axons were cut with the tungsten needle after 3–4 days *in vitro*, and the severed bulbs were then studied 2–4 days later. Once again, g_{Na} was normal in 10 severed bulbs, and no bulbs survived longer than 4 days in the absence of the cell body.

g_{Na} in both axonal bulbs and somata is thus reduced only at a final catastrophic stage 4–5 days after cessation of protein synthesis and/or delivery. There is no sign of a gradual decrease in g_{Na} before this time. Under these conditions Na^+ channels have a long lifetime in both cell domains relative to the time of measurement (4 days). Accelerated degradation in the soma is therefore not likely to underlie the large difference in somatic and axonal g_{Na} evident after only 1–3 days *in vitro*.

Selective inhibition of high-level axonal Na^+ channel expression

Selective delivery of Na^+ channels to the axonal compartment of GFL neurones is another possible mechanism for maintaining the highly polarized channel distribution, and a large number of complex processes are undoubtedly involved. Posttranslational modifications of membrane proteins are potentially a source of sorting or targeting information, and we have begun exploring this idea by focusing on glycosylation.

Tunicamycin prevents cotranslational glycosylation of membrane proteins and effectively inhibits expression of functional Na^+ channels in cultured chick muscle (Bar-Sagi and Prives 1983) and neuroblastoma cells (Waechter *et al.* 1983; Smith-Maxwell 1986). In the case of squid GFL neurones tunicamycin specifically interferes with high-level axonal expression of Na^+ channels.

Figure 13.12(a) illustrates the time course of g_{Na} development in cells cultured in tunicamycin. After 5 days, g_{Na} is normal in amount (and properties) in axonless cells (triangles) but is severely reduced in axonal bulbs (circles) in comparison to control data (smooth curves). The bulbar Na^+ channels that are missing are not expressed in the somata of tunicamycin-treated cells [Fig. 13.12(b)]. Total Na^+ conductance (bulb + soma) in 11 treated cells (days 4–6) was 138 ± 21 nS, whereas the control value is significantly greater (288 ± 63 nS). Thus, spatially appropriate, high-level expression does not occur if glycosylation is blocked, but inappropriate somatic expression is unaffected.

Specificity of the tunicamycin effect on axonal g_{Na} is also indicated by several other observations. First, K^+ conductance density in axonal bulbs is not reduced by tunicamycin (Gilly *et al.* 1990). Second, growth of axonal bulbs is not impeded in the poisoned cultures [Fig. 13.8(c)]. Third, biochemical analysis of lobes cultured in tunicamycin indicates that total glycoprotein synthesis is inhibited by more than 90 per cent after 2–3 days [Fig. 13.13(a)], whereas protein synthesis is only moderately reduced [Fig. 13.13(b)]. Figure 13.13(a) also shows the effect of tunicamycin on the daily rate of bulbar g_{Na} increase as determined in voltage-clamp experiments. Similarity in the time course for inhibition of g_{Na} appearance and glycoprotein synthesis strongly suggests that lack of glycosylation of the Na^+ channel itself or some accessory protein is responsible for disrupting axonal Na^+ channel expression.

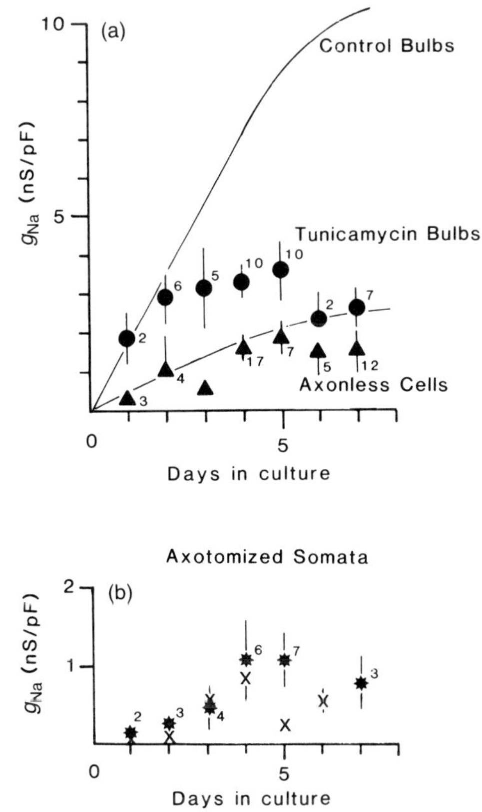

Fig. 13.12. Tunicamycin selectively inhibits development of g_{Na} in axonal bulbs. (a) g_{Na} measured in cells cultured in tunicamycin (1–2 μg/ml) is depressed after day 2 in axonal bulbs (circles) but not in cell bodies of axonless cells (triangles). Control data follows the smooth curves derived from Fig. 13.10. (b) The small g_{Na} found in acutely axotomized somata of cells that supported axonal bulbs is not affected in tunicamycin-treated cultures. Asterisks represent tunicamycin data; crosses are control data replotted from Fig. 13.10. Means ±1 SEM and number of 'cells' are indicated.

Spatially inappropriate expression in axonless cell bodies is insensitive to the status of the glycosylation machinery. This suggests that these channels, although functionally normal in both control and tunicamycin-treated cells, have escaped the control processes that strictly regulate Na^+ channel trafficking *in vivo* and *in vitro*.

Conclusions and future directions

At this point several conclusions emerge that act to guide our thinking about future avenues to follow. On the most general level, it appears that the squid

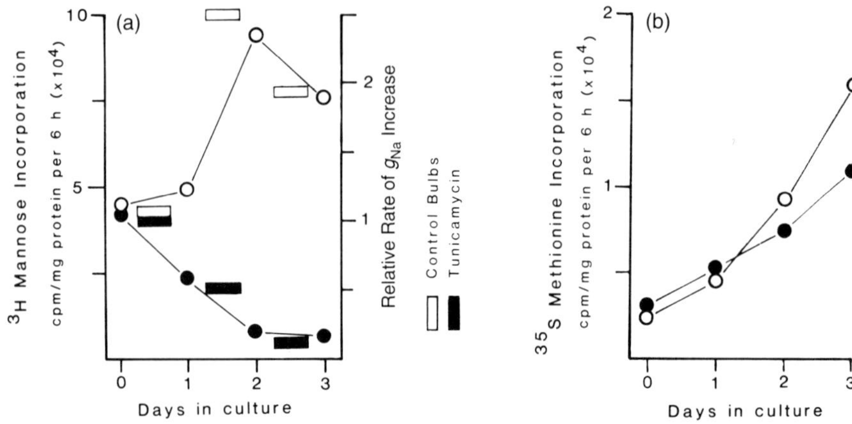

Fig. 13.13. Tunicamycin (2 μg/ml) inhibits glycosylation of proteins and the rate of g_{Na} development in parallel, but has little effect on total protein synthesis. (a) Glycoprotein synthesis, measured as incorporation of [^{3}H]mannose into TCA-precipitable protein (Gilly *et al.* 1990) is reduced in the presence of tunicamycin (filled circles) to less than 10 per cent of control values (open circles) by 2 days *in vitro*. Open (control) and filled bars show the identical effect of tunicamycin on the rate of g_{Na} increase in axonal bulbs (computed from daily differences of data in Fig. 13.12 and normalized to the changes between days 0 and 1). (b) Tunicamycin (closed circles) shows little effect on total protein synthesis, measured as [^{35}S]methionine incorporation into protein, at times when glycosylation is eliminated. Open circles are control data. Metabolic labelling with both isotopes was carried out in cultures of whole giant fibre lobes maintained *in vitro* for the indicated number of days and then labelled for a 6-h period in a L-15 medium containing 1 μM cold methionine.

GFL neurone–giant axon system is indeed a useful model system for studying the cell biology behind Na^+ channel sorting and trafficking. GFL neurones *in vitro* continue to maintain their normal, highly polarized, Na^+ channel distribution for many days. More specifically, we do not believe this reflects a difference in channel lifetime in the axon versus cell body based on work described in this chapter; but this conclusion must be further substantiated by other approaches. The idea of selective targeting and delivery of Na^+ channels to the axonal domain is attractive, and our results with tunicamycin are intriguing. What are the important differences between inappropriate expression of Na^+ channels in cell bodies and appropriate expression in axonal membrane, and how does glycosylation play a role in the latter? These are questions we intend to pursue.

Although we have made valuable progress with this system, we must develop new tools in order to attack the problems of interest. Voltage-clamp assays for Na^+ channel expression are extremely sensitive and superb for identifying functional properties, but functional channels are *all* that they see. The sorting and trafficking machinery for these channels is hidden inside the cell, and all channels other than those functioning in the surface membrane are invisible to us at present. Obviously we must develop biochemical probes to detect Na^+ channel protein before we can intelligently tinker with the relevant machinery.

Development of antibodies that recognize squid Na^+ channels is presently under way. Our approach has been to clone a fragment of a cDNA corresponding to a Na^+ channel mRNA isolated from GFL (Rosenthal and Gilly 1993) and to express the corresponding peptide in a bacterial fusion protein as a source of antigen. This work has gone well, and useful antibodies should soon be available. At that time we will be in a position to characterize the biosynthetic pathway of squid Na^+ channels and study glycosylation of the channel itself and how it relates to control of spatial distribution.

Suitable antibodies will also allow us to take further advantage of many features of the GFL neurone–giant axon system outlined in this chapter. Thus, we hope to develop methods for mapping expressed channels using immunostaining and imaging to complement patch-clamp experiments, and for immunoprecipitating Na^+ channel-containing vesicles from extruded axoplasm. This unique motor neurone will continue to lend itself to approaches impossible anywhere else.

Although this paper has concentrated on Na^+ channels, the system we describe should be equally valuable for studying the cell biology behind sorting and trafficking of other important membrane proteins. K^+ channel subtypes may be preferentially expressed in axons versus cell bodies. Ca^{2+} channels are present in the soma, virtually absent in the giant axon, and undoubtedly present at high density in the motor endings. Glutamate receptors are probably highly localized in the postsynaptic dendrites at the giant synapse. Cation pumps and anion transporters may also be differentially expressed in different domains. Advantages of this giant cell should benefit studies of these molecules as well.

Most fundamental of the important features displayed by squid GFL cells is that they are real neurones with a unique identity that reside in one of the most neurologically sophisticated invertebrates known. We can study the properties of cell bodies, dendrites, and axons of GFL neurones in preparations that reflect the operational status of these motor neurones in the living animal. This not only provides a valuable check on the fidelity of our manipulated neurones *in vitro*, but can also reveal how this giant motor neurone functions *in vivo* and how function and cellular properties change during development or in response to injury. Such complex aspects of neuronal specificity are moulded by basic processes like Na^+ channel sorting and trafficking, and we are just beginning to attack the squid giant axon pathway in these areas (Otis and Gilly 1990; Gilly *et al.* 1991).

The cell biological principles by which this friendly but simple-minded giant maintains and modifies its Na^+ channel distribution and other aspects of its molecular personality will almost certainly be relevant to the operation of other neurones, regardless of how complex they might be or in what brain they reside.

Acknowledgements

We thank the Wellcome Trust for travel grants for M.T.L. and J.R. to attend the cephalopod neurobiology symposium. This work was supported by the Whitehall Foundation, US National Science Foundation, and the National Institutes of Health.

References

Baines, A. J. (1990). Ankyrin and the node of Ranvier. *Trends in Neuroscience*, **13**, 119–21.

Bar-Sagi, D. and Prives, J. (1983). Tunicamycin inhibits the expression of surface Na^+ channels in cultured muscle cells. *Journal of Cellular Physiology*, **11**, 77–81.

Baas, P. W., Deitch, J. S., Black, M. M., and Banker, G. A. (1988). Polarity orientation of microtubules in hippocampal neurons: uniformity in the axon and nonuniformity in the dendrite. *Proceedings of the National Academy of Sciences, USA*, **85**, 8335–9.

Bezanilla, F. and Correa, A. M. (1991). Single sodium channels in high internal sodium in the squid giant axon. *Biophysical Journal*, **59**, 12a.

Brady, S. T., Lasek, R. J., and Allen, R. D. (1985). Video microscopy of fast axonal transport in extruded axoplasm: a new model for study of molecular mechanisms. *Cell Motility and the Cytoskeleton*, **5**, 81–101.

Breitfeld, P. P., McKinnon, W. C., and Mostov, K. E. (1990). Effect of nocodazole on vesicular traffic to the apical and basolateral surface of polarized MDCK cells. *Journal of Cell Biology*, **109**, 475–86.

Brismar, T. and Gilly, W. F. (1987). Synthesis of sodium channels in the cell bodies of squid giant axons. *Proceedings of the National Academy of Sciences, USA*, **84**, 1459–63.

Dotti, C. G. and Simons, K. (1990). Polarized sorting of viral glycoproteins to the axon and dendrites of hippocampal neurons in culture. *Cell*, **62**, 63–72.

Gilbert, D. L., Adelman, W. J., and Arnold, J. M. (ed.) (1990). *Squid as experimental animals*. Plenum Press, New York.

Gilly, W. F. and Armstrong, C. M. (1984). Threshold channels—a novel type of sodium channel in squid giant axon. *Nature*, **309**, 448–50.

Gilly, W. F. and Brismar, T. (1989). Properties of appropriately and inappropriately expressed sodium channels in squid giant axon and its somata. *Journal of Neuroscience*, **9**, 1362–74

Gilly, W. F., Lucero, M. T., and Horrigan, F. T. (1990). Control of the spatial distribution of sodium channels in giant fiber lobe neurons of the squid. *Neuron*, **5**, 663–4.

Gilly, W. F., Hopkins, B., and Mackie, G. O. (1991). Development of giant motor axons and neural control of escape responses in squid embryos and hatchlings. *Biological Bulletin*, **180**, 209–20.

Griffiths, G. and Simons, K. (1986). The trans Golgi network: sorting at the exit site of the Golgi complex. *Science*, **234**, 438–43.
Llano, I. and Bookman, R. J. (1986). Ionic conductances of squid giant fiber lobe neurons. *Journal of General Physiology*, **88**, 543–69.
Martin, R. and Miledi, R. (1986). The form and dimensions of the giant synapse of squids. *Philosophical Transactions of the Royal Society of London, Series B*, **312**, 355–77.
Matteson, D. R. and Armstrong, C. M. (1982). Evidence for a population of sleepy sodium channels in squid axon at low temperature. *Journal of General Physiology*, **79**, 739–58.
Miledi, R. (1967). Spontaneous synaptic potentials and quantal release of transmitter in the stellate ganglion of the squid. *Journal of Physiology*, **192**, 379–406.
Otis, T. S. and Gilly, W. F. (1990). Jet-propelled escape in the squid *Loligo opalescens*: concerted control by giant and non-giant motor axon pathways. *Proceedings of the National Academy of Sciences, USA*, **87**, 2911–15.
Pfeffer, S. R. and Rothman, J. E. (1987). Biosynthetic protein transport and sorting by the endoplasmic reticulum and Golgi. *Annual Reviews of Biochemistry*, **57**, 829–52.
Rosenthal, J. J. C. and Gilly, W. F. (1993). Amino acid sequence of a putative sodium channel expressed in the giant axon of the squid *Loligo opalescens*. *Proceedings of the National Academy of Sciences, USA*, **90**, 10026–30.
Schnapp, B. J. and Reese, T. S. (1986). New developments in understanding rapid axonal transport. *Trends in Neurosciences*, **9**, 155–62.
Simons, K. and Wandinger-Ness, A. (1990). Polarized sorting in epithelia. *Cell*, **62**, 207–10.
Smith-Maxwell, C. J. (1986). Sodium channel current is decreased by the glycosylation inhibitor tunicamycin. *Biophysical Journal*, **49**, 41a.
Sossin, W. H., Fisher, J. M., and Scheller, R. H. (1990). Sorting within the regulated secretory pathway occurs in the *trans*-Golgi network. *Journal of Cell Biology*, **110**, 1–12.
Vandenberg, C. A. and Bezanilla, F. (1991). Single-channel, macroscopic, and gating currents from sodium channels in the squid giant axon. *Biophysical Journal*, **60**, 1499–510.
Waechter, C. J., Schmidt, J. W., and Catterall, W. A. (1983). Glycosylation is required for maintenance of functional sodium channels in neuroblastoma cells. *Journal of Biological Chemistry*, **258**, 5117–23.
Wandinger-Ness, A., Bennett, M. K., Antony, C., and Simons, K. (1990). Transport vesicles mediate the delivery of plasma membrane proteins to the apical and basolateral domains of MDCK cells. *Journal of Cell Biology*, **111**, 987–1000.
Wonderlin, W. F. and French, R. J. (1991). Ion channels in transit: voltage-gated Na and K channels in axoplasmic organelles of the squid, *Loligo pealei*. *Proceedings of the National Academy of Sciences, USA*, **88**, 4391–5.

PART III

Axon–Schwann cell interaction

14

Electrophysiology of squid Schwann cells

N. JOAN ABBOTT, E. R. BROWN, Y. PICHON, and FUMIO KUKITA

Summary

Schwann cells of the squid giant nerve fibre form a monolayered sheath around the axon. The accessibility and size of the cells permits microelectrode impalement. The membrane potential is around −40 mV at rest, the result of a membrane predominantly permeable to K^+. The Schwann cell has a lower internal $[K^+]$ and higher $[Na^+]$ than the giant axon, which helps explain the low membrane potential. Resting Cl^- permeability has not been accurately measured, but may be significant; Cl^- distribution across the membrane appears to be close to equilibrium. The input impedance of squid Schwann cells is between 3 and 13 MΩ. Calculations of membrane resistance depend on assumptions about the effective surface area of the membrane, but figures of 1.2–4.8 kΩ.cm² have been derived, compared with about 1 kΩ.cm² for the axon membrane. Schwann cells are weakly dye- and electrically-coupled in small groups, but the nature of the junctions responsible is unclear. Current injection provides evidence of non-ohmic membrane behaviour, and 'spike-like' transients of potential to depolarizing current steps. Axonal stimulation or addition of glutamate or carbachol causes a long-lasting hyperpolarization of the Schwann cell, associated with increased K^+-selectivity of the membrane. Transient hyperpolarizations are often seen during this activation, and appear to be due to opening of calcium-activated K^+ channels. The membrane properties may underlie important activities of the Schwann cell including periaxonal ion homeostasis and axon–Schwann cell signalling. The similarities in membrane properties with vertebrate Schwann cells suggest that axon-associated peripheral glial cells may function on similar general principles in vertebrates and invertebrates.

Schwann cells are the class of glial cells of the peripheral nervous system associated with axons. The myelin-forming Schwann cells of large-diameter vertebrate axons have a clear physiological function important for axonal physiology. The smallest axons are surrounded in bundles by non-myelinating Schwann cells, and here the function is less clear, although a role in axonal nourishment and homeostasis of the local environment has been proposed. Among invertebrates, while bundles of small axons are again surrounded by an enveloping glial/Schwann cell, the glial ensheathment of giant axons is more elaborate, ranging from a 1:1 relationship, with the glial cell taking

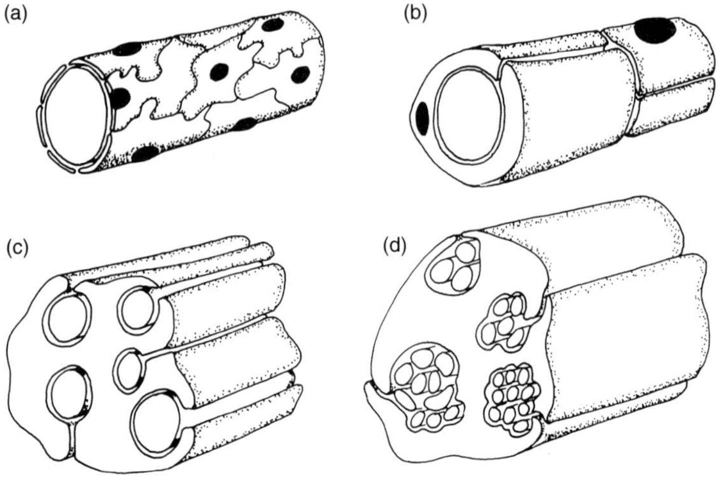

Fig. 14.1. Diagram illustrating the four types of ensheathment of axons by Schwann cells in the squid stellar nerve. (a) The giant nerve fibre with the axon surrounded by a epithelial-like monolayer of several Schwann cells; (b) a medium-sized axon (1.5–10 μm diameter) ensheathed by a single Schwann cell; (c) several small (0.5–1 μm diameter) axons ensheathed by a single Schwann cell. In (d) bundles of the smallest axons (<0.5 μm diameter) are ensheathed by a single Schwann cell. (Redrawn with permission from G. M. Villegas and Villegas (1968).)

multiple turns around the axon (crustaceans, Lane and Abbott 1975; insects, Lane and Treherne 1973) to a monolayer sheath formed by numerous Schwann cells as in squid (G.M. Villegas and Villegas 1968, 1984; Fig. 14.1). Damage to giant axons in crayfish central nervous system (CNS) connectives is followed by hypertrophy of the adaxonal glia and regeneration of the axon, evidence for a trophic role of glial cells in nourishing the axon (Bittner 1981).

The squid giant axon shows a complex system for chemical signalling from axon to Schwann cell (see Chapter 15), which may be an important way for the Schwann cells to match their activity to the requirements of the axon. As the signalling involves modulation of the electrophysiological properties of the Schwann cell, a complete understanding of the signalling requires consideration of these electrical properties. The electrophysiological properties also determine the mechanisms available for ionic homeostasis of the periaxonal microenvironment. This chapter reviews current understanding of the resting and activated properties of the Schwann cell membrane, and the implications for signalling and ionic homeostasis.

Preparations and methods

Three squid species have been extensively used for electrophysiological studies: the tropical squid *Sepioteuthis sepioidea*, and the European species

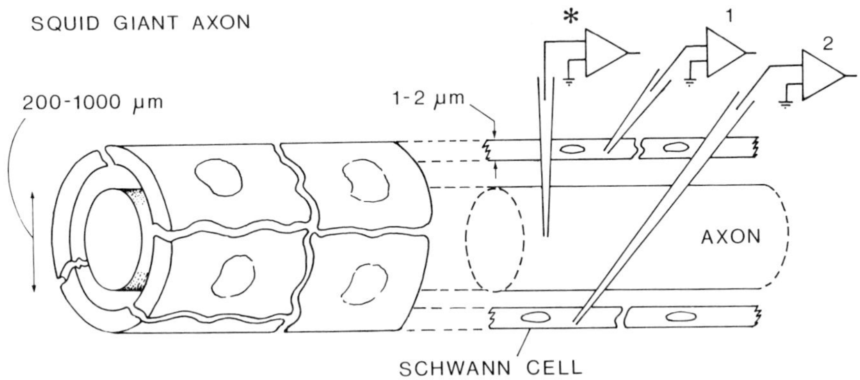

Fig. 14.2. Techniques for microelectrode recording from squid Schwann cells. 1, Impalement of Schwann cells on the upper surface of the giant axon is easiest in *Sepioteuthis* and *Alloteuthis*. 2, Brief sampling from Schwann cells below the axon, with withdrawal of the electrode back into the axon between impalements, is possible in *Sepioteuthis, Loligo*, and *Alloteuthis*. Both methods 1 and 2 allow simultaneous recording from the axon (*). In method 3 (not shown), the axon is slit open longitudinally, pinned out slightly stretched, and impalements made into the Schwann cell monolayer from the axolemmal side.

Alloteuthis subulata and *Loligo forbesi*. The individual Schwann cells are all about 1 μm thick, so microelectrode penetration is difficult but possible. Three methods have been used: continuous recording from Schwann cells around intact axons by maintained impalement (*Sepioteuthis* and *Alloteuthis*); sampling of membrane potentials from a population of Schwann cells by transient impalements (all species); and continuous recording from Schwann cells in a cut-open axon preparation (mainly *Loligo*) (Fig. 14.2).

Ionic permeability and transport of the resting membrane

Membrane potential

An early study by J. Villegas and co-workers (1963) with continuous recording showed three steps of potential as the (3 M KCl) microelectrode was advanced through the axon sheath (Fig. 14.3). By comparison with histology, together with ejection of carmine from the electrode to mark position, it was concluded that the ~−20 mV level corresponded to the outer or 'endoneurial' cells in the sheath, ~−40 mV to Schwann cells, and ~−60 mV to the axon. Subsequent studies used a sampling method to assay membrane potential and its modulation by chemical agents (see below), and the potential in the resting condition was consistently found to be close to −40 mV, in *Sepioteuthis*, *Alloteuthis*, and *Loligo*. Studies with continuous recording using

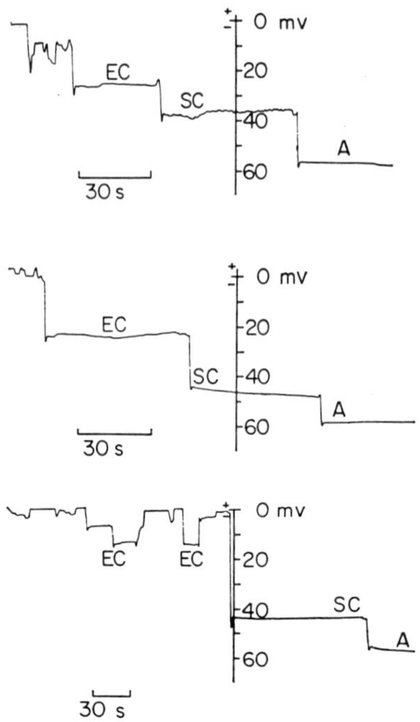

Fig. 14.3. Steps in membrane potential recorded by microelectrodes advanced into three intact *Sepioteuthis* giant axons from the outer surface. The first stable potential corresponds to 'endoneurial' cells (EC), the second to Schwann cells (SC), and the third to the axon (A). (Taken with permission from J. Villegas *et al.* (1963).)

3 M KCl microelectrodes in *Alloteuthis* demonstrated a larger range of potentials in Schwann cells. There was a clustering of potentials at about −30, −40, and −60 mV, with the largest potentials being about −70 mV (Abbott *et al.* 1990; Pichon *et al.* 1990). Although some cells showed relatively stable potentials following penetration, many cells initially at about −40 mV gradually hyperpolarized with time.

The most complete studies have been done on the Schwann cells of the split-open sheath of *Loligo*, which offers more stable conditions for intracellular recording (Brown 1991). The histogram of Schwann cell membrane potential distribution shows a peak at about −40 mV, and again a gradual hyperpolarization with time (Brown *et al.* 1991; Fig. 14.4). The hyperpolarization no longer occurs in Ca^{2+}-free solution, suggesting that it is a calcium-dependent process. The hyperpolarization following impalement is not blocked by external Co^{2+}, showing it is not due to opening of voltage-dependent Ca^{2+} channels (Brown *et al.* 1991). Taken together, the results suggest that the normal resting potential of the Schwann cells is about −40 mV, but that a rise in intracellular [Ca^{2+}] may cause a progressive hyperpolarization in dissected preparations. The mechanism could involve a leak of Ca^{2+} into the cell during

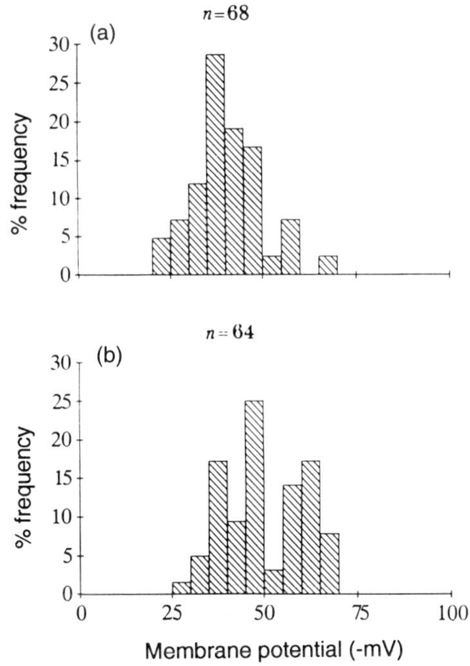

Fig. 14.4. Frequency distribution of Schwann cell membrane potentials, recorded during impalements lasting 2–10 s in a split-open axon preparation of *Loligo forbesi*. (a) At time 0–10 min, 14°C; (b) at 55–65 min after the start of the experiment, the preparation having been allowed to warm to 23°C between 35 and 55 min. Note the hyperpolarizing shift of the frequency distribution. (Taken with permission from Brown *et al.* (1991).)

dissection or microelectrode penetration, followed by opening of Ca^{2+}-activated K^+ channels, possibly accompanied by elevation of intracellular $[K^+]$ by leakage from the KCl microelectrode. The membrane potentials recorded with lithium acetate electrodes are generally more stable and slightly more negative than with KCl electrodes (Brown and Abbott, in preparation; see below).

Ionic selectivity of the resting membrane

J. Villegas and co-workers (Villegas 1968; Villegas *et al.* 1968) used the sampling method to record resting membrane potentials of *Sepioteuthis* Schwann cells in external solutions of different $[K^+]$, and found a 45-mV change per decade change of $[K^+]$ [Fig. 14.5(a)] compared with the 59-mV change expected for a perfect K^+ electrode, indicating a significant permeability to other ions. Large decreases in $[Cl^-]$ and $[Na^+]$ caused only a

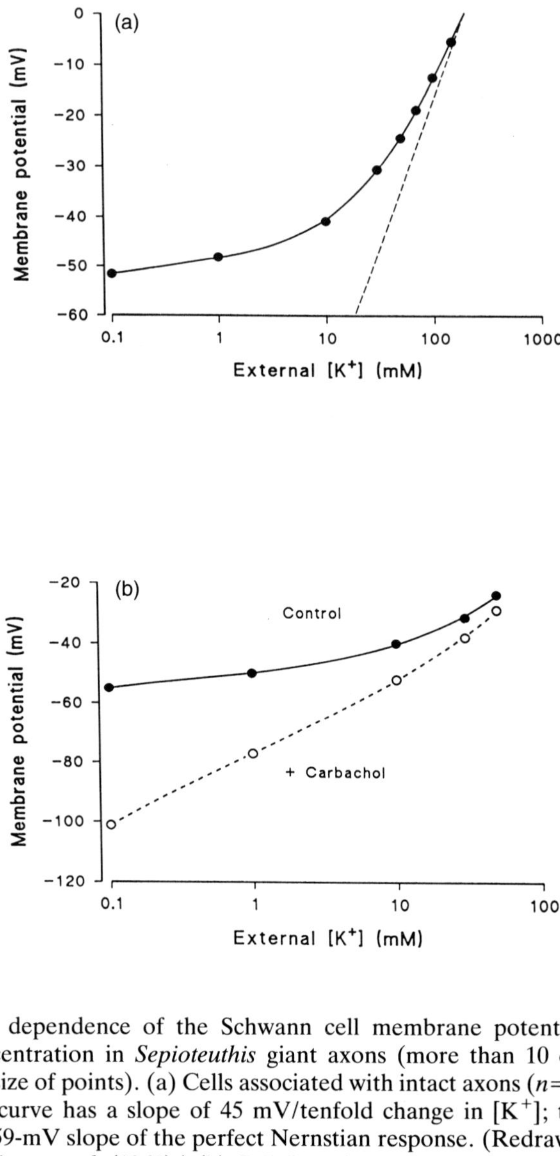

Fig. 14.5. The dependence of the Schwann cell membrane potential on external potassium concentration in *Sepioteuthis* giant axons (more than 10 cells per point, SEM less than size of points). (a) Cells associated with intact axons (n=20). The linear portion of the curve has a slope of 45 mV/tenfold change in [K^+]; the dashed line represents the 59-mV slope of the perfect Nernstian response. (Redrawn with permission from J. Villegas *et al.* (1968).) (b) Cells in split-open axon preparations, in control solutions (filled circles), and following addition of 10^{-6} M carbachol to activate the cholinergic receptors (open circles). Carbachol caused an increase in the slope of the line, compatible with an increase in the K^+ selectivity of the membrane. (Redrawn with permission from J. Villegas (1974).)

1–3 mV change in the Schwann cell membrane potential (J. Villegas *et al.* 1968) and this was taken to indicate negligible permeability to Cl^- or Na^+. However, a membrane highly permeable to Cl^- would show no change in steady-state potential in different $[Cl^-]$, if chloride was in equilibrium at the resting potential (J. Villegas *et al.* 1965, 1968).

In split-open axons of *Loligo*, the Schwann cell membrane potential showed a 52-mV slope per decade change of $[K^+]$ above 20 mM (Brown and Abbott, in preparation). In 10 mM $[K^+]$, reduction of $[Cl^-]$ from 612 to 15 mM caused a slight depolarization, while reduction of $[Na^+]$ from 470 to 141 mM caused a few millivolts hyperpolarization. A permeability ratio P_{Na}:P_K of 0.01–0.02 could explain the observations, but the method is not suitable for determining P_{Cl}:P_K.

Contribution of the electrogenic Na^+ pump

Blocking the Na^+ pump with K-strophanthoside caused an initial hyperpolarization of the membrane potential of *Sepioteuthis* Schwann cells and later depolarization (J. Villegas 1968). The hyperpolarization was attributed to inhibition of an inwardly directed active K^+ transport. However, in *Loligo* Schwann cells, rapid change from a low to a high $[K^+]$ solution caused a transient hyperpolarization (Brown, in preparation), consistent with activation of an electrogenic Na^+K^+ATPase. If the Schwann cell Na^+ pump behaves like that of the squid axon, it would be expected to be maximally activated at 100 mM external $[K^+]$, with a half-saturation concentration of ~20 mM K^+ (Baker *et al.* 1969).

Intracellular ion concentrations and equilibrium potentials

If the isolated cut-open axon preparation is incubated in a non-electrolyte to fill the extracellular spaces, then ion analysis of the sheath will allow calculation of the Schwann cell intracellular ion concentrations. In studies in *Sepioteuthis*, J. Villegas and co-workers found values for intracellular ions of 220 mM for $[K^+]$, 312 mM for $[Na^+]$, and 167 mM for $[Cl^-]$. The calculated equilibrium potentials (E_{ion}), assuming absence of intracellular binding, were: E_{Na}, +9 mV; E_K, −79 mV; and E_{Cl}, −32 mV (Villegas 1984).

In experiments on split-open *Loligo* axons, plots of membrane potential against log external $[K^+]$ have been used to calculate the internal $[K^+]$ (the concentration at which membrane potential equals zero; Brown and Abbott, in preparation). The value is around 100 mM. Assuming that intracellular cations are predominantly K^+ and Na^+, this leaves about 450 mM to be accounted for by Na^+. If E_{Cl} were more positive than the resting membrane potential, removing Cl^- from the external solution should cause hyperpolarization of the membrane, which was seen in *Alloteuthis* using the sampling method (Lieberman and Sanzenbacher 1992), but not using continuous recording in *Loligo* (Brown and Abbott, in preparation). The *Loligo* observations are compatible with $E_{Na} \sim +1$ mV, $E_K \sim -58$ mV, and E_{Cl} = membrane potential (i.e. at equilibrium).

Membrane resistance

J. Villegas (1972) used an intracellular microelectrode to inject current in *Sepioteuthis* Schwann cells, and recorded the resulting voltage deflection. The current–voltage (*I–V*) plot was linear (Fig. 14.6), and gave a mean input resistance of 3 MΩ. Assuming a value of 4×10^{-4} cm^2 for the cell membrane surface area, the calculated membrane resistance (R_m) was 1.2 KΩ cm^2. Brown (1991) repeated the experiment on Schwann cells in the split-open giant axon preparation of *Loligo*, and again found a linear *I–V* plot, with calculated input impedance of 12.5 MΩ. Values for intact *Alloteuthis* axons were similar. Estimation of R_m was more difficult. Dye injection showed that the *Loligo* Schwann cells have an elongated spindle shape, with fine terminal processes (Brown *et al.* 1991; Fig. 14.7). Detailed morphometric analysis of *Alloteuthis* Schwann cells showed a roughly rhomboidal cell body region, around 70 μm × 20 μm × 1 μm thick, and the total membrane surface area per cell was calculated to be 3.14×10^{-4} cm^2 (Brown and Abbott 1993). However, it is not known whether current injection causes a change in membrane potential mainly in the cell body region, or uniformly throughout the elongated process. The calculated R_m was in the range 3.7–4.8 KΩ cm^2, assuming that the whole membrane is involved. No allowance was made for

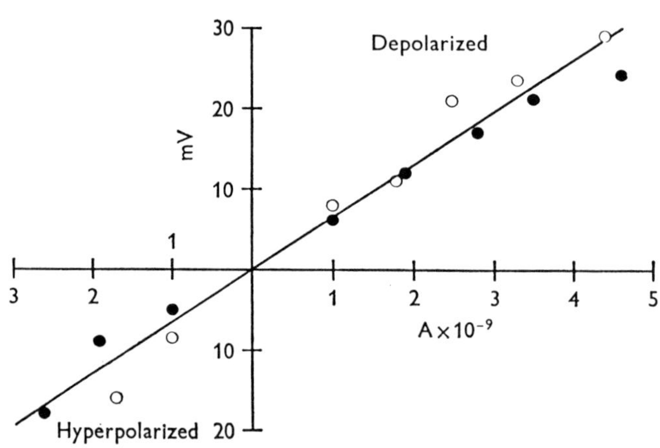

Fig. 14.6. Current–voltage relation of a *Sepioteuthis* Schwann cell, measured with a single intracellular microelectrode, in a split-open axon preparation. The observations were made in two successive runs of current pulses (different symbols) injected into a single Schwann cell. Line drawn by eye. (Taken with permission from J. Villegas (1972).)

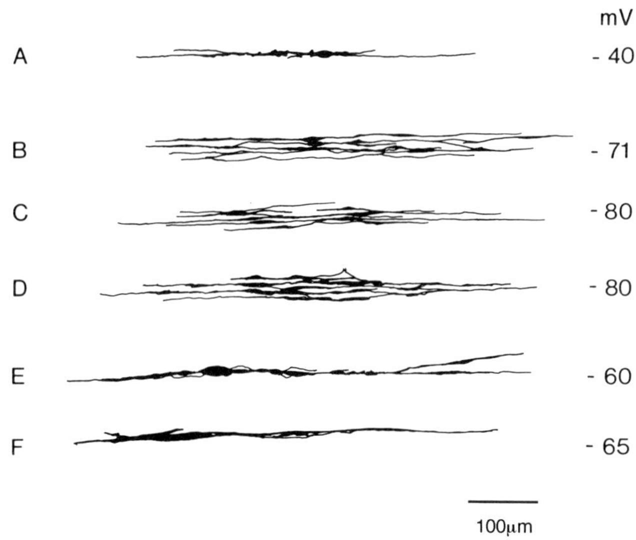

Fig. 14.7. Tracings from Lucifer Yellow (A–E) and horseradish peroxidase (HRP; trace F) fills of Schwann cells in split open giant axons of *Loligo forbesi*. The cell membrane potential is shown at the right. Dye injection into ~−40 mV cells generally filled a single cell (trace A), whereas higher membrane potential cells showed dye coupling among a small group of cells (traces B, C, D). Cell E was in a preparation previously treated with 2 mM octanol to uncouple gap junctions; dye is confined to a single cell. HRP fills (trace F) always showed a single cell, evidence that the cell:cell junctions are impermeable to HRP. (Taken with permission from Brown *et al.* (1991).)

the effect of the membranes of the glial tubular system (Zwahlen *et al.* 1988; Brown and Abbott 1993). For comparison, the R_m of the squid axon membrane is about 1 KΩ cm^2.

Electrical coupling

Injection of Lucifer Yellow via the microelectrode demonstrates clusters of dye-coupled Schwann cells, more obvious at later times and in hyperpolarized cells in split-open axon preparations (Brown *et al.* 1991). Injection of current into one Schwann cell while recording from another cell shows electrical coupling between neighbouring Schwann cells (Brown *et al.* 1991; Fig. 14.8), confirming that the cells are linked by junctions permeable to both dyes and small ions but, as the coupling is relatively weak, the junctional resistance must be high. At the electron microscope level, close membrane contacts are seen between Schwann cells in thin-sectioned material, but these resemble *maculae occludentes* rather than classical gap junctions (Brown and Abbott 1993). A full analysis using extracellular lanthanum in thin-sectioned material,

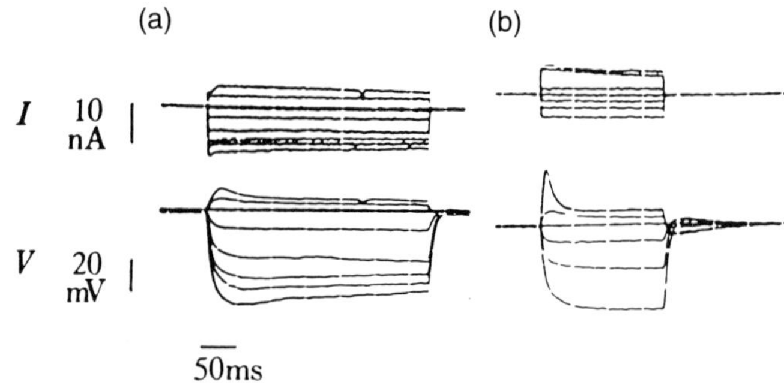

Fig. 14.8. Evidence for electrical coupling between Schwann cells. Oscilloscope traces showing *I*, current injected into one cell, and *V*, voltage monitored in a nearby cell. The electrodes were 60 μm apart (laterally) in both (a) and (b). Note the non-ohmic ('spikelike') response at the start of the depolarizing current injection, and the rectification of the voltage traces. From Brown *et al.* (1991) with permission.

and freeze-fracture examination to show intramembranous particle organization, will be necessary to reveal the junctions responsible for the dye and electrical coupling.

The electrical coupling studies revealed additional interesting non-ohmic behaviour of the membranes, with a 'spike-like' transient superimposed on the depolarizing deflection caused by current injection (Brown *et al.* 1991; Fig. 14.8). J. Villegas *et al.* (1976) reported that veratrine and grayanotoxin caused a depolarization of the *Sepioteuthis* Schwann cells, which could be reversed by reducing external [Na^+] or adding tetrodotoxin, but no evidence for voltage-dependent Na^+ channels was found. However, fast voltage steps would be needed to show such channels. If voltage-dependent Na^+ or Ca^{2+} channels are present, they could play important roles in modulation and amplification of the axon–Schwann cell signalling system (see below).

Ionic selectivity of the activated Schwann cell membrane

The Schwann cell hyperpolarization that results from axonal stimulation, and that can be mimicked by application of carbachol, appears to involve at least three steps, the first caused by an axonal signal acting on the Schwann cell, the second by acetylcholine (ACh) released from the Schwann cell acting back on Schwann cell ACh autoreceptors, and the third by elevation of intracellular cyclic AMP (cAMP) (see Chapter 15). Ionic events in the Schwann cell membrane associated with each step are not well understood. The final hyperpolarization depends on activation of ACh receptors, because it can be blocked by α-bungarotoxin and d-tubocurarine, blockers of nicotinic ACh receptors. cAMP is implicated in the mechanism, because the hyperpolarization can be potentiated by agents that cause elevation of cAMP (forskolin, theophylline, cAMP analogues; Evans *et al.* 1985).

Given that E_K is generally 15–20 mV more negative than the resting membrane potential, hyperpolarization could be achieved by a rise in K^+ conductance (g_K), a fall in Na^+ conductance (g_{Na}), a fall in g_{Cl} if E_{Cl} is more positive than the membrane potential E_m, or activation of the electrogenic Na^+ pump. The critical experiments to distinguish between these possibilities have not yet been done, but some studies suggest the involvement of K^+ channels. The slope of the relation between Schwann cell membrane potential and log [K^+] in *Sepioteuthis* is steeper following application of carbachol [J. Villegas 1974; Fig. 14.5(b)], implying a more K^+-selective membrane in the activated condition. Following stimulation of the axon in *Alloteuthis* preparations, the Schwann cell membrane potential, in addition to showing the usual slowly developed hyperpolarizations, also shows transient membrane potential changes, most commonly hyperpolarizations (Fig. 14.9), but occasionally including depolarizing components (Abbott *et al.* 1990). Repeated transients were seen in seven of 30 preparations following axonal stimulation, and also occasionally in the split-open *Loligo* axon preparation on application of carbachol and glutamate (Brown 1991). The hyperpolarizing transients were reduced but not abolished by removal of external calcium, and increased in amplitude on return to normal solution. Opening of Ca^{2+}-activated K^+ channels was suggested. It remains to be seen whether the cAMP-dependent hyperpolarization is produced by this mechanism.

The candidates for the axonal signal triggering potential changes in the Schwann cell include K^+ (released during each action potential), glutamate, and peptides. In an early review, axonal activation was illustrated as causing a biphasic response in the Schwann cell, an initial depolarization, then a hyper-

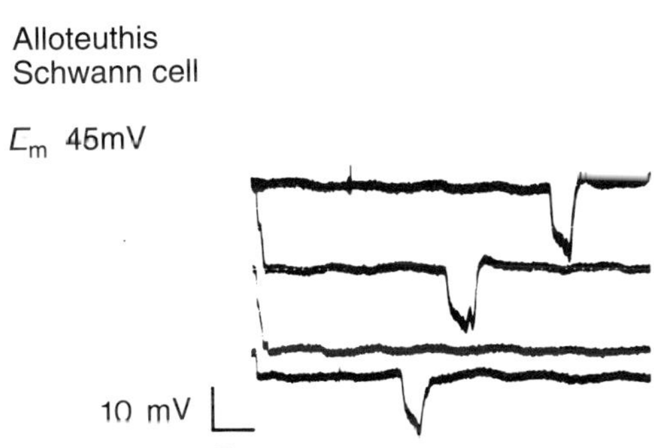

Fig. 14.9. Four consecutive oscilloscope sweeps from a continuous recording of Schwann cell membrane potential in an *Alloteuthis* giant axon (baseline −45 mV) showing spontaneous transient hyperpolarizations, each lasting about 4 ms. The transients were triggered by a period of axonal stimulation (30 s, 50 Hz), and continued for more than an hour after cessation of stimulation (Abbott *et al.* 1990).

polarization (fig. 1 in J. Villegas 1981); the depolarization could be due to K^+. However, the depolarization does not appear in most experimental traces (e.g. J. Villegas 1972). More recently, it has been shown that the Schwann cell hyperpolarization (with intact cholinergic system) can be blocked by 2-aminophosphonobutyrate (2-APB or AP-4), a broad-spectrum glutamate-receptor antagonist, and mimicked by application of glutamate, quisqualate, or kainate, but not N-methyl-D-aspartate (NMDA) (Lieberman *et al.* 1989). This suggests that glutamate or a related substance is the primary signal from the axon. In some experiments it was possible to detect an early depolarization in response to glutamate application, but this was not an invariable finding, and blocking of the depolarization by using Na^+-free solution did not stop the glutamate-induced hyperpolarization, showing that the depolarization is not a prerequisite for the release of ACh (Lieberman and Sanzenbacher 1992). A possible sequence is release of glutamate from the axon by electrical activity (e.g. depolarization-mediated reversal of Na^+-dependent glutamate transporter, cf. Szatkowski *et al.* 1990), acting on Schwann cell glutamate receptors to cause a rise in intracellular [Ca^+] without depolarization (by a process possibly analogous to that of the vertebrate metabotropic glutamate receptor, involving inositol 1,4,5-trisphosphate (IP_3)), and release of ACh by a Ca^{2+}-mediated process. However, the detailed steps in the sequence require further investigation.

Relevance to physiological function

The electrophysiological properties of the Schwann cell are of relevance not only to the processes underlying axon–Schwann cell signalling, but also to other documented functions of the Schwann cells, in particular their role in regulation of the [K^+] and [Na^+] in the periaxonal space (see Chapter 16). Thus a significant K^+ and Cl^- permeability of the resting Schwann cell membrane would permit net KCl uptake following axonal activity, and the presence of an electrogenic Na^+ pump, activated by a rise in external [K^+], would also contribute to the ion-regulatory capacity of the Schwann cell. The fact that Na^+ is close to equilibrium across the Schwann cell membrane at rest may be significant here; activation of the pump by elevation of extracellular [K^+] would exchange K^+ for Na^+, and the high intracellular [Na^+] would then effectively permit the cell to act as a cation reservoir, with a large capacity for taking up K^+ and releasing Na^+, the combination required for ionic homeostasis in the periaxonal space during axonal activity.

The functional significance of the electrical coupling between Schwann cells is less clear. It would serve to integrate axon–Schwann cell signalling and metabolic coupling by permitting a more uniform response in the Schwann cells to axonal stimulation, regardless of the receptor distribution (i.e. even if signalling is confined to a relatively few sites on the axon or Schwann cell layer). The possible advantage of the electrical coupling for K^+ regulation by 'spatial buffering' has been considered (see Chapter 16). However, since the spindle-shaped morphology of the Schwann cell, running parallel to the axonal axis, is not advantageous for spatial buffering, further elongation of the network by electrical coupling would add no benefit.

Comparison with vertebrate Schwann cells

Vertebrate CNS glia (astrocytes, oligodendrocytes), whether freshly isolated, cultured, or studied in brain slices, generally show high membrane potentials, typically −70 to −90 mV (Kettenmann *et al.* 1983; Ballanyi *et al.* 1987). It is therefore necessary to consider whether the membrane potential recorded in squid Schwann cells could be artefactually low, perhaps because of electrical shunting by membrane damage caused by microelectrode penetration of these very small cells. Several lines of evidence suggest that close to −40 mV is the real potential in squid Schwann cells. The mean E_m is very consistent between species, in spite of the use of a variety of microelectrode methods and Schwann cell preparations (associated with an intact axon or in split-axon preparations). The near-Nernstian response of the membrane to K^+ argues against significant shunting by electrode damage. The intracellular $[K^+]$ calculated from the Nernst plot (which would be influenced by electrical shunting if present) is close to that determined independently by ionic analysis (J. Villegas 1984). The brief and reproducible hyperpolarization caused by axonal stimulation or agonist application would be difficult to explain if there were significant shunting; withdrawal of the electrode and reinsertion during the hyperpolarization records the same potential, again suggesting minimal electrode damage (Brown and Abbott 1992; Fig. 14.10).

There is a body of experimental evidence showing that vertebrate Schwann cells, once differentiated, also maintain low membrane potentials close to −40 mV (Ritchie 1992). The similarity between invertebrate and vertebrate Schwann cells, and difference from CNS glia, could indicate a difference in physiological function between central and peripheral glia, linked to the membrane potential (Barres 1991; Ritchie 1992). If the low membrane potential of vertebrate Schwann cells is produced by similar mechanisms to those of the squid Schwann cells, the argument concerning the relevance to ionic homeostasis could also be applied to the vertebrate system. The similarity of axonal function in vertebrates and invertebrates, with axonal conduction dependent on all-or-none action potentials with prominent Na^+ and K^+ currents, would suggest that the Schwann cells should also have similar mechanisms for ion regulation. The comparison with the squid suggests that similar mechanisms are indeed likely.

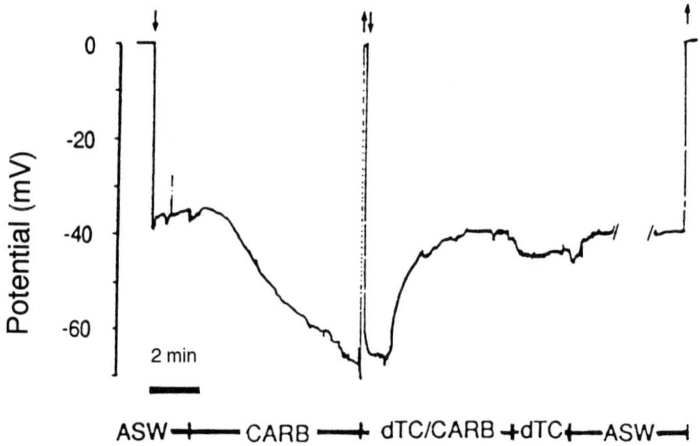

Fig. 14.10. Continuous record of Schwann cell membrane potential in a split-open giant axon preparation from *Loligo forbesi*. The cell was impaled in artificial sea water (ASW). Carbachol 10^{-6} M (CARB) caused a hyperpolarization, blocked by d-tubocurarine 10^{-7} M (dTC). Reinsertion of the electrode after its removal during the carbachol application (arrows in centre) still recorded the same potential, indicating that shunting the membrane potential by electrode damage was minimal. (Taken with permission from Brown and Abbott (1992).)

Conclusions

This survey of studies on the electrophysiology of squid Schwann cells shows a general consistency between observations in *Sepioteuthis*, *Alloteuthis*, and *Loligo*, good evidence for the generality of the phenomena. The documented membrane properties of the Schwann cells would make for efficient periaxonal ion regulation. Further characterization of the membrane ion channels responsible for the resting and activated electrical behaviour of the membranes requires rigorous study using the patch-clamp technique.

Acknowledgements

We are grateful to the Director and staff of the Marine Biological Association Laboratory for facilities and squid, and to the SERC, Wellcome Trust, and Royal Society for support.

References

Abbott, N. J., Pichon, Y., and Brown, E. R. (1990). Long-term microelectrode recording from Schwann cells surrounding the isolated giant axon of the squid; steady and oscillating potentials. *Journal of Physiology*, **430**, 128P.

Baker, P. F., Blaustein, M. P., Keynes, R. D., Manil, J., Shaw, T. I., and Steinhardt, R. A. (1969). The ouabain-sensitive fluxes of sodium and potassium in squid giant axons. *Journal of Physiology*, **200**, 459–96.

Ballanyi, K., Grafe, P., and ten Bruggencate, G. (1987). Ionic activities and potassium uptake mechanisms of glial cells in guinea-pig olfactory cortex slices. *Journal of Physiology*, **383**, 45–67.

Barres, B. A. (1991). New roles for glia. *Journal of Neuroscience*, **11**, 3685–94.

Bittner, G. D. (1981). Trophic interactions of CNS giant axons in crayfish. *Comparative Biochemistry and Physiology*, **68A**, 299–306.

Brown, E. R. (1991). Axon–Schwann cell interaction in the squid. Unpublished PhD thesis, University of London.

Brown, E. R. and Abbott, N. J. (1992). Membrane potential in squid Schwann cells: stable impalements in the split open giant axon preparation of *Loligo*. *Journal of Physiology*, **446**, 320P.

Brown, E. R. and Abbott, N. J. (1993). Ultrastructure and permeability of the Schwann cell layer surrounding the giant axon of the squid. *Journal of Neurocytology*, **22**, 283–98.

Brown, E. R., Bone, Q., Ryan, K. P., and Abbott, N. J. (1991). Morphology and electrical properties of Schwann cells around the giant axon of the squids *Loligo forbesi* and *Loligo vulgaris*. *Proceedings of the Royal Society of London, Series B*, **243**, 255–62.

Evans, P. D., Reale, V., and Villegas, J. (1985). The role of cyclic nucleotides in modulation of the membrane potential of the Schwann cell of squid giant nerve fibre. *Journal of Physiology*, **363**, 151–67.

Kettenmann, H., Sonhof, U., and Schachner, M. (1983). Exclusive potassium dependence of the membrane potential in cultured mouse oligodendrocytes. *Journal of Neuroscience*, **3**, 500–5.

Lane, N. J. and Abbott, N. J. (1975). The organization of the nervous system in the crayfish *Procambarus clarkii*, with emphasis on the blood–brain interface. *Cell and Tissue Research*, **156**, 173–87.

Lane, N. J. and Treherne, J. E. (1973). The ultrastructural organization of peripheral nerves in two insect species (*Periplaneta americana* and *Schistocerca gregaria*). *Tissue and Cell*, **5**, 703–14.

Lieberman, E. M. and Sanzenbacher, E. (1992). Mechanisms of axon-to-Schwann cell signaling in the squid. *Neuroscience*, **47**, 931–9.

Lieberman, E. M., Abbott, N. J., and Hassan, S. (1989). Evidence that glutamate mediates axon-to-Schwann signaling in the squid. *Glia*, **2**, 94–102.

Pichon, Y., Abbott, N. J., and Brown, E. R. (1990). Long term recordings from Schwann cells surrounding squid axons reveal multiple resting states. *Biophysical Journal*, **57**, 532a.

Ritchie, J. M. (1992). Voltage-gated ion channels in Schwann cells and glia. *Trends in Neurosciences*, **15**, 345–51.

Szatkowki, M., Barbour, B., and Attwell, D. (1990). Non-vesticular release of glutamate from glial cells by reversed electrogenic glutamate uptake. *Nature*, **348**, 443–6.

Villegas, G. M. and Villegas, R. (1968). Ultrastructural studies of the squid nerve fibers. *Journal of General Physiology*, **51**, 44s–60s.

Villegas, G. M. and Villegas, R. (1984). Squid axon ultrastructure. *Current Topics in Membranes and Transport*, **22**, 3–37.

Villegas, J. (1968). Transport of electrolytes in the Schwann cell and location of sodium by electron microscopy. *Journal of General Physiology*, **51**, 61s–71s.

Villegas, J. (1972). Axon–Schwann cell interaction in the squid nerve fibre. *Journal of Physiology*, **25**, 275–96.
Villegas, J. (1974). Effects of acetylcholine and carbamylcholine on the axon and Schwann cell electrical potentials in the squid nerve fibre. *Journal of Physiology*, **242**, 647–59.
Villegas, J. (1981). Axon/Schwann-cell relationships in the giant nerve fibres of the squid. *Journal of Experimental Biology*, **95**, 135–51.
Villegas, J. (1984). Axon–Schwann cell relationship. *Current Topics in Membranes and Transport*, **22**, 547–71.
Villegas, J., Villegas, L., Gimenez, M., and Villegas, G. M. (1963). Schwann cell and axon electrical potential differences. Squid nerve structure and excitable membrane location. *Journal of General Physiology*, **46**, 1047–64.
Villegas, J., Villegas, L., and Villegas, R. (1965). Sodium, potassium, and chloride concentrations in the Schwann cell and axon of the squid nerve fiber. *Journal of General Physiology*, **49**, 1–7.
Villegas, J., Villegas, R., and Gimenez, M. (1968). Nature of the Schwann cell electrical potential. Effects of the external ionic concentration and a cardiac glycoside. *Journal of General Physiology*, **51**, 47–64.
Villegas, J., Sevcik, C., Barnola, F. V., and Villegas, R. (1976). Grayanotoxin, veratrine, and tetrodotoxin-sensitive sodium pathways in the Schwann cell membrane of squid nerve fibres. *Journal of General Physiology*, **67**, 369–80.
Zwahlen, M. J., Sandri, C., and Greeff, N. J. (1988). Transglial pathway of diffusion in the Schwann sheath of the squid giant axon. *Journal of Neurocytology*, **17**, 145–59.

15

The pharmacology of receptors present on squid giant axon Schwann cells

PETER D. EVANS, VINCENZINA REALE, ROSA MARIA MERZON, and JORGE VILLEGAS

Summary

The adaxonal Schwann cells of the squid giant axon possess a range of receptors for classical neurotransmitters. When activated these receptors either alter the membrane potential of the Schwann cell or modulate its responsiveness to other inputs. A complex cascade of events is involved in signalling between the giant axon and its surrounding Schwann cells, involving the release of glutamate from the axon, to act on receptors on the Schwann cells which mediate a release of endogenous acetylcholine. The latter molecule then feeds back to activate nicotinic cholinergic receptors on the Schwann cells which initiate the production of a long-lasting hyperpolarization of the Schwann cell membrane. The modulation of this cascade of events by a range of biogenic amines and neuropeptides will be reviewed. Recent evidence for the presence of both *N*-methyl-D-aspartate (NMDA) and non-NMDA type glutamate receptors on the Schwann cells will be discussed, together with evidence for the presence of an active uptake system for L-glutamate in this preparation. The co-release of acetylcholine and a vasoactive intestinal polypeptide (VIP)-like peptide from the Schwann cells will also be considered. Finally, the physiological roles of the different receptor systems present on the Schwann cells will be discussed.

Neuronal–glial cell signalling mechanisms require the presence of receptors on glial cells for the signal molecules released from the neurone. Although an ever-increasing number of receptors for neurotransmitter molecules are being described on glial cells from a number of different sources (see Kimelberg 1988), very few preparations exist in which the anatomical relationships between the neurones and their surrounding glia are maintained. Cephalopod neurobiology has, however, provided us with one of the few exceptions. In the isolated giant axon from the stellate nerve of the squid it is possible to study glial functioning in an intact preparation where the anatomical relationships between the giant axon and its surrounding adaxonal Schwann (glial) cells are preserved. Most of the work to date has been performed on the tropical squid, *Sepioteuthis sepioidea* (see Villegas 1973, 1974, 1975, 1981, 1984; Evans *et al.* 1991*a*), but similar approaches are now being developed for other species of squid, such as *Alloteuthis subulata* and *Loligo forbesi* (Lieberman *et al.* 1989; Brown *et al.* 1991).

Early work on signalling between the giant axon and its surrounding Schwann cells in the tropical squid provided evidence for a complex cascade of events being initiated after axonal stimulation that involved nicotinic acetylcholine receptors and possibly glutamate receptors on the Schwann cells (Villegas 1973, 1974, 1975, 1978*a*,*b*). On the basis of the sensitivity of the Schwann cell membrane potential to exogenously applied glutamate, it was proposed that glutamate might be the messenger molecule released from the giant axon to activate the Schwann cell system (Villegas 1978*a*,*b*).

This suggestion has received further support recently by the demonstration of specific glutamate receptors on the squid Schwann cells (Lieberman *et al.* 1989; Evans *et al.* 1991*b*). The activated Schwann cells were then shown to release endogenous acetylcholine by a calcium-dependent mechanism (Villegas and Jenden 1979). Thus, unlike Schwann cells at the vertebrate neuromuscular junction, which only synthesize and release acetylcholine after denervation (Dennis and Miledi 1974), the squid adaxonal Schwann cells appear to release acetylcholine as part of the neural signalling cascade, perhaps to amplify the strength of the signal generated by axonal stimulation of the Schwann cells or to increase its spatial spread. The acetylcholine released from the Schwann cells then feeds back on to the Schwann cells to activate nicotinic cholinergic receptors whose activation leads to the production of a longlasting hyperpolarization of the Schwann cell membrane (Villegas 1974, 1975). This effect is unusual since nicotinic cholinergic receptors usually mediate rapid effects via the activation of the ion channel formed by the receptor complex itself.

Recent evidence indicates that the second messenger cyclic adenosine monophosphate (cAMP) is involved in the response of the Schwann cells to nicotinic cholinergic receptor activation (Evans *et al.* 1985). Evidence in support of the latter suggestion is that compounds that increase cAMP levels by bypassing the receptor activation stage, such as phosphodiesterase inhibitors, forskolin, and cAMP analogues, can all mimic the action of activation of the receptors by producing a longlasting hyperpolarization. In addition, the actions of carbachol can be potentiated in the presence of low doses of phosphodiesterase inhibitors and forskolin. Recently, the actions of forskolin in a number of preparations were demonstrated not to be due to adenylate cyclase activation (Laurenza *et al.* 1989). However, the actions of forskolin on adenylate cyclase and on other sites of action can be differentiated by the use of the forskolin analogue 1,9 dideoxyforskolin, which does not activate the adenylate cyclase system. We have demonstrated that 1,9 dideoxyforskolin, at concentrations up to 2.5×10^{-5} M, neither induces hyperpolarization of the Schwann cell membrane nor potentiates the effect of a 1-min pulse of 10^{-7} M carbachol in this preparation (Evans, Reale, Merzon and Villegas, unpublished observation). Thus it is likely that the actions of forskolin on the Schwann cells are mediated by activation of the adenylate cyclase system.

Direct receptor-mediated activation of adenylate cyclase would be the simplest explanation of the results just described, but they are also compatible with more complex modes of adenylate cyclase activation (Evans *et al.* 1985). Recent evidence, however, suggests that these results may not be as unusual as first thought. Evidence is now accumulating for the existence of a wide range of distinct subtypes of nicotinic receptor (Chiappinelli 1985; Boulter *et al.* 1987) even in the squid nervous system (Chen *et al.* 1988). In addition, nicotinic cholinergic receptors have been reported to increase cAMP levels in chick embryo myocytes (Zani *et al.* 1986). Furthermore, recent evidence shows that nicotinic cholinergic receptors can induce a prolonged hyperpolarization of neurones in the rat dorsolateral septal nucleus (Wong and Gallagher 1989).

Modulation by aminergic and peptidergic receptors

The complex multistep interaction process between the squid giant axon and the Schwann cells is modulated by a range of aminergic and peptidergic receptors found on the Schwann cells. Table 15.1 shows that the biogenic amine, octopamine, was the most potent of a range of aminergic substances tested for their ability to hyperpolarize the membrane of the Schwann cell (Reale *et al.* 1986). The response had many of the same pharmacological characteristics of octopamine receptor-mediated effects found in other

Table 15.1 Actions of amines on Schwann cell membrane potential

Amine	Hyperpolarizing response ± SE (mV)	*n*
D(−)-Octopamine	12.3 ± 0.3	3
DL-Octopamine	8.8 ± 0.3	17
Tyramine	8.6 ± 0.3	8
L(+)-Octopamine	4.0 ± 0.1	3
L-Adrenaline	3.2 ± 0.6	5
L-Noradrenaline	2.8 ± 0.5	4
Phenylethanolamine	2.0, 3.0	2
5-Hydroxytryptamine	2.3 ± 0.3	3
DL-Synephrine	0.8 ± 0.4	8
Dopamine	0	2
L-Tyrosine	0	2

The results are expressed as the mean ± SE of the maximum hyperpolarization produced by the introduction of a 1-min pulse of the amine into the superfusate at a concentration of 10^{-5} M. *n* = the number of observations for each amine on different nerve fibres. Results taken from Reale *et al.* (1986).

systems in terms of stereospecificity, amine specificity, and interactions with a range of agonists and antagonists. Subthreshold doses of DL-octopamine (10^{-8} M), which alone produced no direct effect on the membrane potential of the Schwann cell, were able to potentiate the effects of submaximal doses of carbachol and of vasoactive intestinal peptide (VIP). Since octopamine is only found at very low levels in squid stellate nerve (Juorio and Molinoff 1974), octopamine may be released into the squid haemolymph at times of stress, as in other invertebrates. Thus squid haemolymph levels would be elevated at the same time as that at which the squid would be likely to exhibit high levels of activity in the giant axons associated with the escape response.

The membrane potential of the Schwann cells is also modulated by a range of neuropeptide receptors that show many pharmacological similarities with their vertebrate counterparts. Table 15.2 shows the range of neuropeptides that are capable of directly hyperpolarizing the Schwann cell membrane (Evans *et al.* 1986). VIP is a very potent agent at inducing a longlasting hyperpolarization of the Schwann-cell membrane, with a threshold occurring between 10^{-10} and 10^{-9} M. Its effects are mimicked by peptide histidine isoleucine (PHI) and by VIP_{10-28}, but not by glucagon and secretin. The VIP responses can also be blocked competitively and reversibly by a range of VIP antagonists, such as (pCl-D-Phe^{6}, Leu^{17}) VIP and by (N-Ac-Tyr^{1}, D-Phe^{7})-GRF (1–29) amide (Evans and Villegas 1988). The VIP responses in this preparation may be mediated via the activation of an inositol phosphate pathway.

Table 15.2 The hyperpolarizing actions of a range of neuropeptides on Schwann-cell membrane potential

Neuropeptide	Hyperpolarizing response ± SE (mV)	*n*
VIP and related peptides		
VIP	10.9 ± 0.2	25
PHI	8.6 ± 0.9	4
VIP_{10-28}	7.0 ± 1.2	3
Secretin	0	4
Glucagon	0	4
Other peptides		
Substance P	10.0 ± 0.3	10
Somatostatin	8.7 ± 0.5	5
APP	5.3 ± 0.5	4
[Met]enkephalin	0	4

The results are expressed as the mean ± SE of the mean of the maximum hyperpolarization produced by the introduction of a 1-min pulse of the neuropeptide into the superfusate at a concentration of 10^{-7} M. n = the number of observations for each peptide. Results taken from Evans *et al.* (1986).

Longlasting hyperpolarizations of the Schwann cell membrane potential can also be produced by a range of other neuropeptides including, substance P, somatostatin, and avian pancreatic polypeptide (Evans *et al.* 1986). The actions of substance P in this preparation appear to be mediated by two distinct classes of receptor, one specific for the N-terminus and the other for the C-terminus of the peptide (Evans *et al.* 1990). Evidence has also been obtained for the presence of a separate set of receptors activated by bradykinin, a neuropeptide with the same N-terminal pair of amino acids (Arg–Pro) as substance P. The non-subtype specific substance P antagonist D-Arg^{1}-D-Trp^{7,9}, Leu^{11} substance P (spantide) was a potent blocker of the effects of substance P in this preparation, but it did not block the effects of VIP, carbachol, or octopamine. [Met]enkephalin does not produce any direct hyperpolarizations of the Schwann cell membrane at concentrations up to 10^{-7} M. However, at this concentration it blocks the hyperpolarizing action of a range of agonists including VIP, octopamine, carbachol, substance P, and somatostatin, as well as the hyperpolarization of the Schwann cells induced by stimulating the giant axon for 1-min periods at 100 Hz. At a concentration of 10^{-8} M the effect of [Met]enkephalin can be partially blocked by the presence of 10^{-5} M naloxone. The mode of action and physiological significance of the actions of the activation of the [Met]enkephalin-activated receptors are unknown.

The neuropeptides, VIP and substance P, as well as producing direct hyperpolarizations of the Schwann cell membrane potential, are also capable of modulating the responses of the Schwann cell to other agonists. Thus, subthreshold doses of VIP can potentiate the effects of both carbachol and octopamine, whereas subthreshold doses of substance P can potentiate the effects of VIP, but not those of carbachol and octopamine.

An extensive array of receptors has been demonstrated pharmacologically to be present in the Schwann cells of the squid giant axon. The goal is now to find physiological roles for these receptors at appropriate times during the behaviour of the animal.

Multiple classes of glutamate receptor

The application of L-glutamate to the squid giant axon–Schwann cell preparation can produce three different responses in the membrane potential of the Schwann cell of the tropical squid, *Sepioteuthis sepioidea*, which appear to be mediated by three pharmacologically distinct classes of receptor (Fig. 15.1) (Evans *et al.* 1991*b*). A class of non-*N*-methyl-D-aspartate (non-NMDA) type receptors mediates a rapid release of acetylcholine from the Schwann cells themselves that leads to the development of a rapid and long lasting hyperpolarization (Villegas 1972, 1973, 1975; Lieberman *et al.* 1989). These receptors were maximally sensitive to L-glutamate itself with a threshold between

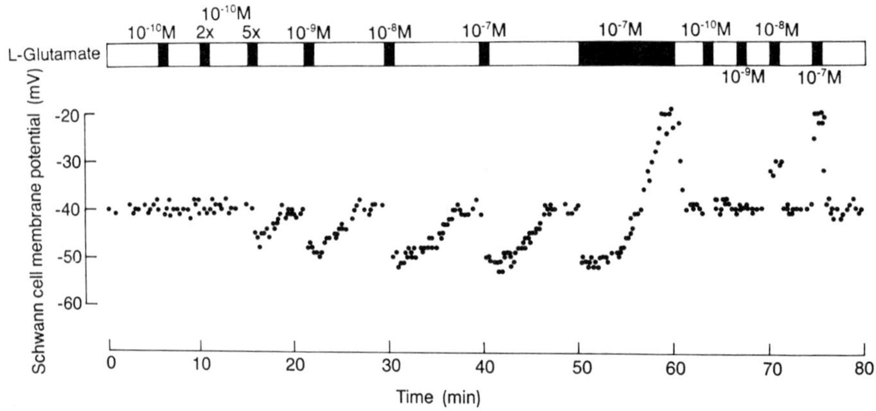

Fig. 15.1. Effect of 1-min pulses of various concentrations of L-glutamate and a 10-min pulse of 10^{-7} M L-glutamate (solid bars) on the Schwann cell membrane potential. Each point represents the potential difference recorded in a different cell. (From Evans *et al.* 1991*b*.)

2×10^{-10} and 5×10^{10} M and were more sensitive to quisqualate and kainate than to NMDA. They were blocked by high doses of the general non-NMDA antagonist glutamate diethyl ester (GDEE) (10^{-4} M) (Lieberman *et al.* 1989) but not by the specific NMDA receptor antagonist DL-2-amino-5-phosphonovaleric acid (APV). They appear to express some similarities with the metabotropic glutamate receptors which mediate their actions via the activation of G proteins and a stimulation of inositol phospholipid metabolism rather than through a direct ion channel interaction (Sugiyama *et al.* 1987, 1989). Thus they are activated by quisqualate but not by α-amino-3-hydroxy-5-methyl-4-isoxazoleproprionic acid (AMPA) (Lieberman *et al.* 1989). An additional similarity is that they are blocked by DL-2-amino-4-phosphonobutyric acid (APB), which conventionally acts like an agonist of some vertebrate glutamate receptor subtypes (Monaghan *et al.* 1989). APB also blocks the actions of non-NMDA type glutamate receptors in several cephalopod muscle preparations (Bone and Howarth 1980; Florey *et al.* 1985). Further, this class of squid Schwann cell glutamate receptor has been postulated to be involved in increasing intracellular Ca^{2+} levels to mediate the release of acetylcholine (Villegas 1978*b*; Evans *et al.* 1991*a*), but it is not currently known if these receptors activate an inositol trisphosphate mediated release of intracellular Ca^{2+} or an increase in Ca^{2+} permeability of the cell membrane.

The lack of involvement of an ion-channel-mediated event in the activation of these receptors is suggested in the tropical squid preparation, since no change in membrane potential can be observed in response to the application of short pulses of glutamate after blockade of the cholinergic receptors with α-bungarotoxin (Evans, Reale, Merzon, and Villegas, unpublished observations). However, in another species of squid, *Alloteuthis subulata*, blockade of the cholinergic receptors with tubocurare reveals the presence of a glutamate-induced depolarization (Lieberman *et al.* 1989). Another difference between the glutamate-receptor-mediated effects in these two species of squid is that aspartate was a much more powerful agonist of these receptors in *Alloteuthis* than in *Sepioteuthis*. Thus the receptors mediating this effect on the adaxonal glial cells of the squid may exhibit species differences.

The slow depolarization of the squid Schwann cells induced by long pulses of L-glutamate is likely to be mediated via the activation of NMDA-type glutamate receptors (Evans *et al.* 1991*b*). This is the first demonstration of NMDA-type glutamate receptors in any glial cell preparation. This glutamate effect is blocked specifically by APV but not by APB (Evans *et al.* 1991*b*). It is not blocked by GDEE (10^{-6} M) and has a threshold for L-glutamate of between 10^{-9} and 10^{-8} M (unpublished observations), while NMDA has a lower threshold of between 5×10^{-10} and 10^{-9} M for the induction of this effect (Evans *et al.* 1991*b*). This receptor in the squid is not activated by quisqualate or kainate. Since Ca^{2+} is needed to initiate and maintain this slow depolarization (Villegas 1978*a,b*; Evans *et al.* 1991*b*), as with NMDA receptors in neurones (Malenka *et al.* 1989), the NMDA receptors on the squid Schwann cell also mediate an increase in calcium permeability. It has been postulated that NMDA receptors in vertebrate preparations can only be activated in the presence of low doses of glycine, which acts as a co-agonist (see Thomson 1989). Thus the squid Schwann cell NMDA receptors may differ from classical vertebrate NMDA receptors in this respect. Another apparent difference in the squid and vertebrate NMDA receptors is that the squid receptor can be activated in artificial sea water containing 53 mM Mg^{2+}, whereas the vertebrate receptor is blocked in a voltage-sensitive way by 1 mM Mg^{2+}. This may well represent an adaptation of the squid NMDA receptor to a marine environment.

The fast depolarizing responses to L-glutamate induced after the activation of the NMDA receptors appear to be mediated by a second class of NMDA receptor. They are specifically activated by NMDA with a threshold between 10^{-10} and 10^{-9} M in contrast to L-glutamate which had a threshold between 10^{-9} and 10^{-8} M (Evans *et al.* 1991*b*) and they are not activated by quisqualate or kainate up to 10^{-5} M. However, the properties of these receptors differ from those of the NMDA receptors mediating the slow depolarization since they are not blocked by APV but are blocked after exposure of the preparation to GDEE (10^{-6} M). Normally, NMDA receptors are associated with the production of slowly induced, longlasting effects. However, a few rapid actions mediated by NMDA receptors have been described in some vertebrate preparations (Mody and Heinemann 1987; D'Angelo *et al.* 1990). Thus a subset of NMDA receptors on the squid Schwann cell may be able to transiently induce rapid depolarizing changes in membrane potential, after the properties of the Schwann cells have been modified by the activation of the NMDA receptors mediating the slow depolarization.

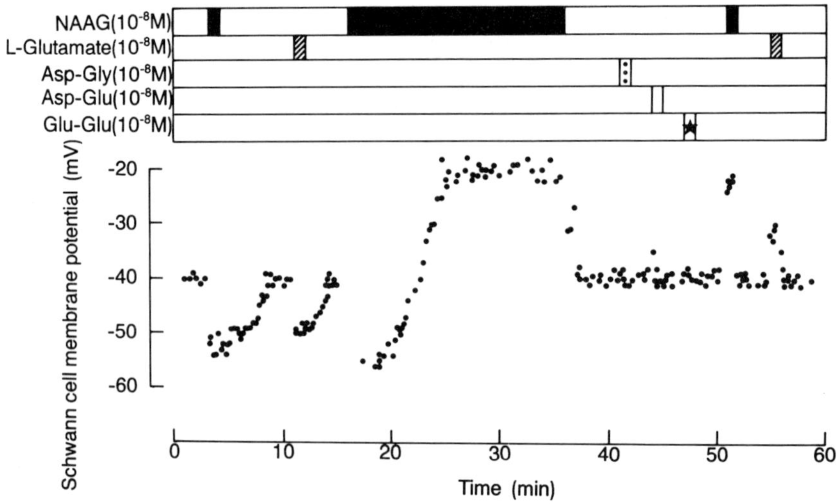

Fig. 15.2. The effect of a 1-min pulse of 10^{-8} M *N*-acetylaspartyl glutamate (NAAG) (solid bars) before and after a 20-min pulse of NAAG. The figure also shows the effect of 1-min pulses of 10^{-8} M L-glutamate (hatched bar), β-Asp–Gly (stippled bar), Asp–Glu (open bar), and Glu–Glu (starred bar) after the induction of a slow depolarization by the prolonged pulse of NAAG. Each point represents the potential difference recorded in a different Schwann cell. (From Evans *et al.* 1992*a*.)

We have recently found that the dipeptide *N*-acetylaspartyl glutamate is a potent agonist of all three glutamate receptor types present on the squid Schwann cells (Fig. 15.2) (Evans *et al.* 1992*a*). This peptide had previously been shown to be localized selectively in various regions of the vertebrate central nervous system and to be capable of activating only a portion of the glutamate receptors present (Zaczek *et al.* 1983; Moffett *et al.* 1990). A range of other dipeptides containing excitatory amino acids were also tested but they were much less effective. These findings suggest that serious consideration should be given to the possibility that the active signal released from the giant axon to stimulate the Schwann cell system in the squid could well be a dipeptide containing excitatory amino acids.

Glutamate uptake system

Evidence is accumulating in the giant axon–Schwann cell system of the tropical squid for the modulation of the effectiveness of the glutamate signal released from the axon by the actions of a high-affinity uptake system for glutamate. The presence of a glutamate-uptake system has been reported in the axolemma of the squid giant axon (Baker and Carruthers 1984), but has also been reported in the Schwann cell plasma membrane (see Villegas 1984). Schwann cells present in the axoplasm-free sheaths of split giant axon preparations, when exposed to low external concentrations of [^{14}C]-L-glutamate in the presence of external Na^+, are able to accumulate radioactively labelled material in their interior by means of an apparently high-affinity uptake mechanism (Blanco and Villegas, unpublished). In functional terms the presence of a glutamate uptake blocker (*p*-chloromercuriphenylsulfonic acid, PCMS, 100 μM) can reduce the threshold for the production of a rapid hyperpolarization of the Schwann cell membrane induced by exposure to a short 1-min pulse of L-glutamate from 2×10^{-10} to 5×10^{-10} M under control conditions to between 10^{12} and 5×10^{12} M in the presence of the blocker (Fig. 15.3) (Evans *et al.* 1992*b*). In addition, in the presence of the blocker, stimulation of the giant axon at 100 Hz for 5 min can induce a subsequent depolarization of the Schwann cell membrane (Fig. 15.4), which is due to the activation of the NMDA receptors in this preparation since short 1-min pulses of 100 Hz stimulation and short 1-min pulses of L-glutamate given subsequently now induce rapid depolarizations (Evans *et al.* 1992*b*). Thus it would seem that during stimulation of the giant axon under normal control conditions not enough glutamate accumulates to activate the NMDA receptors which have a higher threshold than the non-NMDA receptors in this preparation. However, in the presence of the uptake inhibitor the activity of the glutamate uptake system is reduced sufficiently for the extracellular level of axonally released glutamate to exceed the threshold for the activation of the NMDA type glutamate receptors in this preparation.

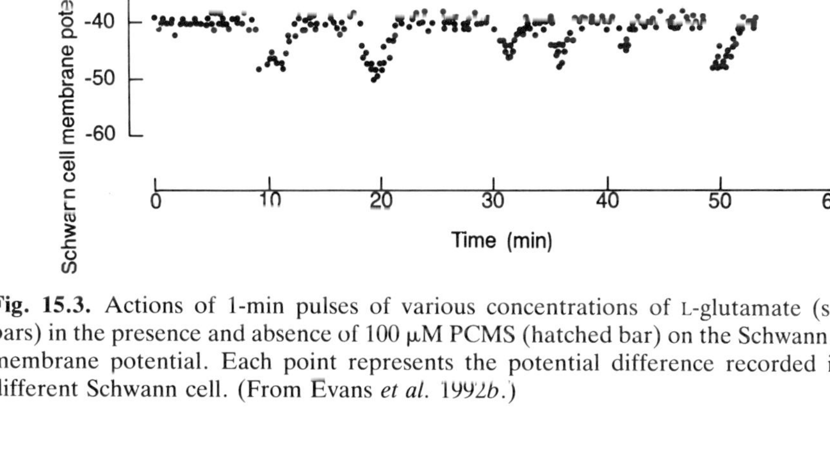

Fig. 15.3. Actions of 1-min pulses of various concentrations of L-glutamate (solid bars) in the presence and absence of 100 μM PCMS (hatched bar) on the Schwann cell membrane potential. Each point represents the potential difference recorded in a different Schwann cell. (From Evans *et al.* 1992*b*.)

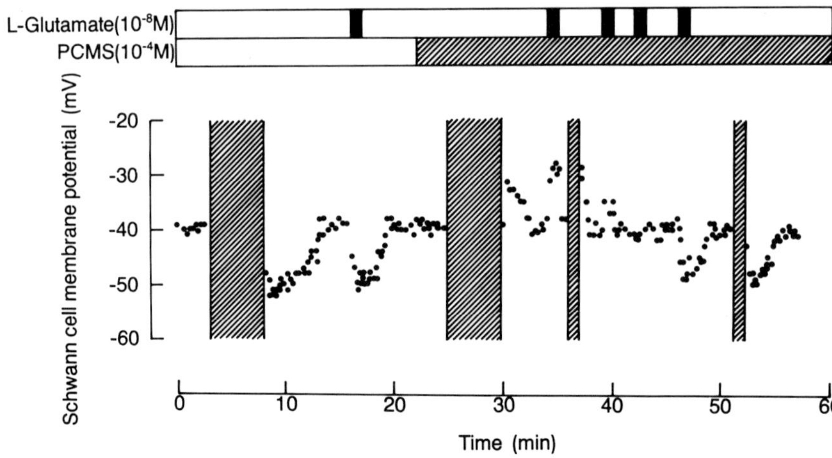

Fig. 15.4. The effect of stimulating the giant axon for 5 min at 100 Hz in the absence and presence of 100 μM PCMS (hatched bar) on the Schwann cell membrane potential. In the presence of the glutamate uptake inhibitor, both 1-min pulses of 10^{-8} M L-glutamate and 1-min periods of giant axon stimulation at 100 Hz produce rapid depolarizing effects after the 5-min period of giant axon stimulation which itself produces a depolarization. These effects gradually change to hyperpolarizations 15–20 min after the end of the prolonged period of giant axon stimulation. Each point represents the potential difference recorded in a different Schwann cell. (From Evans *et al.* 1992*b*.)

The functional significance of the presence of a very active glutamate uptake system in a preparation where glutamate is used as a signal molecule from the giant axon to activate a complicated cascade of events in the Schwann cell leading to a change in the potassium permeability of the cells when the giant axon is active is unclear. The uptake system could be present to protect the neuronal and glial membranes from continuous exposure to the levels of free L-glutamate in the squid haemolymph, so that glutamate-induced changes in the Schwann cell properties could be more closely matched to patterns of axonal stimulation. However, the presence of such an active glutamate uptake system means that the Schwann cells would have a reduced sensitivity to glutamate released from the giant axon during activity, unless some form of spatial separation occurs between the release and uptake sites. Another interesting possibility to consider, however, is that some of the recently discovered aminergic and peptidergic modulators of the squid giant axon-Schwann cell pathway (see above) may be capable of modulating the sensitivity of the Schwann cell system to released glutamate by changing the effectiveness of the glutamate uptake system in this preparation. Indeed, a monoaminergic-receptor-mediated modulation of the kinetics of a high-affinity uptake system for glutamate into astrocytes in primary culture has already been reported (Hansson and Rönnbäck 1991). In addition, NMDA receptor activation in vertebrates by glutamate stimulates phospholipase A_2 to release arachidonic acid which produces a prolonged inhibition of glutamate uptake into glial (Müller) cells of the salamander retina (Barbour *et al.* 1989). Thus the possibility of modulation of the activity of the glutamate uptake system requires further investigation in the Schwann cells of the squid giant axon.

Co-release of acetylcholine and a VIP-like peptide

The Schwann cell of the giant axon of the tropical squid responds to a short 1-min burst of axonal stimulation with a rapid hyperpolarization of the membrane that can be blocked by 10^{-8} M α-bungarotoxin (Villegas 1975). However, if the experiment is repeated with the 100-Hz stimulation being given in the presence of 10^{-8} M DL-octopamine, a further hyperpolarizing response is initiated that can be blocked by a series of specific VIP antagonists (Evans *et al.* 1986; Evans and Villegas 1988). This suggests that axonal stimulation releases both a cholinergic and a peptidergic component in this preparation. Previous evidence (Villegas and Jenden 1979) suggests that the cholinergic component is due to the release of acetylcholine from the Schwann cell itself.

The axonal activation stage in the above experiments can be bypassed by directly activating the Schwann cells by the application of pulses of L-glutamate. Figure 15.5 shows that the effect of 1 min pulses of either 10^{-9} or 10^{-6} M L-glutamate on the Schwann cell membrane potential are blocked after exposure of the preparation to 10^{-8} M α-bungarotoxin. However, if the 1-min pulse of 10^{-9} M L-glutamate is repeated in the presence of a subthreshold concentration of DL-octopamine (10^{-8} M) a further hyperpolarization can be revealed, which is blocked by the VIP blocking agent (pCl-D-Phe^6, Leu^{17}) VIP (10^{-9} M). This experiment suggests that the VIP-like peptide in this preparation is also released from the Schwann cells themselves along with acetylcholine.

Physiological role

The idea that glutamate may be released non-synaptically by axons to act as a signal to modulate the properties of their surrounding glial cells was first proposed by Villegas (1978*a*,*b*), but it may also be true for vertebrate glial cells (see Barres 1989). The non-NMDA glutamate receptors mediating the co-release of acetylcholine and the VIP-like peptide from the Schwann cells are extremely sensitive to L-glutamate (threshold between 2×10^{-10} and 5×10^{-10} M) and are likely to be activated by glutamate released non-synaptically

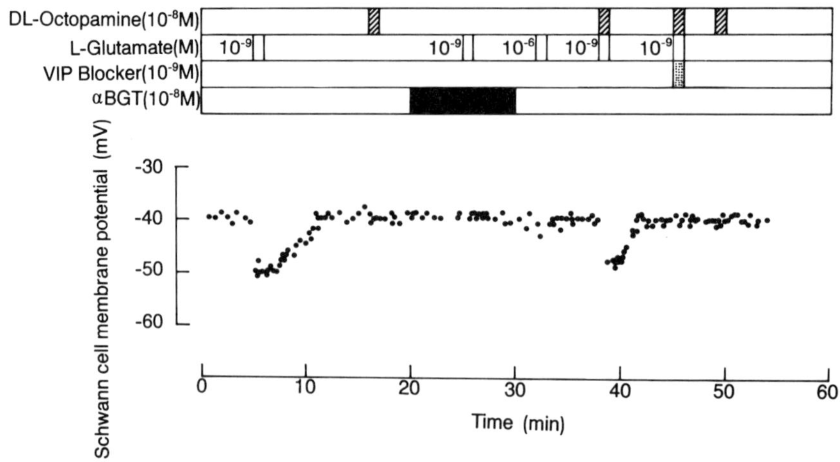

Fig. 15.5. Release of a VIP-like peptide in response to a 1-min pulse of 10^{-9} M L-glutamate in the presence of 10^{-8} M DL-octopamine after the cholinergic component of the response has been blocked by exposure of the preparation to 10^{-8} M α-bungarotoxin (αBGT). The response to the release of the endogenous peptide is blocked in the presence of 10^{-9} M of the VIP blocking agent (pCl-D-Phe^{6}, Leu^{17}) VIP. Each point represents the potential difference recorded in a different Schwann cell.

from the squid giant axon by stimulation at high frequencies (Lieberman *et al.* 1989; Evans *et al.* 1991*b*). The activation of these receptors turns on a complex multistep process that amplifies the original signal and leads to an increase in the permeability of the Schwann cell membrane to potassium ions at times when the squid is likely to be using its giant fibre-mediated escape response pathway (Villegas 1981; Villegas *et al.* 1988; Evans *et al.* 1991*a*). One possible function for this system would be to buffer any accumulation of potassium ions in the adaxonal space that might interfere with the conduction of nerve impulses along the giant axons. This whole multistep pathway appears capable of being modulated at many points to suit the minute by minute behavioural needs of the animal. It is now clear that the observed cholinergic response initiated by axonal stimulation is in fact already potentiated by the actions of the co-release subthreshold levels of the endogenous VIP-like peptide. It is worth speculating that, just as for co-release from neuronal systems, different patterns of stimulation of the Schwann cell system may result in the release of different ratios of the cholinergic and peptidergic components. By analogy with other invertebrate systems, it seems likely that octopamine, released into the haemolymph of the squid under stressful circumstances that are likely to activate the escape response, will also potentiate several components of this signalling system. However, at present, functional roles for the various other peptidergic modulatory systems present on the Schwann cells are not known.

The NMDA-type glutamate receptors present in this preparation have higher thresholds for activation by L-glutamate (between 10^{-9} and 10^{-8} M) and are thus only likely to be activated *in vivo* under extreme conditions of prolonged activity of the giant axon at high frequency, unless some form of inhibiting modulation of the high-affinity glutamate uptake system can reduce this threshold under certain behavioural circumstances. The activation of the NMDA receptors associated with the slow depolarization develops slowly during a prolonged exposure to glutamate and initiates a longlasting change in the responsiveness of the Schwann cell to further applications of glutamate. The slow depolarization above the resting membrane potential is not, however, a prerequisite for the induction of the change in responsiveness. Thus the Schwann cells of the squid giant axon possess NMDA receptors that can lead to long-term changes in responsiveness as do an increasing number of different neuronal systems (Kauer *et al.* 1988*a*, *b*; Davies *et al.* 1989; Malenka *et al.* 1989; Watkins and Collingridge 1989).

Acknowledgements

The work reported in this chapter was partially supported by CONICIT research grants S1-0808 and S1-1176, by British Council travel grants awarded to P.D. Evans, V. Reale, and J. Villegas for exchange visits between our laboratories, by grants from the Marshall and Orr Bequest of the Royal Society towards the cost of consumables for part of the work, and by the collaborative support of Fundación para la Ciencia 'José Gregorio Hernández' (FUNDACIENCIA) and its Estación de Investigaciones Marinas de Mochima.

References

Baker, P. F. and Carruthers, A. (1984). Transport of sugars and amino acids. *Current Topics in Membranes and Transport*, **22**, 91–130.

Barbour, B., Szatkowski, M., Ingledew, N., and Attwell, D. (1989). Arachidonic acid induces a prolonged inhibition of glutamate uptake into glial cells. *Nature*, **342**, 918–20.

Barres, B. A. (1989). A new form of transmission? *Nature*, **339**, 343–4.

Bone, Q. and Howarth, J. V. (1980). The role of L-glutamate in neuromuscular transmission in some molluscs. *Journal of the Marine Biological Association of the United Kingdom*, **60**, 619–26.

Boulter, J., Connolly, J., Deneris, E., Goldman, D., Heinemann, S., and Patrick, J. (1987). Functional expression of two neuronal nicotinic acetylcholine receptors from cDNA clones identifies a gene family. *Proceedings of the National Academy of Sciences, USA*, **84**, 7763–7.

Brown, E. R., Bone, Q., Ryan, K. P., and Abbott, N. J. (1991). Morphology and electrical properties of Schwann cells around the giant axon of the squids *Loligo forbesi* and *Loligo vulgaris*. *Proceedings of the Royal Society of London, Series B*, **243**, 255–62.

Chen, S-J., Spathis, R., and Schmidt, J. (1988). Binding sites for [^{3}H]-acetylcholine and ^{125}-I-α-bungarotoxin in the optic ganglion of *Loligo pealii*. *Comparative Biochemistry and Physiology*, **90C**, 317–23.

Chiappinelli, V. A. (1985). Actions of snake venom toxins on neuronal nicotinic receptors and other neuronal receptors. *Pharmacological Therapeutics*, **31**, 1–32.

D'Angelo, E., Rossi, P., and Garthwaite, J. (1990). Dual-component NMDA receptor currents at a single central synapse. *Nature*, **346**, 467–70.

Davies, S. N., Lester, R. A. J., Reymann, K. G., and Collingridge, G. L. (1989). Temporally distinct pre- and post-synaptic mechanisms maintain long-term potentiation. *Nature*, **338**, 500–3.

Dennis, M. J. and Miledi, R. (1974). Electrically induced release of acetylcholine from denervated Schwann cells. *Journal of Physiology*, **237**, 431–52.

Evans, P. D. and Villegas, J. (1988). The action of vasoactive intestinal peptide antagonists on peptidergic modulation of the squid Schwann cell. *Journal of Experimental Biology*, **138**, 259–69.

Evans, P. D., Reale, V., and Villegas, J. (1985). The role of cyclic nucleotides in modulation of the membrane potential of the Schwann cell of the squid giant nerve fibre. *Journal of Physiology*, **363**, 151–67.

Evans, P. D., Reale, V., and Villegas, J. (1986). Peptidergic modulation of the membrane potential of the Schwann cell of the squid giant nerve fibre. *Journal of Physiology*, **379**, 61–82.

Evans, P. D., Reale, V., Merzon, R. M., and Villegas, J. (1990). Substance P modulation of the membrane potential of the Schwann cell of the squid giant nerve fibre. *Glia*, **3**, 393–404.

Evans, P. D., Reale, V., Merzon, R. M., and Villegas, J. (1991*a*). Mechanisms of axon–Schwann cell signalling in the squid nerve fibre. *Annals of the New York Academy of Sciences*, **633**, 434–47.

Evans, P. D., Reale, V., Merzon, R. M., and Villegas, J. (1991*b*). *N*-methyl-D-aspartate (NMDA) and non-NMDA type glutamate receptors are present on squid giant axon Schwann cells. *Journal of Experimental Biology*, **157**, 593–600.

Evans, P. D., Reale V., Merzon, R. M., and Villegas, J. (1992*a*). *N*-methyl-D-aspartate (NMDA) and non-NMDA (Metabotropic) type glutamate receptors modulate the membrane potential of the Schwann cell of the squid giant nerve fibre. *Journal of Experimental Biology*, **173**, 229–49.

Evans, P. D., Reale, V., Merzon, R. M., and Villegas, J. (1992*b*). The effect of a glutamate uptake inhibitor on axon–Schwann cell signalling in the squid giant nerve fibre. *Journal of Experimental Biology*, **173**, 251–60.

Florey, E., Dubas, F., and Hanlon, R. T. (1985). Evidence for L-glutamate as a transmitter substance of motorneurons innervating squid chromatophore muscles. *Comparative Biochemistry and Physiology*, **82C**, 259–68.

Hansson, E. and Rönnbäck, L. (1991). Receptor regulation of the glutamate, GABA and taurine high-affinity uptake into astrocytes in primary culture. *Brain Research*, **548**, 215–21.

Juorio, A. V. and Molinoff, P. B. (1974). The normal occurrence of octopamine in neural tissues of the Octopus and other cephalopods. *Journal of Neurochemistry*, **22**, 271–80.

Kauer, J. A., Malenka, R. C., and Nicoll, R. A. (1988*a*). NMDA application potentiates synaptic transmission in the hippocampus. *Nature*, **334**, 250–2.

Kauer, J. A., Malenka, R. C., and Nicoll, R. A. (1988*b*). A persistent postsynaptic modification mediates long-term potentiation in the hippocampus. *Neuron*, **1**, 911–17.

Kimelberg, H. K. (ed.) (1988). *Glial cell receptors*. Raven Press, New York.
Laurenza, A. S., Sutkowski, E. K., and Seamon, K. B. (1989). Foskolin: a specific stimulator of adenylyl cyclase or a diterpene with multiple sites of action? *Trends in Pharmacological Sciences*, **10**, 442–7.
Lieberman, E. M., Abbott, N. J., and Hassan, S. (1989). Evidence that glutamate mediates axon-to-Schwann cell signalling in the squid. *Glia*, **2**, 94–102.
Malenka, R. C., Kauer, J. A., Perkel, D. J., and Nicoll, R. A. (1989). The impact of postsynaptic calcium on synaptic transmission—its role in long-term potentiation. *Trends in Neurosciences*, **12**, 444–50.
Mody, I. and Heinemann, U. (1987). NMDA receptors of dendate gyrus granule cells participate in synaptic transmission following kindling. *Nature*, **326**, 701–4.
Moffett, J. R., Williamson, L., Palkovits, M., and Namboodiri, M. A. A. (1990). *N*-acetylaspartylglutamate: a transmitter candidate for the retinohypothalmic tract. *Proceedings of the National Academy of Sciences, USA*, **87**, 8065–9.
Monaghan, D. T., Bridges, R. J., and Cotman, C. W. (1989). The excitatory amino acid receptors: their classes, pharmacology and distinct properties in the function of the central nervous system. *Annual Review of Pharmacology and Toxicology*, **29**, 365–402.
Reale, V., Evans, P. D., and Villegas, J. (1986). Octopaminergic modulation of the membrane potential of the Schwann cell of the squid giant nerve fibre. *Journal of Experimental Biology*, **121**, 421–43.
Sugiyama, H., Ito, I., and Hirono, C. (1987). A new type of glutamate receptor linked to inositol phospholipid metabolism. *Nature*, **325**, 531–3.
Sugiyama, H., Ito, I., and Watanabe, M. (1989). Glutamate receptor subtypes may be classified into two major categories: a study on *Xenopus* oocytes injected with rat brain mRNA. *Neuron*, **3**, 129–32.
Thomson, A. M. (1989). Glycine modulation of the NMDA receptor/channel complex. *Trends in Neurosciences*, **12**, 349–53.
Villegas, J. (1972). Axon–Schwann cell interaction in the squid nerve fibre. *Journal of Physiology*, **225**, 275–96.
Villegas, J. (1973). Effects of tubocurarine and eserine on the axon Schwann cell relationship in the squid nerve fibre. *Journal of Physiology*, **232**, 193–208.
Villegas, J. (1974). Effects of acetylcholine and carbamylcholine on the axon and Schwann cell electrical potentials in the squid nerve fibre. *Journal of Physiology*, **242**, 647–59.
Villegas, J. (1975). Characterization of acetylcholine receptors in the Schwann cell membrane of the squid nerve fibre. *Journal of Physiology*, **249**, 679–89.
Villegas, J. (1978*a*). Cholinergic properties of satellite cells in the peripheral nervous system. In *Dynamic properties of glia cells* (ed. E. Schoffeniels, G. Franck, L. Hertz, and D. B. Tower), pp. 207–15. Pergamon Press, Oxford.
Villegas, J. (1978*b*). Cholinergic systems in axon–Schwann cell interactions. *Trends in Neurosciences*, **1**, 66–8.
Villegas, J. (1981). Axon/Schwann cell relationships in the giant nerve fibre of the squid. *Journal of Experimental Biology* **95**, 135–51.
Villegas, J. (1984). Axon–Schwann cell relationship. *Current Topics in Membranes and Transport*, **22**, 547–71.
Villegas, J. and Jenden, D. J. (1979). Acetylcholine content of the Schwann cell and axon in the giant nerve fibre of the squid. *Journal of Neurochemistry*, **32**, 761–6.
Villegas, J., Evans, P. D., and Reale, V. (1988). Electrophysiology of Schwann cell

receptors. In *Glial cell receptors* (ed. H. K. Kimmelberg), pp. 141–57. Raven Press, New York.

Watkins, J. C. and Collingridge, G. L. (ed.) (1989). *The NMDA receptor*. IRL Press, Oxford.

Wong, L. A. and Gallagher, J. P. (1989). A direct nicotinic receptor-mediated inhibition recorded intracellularly in vitro. *Nature*, **341**, 439–42.

Zaczek, R., Koller, K., Cotter, R., Heller, D., and Coyle, J. T. (1983). *N*-acetyl-aspartylglutamate: an endogenous peptide with high affinity for a brain 'glutamate' receptors. *Proceedings of the National Academy of Sciences, USA*, **80**, 1116–19.

Zani, B. M., Grassi, F., Molinano, M., Monaco, L., and Eusebi, F. (1986). Cyclic AMP regulates the life time of acetylcholine-activated channels in cultured myotubes. *Biochemical and Biophysical Research Communications*, **140**, 243–9.

16

Periaxonal ion regulation in the squid

Y. PICHON, N. JOAN ABBOTT, E. R. BROWN, ISAO INOUE, and PATRICIA A. REVEST

Summary

In the majority of animal groups, axons release K^+ and take up Na^+ during action potential production, with a tendency for an accumulation of K^+ and depletion of Na^+ in the narrow extracellular spaces adjacent to the axon. Homeostatic mechanisms are present to reduce the severity of these changes, and so limit their undesirable effects on neuronal function. This chapter reviews studies in squid giant axons, to show that under normal physiological conditions the Schwann cell sheath around the axon has powerful mechanisms for regulating the $[K^+]$ in the periaxonal space, using a combination of passive diffusion (particularly via the transcellular glial tubular system and across its membranes) and carrier-mediated transport. The significance of this regulation for the normal function of the giant axon system in swimming and escape responses is discussed.

Need for periaxonal ion regulation

Axons in vertebrates and invertebrates use a sequence of voltage-dependent ionic currents to generate propagating action potentials, generally an early inward Na^+ current followed by a late outward K^+ current (Hodgkin and Huxley 1952; Hodgkin 1958; Fig. 16.1). If the axon is surrounded by a narrow interstitial space, there is a tendency for K^+ accumulation and Na^+ depletion in the space during activity. Severe K^+ accumulation and Na^+ depletion will impair axonal function, by causing depolarization, reducing the electrochemical gradient driving Na^+ entry, and interfering with Na^+ current (Adelman and Palti 1969) and closing of K^+ channels (Swenson and Armstrong 1981; Clay 1986). The axonal resting potential is relatively insensitive to small changes of $[K^+]$ in the physiological range, because the significant Na^+ permeability of the membrane flattens the Nernst plot of membrane potential versus log $[K^+]$ in this range (Curtis and Cole 1942; Baker *et al.* 1962*b*; Fig 16.2). However, although moderate K^+ accumulation around large axon segments has little effect on the resting potential, such accumulation can still influence action potential generation and conduction because of its effects on

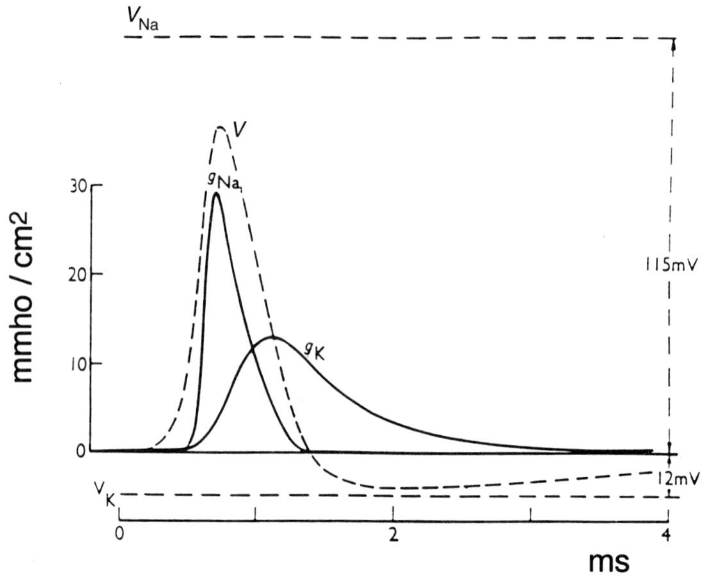

Fig. 16.1. Theoretical action potential (V) and conductance (g) changes obtained by numerical solution of the Hodgkin–Huxley equations, using constants derived from voltage-clamp experiments in *Loligo forbesi* at 18.5 °C. Note that, as the undershoot of the action potential approaches the potassium equilibrium potential, V_K, the membrane behaves like a K^+-selective electrode at this point. The size of the undershoot is then a measure of the $[K^+]$ in the periaxonal space. (Taken with permission from Hodgkin and Huxley (1952).)

voltage-gated Na and K channels. The situation is particularly critical around axon branch points, in zones of restricted extracellular space, at low temperatures, and in hypoxia, conditions in which significant reductions in the safety factor for conduction can occur (reviewed in Smith 1980). It would be predicted that the neural processes underlying escape behaviour, especially the giant fibre activation of the mantle muscle, would require efficient K^+ homeostasis. This chapter reviews the observations on squid giant axons that demonstrate effective ion regulation under physiological conditions, and experiments designed to show the mechanisms responsible.

Potassium accumulation

Early experiments: significant K^+ accumulation in *Loligo* axons

The third-order giant axon of the squid stellar nerve runs in a bundle together with many small axons. The giant axon is surrounded by a closely apposed layer of Schwann (glial) cells that separate the axon from the bulk interstitial

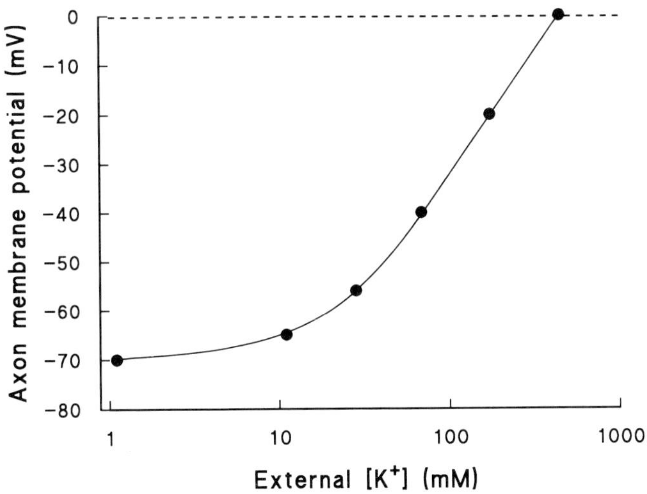

Fig. 16.2. Dependence of the resting membrane potential on external [K^+] in giant axon of *Loligo pealei*. At high K^+ the curve is a straight line, with a slope close to 58 mV per decade change in [K^+] (Nernst equation), but in the physiological range (~10mM [K^+]) the potential is relatively insensitive to K^+ concentration, because of the contribution of Na^+ permeability. (Redrawn with permission from Curtis and Cole (1942), using a −14 mV correction for electrode junction potential.)

space of the axon bundle (Geren and Schmitt 1954; G. M. Villegas and Villegas 1960, 1984). In a classic study at the Marine Biological Association Laboratory, Plymouth, on axons of the squid *Loligo forbesi*, Frankenhaeuser and Hodgkin (1956) showed that the action potential undershoot could be used as a measure of [K^+] bathing the axonal surface, and that progressive decline of undershoots during a train of action potentials indicated K^+ accumulation in a space adjacent to the axon (Fig. 16.3). Calibration studies exposing the axon to different external K^+ concentrations showed an approximately linear relation between undershoot amplitude and [K^+] in the physiological range, with a 1 mV change in undershoot corresponding to approximately 1 mM change in [K^+] (Fig. 16.4). For a seven action potential train at 50 Hz, K^+ accumulation could be described by a single exponential of time constant τ (mean 58 ms), and the mean rise in [K^+] immediately after a single spike was 1.6 mM. The results could be fitted by a formal model in which K^+ accumulates in a narrow space of width θ outside the axon, from which it is cleared by a pathway of permeability P [Fig. 16.5(a)]. θ was given by the K^+ leakage per impulse divided by the rise in [K^+], and averaged 27 nm, while P could be calculated from θ/τ, giving a mean value of 6×10^{-5} cm/s. The clearance mechanism had a temperature coefficient (Q_{10}) of 1.3, compatible with a passive diffusion process. The calculated values of θ and P were close to estimates based on the morphology of the Schwann cell sheath, θ being

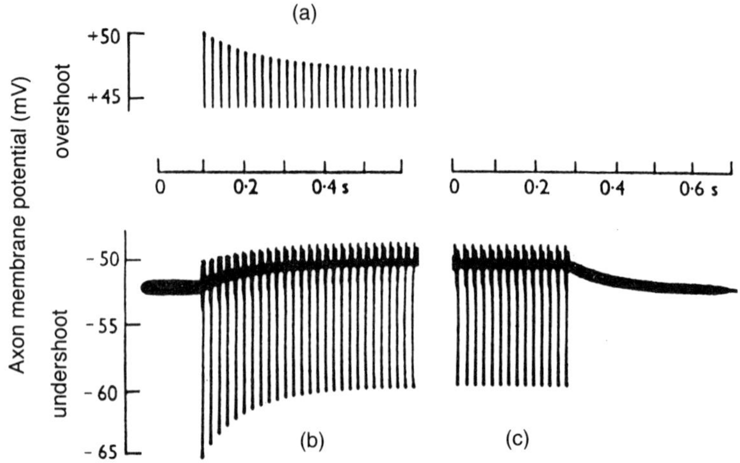

Fig. 16.3. Records of the peak action potential overshoots (a), and the undershoots at the beginning (b) and end (c) of a train of action potentials at 50 Hz, lasting around 1 s (note break in record). *Loligo forbesi*, temperature 19.8°C, $[K^+]$ 10 mM. The polarity of potentials follows the modern convention—depolarization upwards, hyperpolarization downwards—the inverse of the original. (Taken with permission from Frankenhaeuser and Hodgkin (1956).)

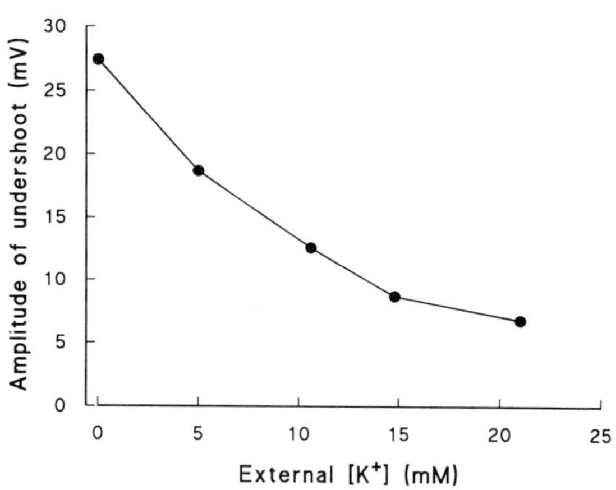

Fig. 16.4. Relation between the amplitude of the action potential undershoot and external $[K^+]$ in axons from *Loligo forbesi*, a calibration curve that can be used to deduce changes in periaxonal $[K^+]$ during a train of action potentials. In the physiological range (around 10 mM), a 1-mV change in undershoot corresponds to *ca.* 1 mM change in $[K^+]$. (Taken with permission from Frankenhaeuser and Hodgkin (1956).)

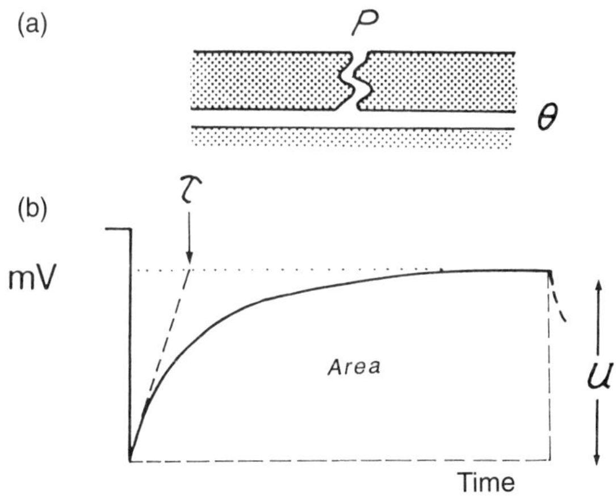

Fig. 16.5. Diagram to show formal model of the axon surface, developed by Frankenhaeuser and Hodgkin (1956) in which K^+ accumulates in a narrow space of width θ, and escapes across the Schwann cell sheath by a pathway of permeability *P*. (b) The 'envelope' of action potential undershoots during a train of action potentials, here shown as a single exponential, of time constant τ and coefficient *U* (Astion *et al*. 1988). The area under the curve is a measure of the degree of K^+ accumulation during the train (Abbott *et al*., in preparation).

similar to the width of the periaxonal space (~10 nm), and *P* approximately as expected if K^+ clearance occurred by diffusion through intercellular (mesaxonal) clefts across the Schwann cell layer (Geren and Schmitt 1954). A few voltage clamp experiments reported in the same study showed K^+ accumulation during pulses of greater than 4 ms duration, as judged from tail currents and sag of the steady-state K^+ current I_K. Most published voltage-clamp studies since the 1960s have shown appreciable periaxonal K^+ accumulation (e.g. Haydon *et al*. 1984; Fig. 16.6).

Modification of K^+ accumulation by osmotic manipulation

K^+ accumulation creates problems in voltage-clamp experiments by causing inactivation of Na^+ currents (Adelman and Palti 1969, 1972) and inaccuracy in measurement of K^+ conductance g_K (Adelman *et al*. 1973). Adelman and Fitzhugh (1975) introduced modified Hodgkin–Huxley equations describing membrane excitability, to allow for the effects of the accumulation. The severity of K^+ accumulation could be reduced by using hypertonic solutions to cause cell shrinkage (Adelman *et al*. 1973), for example 1.31 compared with normal 1.08 Osmol./l. This manoeuvre caused an increase of 194 per cent in θ. Subsequent morphometric study at the electron microscopic (EM) level showed a 147 per cent increase in the measured periaxonal space width, and 82 per cent increase in mesaxonal cleft width (Adelman *et al*. 1977). The

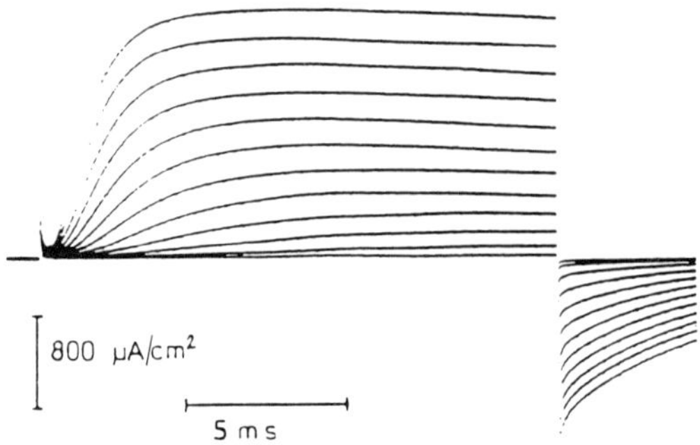

Fig. 16.6. Current traces from voltage-clamped axon of *Loligo forbesi* showing K^+ accumulation, seen in the sag of the steady-state outward K^+ current, and inwardly directed tail currents. Tetrodotoxin was used to block the Na^+ current; holding potential −60 mV; temperature 6°C. (Taken with permission from Haydon *et al.* (1984).)

discrepancies between the widths of the anatomical and physiological periaxonal space (~10 nm in EM, θ = 30 nm from physiological estimates) were taken to indicate that a proportion of the physiological periaxonal space included mesaxonal cleft, but no explanation was offered for the smaller percentage changes seen in the EM compared with the physiological estimates. More recently, Kukita and co-workers (Kukita and Yamagishi 1983; Kukita 1988) have used an outward water flow across the axon membrane to overcome K^+ accumulation, by generating a gradient of osmolarity between axon interior and external solution (Fig. 16.7).

Little K^+ accumulation in dissected fresh *Loligo* axons

Loligo forbesi used in Plymouth are often caught by trawling some distance offshore, so that by the time animals reach the laboratory some may be in poor condition. However, the axon membrane is relatively robust, so voltage-clamp studies of axonal currents can be done even on axons from animals decapitated at sea whose mantles have been kept on ice. Y. Pichon introduced jigging as a method of *Loligo* catching in Roscoff, France, so that animals were in good condition when caught, and survived for several months in a deep outside sea water tank. Voltage-clamp studies on axons from these healthy animals showed some variation in K^+ accumulation at 2°C, it being possible to divide axons into groups A, B, and C, depending on the level of K^+ accumulation (Pichon *et al.* 1987; Larmet and Pichon 1988; Fig. 16.8). Type A axons, which showed significant accumulation, were usually old or damaged axons; type C axons showed effectively no accumulation; type B were intermediate. Fresh axons showed type B or type C behaviour. Because of the electrokinetic effects of an imposed voltage on K^+ movement in the

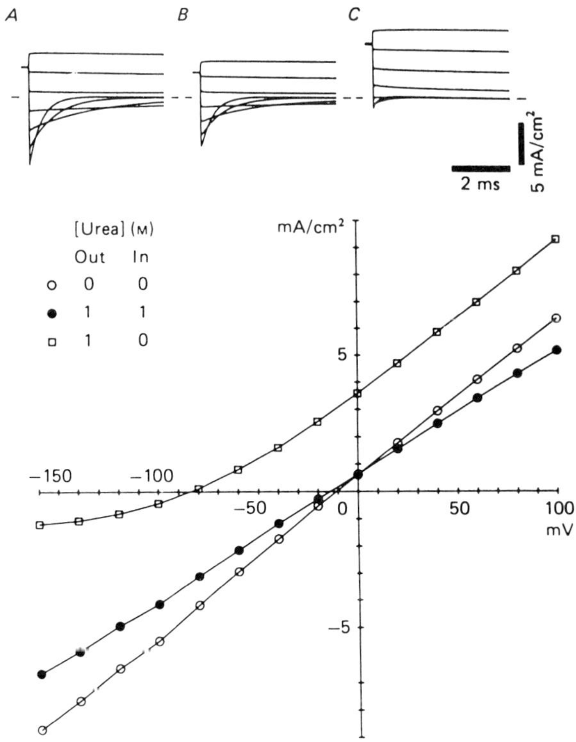

Fig. 16.7. The use of an osmotic water flow to reduce periaxonal K^+ accumulation. The effects of an osmotic gradient on tail currents (above) and instantaneous *I–V* relations (below), the latter derived from a depolarization to 50 mV for 10 ms. Changing both external and internal solutions from isotonic (A, open circle) to hypertonic (B, closed circle) had little effect, whereas changing an internal hypertonic solution to an isotonic solution (C, open square) caused reduction in tail currents and a shift of the *I–V* relation, with clear outward rectification. Families of current traces (above) were recorded at potentials between −160 and 80 mV in 40 mV steps. (Taken with permission from Kukita (1988).)

region of the membrane (see the 'Appendix' to this chapter), it would be useful to extend these experiments with longer pulses (cf. Chapter 5, Fig. 5.3) for a more complete analysis of K^+ accumulation and clearance.

Little K^+ accumulation in *Alloteuthis* axons *in situ*

The small squid *Alloteuthis subulata*, available in Plymouth, is particularly suitable for studies on axons *in situ*. The mantle is relatively transparent, so axons can be impaled *in situ* without dissection. There was little decline in undershoots during a burst of action potentials in fresh preparations, showing

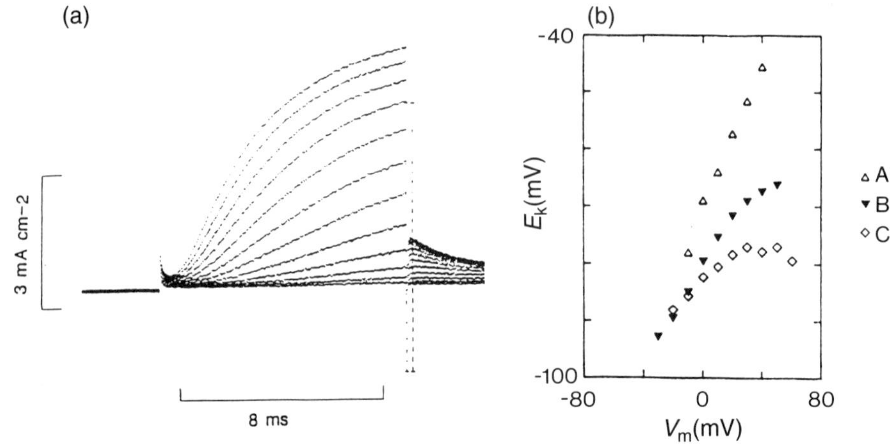

Fig. 16.8. Voltage-clamp studies of fresh axons from *Loligo forbesi*. (a) Family of K^+ currents: holding potential, −70 mV; 10-mV depolarizing steps; temperature, 2°C. The Na^+ current was blocked by tetrodotoxin. Note that, although the K^+ current is larger than in Fig. 16.6, there is less K^+ accumulation (outward tail currents, no sag of steady-state outward currents). (b) Characteristic potassium equilibrium potential (E_K) versus membrane potential (V_m) curves for three categories of giant axons: A, rapidly accumulating (usually old or damaged axons); B, slowly accumulating (usually fresh axons); and C, non-accumulating (fresh axons). Note that at the resting level, E_K is the same in all cases (around −90 mV), suggesting that the K^+ leak from the axons is compensated by effective clearance. (Taken with permission from Pichon *et al.* (1987).)

little K^+ accumulation; accumulation gradually increased with time (Abbott *et al.* 1988; Fig. 16.9).

Relation of K^+ accumulation to the physiological state of Schwann cells

It proved possible to prepare isolated *Alloteuthis* axons showing little K^+ accumulation, by careful dissection without stretching or drying, and by keeping the preparation at < 15°C (Abbott *et al.* 1988). Electron microscopic examination of axons that had been studied electrophysiologically showed a correlation between Schwann cell morphology and degree of accumulation. Non-accumulating axons showed Schwann cells with well preserved membranes and organelles, while accumulating axons showed disrupted Schwann cells with ruptured membranes and vacuolated cytoplasm (Brown 1991, 1993). It was concluded that healthy axons have effective mechanisms for regulating $[K^+]$ in the periaxonal space, and that the geometry of the space, and effectiveness of K^+ clearance, depend on the physiological state of the associated Schwann cells.

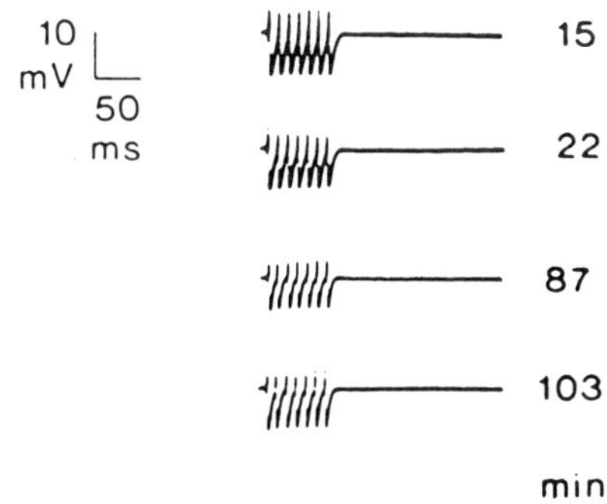

Fig. 16.9. *In situ* axon of *Alloteuthis subulata*; action potential undershoots during 70-ms stimulation at 100 Hz, 18°C. There was little undershoot decline during the burst (little K^+ accumulation) even after 103 min without vascular perfusion. (Taken with permission from Abbott *et al.* (1988).)

K^+ homeostasis by Schwann cells

Passive and carrier-mediated processes

EM morphometric study of the squid giant axon sheath has shown that the plasmalemmal surface areas of the axon and Schwann cell are in the ratio of approximately 1:10 in *Loligo* (Brown *et al.* 1991) and 1:4 in *Alloteuthis* (Brown and Abbott 1993). In the discussion that follows, it will be assumed that Schwann cell mechanisms contribute the major fraction of K^+ clearance, reflecting their advantageous membrane area. The role of the axon in its own homeostasis will be discussed later.

Passive processes

The early studies by Frankenhaeuser and Hodgkin (1956) on *Loligo* axons indicated that passive diffusion via an extracellular route could account for the rate of K^+ clearance observed. In the light of later experience, it seems likely that these axons showed unphysiological levels of K^+ accumulation. It was therefore useful to re-examine the mechanisms for clearance using *Alloteuthis*, in which healthy Schwann cells and low accumulation could be preserved.

If extracellular diffusion accounts for a major fraction of K^+ clearance, it should be possible to change accumulation by osmotic manipulation, hypertonic solutions reducing accumulation (by shrinking Schwann cells), and hypotonic solutions increasing it (Schwann cell swelling). Astion *et al.* (1988) showed that, in control *Alloteuthis* axons kept at 12°C and stimulated for 200 ms at 60 Hz, K^+ accumulation was relatively small (typically < 1 mM change during the action potential train). The action potential undershoots declined along a curve that could be fitted by a single exponential, and from the

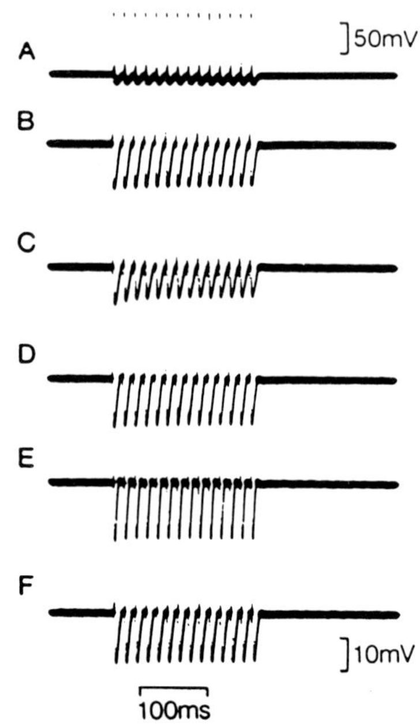

Fig. 16.10. Isolated axon of *Alloteuthis subulata* showing the effect of changing osmolarity. Trace A shows a train of action potentials in artificial sea water (ASW) at 10–12°C and trace B shows the undershoots at higher amplification (little accumulation). Trace C, undershoots after 6 min in hypotonic (80 per cent) ASW; trace D, 6 min after return to normal ASW; trace E, after 6 min in hypertonic (+ 17 per cent) ASW; trace F, 6 min after return to normal ASW. Only with cell swelling caused by hypotonic solution is K^+ accumulation marked. (Taken with permission from Astion *et al.* (1988).)

observed time constant τ and coefficient U, applying the formal Frankenhaeuser and Hodgkin model, θ and P could be calculated. Osmotic manipulation had the predicted effect on K^+ accumulation (Fig. 16.10). Although the results were qualitatively similar to those previously observed, the calculated values of θ and P of the least accumulating axons were much larger, respectively, 6 and 11 times the mean values reported by Frankenhaeuser and Hodgkin (1956). If K^+ clearance occurs via an extracellular route, the results imply a very large periaxonal space and open pathway across the Schwann cell layer.

Alternatively, passive clearance may include a diffusive component across Schwann cell membranes, in which case the calculated θ will be a 'virtual' space with contribution from the Schwann cell volume. Preliminary study of rapidly frozen axons indicates that the periaxonal space width is similar to that in conventional fixation (Brown 1991), making it unlikely that extracellular diffusion can account for the K^+ clearance in the least accumulating

Alloteuthis axons. Cell penetration via K^+ and Cl^- channels (see below, and Chapter 14, this volume) may be responsible.

Carrier-mediated processes

Examination of trains of action potentials lasting 5–10 s in *Alloteuthis* showed that the undershoots declined along a complex time course, which required at least three exponentials to fit it (Fig. 16.11), i.e. at longer times additional mechanisms contributing to accumulation/clearance could be detected (Abbott *et al.* 1989). As an electrogenic $Na^+ K^+$ATPase is likely to be present in the Schwann cell membrane, and as a bumetanide-sensitive $Na^+–K^+–2Cl^-$ co-transporter has been implicated in other glial ion regulatory systems (Kimelberg and Frangakis 1985; Walz and Hinks 1985; Hassan and Lieberman 1988), experiments were designed to test the effect of transport inhibitors on K^+ accumulation. Osmotic agents expected to modulate extracellular clearance routes were also tested. Curves of peak undershoots versus time were analysed, to give τ and U values for each of the three exponential components (fast, medium, and slow), and 'virtual' θ and P values calculated for each. Osmotic agents had most effect on the fast component, as expected if this component includes passive diffusion via an extracellular route. Ouabain

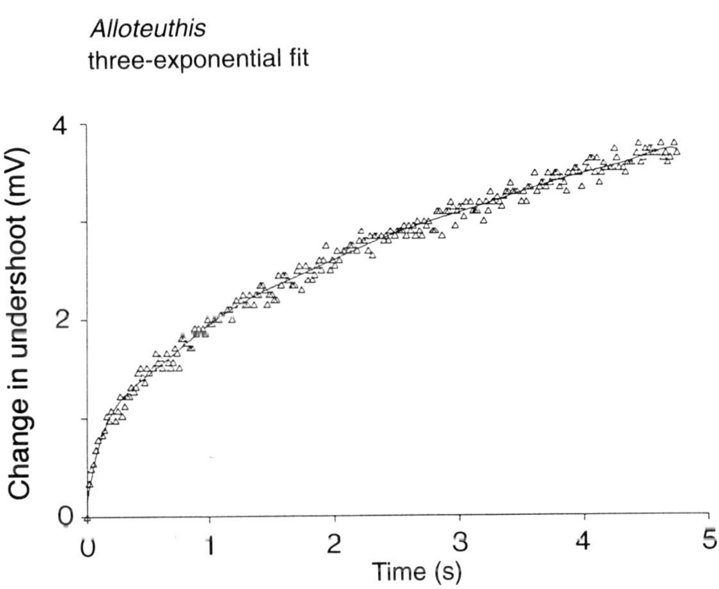

Fig. 16.11. Isolated axon of *Alloteuthis subulata*, 12°C. The triangles are individual action potential undershoots during a 5-s stimulation period at 50 Hz, expressed as change compared with the initial value. Three exponentials are required to fit the curve. (Abbott *et al.*, in preparation.)

and some effect on the fast component, which could be due to cell swelling resulting from blocking the Na^+ K^+ pump. Bumetanide and ouabain had their major effects on the medium and slow components of accumulation, evidence for the contribution of both Na^+ K^+ ATPase and Na^+–K^+–$2Cl^-$ co-transporter to the clearance process (Abbott *et al.* 1989 and in preparation).

Nature of extracellular routes involved in K^+ clearance

The earlier studies of Frankenhaeuser and Hodgkin (1956), Adelman *et al.* (1977), and Astion *et al.* (1988) were interpreted in terms of K^+ clearance via extracellular diffusion through the clefts between Schwann cells. In analogous experiments on giant axons from the crayfish ventral nerve cord, much more rapid K^+ clearance was seen, and attributed to a system of transcellular tubules acting as a short cut across the adaxonal glial cells (Shrager *et al.* 1983). Although a similar tubular structure in squid Schwann cells was described 20 years ago (Hama 1962), it was subsequently ignored. It was illustrated in a 1984 review (G. M. Villegas and Villegas 1984), and subjected to morphometric analysis by Zwahlen *et al.* (1988) (Fig. 16.12). The latter authors also showed a single-stranded junction at the axonal end of the intercellular cleft. The morphometric analysis suggested that the glial tubular system was likely to be the predominant extracellular route for ion flux, and could account for the series resistance measured under voltage and current clamp (cf. Binstock *et al.* 1975).

Recently, a detailed morphometric and tracer study of the Schwann cell

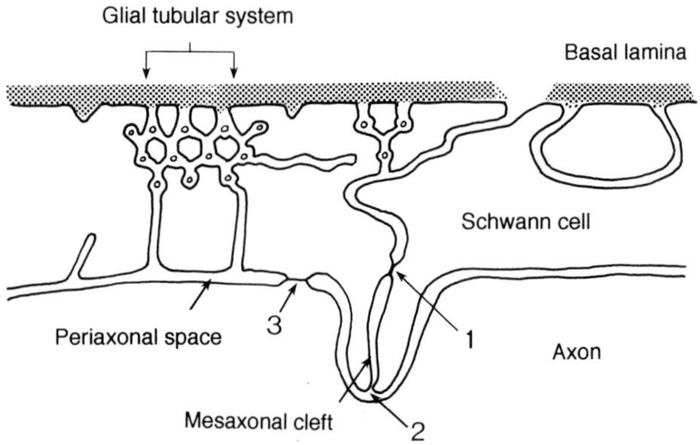

Fig. 16.12. Diagram of extracellular pathways across the Schwann cell layer, based on studies by Zwahlen *et al.* (1988) and Brown and Abbott (1993). The extensive 'glial tubular system' (or lattice) towards the basal side of the Schwann cell, connecting to the periaxonal space, is shown. Morphometric and tracer analysis suggests that the tubular system is the predominant clearance pathway for K^+. Close membrane apposi-tions between Schwann cells (1,2) and between Schwann cell and axon (3) are indicated.

sheath of *Alloteuthis* has been done (Brown and Abbott 1993). Both the cleft and tubular spaces are penetrated by electron-dense tracers (lanthanum, tannic acid, horse-radish peroxidase) and are hence potential routes for ion flux, but the quantitative analysis shows a much greater axon-facing surface area occupied by tubular openings than by cleft openings (0.26 versus 0.02 per cent area, respectively). Moreover, in order to explain the low series resistance (*ca.* 1 Ω cm^2; Binstock *et al.* 1975), it is necessary to include the tubular pathway. The conclusion is that the K^+ clearance component due to passive extracellular diffusion is predominantly via the transcellular tubules. The tubules have a relatively straight section perpendicular to the surface membrane at the axon-facing end, and a more highly branched 'lattice' arrangement at the basal lamina end, which would be an effective way of removing K^+ quickly from the axon surface into a more peripheral 'sink', yet not distributing it so far from the axon as to hinder recycling. Osmotic manipulation of the bathing solution causes greater change in the width of the tubular system than in the width of the mesaxonal cleft (Abbott and Brown 1989; Brown 1991), suggesting that the reduction in K^+ accumulation seen in hypertonic solutions is mainly due to increased permeability of the tubular route.

Role of Schwann cell membrane conductance in K^+ clearance

The electrophysiological analysis of the Schwann cell properties shows a relatively K^+-selective membrane (J. Villegas 1974; Brown 1991), possibly associated with high chloride permeability (Lieberman and Sanzenbacher 1992). If K^+ and Cl^- channels are present in the tubular membranes, this would also be an efficient way of increasing the surface area for uptake in the Schwann cells. Calculations by I. C. H. Smith (in preparation) show that, if the Schwann cell tubular membrane has a K^+ permeability similar to that of the axon ($P_K = 1\text{–}10 \times 10^{-5}$ cm/s at -40 mV membrane potential), it would allow between 58 and 98 per cent of K^+ cleared by the tubules to be taken up into the Schwann cells. The lateral branches of the lattice would substantially enhance the K^+ capture and speed of dispersal. As any net KCl uptake would be accompanied by osmotic movement of water, there would be an associated cell swelling. At longer times this would interfere with the efficiency of the extracellular clearance route and make carrier-mediated uptake more important. The progressive increase in accumulation during longer pulse trains (Abbott *et al.* 1989) (rather than establishment of the expected steady state) may be the result of volume changes in the tubular network.

Can Schwann cells act as 'spatial buffers' for K^+?

The 'spatial buffer' is a documented mechanism by which K^+ can be cleared from sites of activity in vertebrate neural tissue (Orkand *et al.* 1966). By this mechanism, a glial cell or syncytium showing high K^+ selectivity is able to move K^+ in at one extremity, out at another, driven by an extracellular gradient of $[K^+]$ and hence membrane potential. The mechanism has been demonstrated in cerebral cortex (Gardner-Medwin 1983*a,b*, 1986; Gardner-Medwin and Nicholson 1983) and retina (Newman *et al.* 1984; Newman 1986). We have considered whether squid Schwann cells could operate such a mechanism. The conclusion is that the cell geometry is not suitable; the short distance across the Schwann cell sheath ($< 6\ \mu m$) together with the longitudinal orientation of cells along the axon (leading to relatively uniform exposure of the Schwann cell membranes to K^+ during a burst of action potentials), and the high internal resistance of the fine processes and coupling junctions between cells (Brown *et al.* 1991), makes spatial buffering unlikely. However, the elongated Schwann cell shape, with the majority of fine processes arranged as a feltwork on the basal lamina side (Brown and Abbott 1993), would have advantages in increasing the membrane surface area, and hence concentrating K^+ uptake channels and transporters towards the basal side of the Schwann cell. This would keep the K^+ 'sink' away from the periaxonal space.

Role of the Na^+ pump in K^+ and Na^+ homeostasis

Electrophysiological studies of *Loligo* Schwann cells in split-open axon preparations show transient hyperpolarizations on moving from low to high external $[K^+]$ (Brown, in preparation), the opposite behaviour to that expected from the Goldman equation (depolarization), but compatible with activation of the electrogenic Na^+ K^+ ATPase. The resting Schwann cell appears to have a high intracellular $[Na^+]$, and low intracellular $[K^+]$ (J. Villegas 1968, 1974; Brown 1991; see also Chapter 14). Activation of the pump by elevation of periaxonal $[K^+]$ will lead to Na^+–K^+ exchange, with efflux of Schwann cell Na^+, effectively counteracting the K^+ accumulation and Na^+ depletion caused by a train of action potentials. The high intracellular $[Na^+]$ can be seen as providing an effective cation reserve, with a large capacity to take up K^+, and supplying Na^+ in exchange to maintain the action potential. The increased K^+ accumulation seen in the presence of ouabain (Abbott *et al.* 1989) shows the importance of this mechanism, particularly during longer trains of action potentials.

Role of axonal membrane in ion homeostasis

In the long term, the axon must be able to regain its resting ionic status by active ion uptake. There is a great deal of information on the activity of the axonal Na^+ K^+ ATPase, which mediates electrogenic, ATP-dependent Na^+ efflux in exchange for K^+ uptake, with a stoichiometry of 3 Na^+: 2 K^+: 1 ATP molecule hydrolysed (Baker *et al.* 1969; Beaugé 1984; De Weer *et al.*, 1988; see also Chapter 4). The axon membrane also contains the Na^+–K^+–$2Cl^-$ co-transporter (Russell 1976, 1983; Altamirano and Russell 1987; Altamirano *et al.* 1988), Na^+–Ca^{2+} exchanger (DiPolo *et al.* 1985), and several mechanisms for pH regulation (Boron *et al.* 1988; Boron and Knakal 1992). It should eventually be possible to produce a temporal and spatial balance sheet that shows the contribution of the Schwann cell and axonal transport mechanisms to ionic regulation under steady-state conditions, and during periods of axon stimulation.

Physiological significance of periaxonal ion homeostasis

The introductory section referred to the safety factor for conduction as one critical feature requiring good periaxonal ion homeostasis. Recent studies have shown that the squid can vary the force produced by giant-axon-mediated mantle contraction during swimming (Otis and Gilly 1990). Furthermore, the force produced by the mantle muscle depends on the frequency and burst length of action potentials in the giant axon, with a train of five action potentials giving a 20 per cent greater muscle tension than a train of four (Brown 1991; Brown and Bone 1991). Thus K^+ accumulation causing failure of a single action potential in a short train could have a significant effect on muscle performance in swimming and escape responses.

Most studies of isolated axons have only considered the consequences of K^+ accumulation caused by axonal activity, but, where the axons run through mantle muscle, they may also be exposed to elevated $[K^+]$ due to muscle action potentials. Because of the relatively low vascularity of mantle muscle in cephalopods (Bone *et al.* 1981; Abbott and Bundgaard 1987; see also Chapter 28), K^+ accumulation is likely to be severe under conditions of repeated escape jetting and strong swimming. It is interesting that the Na^+ K^+ ATPase of the squid axon saturates at around 100 mM $[K^+]$, with half-saturation at about 10 mM $[K^+]$ (Baker *et al.* 1969), suggesting that the transporter may have to cope with K^+ concentrations up to 20–30 mM under physiological conditions. Taken together these observations indicate that good periaxonal K^+ homeostasis, and a mechanism matching K^+ removal with Na^+ supply, would have powerful selective advantage.

Conclusion

Survey of studies using microelectrodes and voltage clamp has shown that the squid giant axon has efficient methods for periaxonal ion regulation, and consideration of the relative membrane area available and fine structural studies suggest that the Schwann cells are the major site of ion regulation during trains of action potentials. The transcellular tubular system of the Schwann cell layer provides a route for extracellular diffusion, which is an efficient way of removing K^+ from the axon surface at short times. The geometry of the tubular system and of the fine Schwann cell processes suggests that the tubular/cleft system acts as a K^+ 'sink' towards the basal lamina side of the cell. Ion channels (especially K^+, Cl^-) in the Schwann cell membranes are likely to contribute to the 'passive' component of K^+ removal, while carrier-mediated transport (Na^+ K^+ ATPase and Na^+–K^+–$2Cl^-$ co-transporter) appear as components with longer time constants. The Schwann cell Na^+ pump and the resting ion distributions (low [K^+] and high [Na^+] intracellularly) make the cell an effective exchange reserve for Na^+ and K^+. Final recycling of ions depends on axonal transport mechanisms. Efficient periaxonal regulation for K^+ and Na^+ is important for faithful conduction at vulnerable points such as axon branches, and for reliable control of effector organs including muscle.

Appendix: critique of experimental techniques

The majority of experiments on periaxonal K^+ regulation in the squid have used either microelectrodes or voltage-clamp techniques to examine K^+ accumulation. Since it is helpful to be able to compare results from the two techniques, and since there are possible artefacts associated with both, a critique follows.

Microelectrode methods

Axon cleaning

Microelectrodes were used in the earliest studies of K^+ accumulation. Where a cannulating electrode is used to measure membrane potential and record undershoots (Frankenhaeuser and Hodgkin 1956), it is first necessary to clean the axon to improve the optics for cannula placement. As this operation may take 30–45 min and Schwann cells may be damaged by the cleaning, K^+ regulation may be compromised. K^+ accumulation has been studied in *Alloteuthis* axons using fine glass microelectrodes (10–50 MΩ) and, as the transparency of the preparation permits accurate impalement with minimal dissection, these preparations generally show healthy Schwann cells and little K^+ accumulation at the start of an experiment (Abbott *et al.* 1988; Astion *et al.* 1988). Schwann cell deterioration and K^+ accumulation occurs progressively in dissected axons, but can be minimized by working in flowing solutions at temperatures less than 15°C. In theory, microelectrode penetration through the sheath to the axon interior should damage Schwann cells and hence K^+ regulation in that region, but in practice this is not a problem, presumably because of membrane sealing around the electrode shank. The time taken between axon dissection and microelectrode penetration can be as little as 5 min, allowing approximately 1 h for experimentation without significant deterioration.

Calibration

Calibration experiments by Frankenhaeuser and Hodgkin (1956) showed that, in the range 5–30 mM [K^+], the membrane potential at the peak of the action potential undershoot was a nearly linear function of [K^+], with about 1 mV decline caused by a 1 mM rise in [K^+] (Fig. 16.4). This permits simple calculation of periaxonal [K^+] from potential recordings. However, the [K^+] can be estimated at only a single time point per action potential.

Quantitative analysis

For the quantitative analysis of K^+ accumulation and clearance, it is necessary to know how much K^+ comes out of the axon during each action potential. A figure of 7.5 pmol/cm^2 at 11 °C (Landowne and Scruggs 1976) has been used in analysis of *Alloteuthis* data. As the amount of K^+ lost per action potential depends on the K^+ conductance g_K and on the electrochemical gradient for K^+, it is dependent on the resting potential and the amplitude of the action potential. Consistent results have been obtained by using healthy axons with resting potentials greater than −60 mV and action potentials greater than 100 mV.

For studies using trains of action potentials, it is assumed that the amount of K^+ lost per action potential is constant. This assumption cannot easily be tested, which gives some uncertainty to quantitative analysis. However, if experimental conditions are kept constant (freshness of animals, dissection, temperature, duration of experiment), then it should be possible to compare the effects of different treatments on K^+ accumulation. This is the approach adopted by Astion *et al.* (1988) and Abbott *et al.* (1989).

Voltage-clamp methods

In theory, voltage-clamp methods should be superior to microelectrode methods for investigating periaxonal K^+ regulation, as it is possible to derive absolute values for K^+ current and conductance (I_K and g_K), and to control I_K by control of the internal solution and pulse protocol (Adelman and Gilbert 1990; Mullins and Brinley 1990). In practice, there are several possible artefacts associated with the method.

Axon cleaning

Fine cleaning of the axon to give good optics for cannulation/perfusion and efficient voltage control by removal of surface fibres has been shown to damage Schwann cells (Brown 1993).

Perfusion

Two methods are commonly used for removing the axoplasm for internal perfusion: the roller method in which the axon is squeezed between a rubber roller and mat, then re-inflated with saline solution (Baker *et al.* 1961, 1962a); and the Tasaki method, in which an axial pipette is gradually advanced, sucking out axoplasm to make a hollow tube, and this is filled with saline solution (Oikawa *et al.* 1961). In both cases, the mechanical distortion of the sheath damages Schwann cells (Baker *et al.* 1962*a*). Perfused axons permit selection of an appropriate perfusion medium and pressure to reduce K^+ accumulation (Kukita 1988), but it is difficult to preserve the normal geometry, and hence to study mechanisms under physiologically relevant conditions. Modifications of the normal bathing medium (e.g. elevation of $[Ca^{2+}]$ from 11 to 20 mM, reduction of $[Na^+]$ to reduce excitability) are frequently made to reduce axonal currents and permit more effective voltage control, but the effect of these manipulations on Schwann cell physiology has not been considered.

Deterioration with time

The procedures involved in cleaning, squeezing/coring, perfusion, and insertion of axial electrodes take time, generally 30–40 min minimum, so the Schwann cells will generally be in worse shape than in microelectrode experiments. The time can be reduced to 15–20 min by omitting perfusion, but in this case the axial electrode deteriorates more rapidly by contact with axoplasm, and needs more frequent cleaning and replacement; the axoplasmic environment is also less efficiently controlled and hence more variable.

Effect of voltage-clamp pulse

The normal physiological condition is for K^+ to be delivered to the periaxonal space in a series of short bursts, by a brief train of action potentials. Voltage-clamp protocols allow control of pulse lengths, with increasing K^+ loading of the periaxonal space with increasing pulse length. As the amount of K^+ lost depends on the g_K, the driving force, and the time, and as g_K is sensitive to the health of the axon and the temperature, it is important to make comparisons under similar conditions.

Measurement of potassium equilibrium potential E_K

The amount of K^+ in the periaxonal space is calculated by an estimate of E_K, for example, by clamping to a range of negative voltages at the end of the depolarizing pulse, and calculating the reversal potential from the tail currents. It is generally necessary to curve fit the tail currents and extrapolate back to zero time (pulse off) to overcome problems associated with the capacitative artefact, but single-point analysis has also been used (Pichon *et al.* 1987). The accuracy of the calculation of E_K is dependent on the accuracy of tail current recording and extrapolation.

Electrokinetic effect of the voltage-clamp pulse

When a voltage gradient is imposed between the axon interior and the space beyond the Schwann cell, as in axial wire voltage clamp, electrochemical theory predicts that K^+ salt will be concentrated at the inner face of the axon membrane, and diluted at the outer surface (Helfferich 1962). The apparent E_K will therefore be artefactually increased (internal negative) by this effect. Hence short pulses will give outward tail currents (interpreted as absence of K^+ accumulation, or even K^+ depletion), while longer pulses will give inward tail currents (interpreted as K^+ accumulation) (Inoue, unpublished). This effect will depend on the intracellular $[K^+]$, the voltage gradient, the pulse duration, and the g_K; hence it is necessary to control for all these parameters when comparing different experiments and preparations. In theory it would be best to use a long enough pulse to reach electrokinetic steady state where the rate of K^+ efflux from the axon balances the rate of K^+ clearance (no change in tail current with pulse length). However, this may require unphysiologically long depolarizing pulses. Nevertheless, it is important to test for the presence of this artefact, by examining E_K as a function of pulse length. Adam (1973) has also considered the effect of electrokinetic volume flow on K^+ distribution in the narrow periaxonal space, and concluded it was small compared with diffusion, and could be ignored.

Conclusion

It is clear that both microelectrode and voltage-clamp methods have advantages and disadvantages. Careful dissection and design of experiments can help to reduce some artefacts such as those due to Schwann cell damage. A combination of both types of study will be necessary to obtain a complete quantitative picture of the role of Schwann cells in periaxonal K^+ regulation.

Acknowledgements

We are grateful to the Director and staff of the Marine Biological Association Laboratory, Plymouth, for facilities and squid, and to Dr I. C. H. Smith, Dr F. Kukita, Professor S. Yamagishi, and Dr A. R. Gardner-Medwin for helpful discussion. The work was supported by grants from The Royal Society, Wellcome Trust, Nuffield Foundation, and SERC.

References

Abbott, N. J. and Brown, E. R. (1989). Effects of osmotic manipulation on extracellular routes for ion regulation in the squid Schwann cell. *Journal of Physiology*, **418**, 16P.

Abbott, N. J. and Bundgaard, M. (1987). Microvessel surface area, density and dimensions in brain and muscle of the cephalopod *Sepia officinalis*. *Proceedings of the Royal Society of London, Series B*, **230**, 459–82.

Abbott, N. J., Lieberman, E. M., Pichon, Y., Hassan, S., and Larmet, Y. (1988). Periaxonal K^+ regulation in the small squid *Alloteuthis*: studies on isolated and *in situ* axons. *Biophysical Journal*, **53**, 275–9.

Abbott, N. J., Pichon, Y., Inoue, I., Brown, E. R., Lieberman, E. M., Revest, P. A., and Smith, I. C. H. (1989). Role of Schwann cells in periaxonal ion regulation in the squid. *Acta Physiologica Scandinavica*, **136** (Suppl. 582), 48.

Adam, G. (1973). The effect of potassium diffusion through the Schwann cell layer on potassium conductance of the squid axon. *Journal of Membrane Biology*, **13**, 353–86.

Adelman, W. J. and Fitzhugh, R. (1975). Solutions of the Hodgkin–Huxley equations modified for potassium accumulation in a periaxonal space. *Federation Proceedings*, **34**, 1322–9.

Adelman, W. J. and Gilbert, D. L. (1990). Electrophysiology and biophysics of the squid giant axon. In *Squid as experimental animals* (ed. D. L. Gilbert, W. J. Adelman, and J. M. Arnold), pp. 93–132. Plenum Press, New York.

Adelman, W. J. Jr and Palti, Y. (1969). The effects of external potassium and long duration voltage conditioning on the amplitude of sodium currents in the giant axon of the squid, *Loligo pealei*. *Journal of General Physiology*, **54**, 589–606.

Adelman, W. J. Jr and Palti, Y. (1972). The role of periaxonal and perineuronal spaces in modifying ionic flow across neural membranes. *Current Topics in Membranes and Transport*, **3**, 199–235.

Adelman, W. J. Jr, Palti, Y., and Senft, J. P. (1973). Potassium ion accumulation in a periaxonal space and its effect on the measurement of membrane potassium ion conductance. *Journal of Membrane Biology*, **13**, 387–410.

Adelman, W. J. Jr, Moses, J., and Rice, R. V. (1977). An anatomical basis for the resistance and capacitance in series with the excitable membrane of the squid giant axon. *Journal of Neurocytology*, **6**, 621–46.

Altamirano, A. A. and Russell, J. M. (1987). 'Reverse' unidirectional fluxes in squid giant axons. *Journal of General Physiology*, **89**, 669–86.

Altamirano, A. A., Breitwieser, G. E., and Russell, J. M. (1988). Vanadate and fluoride effects on Na^+-K^+-Cl^- cotransport in squid giant axon. *American Journal of Physiology*, **254**, C582–C586.

Astion, M. L., Coles, J. A., Orkand, R. K., and Abbott, N. J. (1988). K^+ accumulation in the space between giant axon and Schwann cell in the squid *Alloteuthis*. *Biophysical Journal*, **53**, 281–5.

Baker, P. F., Hodgkin, A. L., and Shaw, T. I. (1961). Replacement of the protoplasm of a giant nerve fibre with artificial solutions. *Nature*, **190**, 657–64.

Baker, P. F., Hodgkin, A. L., and Shaw, T. I. (1962*a*). Replacement of the axoplasm of giant nerve fibres with artificial solutions. *Journal of Physiology*, **164**, 330–54.

Baker, P. F., Hodgkin, A. L., and Shaw, T. I. (1962*b*). The effects of changes in internal ionic concentrations on the electrical properties of perfused giant axons. *Journal of Physiology*, **164**, 355–74.

Baker, P. F., Blaustein, M. P., Keynes, R. D., Manil, J., Shaw, T. I., and Steinhardt, R. A. (1969). The ouabain-sensitive fluxes of sodium and potassium in squid giant axons. *Journal of Physiology*, **200**, 459–96.

Beaugé, L. (1984). Sodium pump in squid axons. *Current Topics in Membranes and Transport*, **22**, 131–75.

Binstock, L., Adelman, W. J., Jr, Senft, J. P., and Lecar, H. (1975). Determination of resistance in series with the membranes of giant axons. *Journal of Membrane Biology*, **21**, 25–47.

Bone, Q., Pulsford, A., and Chubb, A. D. (1981). Squid mantle muscle. *Journal of the Marine Biological Association of the United Kingdom*, **61**, 619–26.

Boron, W. F. and Knakal, R. C. (1992). Na^+-dependent Cl–HCO_3 exchange in the squid axon. *Journal of General Physiology*, **99**, 817–37.

Boron, W. F., Hogan, E., and Russell, J. M. (1988). pH-sensitive activation of the intracellular-pH regulation system in squid axons by ATP-γ-S. *Nature*, **332**, 262–5.

Brown, E. R. (1991). Axon–Schwann cell interaction in the squid. Unpublished PhD thesis, University of London.

Brown, E. R. (1993). K^+ accumulation around the giant axon of the squid; comparison of electrical and morphological measurements. *Japanese Journal of Physiology*, **43 Suppl 1**, S279–S284.

Brown, E. R. and Abbott, N. J. (1993). Ultrastructure and permeability of the Schwann cell layer surrounding the giant axon of the squid. *Journal of Neurocytology*, **22**, 283–98.

Brown, E. R. and Bone, Q. (1991). Repetitive stimulation of the squid giant axon exerts graded control over mantle tension. *Journal of the Marine Biological Association of the United Kingdom*, **71**, 732.

Brown, E. R., Bone, Q., Ryan, K. P., and Abbott, N. J. (1991). Morphology and electrical properties of Schwann cells around the giant axon of the squids *Loligo forbesi and Loligo vulgaris*. *Proceedings of the Royal Society of London*, **B243**, 255–62.

Clay, J. R. (1986). Potassium ion accumulation slows the closing rate of potassium channels in squid axons. *Biophysical Journal*, **50**, 197–200.

Curtis, H. J. and Cole, K. S. (1942). Membrane resting and action potentials from the squid giant axon. *Journal of Cellular and Comparative Physiology*, **19**, 135–44.

De Weer, P., Rakowski, R. F., and Gadsby, D. C. (1988). Voltage dependence of the Na–K pump. *Annual Review of Physiology*, **50**, 225–41.

DiPolo, R., Bezanilla, F., Caputo, C., and Rojas, H. (1985). Voltage dependence of the Na/Ca exchange in voltage-clamped, dialyzed squid axons. Na-dependent Ca efflux. *Journal of General Physiology*, **86**, 457–78.

Frankenhaeuser, B. and Hodgkin, A. L. (1956). The after-effects of impulses in the giant nerve fibres of *Loligo*. *Journal of Physiology*, **131**, 341–76.

Gardner-Medwin, A. R. (1983*a*). A study of the mechanisms by which potassium moves through brain tissue in the rat. *Journal of Physiology*, **335**, 353–74.

Gardner-Medwin, A. R. (1983*b*). Analysis of potassium dynamics in mammalian brain tissue. *Journal of Physiology*, **335**, 393–426.

Gardner-Medwin, A. R. (1986). A new framework for assessment of potassium buffering mechanisms. *Annals of the New York Academy of Sciences*, **481**, 287–302.

Gardner-Medwin, A. R. and Nicholson, C. (1983). Changes of extracellular potassium activity produced by electric current through brain tissue in the rat. *Journal of Physiology*, **335**, 375–92.

Geren, B. B. and Schmitt, F. O. (1954). The structure of the Schwann cell and its relation to the axon in certain invertebrate nerve fibres. *Proceedings of the National Academy of Sciences, USA*, **40**, 863–71.

Hama, K. (1962). Some observations on the fine structure of the giant synapse in the stellate ganglion of the squid, *Doryteuphis bleekeri. Zeitschrift für Zellforschung*, **56**, 437–44.

Hassan, S. and Lieberman, E. M. (1988). Studies of axon–glial interactions and periaxonal K^+ homeostasis—II. The effect of axonal stimulation, cholinergic agents and transport inhibitors on the resistance in series with the axon membrane. *Neuroscience*, **25**, 961–9.

Haydon, D. A., Elliott, J. R., and Hendry, B. M. (1984). Effects of anesthetics on the squid giant axon. *Current Topics in Membranes and Transport*, **22**, 445–82.

Helfferich, F. (1962). *Ion exchange*. McGraw Hill, New York.

Hodgkin, A. L. (1958). Ionic movements and electrical activity in giant nerve fibres. *Proceedings of the Royal Society of London, Series B*, **148**, 1–37.

Hodgkin, A. L. and Huxley, A. F. (1952). A quantitative description of membrane current and its application to conduction and excitation in nerve. *Journal of Physiology*, **117**, 500–44.

Kimelberg, H. K. and Frangakis, M. V. (1985). Furosemide- and bumetanide-sensitive ion transport ratio and volume control in primary astrocyte cultures from rat brain. *Brain Research*, **361**, 125–34.

Kukita, F. (1988). Removal of periaxonal potassium accumulation in a squid giant axon by outward osmotic water flow. *Journal of Physiology*, **399**, 647–56.

Kukita, F. and Yamagishi, S. (1983). Effects of an outward water flow on squid giant axon. *Journal of Membrane Biology*, **75**, 33–44.

Landowne, D. and Scruggs, V. (1976). The temperature dependence of the movement of potassium and chloride ions associated with nerve impulses. *Journal of Physiology*, **259**, 145–58.

Larmet, Y. and Pichon, Y. (1988). Lack of potassium accumulation in fresh axons of *Loligo forbesi. Journal of Physiology*, **396**, 166P.

Lieberman, E. M. and Sanzenbacher, E. (1992). Mechanisms of glutamate activation of axon-to-Schwann cell signaling in the squid. *Neuroscience*, **47**, 931–9.

Mullins, L. J. and Brinley, F. J. (1990). Internal dialysis in the squid giant axon. In *Squid as experimental animals* (ed. D. L. Gilbert, W. J. Adelman, and J. M. Arnold), pp. 93–132. Plenum Press, New York.

Newman, E. A. (1986). Regional specialization of the membrane of retinal glial cells and its importance to K^+ spatial buffering. *Annals of the New York Academy of Sciences*, **481**, 273–85.

Newman, E. A., Frambach, D. A., and Odette, L. L. (1984). Control of extracellular potassium levels by retinal glial cell siphoning. *Science*, **225**, 1174–5.

Oikawa, T., Spyropoulis, C. S., Tasaki, I., and Teorell, T. (1961). Methods for perfusing the giant axon of *Loligo pealei. Acta Physiologica Scandinavica*, **52**, 195–6.

Orkand, R. K., Nicholls, J. G., and Kuffler, S. W. (1966). Effect of nerve impulses on the membrane potential of glial cells in the central nervous system of amphibia. *Journal of Neurophysiology*, **29**, 788–806.

Otis, T. S. and Gilly, W. F. (1990). Jet-propelled escape in the squid *Loligo opalescens*: concerted control by giant and non-giant motor axon pathways. *Proceedings of the National Academy of Sciences, USA*, **87**, 2911–15.

Pichon, Y., Abbott, N. J., Lieberman, E. M., and Larmet, Y. (1987). Potassium homeostasis in the nervous system of cephalopods and Crustacea. *Journal de Physiologie, Paris*, **82**, 346–56.

Russell, J. M. (1976). ATP-dependent chloride influx into internally dialyzed squid giant axons. *Journal of Membrane Biology*, **28**, 335–49.
Russell, J. M. (1983). Cation-coupled chloride influx in squid axon. *Journal of General Physiology*, **81**, 909–25.
Shrager, P., Starkus, J. D., Lo, M.-V. C., and Perrachia, C. (1983). The periaxonal space of crayfish giant axons. *Journal of General Physiology*, **82**, 221–44.
Smith, D. O. (1980). Mechanisms of action potential propagation failure at sites of axon branching in the crayfish. *Journal of Physiology*, **301**, 243–59.
Swenson, R. P. and Armstrong, C. M. (1981). K^+ channels close more slowly in the presence of external K^+ and Rb^+. *Nature* **291**, 427–9.
Villegas, G. M. and Villegas, R. (1960). The ultrastructure of the giant nerve fiber of the squid. Axon–Schwann cell relationship. *Journal of Ultrastructure Research*, **3**, 362–73.
Villegas, G. M. and Villegas, R. (1984). Squid axon ultrastructure. *Current Topics in Membranes and Transport*, **22**, 3–37.
Villegas, J. (1968). Transport of electrolytes in the Schwann cell and location of sodium by electron microscopy. *Journal of General Physiology*, **51**, 61s–71s.
Villegas, J. (1974). Effects of acetylcholine and carbamylcholine on the axon and Schwann cell electrical potentials in the squid nerve fibre. *Journal of Physiology*, **242**, 647–59.
Walz, W. and Hinks, E. C. (1985). Carrier-mediated KCl accumulation accompanied by water movements is involved in the control of physiological levels by astrocytes. *Brain Research*, **343**, 44–51.
Zwahlen, M. J., Sandri, C., and Greeff, N. J. (1988). Transglial pathway of diffusion in the Schwann sheath of the squid giant axon. *Journal of Neurocytology*, **17**, 145–59.

PART IV

Squid giant synapse

17

Synaptic transmission in the squid stellate ganglion

R. LLINÁS and M. SUGIMORI

Summary

Our understanding of the mechanism responsible for depolarization–release coupling at chemical junctions has advanced quite steadily over the last three decades, much of it directly due to the study of the giant synapse in the squid stellate ganglion. From such work, general agreement has been reached regarding several aspects of synaptic transmission: i) Presynaptic membrane depolarization, the most common triggering signal for release, is achieved by an influx of calcium into the presynaptic terminal. ii) Synaptic transmitter is released from specialized sites characterized by the confluence of an intracellular calcium concentration ($[Ca^{2+}]_i$) increase, the biochemical cascade triggered by the calcium, and the intracellular organelles involved in this secretory event. iii) The localization of the intracellular calcium concentration microprofiles that follow voltage-dependent calcium-channel activation is a critical parameter in ensuring that the set of events triggering release occurs with sufficient speed to guarantee a short synaptic delay.

Anatomy

The giant synapse is the last junction in a three-link chain of giant neurones that form the escape system in squid. Its general morphology was described in the classical work of J. Z. Young (1936, 1938, 1939). This chain is a bilaterally symmetrical system that originates with the first-order giant neurones located one on each side of the ventral magnocellular lobe of the brain (Fig. 17.1). Each of the two first-order neurones sends its axon toward the midline, where it fuses with its counterpart to form a cytoplasmic bridge. After forming this bridge the two axons course caudally to make axo-axonic synapses with the second-order giant cells in the dorsal magnocellular lobe. The axons of the second-order neurones in turn leave the brain in the pallial nerves and terminate in the stellate ganglia (found at each side of the dorsomedial region

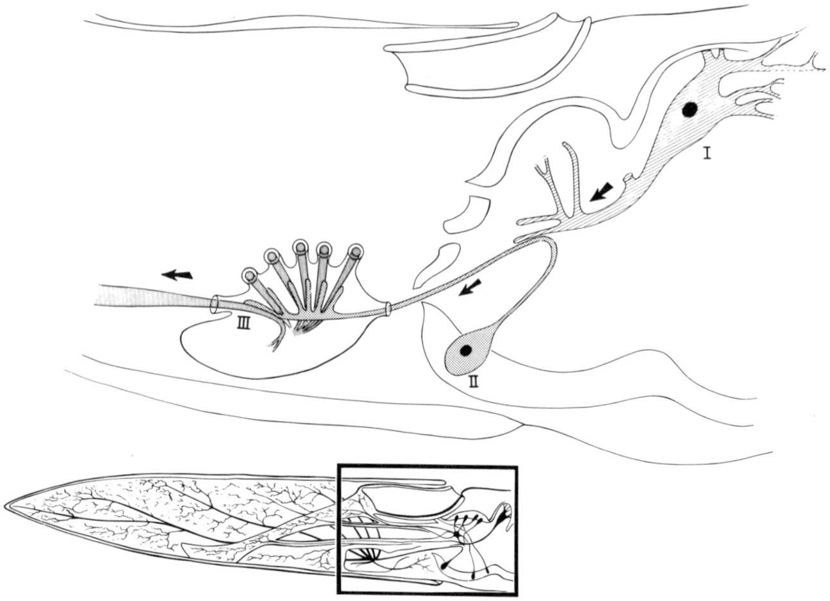

Fig. 17.1. Chain of giant nerve cells in the squid. Synaptic transmission at the giant synapse in the squid is usually initiated when the first-order giant neurone (I) in the palliovisceral ganglion is activated by sensory input. This excites the second-order giant neurone (II) whose axon forms the presynaptic terminal in the stellate ganglion. The stellate ganglia lie in the mantle on either side of the midline. The postsynaptic axons of the giant synapse (III) arise from cell bodies in the stellate ganglion and innervate the musculature of the mantle. The inset illustrates the innervation of the mantle by the branches of the giant axon. (Modified with permission from Young (1939).)

of the mantle). In each stellate ganglion, the axon of the second-order generates eight to 10 presynaptic terminal digits. Each of these terminals contacts a third-order axon, the last link in the chain. This contact takes place via spinal protrusions emerging from the third-order axons, which demarcate the extent of the junction (Young 1973). The third-order axons are the postsynaptic elements of the giant synapses. These large single fibres are formed by the fusion of the axons from many neurones located at the giant axon lobe and each courses radially to innervate the mantle musculature (Fig. 17.1).

In summary then, the giant synapses are axo-axonic junctions between the axon of the second-order giant neurone in the brain and the giant axons of the stellate ganglion. The synapse that has received the most attention is the largest and most distal of these junctions and is commonly referred to as the giant synapse.

The actual contact is made between the smooth surface of the presynaptic

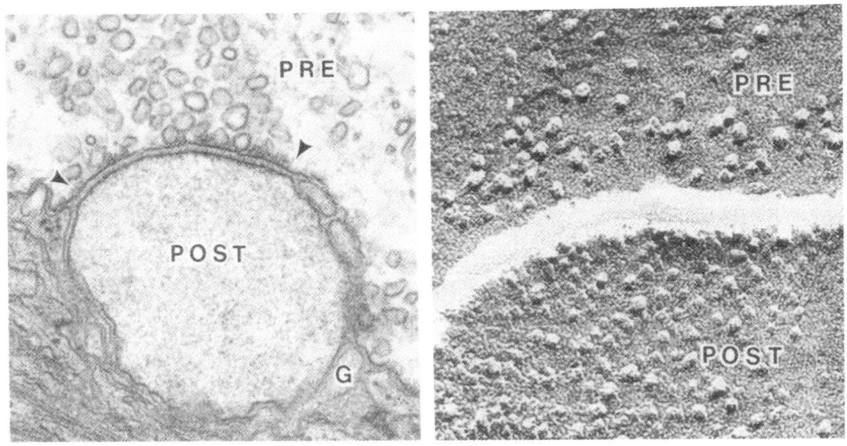

Fig. 17.2. Left: Thin section through an active zone from the giant synpase of *Loligo paelei*. The presynaptic axon (PRE) contacts the postsynaptic axon spine (POST) without an intervening glial element (G). The active zone (arrowheads) is marked by parallel cell membranes separated by a cleft, electron-dense fuzz associated with these membranes, and the presence of synaptic vesicles nearby (× 46 000). Right: Freeze-fracture micrograph through an active zone of a squid giant synapse. Large intramembrane particles occur in patches on the cytoplasmic leaflet of the presynaptic membrane and the external leaflet of the postsynaptic membrane. Densities of these particles were obtained from this and similar micrographs (× 175 000). (Taken with permission from Pumplin *et al.* (1981).)

terminal and spine-like processes that protrude from the postsynaptic axon (Young 1939; Robertson 1953) and envelop the terminal (Young 1973; Pumplin and Reese 1978; Pumplin *et al.* 1981). These dendrite-like processes branch to form the postsynaptic junctions, the final bulbous heads which have a diameter of about 1 μm (Hama 1962; Pumplin and Reese 1978; Fig. 17.2, left panel). The number of contacts between a presynaptic terminal and postsynaptic element has been estimated to be between 4000 and 10 000 (Martin and Miledi 1975; Pumplin and Reese 1978; Pumplin *et al.* 1981).

The site of synaptic contact is limited to the 'head' of each spinous process by the curvature of the postsynaptic elements and the presence of connective tissue surrounding the necks of the postsynaptic spines (Fig. 17.2, left panel). At the ultrastructural level the presynaptic terminal demonstrates the usual synaptic vesicles, which have been observed with both transmission electron microscopy (Hama 1962; Castejon and Villegas 1964; Martin and Miledi 1975; Pumplin and Reese 1978) and freeze-fracture techniques (Pumplin and Reese 1978; Pumplin *et al.* 1981). The vesicles are located immediately opposite the postsynaptic spines (Hama 1962; Pumplin and Reese 1978), suggesting that vesicle fusion occurs at that point.

The active zone

Morphologically, the idea that a well-defined entity is always associated with rapid chemical synaptic transmission was first proposed by Couteaux and Pécot-Dechavassine (1970) for the neuromuscular junction and was named the 'active zone'. The concept further implied that synaptic transmitter was released only at such specialized sites and involved presynaptic and postsynaptic specializations at the apposition points between these elements (Couteaux and Pécot-Dechavassine 1970). This site is the place where synaptic vesicles (De Robertis and Bennett 1955; Palay 1956) are believed to fuse with the plasmalemma (Ceccarelli *et al.* 1972; Heuser *et al.* 1979) in order to release synaptic transmitter. The active zone was also initially demonstrated to be in close proximity to the postsynaptic channels known to respond to iontophoresis of acetylcholine at the neuromuscular junction (del Castillo and Katz 1956).

From a physiological point of view, experimental work initially performed in the squid giant synapse demonstrated that the latency between calcium entry and transmitter release could be as short as 200 μs (Llinás 1977; Llinás *et al.* 1981*a*, 1982). Furthermore, based on those measurements, it was proposed that the active-zone particles observed in freeze-fracture replicas of the presynaptic membrane could represent the sites of calcium entry (Pumplin *et al.* 1981; Fig. 17.2, right panel). Quantification of particle number on freeze-fracture replicas is consistent with the view that these particles are the calcium channels (Pumplin *et al.* 1981), the average number of active zone particles per synapse being 1.1×10^7.

Taken together these findings indicated, as stated above, that the release process had to be encompassed in a morphological unit, to include the sites of calcium entry and the transmitter release sites. Also, employing voltage-clamp techniques, physiological studies of the presynaptic terminal of the squid giant synapse confirmed the classical hypothesis of Katz (1969) that synaptic transmission is triggered by calcium entry.

Initial findings

J. Z. Young was the first to investigate this synapse and to demonstrate that stimulation of the second-order giant fibre resulted in the activation of the tertiary giant axon system (Young 1939). T. H. Bullock (1946, 1948) and Bullock and Hagiwara (1957) further demonstrated the steps in synaptic transmission at the stellate ganglion using intracellular and extracellular recording techniques. This initial work was followed by the classical work of Hagiwara and Tasaki (1958), Takeuchi and Takeuchi (1962), and Miledi and Slater (1966) which established the relation between the amplitude and duration of the presynaptic action potential and transmitter release.

Synaptic transmission in the absence of action potentials

That synaptic transmission may be triggered by direct depolarization of the presynaptic terminal was initially demonstrated in the mid-1960s following sodium conductance block using tetrodotoxin (TTX) (Bloedel *et al.* 1966; Katz and Miledi 1966; Kusano *et al.* 1967). In addition, when the potassium conductance was blocked [by the use of tetraethylammonium (TEA)], synaptic transmission was blocked during direct presynaptic depolarization to membrane potentials positive to +50 mV (the suppression potential), but returned at the end of the pulse (the 'off' response) (Katz and Miledi 1967; Kusano *et al.* 1967). Following this significant set of findings it was demonstrated that presynaptic calcium-dependent spikes could be recorded at the presynaptic terminal following TTX and TEA administration (Katz 1969). That a clear increase in presynaptic intracellular calcium concentration, $[Ca^{2+}]_i$, actually occurs during synaptic transmission was determined by the use of $[Ca^{2+}]_i$ makers such as aequorin (Llinás *et al.* 1972; Llinás and Nicholson 1975) and Arsenazo III (Miledi and Parker 1981; Charlton *et al.* 1982). Moreover, the suppression potential was shown to be produced by a reduction in calcium entry due to the decrease in the calcium driving force (Llinás and Nicholson 1975), and not to a direct effect on the release process itself.

Voltage-clamp studies

The next step in studying the properties of the postsynaptic elements that govern transmitter release was the development of a presynaptic voltage-clamp paradigm (Llinás *et al.* 1981*a*, *b*; Augustine and Charlton 1986; Augustine *et al.* 1988). The methodology employed in these studies is described in detail in the original papers (Llinás *et al.* 1976, 1980, 1981*a*,*b*). After double or triple microelectrode impalement of the presynaptic terminal [Fig. 17.3(a)] and blockage of the voltage-gated sodium and potassium conductances with TTX and TEA + 4-aminopyridine (4-AP), respectively, the membrane potential of the terminal was depolarized using square voltage-clamp steps. The membrane potential was usually held at −70 mV.

A rapid presynaptic depolarization (usually 4–10 ms) triggers the onset of a rather slow inward calcium current that does not inactivate and that reaches a peak whose amplitude varies with the level of depolarization [Fig. 17.3(b) and (c)]. At a depolarization step of 30 mV (from −70 to −40 mV) [Fig 17.3(b)] a small inward current (lowest record) is seen that begins very slowly and shows a close-to-linear increase with time. At the end of the pulse, a fast tail current is observed as shown in Fig. 17.3(d). These transient currents and the postsynaptic potentials that accompany them are blocked by manganese chloride (10 mM) or cadmium chloride (1 mM) (Llinás 1977; Llinás *et al.* 1980) and are absent if calcium is removed from the extracellular medium.

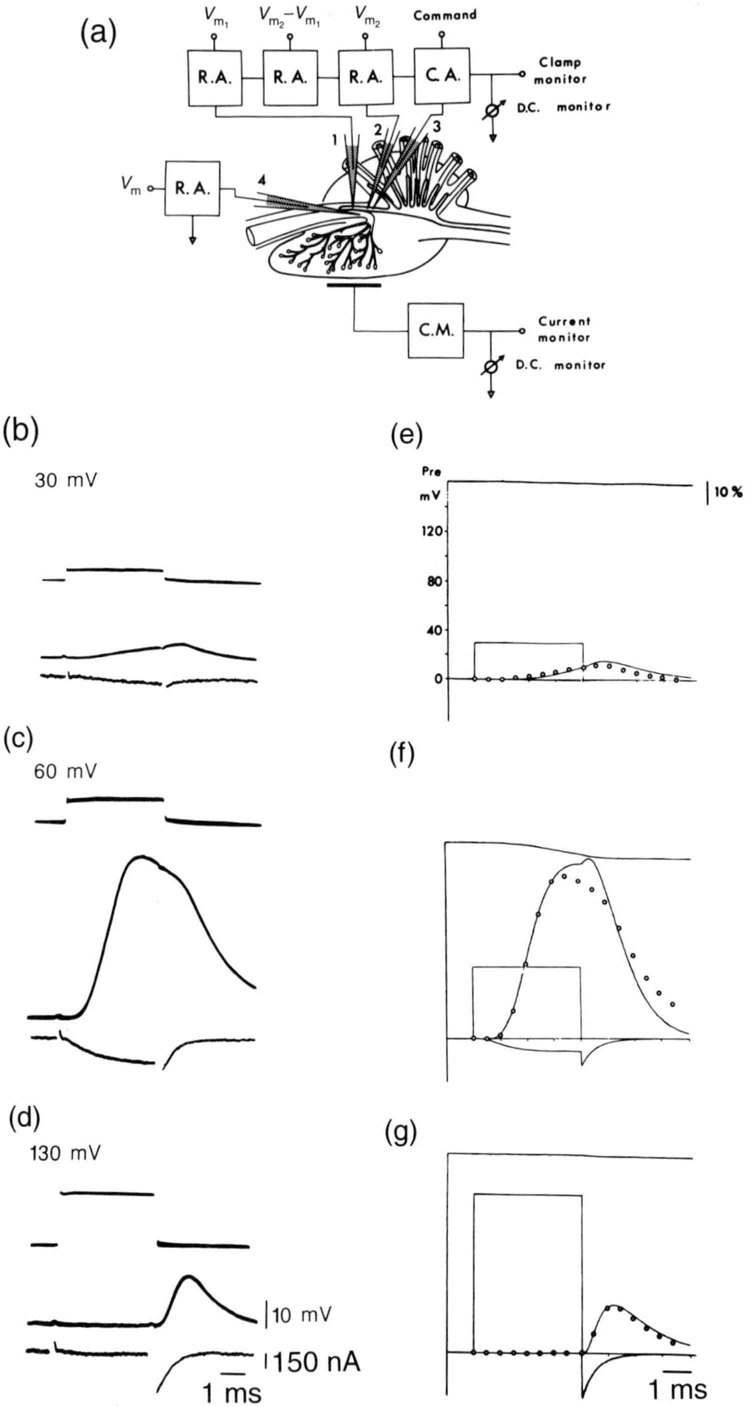
(a)
V_{m_1}
$V_{m_2}-V_{m_1}$
V_{m_2}
Command
R.A.
R.A.
R.A.
C.A.
Clamp monitor
D.C. monitor
1
2
3
4
V_m
R.A.
C.M.
Current monitor
D.C. monitor
(b)
30 mV
(c)
60 mV
(d)
130 mV
10 mV
150 nA
1 ms
(e)
Pre mV
120
80
40
0
10 %
(f)
(g)
1 ms

Fig. 17.3. Synaptic transmission during voltage clamp of presynaptic terminal. (a) Voltage-clamp technique. The presynaptic terminal is impaled with electrodes 1, 2, and 3. The difference between electrodes 1 and 2 ($V_{m2}-V_{m1}$) gives the presynaptic terminal current using recording amplifiers (R.A.) after current injection through the third presynaptic electrode (3) through a feedback command amplifier (C.A.). The postsynaptic response (V_m) is recorded through the R.A. connected to electrode 4 in the postsynaptic terminal. Total current is measured at the indifferent electrode current monitor (C.M.). (b)–(d) Experimental data. Top trace, presynaptic voltage; middle trace, postsynaptic response; lower trace, calcium current. The S shape of the current onset can be seen at 60 mV depolarization from a holding potential of −70 mV. Note the fast tail current and the 'on' and 'off' excitatory postsynaptic potential (EPSP). (e)–(g) Numerical solution to a mathematical model. Top trace, vesicle depletion; other traces as in (b)–(d). Open circles are recorded EPSPs. $[Ca^{2+}]_o$ = 10nM. (Modified with permission from Llinás *et al.* (1981*b*).)

The postsynaptic response associated with this current is illustrated in the middle trace in Fig. 17.3(b)–(d). Note that the postsynaptic response during the pulse also has a close-to-linear rate of rise, and that the excitatory postsynaptic potential (EPSP) that follows the break of the voltage-clamp pulse (off response) is associated with the inward tail current.

The calcium current (I_{Ca}) increases in amplitude with increasing levels of depolarization reaching a maximum in most synapses at 60 mV from rest (−10 mV absolute), a value quite close to that determined by Baker and his colleagues for the squid giant axon using aequorin (Baker *et al.* 1971). At this level, the sigmoidal character of the current onset is evident [Fig. 17.3(c), bottom trace], after which the current reaches a plateau indicating the absence of inactivation, at least for pulses up to 100 ms. A tail current is again apparent at the end of the pulse. Accompanying this increase in I_{Ca} a change is seen in the shape, rate of rise, and amplitude of the EPSP. With a presynaptic depolarization of 60 mV the EPSP reaches a peak, begins to decline during the pulse, and is followed by a secondary transmitter release, the off response [Fig. 17.3(c), arrow].

As the presynaptic voltage step was increased beyond 60 mV, the amplitude of the 'on' presynaptic calcium current decreased while the tail current increased reaching a maximum near 100 mV depolarization (+ 40 mV absolute). At the same time, the amplitude and rate of rise of the 'on' EPSP decreased while the 'off' response increased. The 'on' current continued to decrease beyond 110 mV until the suppression potential (Katz and Miledi 1967) was reached at a depolarization close to 130 mV [Fig. 17.3(d), lowest trace]. At this point, no inward current was seen during the voltage pulse, but a large tail current was evident following the pulse break. Similarly, no synaptic transmission was seen during the pulse, and a sharp, short-latency 'off' postsynaptic response was generated by the tail current. A clear correlation is thus present between the characteristics of the presynaptic I_{Ca} and the postsynaptic response, the 'on' EPSP being related to the I_{Ca} during the pulse and the 'off' EPSP to the tail I_{Ca}.

Synaptic delay

Among the parameters relating calcium entry to transmitter release few have been more important in determining the mechanisms of the release process than the synaptic delay. Indeed, the very rapidity of this phenomenon (100s of μs) probably represents close to the maximum speed of organelle movement. This temporal relationship between depolarization and transmitter release may be divided into two parts. The first, component A, is related to the time required for opening of the calcium channels, whereas the second, component B, relates to the time between calcium entry and the onset of the postsynaptic potential (Fig. 17.4). Examples of how these two components affect synaptic delay can be obtained by comparing the delay for the 'on' and 'off' release as shown in Fig. 17.4(a) and (b). Here, in (a), a presynaptic voltage-clamp step (to -10 mV) demonstrates a characteristic 1-ms delay for the 'on' response (composed of components A and B). The record in Fig. 17.4(b) was taken at the suppression potential and illustrates the tail calcium current and the latency (about 200 μs, comprising the B component alone) for the 'off' synaptic potential generated by this short current injection. The actual synaptic delay may, in fact, be slightly shorter since the falling phase of the pulse is not instantaneous. These two components were measured directly by comparing the onset time of the voltage-clamp pulse to the time of onset for the inward calcium current (giving a value for the A component) and then measuring the delay between the current onset and the initiation of the postsynaptic potential (giving a value for the B component). In short, the synaptic delay was brief enough to indicate that calcium entry should occur, as expected from other considerations, at the active zone and at particular sites within this compartment.

By using a presynaptic voltage-clamp circuit to simulate an action potential, a synaptic potential that precisely mimicked that produced by a natural presynaptic action potential could be generated following block of the sodium and potassium conductances (Llinás *et al.* 1982). From the above it became apparent that the main role of the action potential was to trigger a very well-timed entry of calcium into the presynaptic terminal at the active zone. This process would then activate transmitter release. Indeed, the presynaptic calcium current triggered by an action potential is a tail current and the EPSP an 'off' response.

Profile of the intracellular calcium concentration

At the beginning of the last decade it became apparent that, given the size of the calcium current per action potential, the calcium concentration against

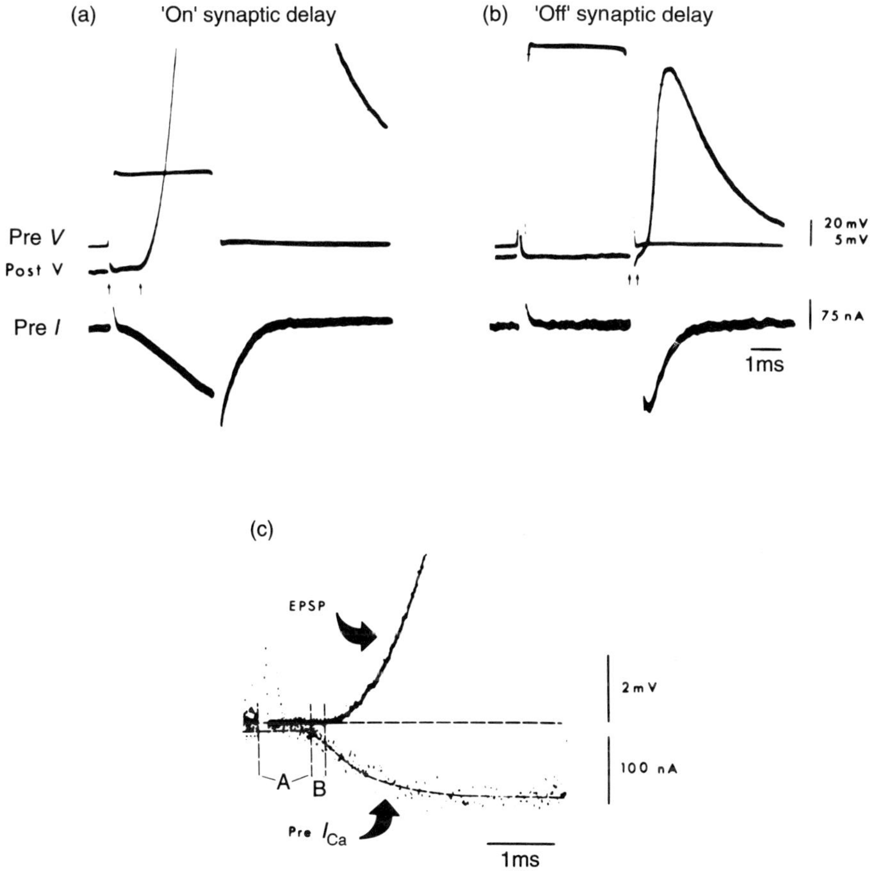

Fig. 17.4. Synaptic delay. (a) The 'on' response is shown for a voltage step of 60 mV from resting (upper trace, Pre *V*). Postsynaptic potential (Post *V*) shows a synaptic delay (arrows) of 1 ms. The presynaptic current (Pre *I*) shows a slow onset and a sharp tail current. (b) The 'off' delay (about 200 μs, marked by arrows) following a depolarization to the suppression potential (140 mV from resting) and the associated tail current. (c) Records showing the delay (A) for the onset of the calcium current (Pre I_{Ca}) and the delay (B) between onset of the calcium current and onset of the postsynaptic response (EPSP). (Taken with permission from Llinás (1977).)

the intracellular membrane could reach rather high levels (Llinás *et al.* 1981*a*). In fact, based on theoretical calculations, the concentration was proposed to be on the order of hundreds of μM during channel activation. This model assumed that calcium entry was not limited to any particular part of the membrane but, rather, occurred throughout the presynaptic terminal. The next step was to place these measurements of calcium current in the context of calcium entry occurring at particular sites.

Spatial and temporal distribution of $[Ca^{2+}]_i$ during the opening of a single-calcium channel

Following the initial observations noted above it became clear that the only appropriate manner in which to study the relation between calcium entry and transmitter release would be to determine as closely as possible the exact level of calcium concentration change at the active zone. To this end a set of calculations and computer modelling were implemented (Chad and Eckert 1984; Simon *et al.* 1984; Fogelson and Zucker 1985; Simon and Llinás 1985) and it was concluded that the calcium concentration immediately against the membrane and in the immediate cytostolic volume could not be determined from a study of the macroscopic voltage-clamp currents measured during synaptic release. It was also determined that, if calcium enters through single channels, as it must, and calcium moves away from the points of entry by diffusion, the calcium isoconcentration contour near each channel at the cytosolic level will have a well-defined tridimensional shape (Simon and Llinás 1985). At the time of maximum ion flow, the ions entering through a single channel would form a hemispherical isoconcentration profile determined by calcium diffusability and calcium buffer mobility within the terminal (Simon and Llinás 1985; Fig. 17.6).

A calculated three-dimensional isoconcentration surface for the calcium concentration during a single channel opening is shown in the left panel of

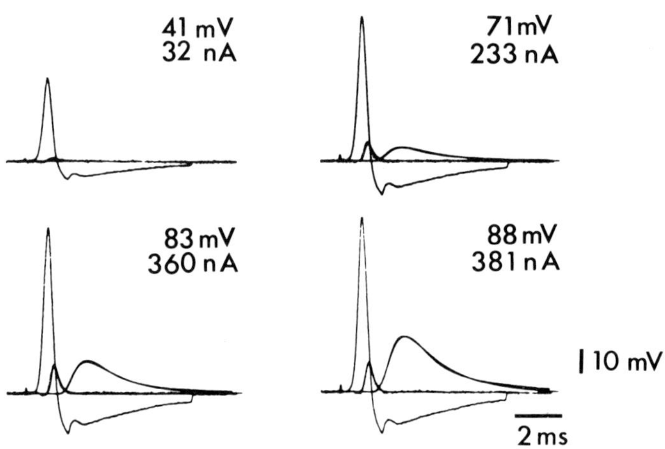

Fig. 17.5. Relationship between amplitude of simulated presynaptic spike (large biphasic potential), I_{Ca} (short monophasic peak), and postsynaptic potential (long monophasic peak, EPSP). The time course and amplitude of I_{Ca} and EPSP evoked by four artificial spikes is shown. The spike amplitude and that of the I_{Ca} is given for each trace. Experimental results for the I_{Ca} and EPSP are superimposed on those generated by the theoretical model. (Modified with permission from Llinás *et al.* (1982).)

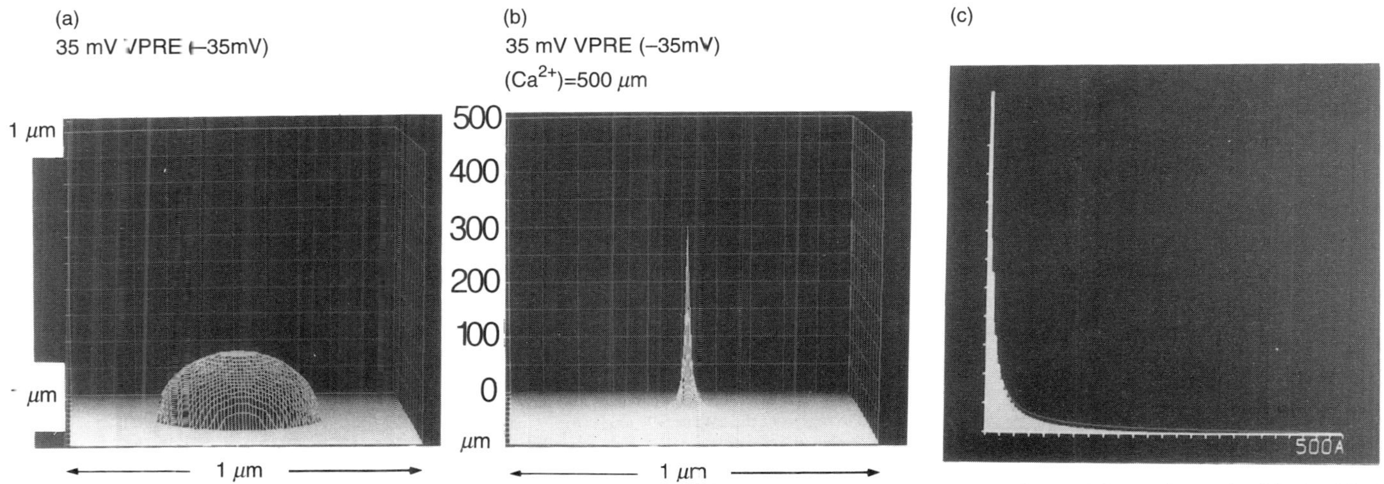

Fig. 17.6. Steady-state distribution of cytosolic calcium, ($[Ca^{2+}]_i$), following the opening of a calcium channel. (a) A 10-μM isoconcentration surface for $[Ca^{2+}]_i$. (b) $[Ca^{2+}]_i$ plotted in the two-dimensional surface immediately under the membrane for depolarization of 35 mV, from a resting potential of −70 mV. (c) $[Ca^{2+}]_i$ as a function of distance from the channel in the absence of a buffer (solid line) and in the presence of a mobile buffer for calcium. (Modified with permission from Simon and Llinás (1985).)

Fig. 17.6. The steady-state $[Ca^{2+}]_i$ in μM (x-axis) in the two-dimensional plane immediately under the membrane (y- and x-axis) is plotted as a function of distance from the channel pore for a 35 mV depolarization. The hemi-cross-section of the calcium concentration profile shown in the centre panel is plotted in the right panel. Over the first few hundred angstroms from the channel pore the concentration is close to steady state within a microsecond of channel opening (Simon and Llinás 1985). Following channel closing, the peak $[Ca^{2+}]_i$ near the channel approaches zero in less than 1 μs. The evaluation of the maximal possible rise of free calcium during calcium entry was calculated to be on the order of 300 μM. This calculation assumed that: (1) the presynaptic terminal had a finite volume; (2) the membrane lacked trans-membrane pumps for the removal of calcium; and (3) the I_{Ca} was 200 nA per terminal (150 $\mu A/cm^2$, the maximum I_{Ca} measured during a pulse).

Calcium concentration microdomains

Experimental and theoretical considerations indicated that calcium concentration peaks or microdomains should occur against the inside of the presynaptic terminal, and that these should last for the duration of the average open time of the channel. This influx would generate a $[Ca^{2+}]_i$ profile as high as several hundred μM near the internal mouth of the calcium channels, which would trigger vesicle fusion and transmitter release in the immediate vicinity of the Ca entry.

This hypothesis was recently tested in collaboration with Robert Silver (Llinás *et al.* 1991) using aequorin, a protein that emits photons in the presence of calcium (Shimomura *et al.* 1962; Llinás *et al.* 1972). In these experiments a hybrid synthetic *n*-aequorin-J (kindly developed by Dr O. Shimomura), with a sensitivity to $[Ca^{2+}]_i$ on the order of 10^{-4}, was injected presynaptically into the squid giant synapse and detected with a photon-counting camera. Upon stimulation, calcium entry triggered photo responses that were imaged as brief light-emitting sites with an average diameter of approximately 0.5 μm. They were found over about 5–10 per cent of the presynaptic area and were restricted to the active zones, demonstrating directly that $[Ca^{2+}]_i$ can reach several hundred μM during presynaptic activity (Fig. 17.7). The rapidly occurring light-emission points were accumulated during several seconds of stimulation, and the spatial and temporal distribution of the events was studied and compared with a second set of such light-emitting points or quantum emission domains (QEDs) obtained from the same terminal some minutes later. This measurement demonstrated the existence of such QEDs and showed that their location was independent of time.

Finally, detailed image analysis of the temporal and spatial dynamics of microdomain QEPs revealed the temporal cycling of active sites from one point of membrane to the next during stimulation. This finding suggests that

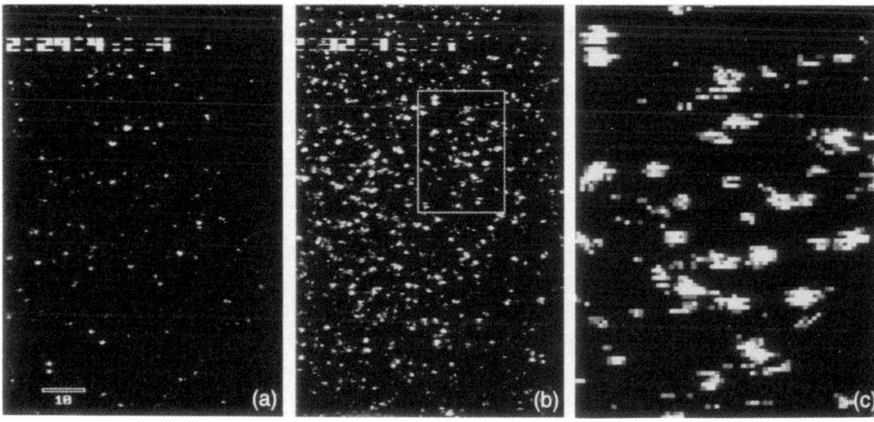

Fig. 17.7. Imagery of Ca^{2+}-dependent aequorin luminescence from presynaptic terminal of the squid giant synapse. Note the discrete sizes of individual blips. (a) A 10-s integration of the presynaptic terminal 30 s prior to stimulation. (b) A 10-s integration of the same region of the presynaptic terminal as shown in (a) 60 s following the onset of tetanic stimulation. Note the increase in the number of blips compared to (a). (c) An enlarged image of calcium microdomains in the region indicated within the inset borders of (b). Scale bar, 10 μm. (Modified with permission from Llinás *et al.* (1991).)

an even more complex modulation of activity may be present in these terminals. Thus, beyond calcium microdomains, other parameters such as the degree of phosphorylation of synapsin I (Llinás *et al.* 1985) may also show dynamic interactions, which would ensure a fine control of transmitter release from one impulse to the next.

Ultimately, the important issue that has emerged from the study of synaptic transmission at the squid giant synapse concerns the mechanisms for transmitter release. Perhaps it is important to remember that even apparently simple biological processes such as the release of synaptic transmitter are only recognized in their extraordinary complexity when viewed in detail. It is obvious that nature writes with fine print and yet, as often happens, such writing is best read in large structures, such as the stellate ganglion of the squid.

References

Augustine, G. J. and Charlton, M. P. (1986). Calcium dependence of presynaptic calcium current and post-synaptic response at the squid giant synapse. *Journal of Physiology*, **381**, 619–40.

Augustine, G. J., Charlton, M. P., and Horn, R. (1988). Role of calcium-activated potassium channels in transmitter release at the squid giant synapse. *Journal of Physiology*, **398**, 149–64.

Baker, P. F., Hodgkin, A. L., and Ridgway, E. B. (1971). Depolarization and calcium entry in squid giant axons. *Journal of Physiology*, **218**, 709–55.

Bloedel, J., Gage, P. W., Llinás, R., and Quastel, D. M. (1966). Transmitter release at the squid giant synapse in the presence of tetrodotoxin. *Nature*, **212**, 49–50.

Bullock, T. H. (1946). A preparation for the physiological study of the unit synapse. *Nature*, **158**, 555–6.

Bullock, T. H. (1948). Properties of a single synapse in the stellate ganglion of squid. *Journal of Neurophysiology*, **11**, 343–64.

Bullock, T. and Hagiwara, S. (1957). Intracellular recording from the giant synapse of the squid. *Journal of General Physiology*, **40**, 565–77.

Castejon, O. J. and Villegas, G. M. (1964). Fine structure of the synaptic contacts in the stellate ganglion of the squid. *Journal of Ultrastructural Research*, **10**, 585–98.

Ceccarelli, B., Hurlbut, W. P., and Mauro, A. (1972). Depletion of vesicles from frog neuromuscular junctions by prolonged tetanic stimulation. *Journal of Cell Biology*, **54**, 30–8.

Charlton, M. P., Smith, S. J., and Zucker, R. S. (1982). Role of presynaptic calcium ions and channels in synaptic facilitation and depression at the squid giant synapse. *Journal of Physiology*, **323**, 173–93.

Chad, J. E. and Eckert, R. (1984). Calcium 'domains' associated with individual channels may account for anomalous voltage relations of Ca-dependent responses. *Biophysical Journal*, **45**, 993–9.

Couteaux, R. and Pécot-Dechavassine, M. (1970). Vésicules synaptiques et poches au niveau des 'zones actives' de la jonction neuromusculaire. *Comptes Rendus Hebdomadaires des Séances de l'Academie des Sciences, D: Sciences Naturelles*, **271**, 2346–9.

De Robertis, E. D. P. and Bennett, H. S. (1955). Some features of the submicroscopic morphology of synapses in frog and earthworm. *Journal of Biophysical and Biochemical Cytology*, **1**, 47–58.

del Castillo, J. and Katz, B. (1956). Localization of active spots within the neuromuscular junction of the frog. *Journal of Physiology* **132**, 630–49.

Fogelson, A. L. and Zucker, R. S. (1985). Presynaptic calcium diffusion from various arrays of single channels: implications for transmitter release and synaptic facilitation. *Biophysical Journal*, **48**, 1003–17.

Hagiwara, S. and Tasaki, I. (1958). A study on the mechanism of impulse transmission across the giant synapse of the squid. *Journal of Physiology*, **143**, 114–37.

Hama, K. (1962). Some observations on the fine structure of the giant synapse in the stellate ganglion of the squid, *Doryteuphis bleekeri. Zeitschrift für Zellforschung* **56**, 437–44.

Heuser, J. E., Reese, T. S., Dennis, M. J., Jan, Y., Jan, L., and Evans, L. (1979). Synaptic vesicle exocytosis captured by quick freezing and correlated with quantal transmitter release. *Journal of Cell Biology*, **81**, 275–300.

Katz, B. (1969). *The release of neural transmitter substances*, The Sherrington Lectures X. Charles C. Thomas, Springfield, Illinois.

Katz, B. and Miledi, R. (1966). Input–output relation of a single synapse. *Nature*, **212**, 1242–5.

Katz, B. and Miledi, R. (1967). A study of synaptic transmission in the absence of nerve impulses. *Journal of Physiology, London*, **192**, 407–36.

Kusano, K., Livengood, D. R., and Werman, R. (1967). Correlation of transmitter release with membrane properties of the presynaptic fiber of the squid giant synapse. *Journal of General Physiology*, **50**, 2579–610.

Llinás, R. R. (1977). Calcium and transmitter release in squid synapse. In *Society for Neuroscience Symposia*, Vol. 2 (ed. W. M. Cowan and J. A. Ferendelli), pp. 139–60. Society for Neuroscience, Bethesda, Maryland.

Llinás, R. and Nicholson, C. (1975). Calcium role in depolarization–secretion coupling: An aequorin study in squid giant synapse. *Proceedings of the National Academy of Sciences, USA*, **72**, 187–90.

Llinás, R., Blinks, J. R., and Nicholson, C. (1972). Calcium transient in presynaptic terminal of squid giant synapse: detection with aequorin. *Science*, **176**, 1127–9.

Llinás, R., Steinberg, I. Z., and Walton, K. (1976). Presynaptic calcium currents and their relation to synaptic transmission: voltage clamp study in squid giant synapse and theoretical model for the calcium gate. *Proceedings of the National Academy of Sciences, USA*, **73**, 2913–22.

Llinás, R., Steinberg, I. Z., and Walton, K. (1980). Transmission in the squid giant synapse: a model based on voltage clamp studies. *Journal de Physiologie, Paris*, **76**, 413–18.

Llinás, R., Steinberg, I. Z., and Walton, K. (1981*a*). Presynaptic calcium currents in squid giant synapse. *Biophysical Journal*, **33**, 289–321.

Llinás, R., Steinberg, I. Z., and Walton, K. (1981*b*). Relationship between presynaptic calcium current and postsynaptic potential in squid giant synapse. *Biophysical Journal*, **33**, 323–52.

Llinás, R., Sugimori, M., and Simon, S. M. (1982). Transmission by presynaptic spike-like depolarization in the squid *Loligo pealii* giant synapse. *Proceedings of the National Academy of Sciences, USA*, **79**, 2415–19.

Llinás, R., McGuinness, T., Leonard, C. S., Sugimori, M., and Greengard, P. (1985). Intraterminal injection of synapsin I or calcium/calmodulin-dependent protein kinase II alters neurotransmitter release at the squid giant synapse. *Proceedings of the National Academy of Sciences, USA*, **82**, 3035–9.

Llinás, R., Sugimori, M., and Silver, R. B. (1991). Imaging preterminal calcium concentration microdomains in the squid giant synapse. *Biological Bulletin*, **181**, 316–17.

Martin, R. and Miledi, R. (1975). A presynaptic complex in the giant synapse of the squid. *Journal of Neurocytology*, **4**, 121–9.

Miledi, R. and Parker, I. (1981). Calcium transients recorded with arsenazo III in the presynaptic terminal of the squid giant synapse. *Proceedings of the Royal Society of London, Series B*, **212**, 197–211.

Miledi, R. and Slater, C. R. (1966). The action of calcium on neuronal synapses in the squid. *Journal of Physiology*, **184**, 473–98.

Palay, S. L. (1956). Synapses in the central nervous system. *Journal of Biophysical and Biochemical Cytology*, **2**, 193–201.

Pumplin, D. W. and Reese, T. S. (1978). Membrane ultrastructure of the giant synapse of the squid *Loligo pealii*. *Neuroscience*, **3**, 685–97.

Pumplin, D. W., Reese, T. S., and Llinás, R. (1981). Are the presynaptic membrane particles the calcium channels? *Proceedings of the National Academy of Sciences, USA*, **78**, 7210–13.

Robertson, J. D. (1953). Ultrastructure of two invertebrate synapses. *Proceedings of the Society for Experimental Biology and Medicine*, **82**, 219–23.

Shimomura, O., Johnson, F. H., and Saiga, Y. (1962). Extraction, purification and properties of aequorin, a bioluminescent protein from the luminous hydromedusan aequorea. *Journal of Cellular Physiology*, **59**, 223–39.

Simon, S. M. and Llinás, R. R. (1985). Compartmentalization of the submembrane calcium activity during calcium influx and its significance in transmitter release. *Biophysical Journal*, **48**, 485–98.

Simon, S. M., Sugimori, M., and Llinás, R. (1984). Modeling of submembranous calcium concentration changes and their relation to rate of presynaptic transmitter release in the squid giant synapse. *Biophysical Journal*, **45**, 264a.

Takeuchi, A. and Takeuchi, N. (1962). Electrical changes in pre- and postsynaptic axons of the giant synapse of *Loligo*. *Journal of General Physiology*, **45**, 1181–94.

Young, J. Z. (1936). Structure of nerve fibers and synapses in some invertebrates. *Cold Spring Harbor Symposia of Quantitative Biology*, **4**, 1–6.

Young, J. Z. (1938). The functioning of the giant nerve fibers of the squid. *Journal of Experimental Biology*, **15**, 170–85.

Young, J. Z. (1939). Fused neurons and synaptic contacts in the giant nerve fibers of cephalopods. *Philosophical Transactions of the Royal Society of London, Series B*, **229**, 465–503.

Young, J. Z. (1973). The giant fibre synapse of *Loligo*. *Brain Research*, **57**, 457–60.

18

Multiple calcium signalling pathways in squid giant presynaptic terminals

GEORGE J. AUGUSTINE, JOACHIM DEITMER, MICHAEL HANS, DIETER SWANDULLA, and KARL ZIPSER

Summary

The fluorescent Ca indicator, fura-2, can be used to measure the changes in presynaptic Ca concentration associated with action potentials. When using either a photomultiplier tube or a video camera to detect fura-2 fluorescence, trains of presynaptic action potentials produce rapid rises in Ca concentration that reach a peak level of a few nM per action potential and that decay very slowly, over hundreds of seconds. However, these measured rises are not those responsible for triggering transmitter release because injection of the Ca buffer, EGTA, blocks these rises in Ca concentration but not transmitter release. EGTA injection blocks synaptic augmentation, a form of synaptic plasticity that increases the amount of release produced by an action potential, suggesting that the measured Ca rises mediate augmentation. Our conclusion is that transmitter release is mediated by a rise in Ca concentration that is too localized to be detected with fura-2 measurements, while augmentation is mediated by a more widespread and detectable Ca signal.

More than three decades of investigation has established that the calcium (Ca^{2+}) ion is an intracellular messenger essential for the triggering of neurotransmitter release at chemical synapses (Katz 1969). In brief, there is evidence that electrical activity in a presynaptic neurone opens voltage-gated Ca^{2+} channels in the terminal region of this neurone. The resulting influx of Ca^{2+} causes a transient rise in intracellular Ca^{2+} concentration, $[Ca^{2+}]_i$, within the terminal and it is this rise in $[Ca^{2+}]_i$ that is thought to be the signal that initiates the exocytotic release of neurotransmitter (Augustine *et al.* 1987; Sudhof and Jahn 1991; Trimble *et al.* 1991).

Despite the acknowledged importance of Ca^{2+} in neurotransmitter release, it has proved experimentally difficult to define the magnitude and time course of the presynaptic $[Ca^{2+}]_i$ signal. Early efforts to examine this signal employed the photoprotein aequorin (Llinás and Nicholson 1975), the metallochromic dye, arsenazo III (Miledi and Parker 1981; Charlton *et al.* 1982; Augustine *et al.* 1985*a*), and Ca^{2+}-sensitive microelectrodes (Deitmer *et al.* 1983). While these studies have amply demonstrated that $[Ca^{2+}]_i$ does indeed rise during presynaptic electrical activity, they lacked adequate precision to define the spatial dimensions of this signal. In this chapter, we consider recent studies that use the fluorescent Ca^{2+} indicator, fura-2 (Grynkiewicz *et al.* 1985), to further characterize presynaptic Ca^{2+} signalling.

Fura-2 photometry reveals longlasting calcium signals

One approach to defining the $[Ca^{2+}]_i$ changes associated with transmitter release is to microinject fura-2 into the presynaptic terminal, excite this dye with a two-wavelength monochromator, and then use a photomultiplier to measure the resultant fluorescence emission. Such measurements indicate that the $[Ca^{2+}]_i$ of a resting presynaptic terminal is on the order of 50–100 nM (Zipser *et al.* 1991). Electrical activity causes Ca^{2+} channels to open and the resultant influx of Ca^{2+} causes $[Ca^{2+}]_i$ to rise transiently (Fig. 18.1). As can be seen in this figure, the changes in $[Ca^{2+}]_i$ produced by a brief train of action potentials (50 Hz, 3 s) are quite long-lasting and require several hundred seconds to fully return to the resting level. The slow decline in $[Ca^{2+}]_i$ is caused by Ca^{2+} being removed from presynaptic cytoplasm via mechanisms such as Na^+/Ca^{2+} exchange (Zipser *et al.* 1991).

The peak magnitude of the $[Ca^{2+}]_i$ changes associated with a train of presynaptic action potentials depends upon the number of action potentials in the train. Although responses to single action potentials are slightly beyond

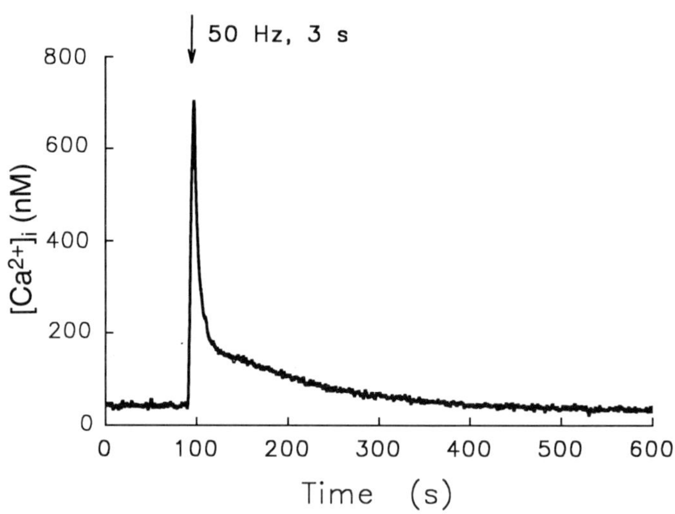

Fig. 18.1. Photomultiplier detection of presynaptic Ca^{2+} signals. A terminal filled with fura-2 was stimulated with a train of action potentials (50 Hz, 3 s) at the time indicated by the arrow. Photomultiplier measurements of fura-2 fluorescence revealed that presynaptic $[Ca^{2+}]_i$ rose rapidly and decayed over hundreds of seconds.

the resolution of the method, the fact that the time course of the $[Ca^{2+}]_i$ change is virtually independent of the Ca^{2+} load (Zipser *et al.* 1991) allows an estimate of the magnitude of this signal from recordings of $[Ca^{2+}]_i$ changes associated with trains of action potentials. By varying the number of action potentials in the train, the relationship between action potential number and resultant change in peak $[Ca^{2+}]_i$ can be quantified. This relationship is approximately linear, with a slope of about 5 nM/action potential. Thus, a single action potential raises average $[Ca^{2+}]_i$ by approximately 5 nM in this presynaptic terminal.

One limitation of such photomultiplier measurements is that they are integrated over the entire volume of the region sensed by the detector. If spatial gradients exist within the volume of the terminal, such integral measurements will not accurately reflect the $[Ca^{2+}]_i$ changes occurring within any single presynaptic compartment. In order to ask whether such gradients exist, we have used video imaging methods to evaluate the spatial distribution of $[Ca^{2+}]_i$ changes associated with presynaptic electrical activity.

Video imaging defines spatial compartments

Conventional video microscopy can be combined with digital image processing technology to provide two-dimensional maps of intracellular $[Ca^{2+}]_i$ (Tsien and Poenie 1986). Details of the methods used to apply such techniques to the squid giant presynaptic terminal may be found in Smith *et al.* (1988) and Swandulla *et al.* (1991). In brief, fura-2 was microinjected into living presynaptic terminals and the resultant fluorescence signals were detected with a video camera attached to a compound microscope. By storing the resultant images on an analogue storage device and performing off-line digital analysis of the images, acquisition rates of 30 images/s were obtained.

Spatial maps of the changes in $[Ca^{2+}]_i$ produced by a 5-s long train of presynaptic action potentials are shown in Fig. 18.2. During the train (top row of images), $[Ca^{2+}]_i$ rises gradually. Throughout the train these rises are spatially non-uniform, being largest in the portion of the presynaptic terminal where active zones, the presynaptic sites of transmitter secretion, are most abundant (Smith *et al.* 1988; Smith and Augustine 1988; Augustine *et al.* 1989; Swandulla *et al.* 1991; Smith *et al.* 1993). However, the lateral gradients of $[Ca^{2+}]_i$ quickly dissipate following a train (bottom row of images in Fig. 18.2), presumably due to diffusion of Ca^{2+} (and perhaps Ca^{2+}-bound fura-2) within the terminal. Just as was observed with photomultiplier measurements, $[Ca^{2+}]_i$ gradually returns to basal levels over many seconds.

A clearer picture of the time course and magnitude of the changes in presynaptic $[Ca^{2+}]_i$ produced by action potentials comes from graphs such as the one shown in Fig. 18.3. This graph shows measurements of the time course of $[Ca^{2+}]_i$ changes within two presynaptic compartments: one compartment,

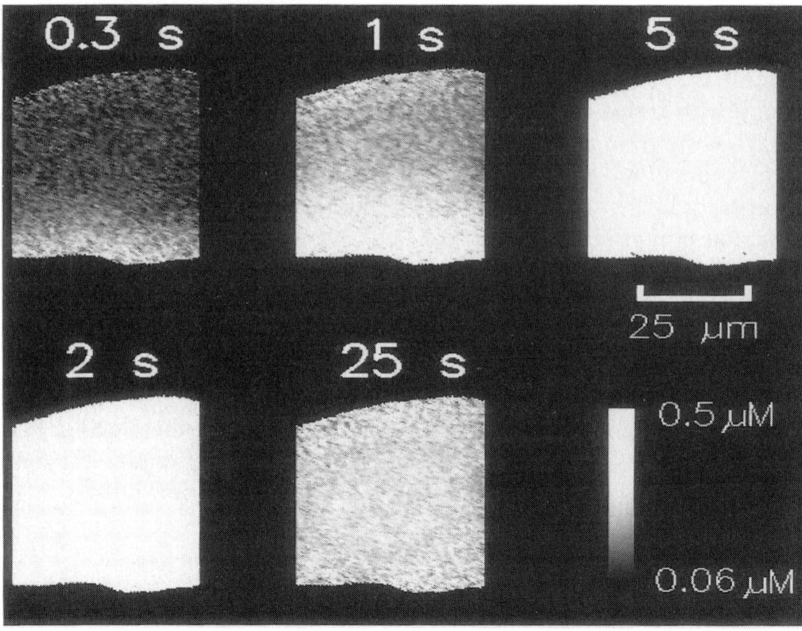

Fig. 18.2. Digital imaging of presynaptic Ca^{2+} signals. A terminal filled with fura-2 was stimulated with a train of action potentials (50 Hz, 5 s) during the time points represented by the top row of images. During this train, substantial gradients of presynaptic $[Ca^{2+}]_i$ are evident, with $[Ca^{2+}]_i$ highest in the portion of the terminal closest to the postsynaptic axon. Following cessation of the action potentials, $[Ca^{2+}]_i$ quickly reached spatial equilibrium (bottom row of images) and then slowly declined over many seconds. (Taken with permission from Swandulla *et al.* (1991).)

near the postsynaptic axon, that is enriched in active zones (labelled 'Near') and another compartment on the other side of the presynaptic terminal, approximately 25 μm away from the active zones (labelled 'Away'). In both compartments, $[Ca^{2+}]_i$ rises to levels on the order of 1 μM or less during the train and subsides very slowly following the train. Note that the $[Ca^{2+}]_i$ values in these two compartments are equal within the first 1–2 s following cessation of electrical activity, meaning that, within this time, equilibration of $[Ca^{2+}]_i$ is complete in the lateral direction. There are also longitudinal $[Ca^{2+}]_i$ gradients, produced by non-uniformities in surface/volume ratio along the length of the terminal, that dissipate more slowly than the lateral gradients (Smith *et al.* 1993).

Several features of the $[Ca^{2+}]_i$ signals measured in the compartment containing active zones suggest that even these spatially resolved measurements do not accurately reflect the $[Ca^{2+}]_i$ signal responsible for triggering transmitter release. First, these signals typically are no larger in magnitude than

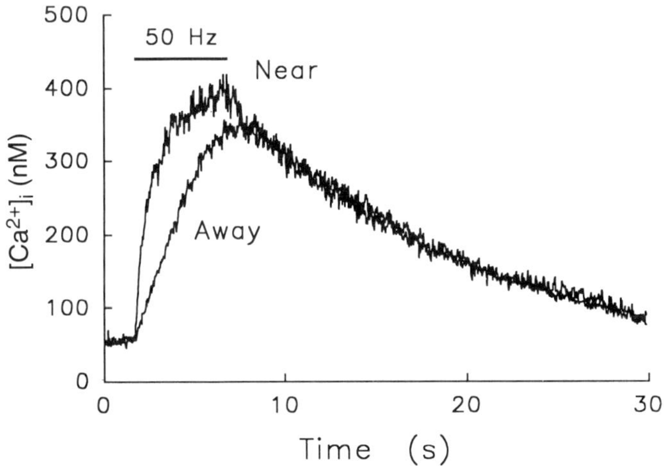

Fig. 18.3. Time course of Ca^{2+} signals measured with digital imaging. $[Ca^{2+}]_i$ changes of two presynaptic compartments—one near the active zones (Near) and another away from the active zones, in the interior of the terminal (Away)—were measured at video frame rates (30 Hz). $[Ca^{2+}]_i$ rose most rapidly and to higher levels in the vicinity of the active zones. (Taken with permission from Swandulla *et al.* (1991).)

the spatially averaged signals measured with a photomultiplier, both amounting to no more than a few nM change in $[Ca^{2+}]_i$ per action potential. Second, their kinetics are too slow: transmitter release is activated and terminated over a period of a few ms (Llinás *et al.* 1981; Augustine *et al.* 1985*b*), while the measured $[Ca^{2+}]_i$ signals rise and decay over many seconds. Thus, even these spatially resolved measurement methods seem inadequate to track the Ca^{2+} signal that triggers transmitter release.

EGTA injection shows that the calcium signal for release is highly localized

The most compelling indication that the $[Ca^{2+}]_i$ rises measured with fura-2 imaging are not those responsible for triggering transmitter release comes from experiments examining the effect of the Ca^{2+} buffer, EGTA, upon the $[Ca^{2+}]_i$ signal measured with fura-2. Microinjection of EGTA into a presynaptic terminal causes the action-potential induced changes in $[Ca^{2+}]_i$, measured with fura-2, to completely disappear (Fig. 18.4; Swandulla *et al.* 1991). In contrast, such injections of EGTA have absolutely no effect on transmitter release evoked by single presynaptic action potentials (Adams *et al.* 1985; Adler *et al.* 1991; Swandulla *et al.* 1991), presumably because the EGTA binds Ca^{2+} too slowly to compete with the Ca^{2+} receptor that triggers release. Whatever the reason for the differential effects of EGTA on release

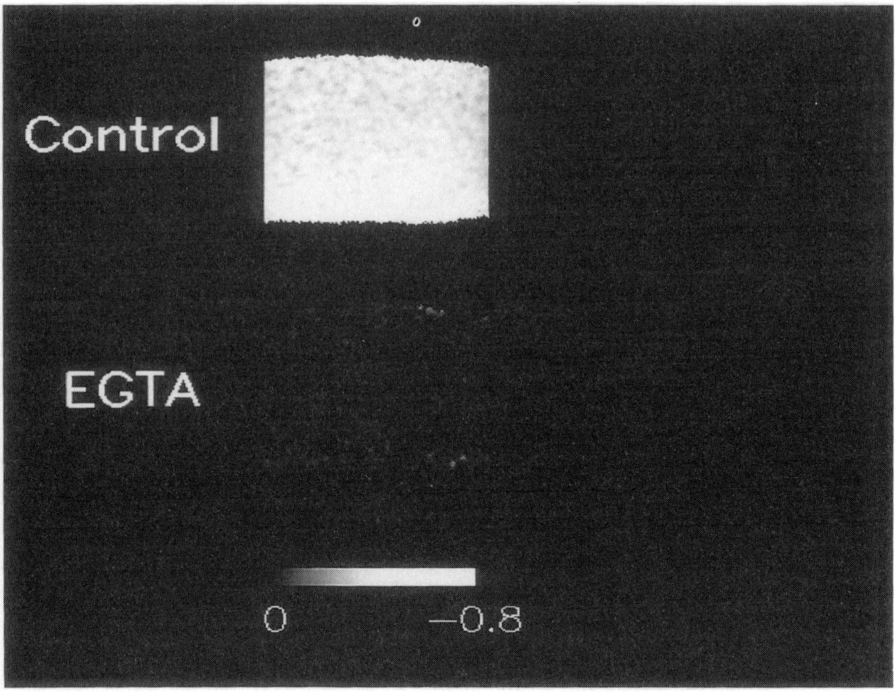

Fig. 18.4. EGTA blocks stimulus-induced presynaptic Ca^{2+} signals. Digital images of changes in fura-2 fluorescence measured before (Control) and following (EGTA) microinjection of EGTA into the presynaptic terminal. Because of changes in the resting $[Ca^{2+}]_i$ level, the scale represents presynaptic $[Ca^{2+}]_i$ levels ranging from 0.19 to 5 μM in the control image and 0.06 to 2.7 μM in the image obtained following EGTA injection. (Taken with permission from Swandulla *et al.* (1991).)

and on the measured $[Ca^{2+}]_i$ signals, this difference proves that the measured $[Ca^{2+}]_i$ signals are not those responsible for release. We imagine that in EGTA-injected terminals the Ca^{2+} signal for release eludes detection because this signal is too spatially localized to be detected by our imaging methods (Swandulla *et al.* 1991). Such acute spatial compartmentalization is also likely to occur in terminals not injected with EGTA (Fogelson and Zucker 1985; Simon and Llinás 1985; Smith and Augustine 1988). In this compartment, local $[Ca^{2+}]_i$ may reach levels on the order of 100 μM or greater (Roberts *et al.* 1990; Adler *et al.* 1991).

A slow calcium signal mediates synaptic augmentation

Our observation that EGTA preferentially blocks the slow $[Ca^{2+}]_i$ signal measured with fura-2 allows EGTA to be used as a probe to address the physiological function of this signal. Because the signal is slow, we have addressed the possibility that it mediates processes slower than neurotransmitter release. In particular, it is possible that this signal is involved in mediating one or more forms of synaptic plasticity; many forms of plasticity have a presynaptic origin and all are slower than the process triggering transmitter release (Zucker 1989). One such form of plasticity is synaptic depression, which develops and decays over a time scale of seconds and acts to decrease the amount of transmitter released by a presynaptic action potential (Kusano and Landau 1975). However, presynaptic microinjection of EGTA has little effect upon the magnitude or time course of synaptic depression (Swandulla *et al.* 1991). Another relatively slow form of plasticity is tetanic and post-tetanic potentiation, which increases the amount of transmitter released by an action potential and consists of various kinetic components with time scales on the order of seconds (Magleby and Zengel 1975). At the squid synapse, the 'augmentation' component of tetanic and post-tetanic potentiation can be evoked by trains of action potentials (Swandulla *et al.* 1991). Presynaptic microinjection of EGTA selectively eliminates this form of plasticity while sparing release evoked by single action potentials (Fig. 18.5). Thus, the slow $[Ca^{2+}]_i$ rise produced by a train of action potentials is responsible for mediating synaptic augmentation.

The quantitative relationship between the level of $[Ca^{2+}]_i$ and the degree of augmentation can be determined from experiments simultaneously employing electrophysiological and optical recording techniques (Swandulla

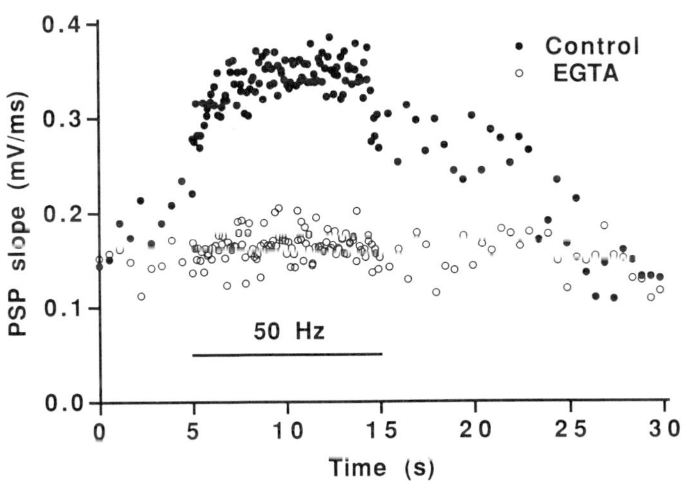

Fig. 18.5. EGTA selectively blocks synaptic augmentation monitored by measurements of the slope of the postsynaptic potential (PSP). A train of presynaptic action potentials (50 Hz) elicits synaptic augmentation in control conditions (solid circles), but augmentation is blocked following presynaptic microinjection of EGTA (open circles). (Taken with permission from Swandulla *et al.* (1991).)

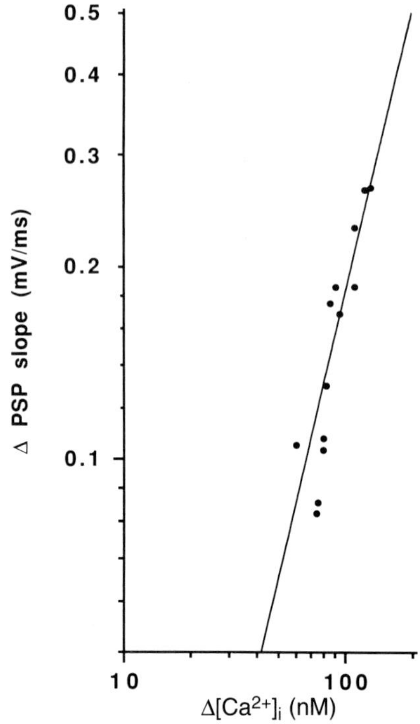

Fig. 18.6. Ca^{2+} requirement for augmentation. The change in $[Ca^{2+}]_i$ produced by a train of presynaptic action potentials (abscissa) is correlated with the resultant degree of augmentation (ordinate) derived from the change in slope of the postsynaptic potential (Δ PSP slope). The resulting relationship is a weak power function (straight line) with an exponent of 1.55. (Taken with permission from Swandulla *et al.* (1991).)

et al. 1991). Such experiments reveal that augmentation occurs with elevations of presynaptic $[Ca^{2+}]_i$ in the sub-micromolar range (Fig. 18.6). This is interesting because such elevations of $[Ca^{2+}]_i$ produce no measurable increases in the basal release occurring between action potentials. Thus, transmitter release and augmentation appear to have different sensitivities to $[Ca^{2+}]_i$.

Conclusions

We have summarized one of a series of experimental iterations aimed at characterizing the presynaptic Ca^{2+} signal for transmitter release. Using the fluorescent Ca^{2+} indicator, fura-2, along with digital imaging microscopy, it is possible to make temporally and spatially resolved measurements of $[Ca^{2+}]_i$ rises associated with neurotransmitter release. Because injection of the Ca^{2+} buffer, EGTA, into presynaptic terminals abolishes measured $[Ca^{2+}]_i$ rises

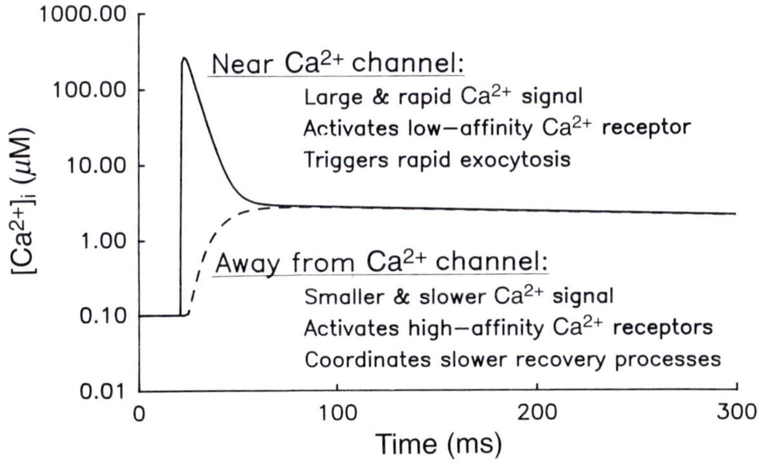

Fig. 18.7. Ca^{2+} signalling pathways in presynaptic terminals. This cartoon schematically illustrates the changes in $[Ca^{2+}]_i$ that we believe to be produced by a single presynaptic action potential. Our concept is that transmitter release is produced by a localized rise in $[Ca^{2+}]_i$ (solid line) that is large and brief. Following termination of the action potential, this rise in $[Ca^{2+}]_i$ spreads throughout the terminal (dashed line), due to diffusion. Because this Ca^{2+} signal is diluted as it spreads throughout the terminal, its magnitude is attenuated and its time course is retarded in comparison to the Ca^{2+} signal that triggers release

but not neurotransmitter release, we must conclude that the measured $[Ca^{2+}]_i$ rises are not those responsible for triggering release. However, because EGTA blocks both the measured rise in $[Ca^{2+}]_i$ and augmentation, a form of synaptic plasticity that modulates the amount of transmitter released by an action potential, this $[Ca^{2+}]_i$ signal may serve to produce augmentation.

Our picture of the Ca^{2+} signalling pathways involved in these two physiological events is shown in the cartoon of Fig. 18.7. The Ca^{2+} signal for neurotransmitter release must be very localized to escape detection in our imaging experiments. Other experiments suggest that this localization occurs as a consequence of a close spatial association between Ca^{2+} channels and the Ca^{2+} receptors that trigger release (Adler *et al.* 1991). Thus, the rise in $[Ca^{2+}]_i$ produced in this highly localized compartment is probably quite large and fast, as illustrated by the solid line in Fig. 18.7. Because of the presumed large magnitude of these changes, as well as the rapid time course of termination of transmitter release (Llinás *et al.* 1981; Augustine *et al.* 1985*b*), it is likely that this $[Ca^{2+}]_i$ rise activates release by binding to receptors with a low affinity for Ca^{2+}. Candidates for such low-affinity receptors include the annexins (Creutz *et al.* 1988) or other vesicle-associated Ca-binding proteins (Petrenko *et al.* 1991). Conversely, the $[Ca^{2+}]_i$ changes occurring at other places within the terminal will be smaller and slower (dashed line in Fig. 18.7). This results from the diffusion of Ca^{2+} ions away from their source, the voltage-gated Ca^{2+} channels, and into the sink of the bulk of the presynaptic cytoplasm. It is these slower and smaller changes in $[Ca^{2+}]_i$ that we are detecting with fura-2 imaging measurements and it is these changes that are responsible for synaptic augmentation. When coupled with recent advances in our understanding of Ca^{2+} regulated phosphorylation of synaptic vesicle proteins (Llinás *et al.* 1991), activation of presynaptic ion channels (Roberts *et al.* 1990; Robitaille and Charlton 1992), and neurotransmitter synthesis, it is possible to propose that this slow Ca^{2+} signal co-ordinates a whole spectrum of reactions that may allow the presynaptic terminal to recover from a bout of transmitter secretion and prepare for future exocytotic activity. Because of the small magnitude of these $[Ca^{2+}]_i$ rises (μM or less) the receptors responsible for these activities are likely to be of higher affinity for Ca^{2+} than those responsible for triggering transmitter release. Candidates for such receptors include calmodulin and its relatives (Kretsinger 1980; Means 1988). Our next experimental challenge is to identify and assign functions to these receptors.

Acknowledgements

This work was supported by NIH grant NS-21624 to GA, DFG funds to JD and DS, a Kuffler Fellowship to DS, and a DOD Fellowship to KZ.

References

Adams, D. J., Takeda, K., and Umbach, J. A. (1985). Inhibitors of calcium buffering depress evoked transmitter release at the squid giant synapse. *Journal of Physiology*, **369**, 145–59.

Adler, E. M., Augustine, G. J., Duffy, S. N., and Charlton, M. P. (1991). Alien intracellular chelators attenuate neurotransmitter release at the squid giant synapse. *Journal of Neuroscience*, **11**, 1496–507.

Augustine, G. J., Charlton, M. P., and Smith, S. J. (1985*a*). Calcium entry into voltage-clamped presynaptic terminals of squid. *Journal of Physiology*, **367**, 143–62.

Augustine, G. J., Charlton, M. P., and Smith, S. J. (1985*b*). Calcium entry and neurotransmitter release at voltage-clamped nerve terminals of squid. *Journal of Physiology*, **367**, 163–81.

Augustine, G. J., Charlton, M. P., and Smith, S. J. (1987). Calcium action in synaptic transmitter release. *Annual Reviews of Neuroscience*, **10**, 633–93.

Augustine, G. J., Buchanan, J., Charlton, M. P., Osses, L. R., and Smith, S. J. (1989). Fingering the trigger for neurotransmitter secretion: studies on the calcium channels of squid giant presynaptic terminals. In *Secretion and its control* (ed. G. S. Oxford and C. M. Armstrong), pp. 203–23. Rockefeller University Press, New York.

Charlton, M. P., Smith, S. J., and Zucker, R. S. (1982). Role of presynaptic calcium ions and channels in synaptic facilitation and depression at the squid giant synapse. *Journal of Physiology*, **323**, 173–93.

Cruetz, C. E., Drust, D. S., Martin, W. H., Kambouris, N. G., Snyder, S. L., and Hamman, H. C. (1988). Calcium-dependent membrane-binding proteins as effectors of secretion in mammalian and fungal cells. In *Molecular mechanisms in secretion* (ed. C. Thorn, M. Treiman, and O. H. Petersen), pp. 575–90. Munksgaard, Copenhagen.

Deitmer, J. W., Augustine, G. J., Zucker, R. S., and Eckert, R. (1983). Stimulus-induced intracellular Ca^{2+} transients in *Aplysia* neurons and squid presynaptic terminals measured with Ca^{2+} sensitive microelectrodes. *Society for Neuroscience Abstracts*, **9**, 1188.

Fogelson, A. L. and Zucker, R. S. (1985). Presynaptic calcium diffusion from various arrays of single channels. *Biophysical Journal*, **48**, 1003–17.

Grynkiewicz, G., Poenie, M., and Tsien, R. Y. (1985). A new generation of Ca^{2+} indicators with greatly improved fluorescence properties. *Journal of Biological Chemistry*, **260**, 3440–50.

Katz, B. (1969). *The release of neural transmitter substances*. Liverpool University Press, Liverpool.

Kretsinger, R. H. (1980). Structure and evolution of calcium-modulated proteins. *CRC Critical Reviews in Biochemistry*, **8**, 119–74.

Kusano, K. and Landau, E. M. (1975). Depression and recovery of transmission at the squid giant synapse. *Journal of Physiology*, **245**, 13–32.

Llinás, R. and Nicholson, C. (1975). Calcium role in depolarization–secretion coupling: an aequorin study in squid giant synapse. *Proceedings of the National Academy of Sciences, USA*, **72**, 187–90.

Llinás, R., Steinberg, I. Z., and Walton, K. (1981). Relationship between presynaptic calcium current and postsynaptic potential in squid giant synapse. *Biophysical Journal*, **33**, 323–52.

Llinás, R., Gruner, J. A., Sugimori, M., McGuinness, T. L., and Greengard, P. (1991). Regulation by synapsin I and Ca^{2+}-calmodulin-dependent protein kinase II of transmitter release in squid giant synapse. *Journal of Physiology*, **436**, 257–82.

Magleby, K. L. and Zengel, J. E. (1975). A quantitative description of tetanic and post-tetanic potentiation of transmitter release at the frog neuromuscular junction. *Journal of Physiology*, **245**, 183–208.

Means, A. R. (1988). Molecular mechanisms of action of calmodulin. *Recent Progress in Hormone Research*, **44**, 223–62.

Miledi, R. and Parker, I. (1981). Calcium transients recorded with arsenazo III in the presynaptic terminal of the squid giant synapse. *Proceedings of the Royal Society of London, Series B*, 212, 197–211.

Petrenko, A. G., Perin, M. S., Davletov, B. A., Ushkaryov, Y. A., Geppert, M., and Sudhof, T. C. (1991). Binding of synaptotagmin to the α-latrotoxin receptor implicates both in synaptic vesicle exocytosis. *Nature*, **353**, 65–8.

Roberts, W. M., Jacobs, R. A., and Hudspeth, A. J. (1990) Colocalization of ion channels involved in frequency selectively and synaptic transmission at presynaptic active zones of hair cells. *Journal of Neuroscience*, **10**, 3664–84

Robitaille, R. and Charlton, M. P. (1992). Presynaptic calcium signals and transmitter release are modulated by calcium-activated potassium channels. *Journal of Neuroscience*, **12**, 297–305.

Simon, S. L. and Llinás, R. R. (1985). Compartmentalization of the submembrane calcium activity during calcium influx and its significance in transmitter release. *Biophysical Journal*, **48**, 485–98.

Smith, S. J. and Augustine, G. J. (1988). Calcium ions, active zones and synaptic transmitter release. *Trends in Neurosciences*, **11**, 458–64.

Smith, S. J., Buchanan, J., Osses, L. R., Charlton, M. P., and Augustine, G. J. (1993). The spatial distribution of calcium signals in squid presynaptic terminals. *Journal of Physiology*, **472**, 573–94.

Smith, S. J., Osses, L. R., and Augustine, G. J. (1988). Fura-2 imaging of localized calcium accumulation within squid 'giant' presynaptic terminals. In *Calcium and ion channel modulation* (ed. A. D. Grinnell, D. Armstrong, and M. B. Jackson), pp. 147–55. Plenum Press, New York.

Sudhof, T. C. and Jahn, R. (1991). Proteins of synaptic vesicles involved in exocytosis and membrane recycling. *Neuron*, **6**, 665–77.

Swandulla, D., Hans, M., Zipser, K., and Augustine, G. J. (1991). Role of residual calcium in synaptic depression and posttetanic potentiation: fast and slow calcium signaling in nerve terminals. *Neuron*, **7**, 915–26.

Trimble, W. S., Linial, M., and Scheller, R. H. (1991). Cellular and molecular biology of the presynaptic nerve terminal. *Annual Reviews of Neuroscience*, **14**, 93–122.

Tsien, R. Y. and Poenie, M. (1986). Fluorescence ratio imaging: a new window into intracellular ionic signaling. *Trends in Biochemical Sciences*, **11**, 450–5.

Zipser, K., Augustine, G. J., and Deitmer, J. (1991). Na/Ca exchange regulates presynaptic calcium levels. *Biophysical Journal*, **59**, 594a.

Zucker, R. S. (1989). Short-term synaptic plasticity. *Annual Reviews of Neuroscience*, **12**, 13–31.

19

Chemical transmission at the squid giant synapse

J. B. MESSENGER, A. DE SANTIS, and D. C. OGDEN

Despite all the work on squid giant fibres since their rediscovery 60 years ago we still know nothing about how they innervate the mantle muscles and do not really understand how they are themselves activated. In particular we do not know the nature of the transmitter(s) at the largest synapse in the animal kingdom: the 'giant synapse' between second- and third-order fibres in the squid stellate ganglion (Fig. 19.1). This chapter discusses some recent experiments suggesting that L-glutamate is very probably an endogenous transmitter at the synapse, including the first demonstration that glutamate, when abruptly released in the synaptic cleft, can elicit action potentials in the third-order giant axon.

There have been several experiments implicating glutamate as a possible transmitter at the synapse (Miledi 1966, 1967, 1969; Kelly and Gage 1969; De Santis *et al.* 1978; Kawai *et al.* 1983; Stanley 1983; Augustine and Eckert 1984;

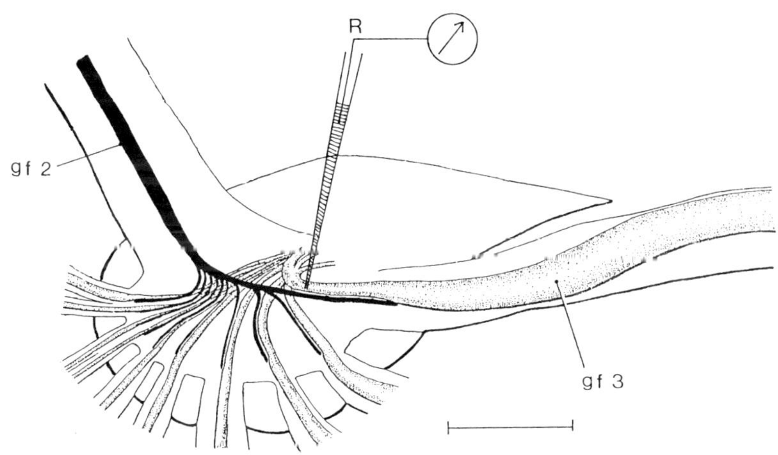

Fig. 19.1. Squid stellate ganglion. The 'giant synapse' investigated here is that between the second-order giant fibre (gf 2, black) and the most medial third order fibre (gf 3). In all experiments a glass microelectrode (R) was inserted into gf 3 as indicated: in the first two series of experiments a suction electrode was used to stimulate the gf 2. Scale bar, 1 mm. (Modified with permission from Young (1939).)

Adams *et al.* 1985; Eusebi *et al.* 1985; Adams and Gillespie 1988). The only other contender, acetylcholine, has been effectively eliminated by the results of Stanley (1984). The evidence for glutamate includes: (1) depolarization of the postsynaptic membrane by iontophoretically applied L-glutamate, kainate, or quisqualate; (2) blockade of the excitatory postsynaptic potential (EPSP) by bath application of glutamate; and (3) irreversible blockade of the EPSP and glutamate-induced depolarization by JSTX, a toxin from Joro spider (*Nephila clavata*) venom. The main objection to glutamate being the endogenous transmitter has always been that the depolarization evoked by glutamate has a different reversal potential from that of the EPSP (Miledi 1969; Llinás *et al.* 1974). An important difficulty here is that the postsynaptic membrane is not uniformly sensitive to iontophoretically applied glutamate or its agonists. There are thousands of sensitive points on the postsynaptic membrane (see Discussion). One consequence of this is that it has proved impossible to elicit action potentials in the giant fibre by iontophoretic application of glutamate. We shall return to these points later.

The starting point for the present series of experiments was the discovery that there are glutamatergic systems in the *Octopus* central nervous system (CNS) (Andrews *et al.* 1983), in the squid chromatophores and mantle muscles (Bone and Howarth 1980; Florey *et al.* 1985; see also Messenger 1991; Cornwell and Messenger, in preparation), and in axon–Schwann cell interaction (Liebermann *et al.* 1989; see also Chapter 15). Allied to this was the sudden appearance of a battery of specific glutamate agonists and antagonists, developed as a result of current interest in glutamate as a major excitatory transmitter in the mammalian CNS (e.g. Watkins and Olverman 1987). A preliminary account of some of these experiments has been given by De Santis and Messenger (1989).

The stellate ganglion

Figure 19.1, which is based on Young's original drawing (1939), shows the main features of the ganglion in *Loligo* and *Alloteuthis*, the two genera used for these experiments. This figure shows 10 third-order giant fibres (the number varies from eight to 12, Martin and Miledi 1986) receiving input from a single second-order giant fibre, which enters the ganglion in the mantle connective and divides to make synaptic contact with each third-order fibre. There is also an accessory second-order fibre, not shown in the figure, which terminates close to the cell bodies in the giant fibre lobe of the ganglion at the so-called 'proximal synapses' (Young 1939). The 'giant synapse' referred to in this chapter is the synapse of the last branch of the second-order axon (the 'pre-axon') with the most medial and thickest of the third-order axons (the 'post-axon'). There is considerable variation in the exact configuration of this synapse but the diagram emphasizes that the synapse extends for about a millimetre: in one specimen of *Loligo vulgaris* (180 mm mantle length) careful measurements based on electronmicrographs gave a value of 1125 μm, and an estimate for the total area of synaptic contacts as 16 312 μm^2 (Martin and Miledi 1986).

In all the experiments reported here an intracellular electrode, filled with 3 M KCl, was routinely inserted into the post-axon in the region of the synapse (for details see De Santis and Messenger 1989). In some experiments, where appropriate, we also impaled the pre-axon. In the first two sets of experiments an extracellular suction electrode was used to stimulate the pallial nerve; in the third set of experiments a second electrode was inserted into the synaptic cleft for iontophoresis.

The experiments

Bath application of new 'glutamate drugs'

With the ganglion immersed in oxygenated filtered sea water (SW) at 14–18 °C the pre-axon was stimulated every 30 s at about 10 times threshold strength in order to produce a stable action potential in the post-axon. The SW was then replaced with a solution (in SW) of the drug being tested and the effects on the post-axon were recorded. Drugs in this and other experiments were supplied by Tocris Neuramin Ltd and Sigma.

We found that L-glutamate and six specific glutamate agonists (four of them novel) all blocked transmission at the giant synapse (Table 19.1). Figure 19.2 shows a typical gradual failing of transmission in the presence of quisqualate. With all seven substances, subsequent washing led to complete recovery of the response. However, as Table 19.1 shows, D-glutamate and L- or D-aspartate did not block transmission, while *N*-methyl-D-aspartic acid (NMDA) either blocked irreversibly or had no effect. There are now numerous glutamate antagonists available and several of these have also been shown to be reversible blockers (Table 19.2). The 'broad-band' antagonists *cis*-2,3-piperidine dicarboxylic acid (*cis*-2,3-PDA) (Davies *et al.* 1982) and γ-D-

Table 19.1 Action of glutamate agonists at squid giant synapse

Reversible blockade	No effect of irreversible blockade
L glutamate	D-glutamate
Kainate	L-, D-aspartate
Quisqualate	NMDA
AMPA	L-AP4
ODAP	
Bromowillardiine	
Domoate	

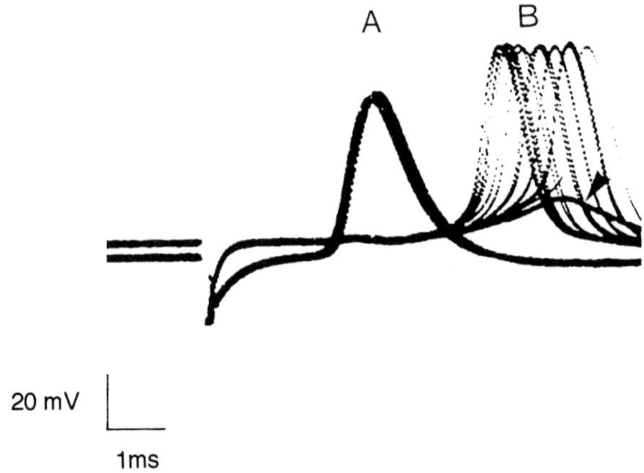

Fig. 19.2. Bath application of quisqualate (1×10^{-4} M) to a ganglion in which the pre-axon and post-axon were impaled and the gf 2 was stimulated every 30 s. The action potential in the pre-axon (A) barely alters but the action potential in the post-axon (B) shifts progressively to the right and fails after 6 min (arrow). The response recovered fully after washing with sea water.

Table 19.2 Action of glutamate antagonists and toxins at squid giant synapse

Antagonist*	Receptor type†	Effect on synapse
γ-DGG	NMDA	Negative or irreversible blockade
CPP	NMDA	Negative
AP5	NMDA	Negative
cis-2,3PDA	'Broad-band'	Blockade
GAMS	K/AMPA‡	Blockade
CNQX	K/AMPA	Blockade
DNQX	K/AMPA	Blockade
Toxin		
JSTX	K/AMPA	Blockade§
Argiotoxin$_{636}$	Quisqualate¶	Irreversible blockade

* γ-DGG, γ-D-glutamylglycine; CPP, 3-((R)-2-carboxypiperazine-4-yl)-propyl-1-phosphonic acid; AP5, D-L-2-amino-5-phosphonopentanoic acid (APV); *cis*-2,3 PDA, *cis*-2, 3-piperidine dicarboxylic acid; GAMS, γ-D-glutamylaminomethylsulfonic acid; CNQX, 6-cyano-7-nitroquinoxaline-2,3-dione (FG9065); DNQX, 6,7-dinotroquinoxaline-2,3-dione (FG9041). JSTX, Joro spider toxin.
† According to the classification of Monaghan *et al.* (1988). NMDA, *N*-methyl-D-aspartic acid; AMPA, α-amino-3-hydroxy-5-methyl-4-isoxazole propionic acid.
‡ Previously quisqualate.
§ Result obtained by Kawai *et al.* (1983).
¶ In insects (Kerry *et al.* 1988).

glutamyl aminomethyl sulfonic acid (GAMS), thought to block kainate and quisqualate receptors in mammals specifically (Davies and Watkins 1985), were very effective blockers of transmission at the giant synapse. Even more effective, however, were the quinoxalinediones, 6-cyano-7-nitroquinoxaline-2,3-dione (FG 9065) (CNQX) and 6,7-dinitroquinoxaline-2,3-dione (FG 9091) (DNQX), which are selective non-NMDA antagonists in mammals (Honoré *et al.* 1988). It is noteworthy that 3-((R)-2 carboxypiperazin-4-yl)-propyl-1-phosphonic acid (CPP) (Davies *et al.* 1986), a highly potent NMDA antagonist in the mammalian CNS, caused no blockade at the giant synapse.

Bath application of argiotoxin$_{636}$

Argiotoxin$_{636}$ is a potent toxin from orb-web spider venom. It is a powerful antiglutamatergic agent that has presumably evolved to paralyse insect prey (Quicke and Usherwood 1990). In the locust it is known to act at quisqualate-type receptors on the muscle (Bateman *et al.* 1985; Kerry *et al.* 1988). We found argiotoxin to be the most potent blocker at the giant synapse that we have yet encountered. It irreversibly blocked transmission even at 10^{-12} M but, with only a limited amount of toxin available to us, we were unable to establish the threshold for its action. As in insects, its activity is proportional to the frequency of stimulation of the presynaptic nerve, a feature that suggests it might act at open channels (Bateman *et al.* 1985) (Fig. 19.3).

Iontophoresis

Using conventional current injection we applied glutamate, kainate, quisqualate, and, for the first time, α-amino-3-hydroxy-5-methyl-4-isoxazole propionic

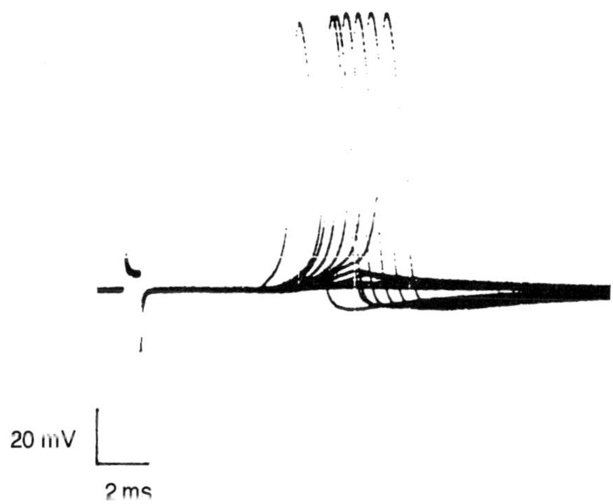

Fig. 19.3. Application of argiotoxin$_{636}$ (1×10^{-10} M) leads to failure of the post-axon action potential after a 4-min stimulation of the pre-axon at 10 Hz. This effect was irreversible.

Fig. 19.4. Depolarization of the postsynaptic membrane (top trace) by iontophoresed quisqualate (1×10^{-2} M) shown by the current pulse in the lower trace.

acid (AMPA) in the region of the postsynaptic membrane. All four substances produced weak depolarization (cf. Miledi 1967, 1969; De Santis *et al.* 1978; Stanley 1983), with quisqualate giving the strongest effect under the conditions tested (Fig. 19.4). Quisqualate, like L-glutamate and kainate, also desensitized the receptors on the postsynaptic membrane. Iontophoretically applied AMPA gave the weakest response of the four substances tested, suggesting that the new classification of glutamate receptor subtypes proposed by Monaghan *et al.* (1989) and based on mammals need not necessarily apply to molluscs.

'Caged' glutamate

The powerful new physiological technique utilizing photolabile 'cages' to render inert an active ion or compound for subsequent rapid release by photolysis (e.g. Gurney and Lester 1987) has already been successfully applied at the giant synapse in a study involving caged calcium (Delaney and Zucker 1990). Having ourselves shown that glutamate photolytically released from 'caged' glutamate expands cephalopod chromatophores (Messenger *et al.* 1991), we decided to apply this technique to the giant synapse.

In the first series of experiments, the sodium salt of N-(1-(2-nitrophenyl) ethoxycarbonyl) glutamate was dissolved in SW and applied to the bath containing the stellate ganglion and left for 20–30 min to allow it to diffuse deep into the ganglion. During this time there was no detectable effect on the resting potential nor any transsynaptic activation. We then discharged a xenon flashlamp delivering a 1-ms pulse of 300–360 nm light at the stellate ganglion. With a 5-mM solution of caged glutamate there was slight depolarization when the lamp was discharged (Fig. 19.5), but with a 50-mM solution the depolarization was sufficiently strong to fire the third-order axon (Fig. 19.6). We estimated that at photolysis only 10 per cent of the glutamate is freed so that, allowing for light lost in the ganglion, the final glutamate

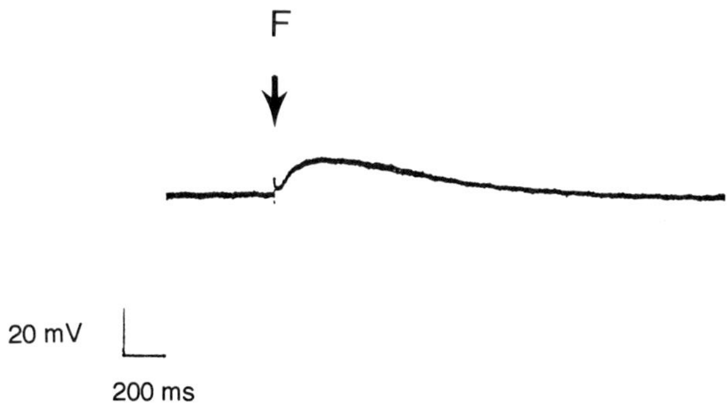

Fig. 19.5. Effect of flash photolysis of glutamate from 5 mM caged glutamate (pH 7.8; 30-min application) on the post-axon membrane potential; a small depolarization is seen. F, Flash.

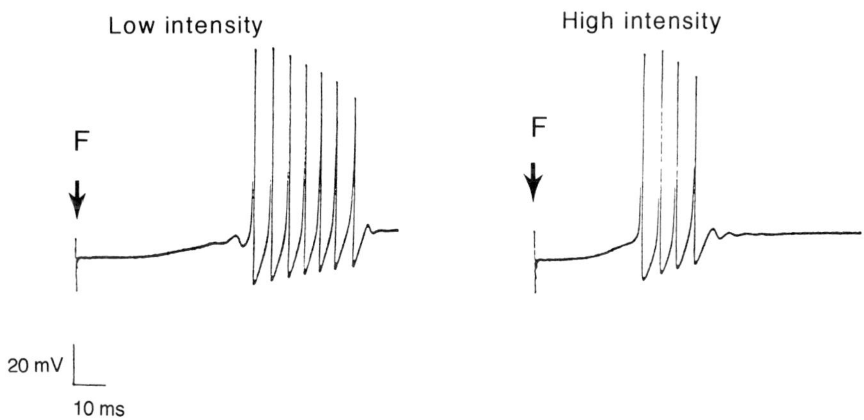

Fig. 19.6. Effect of flash photolysis of glutamate from 50 mM caged glutamate (pH 6.2; 18-min application) on the post-axon membrane potential; action potentials are superimposed on a slow depolarization. F, Flash: 'low' intensity = 40 mJ (300–350 nm); 'high' intensity = 100 mJ (300–350 nm).

concentration was probably less than 1 mM. As can be seen from Fig. 19.6, the photolytic cleavage of this particular caged glutamate was slow (half-time of about 10 ms at pH 6.2), resulting in a slower depolarization than that seen with transsynaptic activation.

In later experiments, flash photolysis of caged glutamate pre-equilibrated in the giant synapse at pH 5.5 caused rapid production of L-glutamate, fast postsynaptic depolarization and generation of postsynaptic action potentials in the third-order giant fibre, further strengthening the case for L-glutamate being an endogenous transmitter at the synapse (Corrie *et al.* 1993).

Discussion

Evaluation of the results

None of these experiments proves that L-glutamate is an endogenous transmitter at the giant synapse but, taken together, they constitute compelling evidence for such an hypothesis.

In the blockade experiments we tested several modern specific glutamate agonists (AMPA, β-*N*-oxalyl-L-α, β-diamino propionic acid (ODAP), bromowillardiine, and domoate) for the first time at the squid giant synapse and found them all to be active. In mammals all these compounds are active at kainate receptors and what, until recently, were termed quisqualate receptors (Monaghan *et al.* 1989). In agreement with this, we found that certain glutamate antagonists were also active at the synapse. Although the NMDA antagonist CPP had no effect, specific non-NMDA antagonists such as GAMS, CNQX, and DNQX, all produced rapid, reversible blocking. In summary, the results show: (1) that specific glutamate agonists desensitize the synapse to the endogenous transmitter(s); and (2) that the postsynaptic membrane is sensitive to non-NMDA glutamate antagonists.

Added to this is the result with argiotoxin$_{636}$. This highly specific antiglutamatergic agent, which enables spiders to paralyse their insect prey, irreversibly blocked transmission at the synapse even at 10^{-12} M. This agrees with the results of Kawai *et al.* (1983) using another toxin from spider venom, JSTX, and again points to there being non-NMDA receptors on the post-axon (see below). It is interesting that with argiotoxin, as in insects (Bateman *et al.* 1985), the speed of blocking is related to the frequency of pre-axon stimulation and therefore to the release of the endogenous transmitter.

The first two series of experiments are open to the criticism that, by using a suction electrode on the pallial nerve, we were stimulating not only the second-order giant fibre but also the accessory giant fibre, so that blockade could have been occurring at the proximal synapses. It will, therefore, be necessary to repeat key experiments using an intracellular stimulating electrode in the second-order giant fibre (Stanley 1990).

The iontophoresis experiments confirm the results of previous authors that L-glutamate, kainate, and quisqualate all depolarize the postsynaptic membrane; this is also true of another non-NMDA agonist, AMPA, although not as strongly as we had anticipated (see below). The level of depolarization we obtained varied considerably according to the position of the electrode at the synapse, as earlier works all report (e.g. Adams and Gillespie 1988). The weak responses we obtained may, therefore, have been due to unfortunate electrode placement; before concluding that AMPA is less active than quisqualate at the postsynaptic membrane more experiments would need to be carried out. However, a more powerful technique for investigating the action of specific neuroactive substances on the post-axon has emerged: flash-photolysis using 'caged' compounds. In these preliminary experiments we allowed caged glutamate to diffuse into the stellate ganglion for 20 min so that it might be assumed to have entered the synaptic cleft and be present at, or close to, many of the synaptic contact areas. We then gave a 1-ms flash to release glutamate *in situ*. The result was a depolarization of the post-axon sufficiently large to produce action potentials (Fig. 19.6). By overcoming desensitization in this way we have shown for the first time that glutamate can activate the third-order giant fibres.

The unusual form of the giant synapse

It is not always recognized that the squid giant synapse is a complex structure. It extends for over a millimetre but the pre- and post-axons are not simply apposed to one another uniformly. As Young's drawing makes clear [Fig. 19.7(a)], the post-axon bears numerous tree-like processes that reach up to and around the pre-axon to make functional contact with it. Martin and Miledi (1986) confirmed this arrangement in their ultrastructural investigation [Fig. 19.7(b)], which included careful measurements based on electron microscope sections. They estimated that there are about 15 000 synaptic contacts and, in one specimen, they estimated the total contact area to be 16 312 μm^2. On average, each synaptic contact was just over 1 μm in diameter but the size ranged from 0.6 to 4 μm in transverse section and from 0.45 to 1.64 μm in longitudinal section. Furthermore, the length of the processes varies from 4 μm centrally to 17 μm laterally. A conspicuous feature of the branching trees with their fine tips is that they are often very narrow at their bases [Fig. 19.7(a)]. Finally, Martin and Miledi (1986) noted that the very regular synaptic gap measures only 12 nm in width.

These features may help to explain some of the anomalous results that we and other workers have obtained with the giant synapse. In the first place, it becomes clear why bath application of drugs requires such a long time to elicit effects, even after the outer capsule of the ganglion has been opened (De Santis and Messenger 1989). Entrance to and exit from the synapse must be impeded by such a narrow cleft. The non-uniform nature of the synapse would also explain why the sensitivity of the postsynaptic membrane to iontophoresed compounds varies so sharply from point to point, as we and others have consistently found. Presumably, maximum sensitivity will be obtained when the pipette tip is at or close to one of the 1-μm synaptic contact areas. This feature also makes it harder to interpret the relative effectiveness of one agonist over another using iontophoresis.

Most important of all, the complex morphology may account for what has always been a very serious objection to L-glutamate being the endogenous transmitter: the discrepancy between the reversal potential measured for

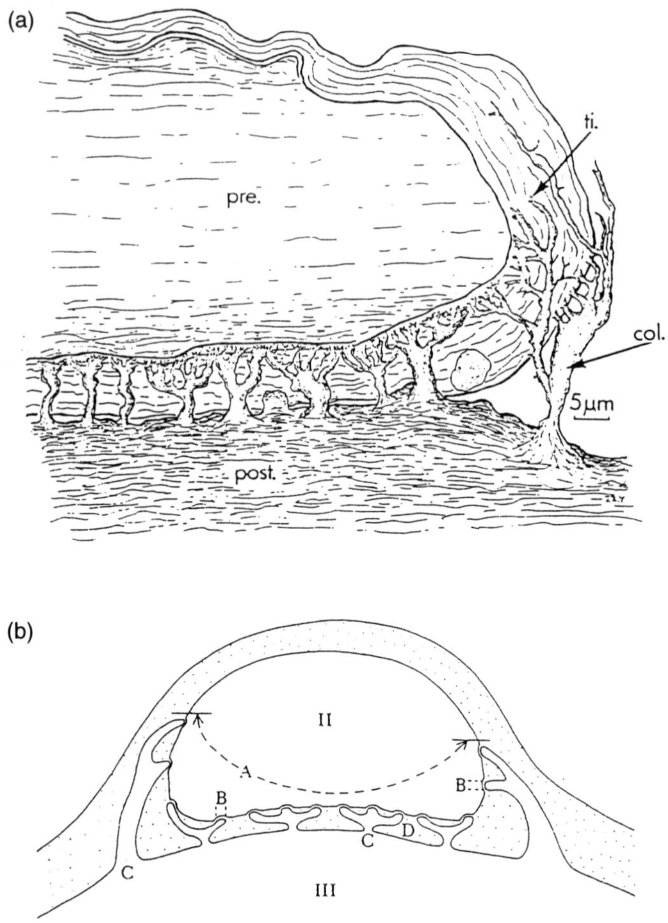

Fig. 19.7. (a) Young's (1973) drawing of part of the giant synapse of *Loligo*. col, Collateral; pre, pre-axon; post, post-axon: ti., tip of collateral. (b) Martin and Miledi's (1986) schematic diagram of the synapse in cross-section. A, Segment of pre-axon (II) with synaptic contacts; B, width of a synaptic contact; C, collateral of post-axon (III); D, sheath around the two giant fibres.

glutamate and that for the endogenous transmitter (Miledi 1969). It could be that simple measurements fail to consider variations in post-synaptic membrane voltage over considerable distances (Dekin 1983). Certainly, as Llinás *et al.* (1974) pointed out, 'for geometric reasons it may be quite difficult to mimic the exact timing and distribution of the natural transmitter release with a point source iontophoretic injection'. The detailed study of Adams and Gillespie (1988) comes to similar conclusions.

Glutamate receptors in cephalopods

These experiments and others also tell us something about the nature of the glutamate receptor subtypes in cephalopods. In mammals the current view (Monaghan *et al.* 1989) seems to be that there are five: NMDA, kainate, AMPA (formerly quisqualate), L(+)-2-amino-4-phosphonobutyric acid (L-AP4), and metabotropic or 1-amino-cyclopentane-1,3-dicarboxylic acid. We have not tested any agonists for metabotropic receptors; and our results with NMDA and L-AP4 were equivocal at the giant synapse. The most potent specific competitive antagonist for NMDA (CPP) did not block transmission, however, and it seems that NMDA receptors may be absent. Experiments with *Octopus* suggest there may be no NMDA receptors centrally (Messenger, in preparation) and experiments with squid chromatophores (Cornwell and Messenger, in preparation; see below) show that they are absent there too. It should be noted, however, that there is evidence for both NMDA and non-NMDA receptors on squid Schwann cells, implicated in axon–Schwann cell signalling (Lieberman *et al.* 1989; see also Chapter 15).

We obtained good evidence for there being receptor sub-types that might be termed kainate in mammals: applications of domoate, bromowillardiine (both agonists) and CNQX, DNQX (both antagonists) led to blockade. There was also evidence for what used to be the quisqualate receptor: quisqualate, ODAP, bromowillardiine, CNQX, and DNQX all caused blockade. So did AMPA but the effect of AMPA when iontophoresed on to the post-axon was certainly not larger than quisqualate. It is worth summarizing here some results with these agonists in another system, *Alloteuthis* chromatophores, where we have immunohistochemical evidence that the innervation may be glutamatergic (see Chapter 22). In the skin preparation we employ (Messenger 1991), topically applied drugs have good access to the chromatophore radial muscles and it is easier to rank the effect of applied drugs. Table 19.3 summarizes our results. The most potent agonist, domoate, is thought to be active at kainate receptors but, clearly, quisqualate is more potent than AMPA. Thus, if there really are two separate receptor subtypes

Table 19.3 Thresholds for squid chromatophore expansion

	Threshold (M)
Domoate	5×10^{-7} to 1×10^{-6}
Quisqualate	2×10^{-5}
L-ODAP	2×10^{-5}
Bromowillardiine	5×10^{-5}
L-glutamate	2×10^{-4}
Kainate	2×10^{-4}
AMPA	5×10^{-4}

on the giant axon, the second one cannot be labelled AMPA. In short, the 'quisqualate' receptor is almost certainly different from the mammalian one. Experiments in which we injected glutamate and its agonists into the blood supplying the brain of the *Octopus* also revealed extreme sensitivity to quisqualate (Andrews *et al.* 1983).

Clearly, cephalopod glutamate receptors are different from mammalian glutamate receptors. We need not be too surprised at this: molluscs and vertebrates have been distinct for about 400 million years (Messenger 1991). Perhaps it is surprising that so many of the new drugs mimicking the action of glutamate are effective at all in cephalopods. Certainly, they are very useful to us and, perhaps, as we accumulate more data about the squid giant synapse, we shall be able to add to the evolutionary debate developing on glutamate receptors.

Future work

To confirm our hypothesis we need to demonstrate that the pre-axon contains L-glutamate in above average levels and ideally to relate glutamate to the vesicles. We need also to demonstrate that, when the pre-axon is stimulated, glutamate is actually released, as Kawagoe *et al.* (1981) have shown at the crayfish neuromuscular junction. We are currently carrying out investigations along these lines. A preliminary immunohistochemical investigation (with J. Storm-Mathisen and O. P. Ottersen of the University of Oslo) has encountered the difficulty that levels of glutamate and some other amino acids in the pre-axon seem to be surprisingly low. Experiments designed to measure the efflux of glutamate from a stimulated pre-axon have yielded data of wide variability. This could be related to the narrowness of the synaptic cleft and its tortuous geometry (see above), which could retard the efflux of the transmitter(s). One way forward here might be to use some of the new glutamate-uptake blockers to enhance 'free' glutamate.

The 'caging' technique should undoubtedly yield important new information about the synapse, for example about the time course of depolarization, threshold levels, equilibrium potentials, and the effectiveness of specific glutamate agonists; experiments with caged compounds are in progress.

Acknowledgements

It is a pleasure to thank the director and staff of the Marine Biological Association Laboratory, Plymouth and the president, director, and staff of the Naples Zoological Station for their support in providing the animals and laboratory facilities for these experiments. We are also deeply indebted to Dr I. Duce for the gift of some argiotoxin$_{636}$ and to Drs D. R. Trentham, J. Corrie, and co-workers for synthesizing and providing us with various types of 'caged' glutamate.

References

Adams, D. J. and Gillespie, J. I. (1988). The actions of L-glutamate at the post-synaptic membrane of the squid giant synapse. *Journal of Experimental Biology*, **140**, 535–48.

Adams, D. J., Takeda, K., and Umbach, J. A. (1985). Inhibitors of calcium buffering depress evoked transmitter release at the squid giant synapse. *Journal of Physiology, London*, **369**, 145–59.

Andrews, P. L. R., Messenger, J. B., and Tansey, E. M. (1983). The chromatic and motor effects of neurotransmitter injections in intact and brain-lesioned *Octopus*. *Journal of the Marine Biological Association of the United Kingdom*, **63**, 355–70.

Augustine, G. and Eckert, R. (1984). Divalent cations differentially support transmitter release at the squid giant synapse. *Journal of Physiology, London*, **346**, 257–71.

Bateman, A., Boden, P., Dell, A., Duce, I. R., Quicke, D. L. J., and Usherwood, P. N. R. (1985). Post synaptic block of a glutamatergic synapse by low molecular weight fractions of spider venom. *Brain Research*, **339**, 237–44.

Bone, Q. and Howarth, J. V. (1980). The role of L-glutamate in neuromuscular transmission in some molluscs. *Journal of the Marine Biological Association of the United Kingdom*, **60**, 619–26.

Corrie, J. T., De Santis, A., Katayama, Y., Khodakhah, K., Messenger, J. B., Ogden, D. C., and Trentham, D. R. (1993). Postsynaptic activation at the squid giant synapse by photolytic release of L-glutamate from a 'caged' L-glutamate. *Journal of Physiology*, **465**, 1–8.

Davies, J. and Watkins, J. C. (1985). Depressant actions of γ-D-glutamylamino-methylsulfonate (GAMS) on amino acid-induced and synaptic excitations in the cat spinal cord. *Brain Research*, **327**, 113–20.

Davies, J., Evans, R. H., Francis, A. A., Jones, A. W., Smith, D. A., and Watkins, J. C. (1982). Conformational aspects of the actions of some piperidine dicarboxylic acids at excitatory amino acid receptors in the mammalian and amphibian spinal cord. *Neurochemical Research*, **7**, 1119–33.

Davies, J., Evans, R. H., Herring, P. L., Jones, A. W., Olverman, H. J., Pook, P., and Watkins, J. C. (1986). CPP, a new potent and selective NMDA antagonist. Depression of central neuron responses, affinity for [^{3}H] D-AP5 binding sites on brain membranes and anticonvulsant activity. *Brain Research*, **382**, 169–73.

Dekin, M. S. (1983). Permeability changes induced by L-glutamate at the crayfish neuromuscular junction. *Journal of Physiology*, **341**, 105–25.

Delaney, K. R. and Zucker, R. S. (1990). Calcium released by photolysis of DM-nitrophen stimulates transmitter release at squid giant synapse. *Journal of Physiology*, **426**, 473–98.

DeSantis, A. and Messenger, J. B. (1989). New evidence that L-glutamate is a transmitter at the squid giant synapse. *Quarterly Journal of Experimental Physiology*, **74**, 219–22.

DeSantis, A., Eusebi, F., and Miledi, R. (1978). Kainic acid and synaptic transmission in the stellate ganglion of the squid. *Proceedings of the Royal Society of London, Series B*, **202**, 527–32.

Eusebi, F., Miledi, R., Parker, I., and Stinnakre, J. (1985). Post-synaptic calcium influx at the giant synapse of the squid during activation by glutamate. *Journal of Physiology*, **369**, 183–97.

Florey, E., Dubas, F., and Hanlon, R. T. (1985). Evidence for L glutamate as a transmitter substance of motoneurons innervating squid chromatophore muscles. *Comparative Physiology and Biochemistry*, **82C**, 259–68.

Gurney, A. M. and Lester, H. A. (1987). Light-flash physiology with synthetic photosensitive compounds. *Physiological Reviews*, **67**, 583–617.

Honoré, T., Davies, S. N., Drejer, J., Fletcher, E. J., Jacobsen, P., Lodge, D., and Nielsen, F. E. (1988). Quinoxalinediones: potent competitive non-NMDA glutamate receptor antagonists. *Science*, **241**, 701–3.

Kawagoe, R., Onodera, K., and Takeuchi, A. (1981). Release of glutamate at the crayfish neuromuscular junction. *Journal of Physiology*, **312**, 225–36.

Kawai, N., Yamagishi, S., Saito, M., and Furuya, K. (1983). Blockade of synaptic transmission in the squid giant synapse by a spider toxin (JSTX). *Brain Research*, **278**, 346–9.

Kelly, J. S. and Gage, P. W. (1969). L-glutamate blockade of transmission at the giant synapse of the squid stellate ganglion. *Journal of Neurobiology*, **1**, 209–19.

Kerry, C. J., Ramsey, R. L., Sansom, M. S. P., and Usherwood, P. N. R. (1988). Single channel studies of non-competitive antagonism of a quisqualate-sensitive glutamate receptor by argiotoxin$_{636}$—a fraction isolated from orb-web spider venom. *Brain Research*, **459**, 312–27.

Lieberman, E. M., Abbott, N. J., and Hassan, S. (1989). Evidence that glutamate mediates axon-to-Schwann cell signaling in the squid. *Glia*, **2**, 94–102.

Llinás, R., Joyner, R. W., and Nicholson, C. (1974). Equilibrium potential for the post-synaptic response in the squid giant synapse. *Journal of General Physiology*, **64**, 519–35.

Martin, R. and Miledi, R. (1986). The form and dimensions of the giant synapse of squids. *Philosophical Transactions of the Royal Society of London, Series B*, **312**, 355–77.

Messenger, J. B. (1991). Transmitters, toxins and phylogeny. *Journal of Zoology*, **223**, 687–94.

Messenger, J. B., Katayama, Y., Ogden, D. C., Corrie, J. E. T., and Trentham, D. R. (1991). Photoloytic release of glutamate from 'caged' glutamate expands squid chromatophores. *Journal of Physiology*, **438**, 293P.

Miledi, R. (1966). Miniature synaptic potentials in squid nerve cells. *Nature*, **212**, 1240–2.

Miledi, R. (1967). Spontaneous synaptic potentials and quantal release of transmitter in the stellate ganglion of the squid. *Journal of Physiology*, **192**, 379–406.

Miledi, R. (1969). Transmitter action in the giant synapse of the squid. *Nature*, **223**, 1284–6.

Monaghan, D. T., Bridges, R. J., and Cotman, C. W. (1989). The excitatory amino acid receptors: their classes, pharmacology, and distinct properties in the function of the central nervous system. *Annual Review of Pharmacology and Toxicology*, **29**, 365–402.

Quicke, D. L. J. and Usherwood, P. N. R. (1990). Spider toxins as lead structures for novel pesticides. In *Safer insecticides: development and use* (ed. E. Hodgson and R. J. Kuhr), pp. 385–452. Dekker, New York.

Stanley, E. F. (1983). Depolarizing and desensitizing actions of glutaminergic and cholinergic agonists at the squid giant synapse. *Biological Bulletin*, **165**, 533.

Stanley, E. F. (1984). The action of cholinergic agonists on the squid stellate ganglion giant synapse. *Journal of Neuroscience*, **4**, 1904–11.

Stanley, E. F. (1990). The preparation of the squid giant synapse for electrophysiological investigation. In *Squid as experimental animals* (ed. D. L. Gilbert, W. J. Adelman, and J. M. Arnold), pp. 171–92. Plenum, New York.

Watkins, J. C. and Olverman, H. J. (1987). Agonists and antagonists for excitory amino acid receptors. *Trends in Neurosciences*, **10**, 265–72.
Young, J. Z. (1939). Fused neurons and synaptic contacts in the giant nerve fibres of cephalopods. *Philosophical Transactions of the Royal Society of London*, *Series B* **229**, 465–503.
Young, J. Z. (1973). The giant fibre synapse of *Loligo*. *Brain Research*, **57**, 457–60.

PART V

Neuromuscular control

20

The structure and physiology of cephalopod muscle fibres

Q. BONE, E. R. BROWN, and M. USHER

Summary

The structure of cephalopod 'skeletal' muscle fibres and radial chromatophore fibres are first described; with the interesting exception of some cross-striated fibres in the tentacles, all are obliquely-striated and of similar structure, divisible into two types according to the volume of the mitochondrial core. In squid mantle, the fibre types are zoned; in the central mitochondria-poor (MP) fibres, mitochondria make up around 6 per cent of the cross-sectional area of the fibre, whereas in the outer layers of mitochondria-rich (MR) fibres, the core occupies around 45 per cent of fibre area. The chromatophore fibres are of the MR type. Cephalopod fibres contract rapidly, but only the responses of the MP type have been examined directly, although what are probably the MR type in the mantle show slower responses. Little is yet known in *Sepia* or squid of the innervation of the mantle circular fibres (which power jetting) but it is probable that the giant axons only supply the MP fibres, and that these are also innervated by small fibres, since escape jets can be driven either by the small fibre system alone or by a combination of small fibres and giant axons. However, the existence of extensive dye-coupling between fibres suggests that the fibres may be electrically-coupled and hence individual fibres are unlikely to be doubly-innervated. The mantle circular fibres and the chromatophore fibres probably are innervated by glutamatergic nerves, whilst the radial fibres of the mantle, the head retractor muscles and some muscles in the tentacles and arms receive a cholinergic innervation. Mechanoreceptors occur in the cephalopod mantle and fins, but their rôle in motor control is not known. Finally, after a brief consideration of muscle biochemistry, some indications are given of the lacunae remaining in our knowledge of cephalopod muscle.

Almost all muscle fibres in cephalopods are quite different in design to those of vertebrates, use different neuromuscular transmitters, and instead of contracting to move skeletal structures, work in opposition to a hydroskeleton, or to other muscle fibres; furthermore, a significant part of the work many of them do is devoted to 'winding up' elastic storage systems. These features make them interesting from a comparative point of view, particularly because cephalopods and fish have long been in direct competition with each other, and have evolved in many instances convergently (Packard 1972). The special features of cephalopods are their remarkable colour changes and their jet-propelled swimming, both of which involve interesting muscular adaptations.

The muscles of only a few living cephalopods have been studied, and even in these, apart from the radial muscle fibres controlling the chromatophores, almost all authors have only examined 'skeletal' muscle fibres, in the arms, tentacles, fins, and mantle. Fortunately, in all the cephalopods (however different) which have been examined, the chromatophore and all 'skeletal' muscle fibres are of similar structure, with only two minor differences, and one striking exception. The muscle fibres of such organs as the hearts, eyes, gills, and ganglia (some ganglia are contractile; Alexandrowicz 1963) have not been considered in this review.

The muscle fibres consist of a central mitochondrial core of varying diameter surrounded by a cortical myofilament zone in which the myofilaments lie in an obliquely striated array (see Fig. 20.1). This organization is slightly different in some *Nautilus* fibres, where there are mitochondria peripherally as well as in the central core (Hochachka *et al.* 1978), and also in fibres of the plexiform layer of the stomach in *Sepia*, where the nucleus is eccentric, and the sparse mitochondria do not form a central core (Amsellem and Nicaise 1980).

There has been some confusion in the literature about the striated nature of cephalopod muscle fibres as Kier (1985) observed, for some earlier workers (e.g. Hanson and Lowy 1957, 1960; Kawaguti and Ikemoto 1957) termed them smooth muscle fibres (although they demonstrated oblique striation). It now seems clear that cephalopod muscle fibres should be classed as striated, like the similarly obliquely striated muscle fibres of annelids, nematodes, and doliolid tunicates. The only exception found so far to the generalization that all cephalopod fibres are obliquely striated is found in the tentacles of squid, where Kier (1985) has shown that almost all the transverse fibres are cross-striated.

No cephalopod muscle fibres are large (maximum diameters of mantle fibres are about 8–10 μm, and most muscle fibres are smaller than this): consequently they do not provide easy 'preparations' for electrophysiological studies, although the chromatophore muscle fibres in particular (which are the most accessible) have been examined in some detail (Florey and Kriebel 1969; Florey *et al.* 1985). Compared with other invertebrates, such as crustacea or insects, relatively little is yet known of cephalopod muscle electrophysiology, pharmacology, or mechanics, and as will become evident in this review, many questions are unresolved.

Organization of cephalopod muscles

Hydrostats

Although some cephalopod muscles insert on skeletal structures (e.g. the head and funnel retractors of *Nautilus*), most operate differently, and are arranged instead either around a fluid-filled hydroskeleton or as muscular hydrostats (muscular organs without an obvious system of skeletal support). In the arms and tentacles, muscle fibres in three planes produce the movements, and there is no fluid-filled hydroskeleton against which they operate, while in the mantle, the fibres are arranged in two planes only: radial fibres thinning the mantle and thus expanding the mantle cavity, opposed by circular fibres contracting against the water within the mantle cavity.

Arms and tentacles

The arms and tentacles of cephalopods are remarkably flexible and extensible, allowing them to catch prey by very rapid movements during which the tentacles may extend up to five times their resting length. The arms and tentacles (and the cirrhi of *Nautilus*) lack a distinct skeletal system, and are composed of closely packed muscles arranged in three planes: longitudinal, transverse, and helical or obliquely circular. There are no significant fluid-filled spaces. The biomechanics of muscular hydrostats of this kind have been considered by Kier and Smith (1985). They point out that given constant volume, the geometrical relation between length and width in these cylinders provides the muscular hydrostat with a system of mechanical amplification during elongation, analogous to a lever system. They also consider how shortening, bending, torsion, and changes in stiffness are brought about in such structures. More recently, the same authors (Kier and Smith 1990) have redescribed the muscles of octopus suckers, and considered their operation as muscular hydrostats.

Mantle

The mantle in *Octopus*, *Sepia*, and squid is a variable volume hydroskeleton surrounded by inner and outer connective tissue layers in which collagen fibres are aligned spirally, forming overlapping helices, in squid with a rather constant fibre angle of $27 \pm 1°$ (Ward and Wainwright 1972). The tunics thus resemble the connective tissue layers of the body walls in other forms with hydrostatic skeletons such as annelids. Clark (1964) has reviewed the geometry of such systems. Within the muscle itself, there are large amounts of collagen fibres disposed in oblique arrays in different planes (Bone *et al*. 1981), which Gosline and Shadwick (1983) have shown to act as an energy storage mechanism, termed the 'mantle spring'. The mantle circular muscles not only power the jet, but also (by storing energy in the 'mantle spring' as they continue to contract near the end of the jet cycle), cause active restoration of original mantle cavity volume. Similar obliquely arrayed connective tissue fibres lie among the radial fibre sheets which divide up the circular muscles, and again, probably act as elastic stores 'wound up' during the hyperinflation caused by radial muscle contraction during respiration and prior to jetting.

Muscle fibre structure

Figure 20.1 summarizes the structure of the muscle fibres described below.

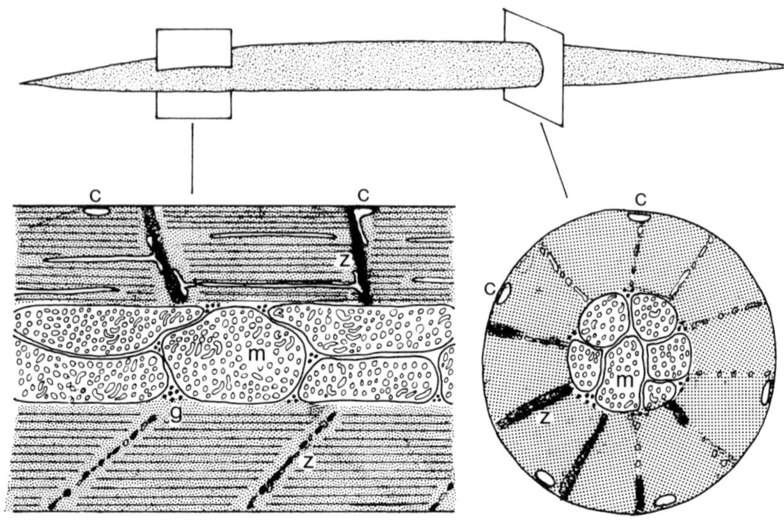

Fig. 20.1. Schematic diagram illustrating structure of the cephalopod muscle fibre. The oblique striations spiral around the fibre, hence in longitudinal section, they present different appearances according to the part of the array sectioned. The central mitochondrial core (m) is surrounded by myofilaments, divided by Z-discs (z). The sarcoplasmic reticulum forms sub-sarcolemmal cisternae (c) from which tubules enter the myofilament zone at Z-level and pass longitudinally. Glycogen rosettes (g) are found around the mitochondria.

Myofilament array of obliquely striated fibres

The earliest workers on the histology of cephalopod mantle muscle, such as Margo (1860) and Boll (1869) recognized that the fibres had a cortical myofibrillar zone surrounding a granular axis in which lay the single nucleus. The excellent studies by Ballowitz (1892) on *Sepiola* and *Eledone*, and those by Marceau (1905, 1906) who examined in addition *Sepia, Loligo*, and *Octopus*, first demonstrated the spiral array of the cortical myofilaments. Marceau dissociated fibres in macerated portions of the mantle to show that they were elongate (2–2.5 mm) and only from 6 to 10 μm in maximum diameter. In the mantle of the small squid *Alloteuthis*, methylene blue-stained fibres are up to 1.4 mm long (Bone *et al.* 1981).

The first electron micrographs (of *Loligo* mantle; Hanson and Lowy 1957;

Kawaguti and Ikemoto 1957) confirmed the spiral nature of the myofibrillar array, and showed that the thick filaments were surrounded by a variable number of thin filaments (often more than six). Later ultrastructural studies with more adequate techniques (those mentioned either being of glycerol-extracted fibres or with poor fixation) showed that the axial granular core consisted of mitochondria with tubular cristae. A brief description of squid mantle muscle fibres was given for *Symplectoteuthis* by Moon and Hulbert (1975); a subsequent detailed description in *Alloteuthis* and *Loligo* was given by Bone *et al.* (1981); and most recently, Kier (1985) has described the muscle fibres of the arms and tentacles in *Loligo*. All are of similar structure, although there are considerable differences in the cross-sectional areas of the mitochondrial core in different fibres. Figure 20.2(A)–(D) shows the structure of the mantle muscle fibres in the small squid *Alloteuthis*. In transverse section, the myofilaments are aligned in bands divided by densities corresponding to Z-lines; this banding pattern results from the obliquity of the myofibrillar array, seen in longitudinal section in Fig. 20.2(B). Since the array spirals round the length of the fibre, a variety of patterns is seen in any longitudinal section (Fig. 20.1). These patterns closely resemble those seen in the small tunicate *Doliolum* where the obliquely striated fibres have a similar spiral array (Bone and Ryan 1974), and are therefore more complex than in other obliquely striated fibres, such as those of annelids and nematodes where the array does not spiral along the fibre (see Lanzavecchia 1981).

In *Loligo* fibres fixed at resting length, the myofilament array is at 6–12° to the longitudinal axis, in contracted fibres 14–18° (Hanson and Lowy 1957). The change in obliquity (and hence spacing of the filament array) during contraction is presumably responsible for the interesting observation that there is a transient increase in opacity of the (extremely transparent) muscle fibres as they contract. Abbott and Lowy (1956) showed that in the mantle of *Loligo* this transient opacity increase began 3 ms before the first sign of tension increase (Fig. 20.3) and was preceded by a small opacity decrease coinciding with the falling phase of the action potential in the stellar nerve stimulated. While the early opacity decrease cannot be ascribed to changes in the obliquity of the array, the major part of the opacity change appears to result from this, the delay between tension increase and the beginning of opacity increase perhaps resulting from compliance due to the large amount of connective tissue within the squid mantle.

Sarcoplasmic reticulum (SR) and fibre diameter

Infoldings of the sarcolemma to form tubules comparable to the vertebrate transverse tubular (T) system are absent in cephalopod muscle fibres. However, subsarcolemmal SR cisternae are prominent around the fibre border, and in some cases regular densities are seen between these cisternae and the

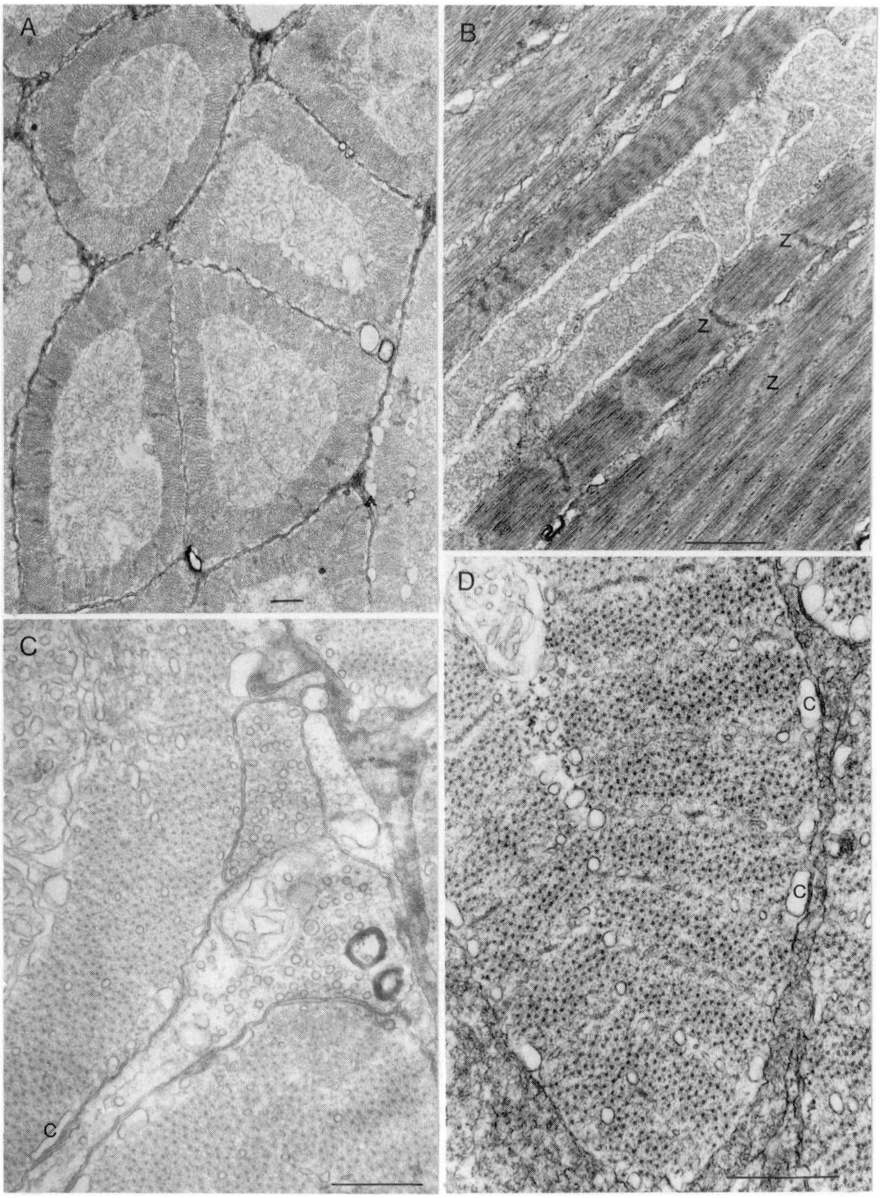

Fig. 20.2. Mantle muscle fibres from *Alloteuthis*. (A) Transverse section of mitochondria-rich fibres, scale bar: 1.0 μm. (B) longitudinal section of mitochondria-rich fibres, scale bar: 1.0 μm. (C) Mitochondria-rich fibres at higher magnification, showing motor terminals with electron lucent vesicles. Scale bar: 0.5 μm. (D) Mitochondria-poor fibre. Scale bar: 0.5 μm. Note conspicuous sub-sarcolemmal cisternae (c) which are probably artefactually swollen; z-discs (z), and in (B), different appearances of myofilament array as it spirals around the fibres.

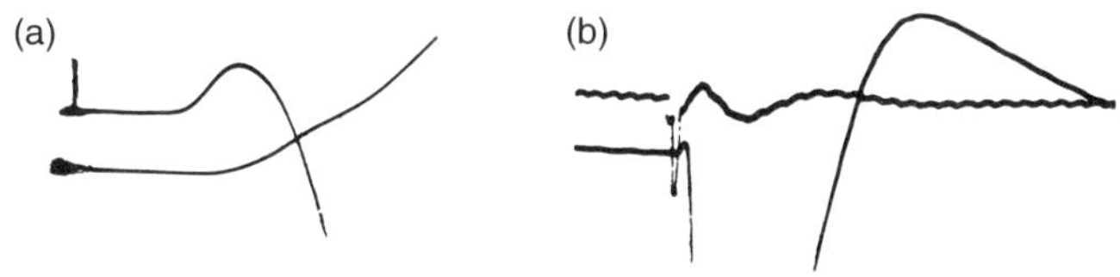

Fig. 20.3. Opacity changes in *Loligo* mantle muscle fibres following single stimuli to stellar nerve. (a) Upper trace opacity change (increase downwards), lower trace tension (increase upwards), sweep duration 50 ms. (b) Upper trace muscle action potential, lower trace opacity change (at lower amplitude than in (a), the latter part of the opacity change record crosses the baseline as it was recorded a.c.-coupled). Sweep duration as (a). (From Abbott and Lowy 1956.)

sarcolemma; this presumably corresponds to the end-feet seen between the SR and T-system in chordate muscle fibres, thought to be formed by ryanodine receptors mediating Ca^{2+} release from the SR (Wagenknecht *et al.* 1989). SR tubules enter the myofilament array at Z-band level and run longitudinally. They link with SR tubules in the central axis of the fibre next to the mitochondrial core. As yet no good estimates are available for the volume of the SR in cephalopod fibres, for although the SR is a prominent feature of all fibres whatever the relative size of the mitochondrial core, fixation problems make the true dimensions of the system difficult to estimate. Thus although the SR appears more prominent in fibres with a small mitochondrial core, such as those seen in Fig. 20.2(d), than in fibres with a larger mitochondrial core as in Fig. 20.2(c), it is not clear whether this difference is real. It might, however, be assumed that the former fibres were more rapidly contracting and hence expected to have a relatively larger SR volume. Moon and Hulbert (1975) estimated that the SR occupied 2–5 per cent of fibre volume in their preparations of fibres with a small mitochondrial core, but noted that the SR in their material was probably swollen due to incorrect osmolarity of the fixative. SR volume in the better-fixed cross-sections of the transverse mitochondria-poor muscle fibres of the arms of *Illex* figured by Kier (1985) is considerably less than this. Cephalopod muscle is curiously difficult to fix, and we have observed even in our 'best' fixations of mantle muscle portions, areas that are apparently well-fixed close to others where osmotic damage is obvious. Whatever the true extent of the SR in different fibres, all are of much smaller diameter than vertebrate fibres. External calcium is required for contraction, since muscle twitches are rapidly and reversibly abolished in Ca^{2+}-free solutions (p. 320). If the calcium required to activate the contractile apparatus enters across the sarcolemma during the muscle action potential, the small diameter of cephalopod fibres may be directly linked to the requirement for rapid diffusion to the regions of the myofibrillar array furthest from the sarcolemma.

Junctions between muscle cells

Various kinds of junction between muscle cells have been observed in cephalopods; in the stomach of *Sepia*, Amsellem and Nicaise (1980) found desmosomes, interdigitations (also seen in squid mantle; Bone *et al.* 1981), and a single example of a possible gap junction. Cloney and Florey (1968) demonstrated gap junctions (then termed close junctions) between the radial muscles of the chromatophores in *Loligo opalescens* as they abut around the chromatophore sac, and subsequently Florey and Kriebel (1969) showed that the fibres were electrically coupled. Gap junctions have not been observed between muscle cells in the arms, tentacles or mantle, but Lucifer Yellow injections of mantle muscle fibres in *Alloteuthis* and *Sepia* show extensive dye-coupling (Brown, unpublished). They are therefore presumably coupled by gap junctions, although two electrode current-injection experiments to demonstrate coupling have so far been unsuccessful owing to the small size of the fibres. Dye-coupling is particularly clear in newly hatched squid and late squid embryos (Fig. 20.4).

Two types of obliquely striated fibres

In the mantle of squid and *Sepia* (Bone *et al.* 1981), and in the funnel and shell retractor muscles of *Nautilus* (Hochachka *et al.* 1978), there are two types of muscle fibre, distinguished by the volume of the mitochondrial core. These were actually first observed in the early study of *Sepia* mantle by Kawaguti and Ikemoto (1957); however, they attributed the difference between them to different states of contraction and did not recognize two fibre types. In *Nautilus* retractor muscle, 90 per cent of the fibres have a mitochondrial core occupying 2 per cent of their cross-sectional area, while in 10 per cent of the fibres, the core (together with some peripherally located mitochondria) occupies 13 per cent of the cross-sectional area. The differences between the two fibre types are more striking in *Sepia* and squid mantle [Fig. 20.5(a)]; in the small squid *Alloteuthis* the mitochondria-rich fibres form a thin inner and outer layer, where the fibres have a mitochondrial core occupying mean values of 45.9 per cent (inner zone) and 47.5 per cent (outer zone). The much

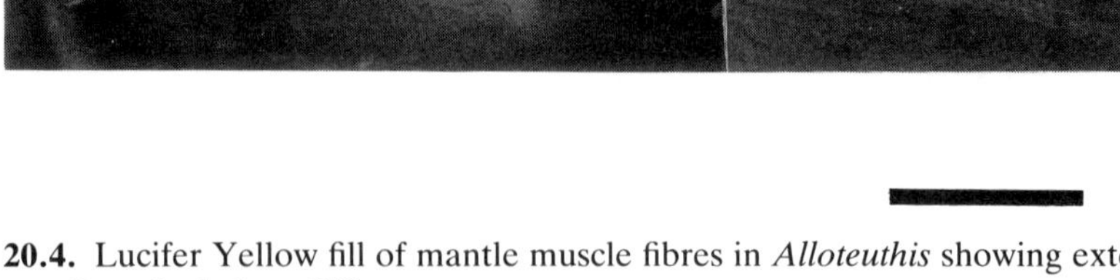

Fig. 20.4. Lucifer Yellow fill of mantle muscle fibres in *Alloteuthis* showing extensive dye-coupling. Scale bar: 200 μm.

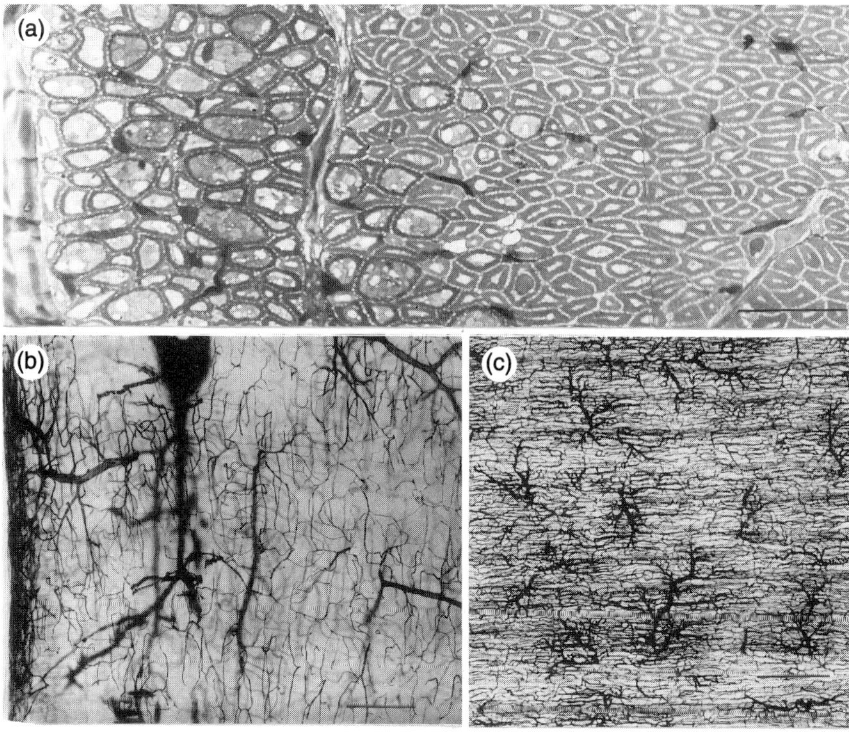

Fig. 20.5. *Alloteuthis* mantle. (a) Longitudinal section of outer region of mantle, showing outer zone of mitochondria-rich larger circular fibres to left, and inner zone of smaller mitochondria-poor circular fibres sectioned transversely. Scale bar: 20 μm. (From Bone *et al.* 1981.) (b) and (c) Thick sections of portions of mantle after injection of vessels with Indian ink. In (b) longitudinal section (circular fibres sectioned transversely) showing much greater density of capillary bed in inner mitochondria-rich layer on left as compared with central zone. Scale bar: 100 μm. In (c), surface view of horizontal section of outer layer. Scale bar: 500 μm (From Bone *et al.* 1981.)

larger central zone consists of fibres with a mean value for the mitochondrial core of only 6.4 per cent. All the radial fibres are similar to those of the central zone. In *Nautilus* retractor muscles, the two fibre types are intermingled in bundles, in contrast to the zoned arrangement within the squid and *Sepia* mantle. This zonation of fibre types (as in fish myotomes) offers the possibility of examining the biochemistry and physiology of the two fibre types separately (e.g. Mommsen *et al.* 1981).

In line with this striking difference in fibre type between the central and inner and outer zones of the mantle, there is an equally striking difference in vascular bed [Fig. 20.5(b) and (c)]. As would be expected, the inner and outer zones, composed of mitochondria rich fibres, have a much higher capillary density than the sparsely supplied central zone. However, even in the inner and outer zones of the mantle, the capillaries (which mainly run longitudinally along the fibres) are seen in transverse section spaced some five to eight muscle fibres apart, so that maximum diffusion distances are 20–30 μm. In the central zone, where the capillaries are spaced 20 muscle fibres or more apart, maximum diffusion distances are over 100 μm. Possibly some of the difficulty in obtaining 'good' fixation of the mantle muscle fibres may result from these relatively large diffusion distances.

Abbott and Bundgaard (1987) extended these observations (made on the small squid *Alloteuthis*) by quantitative measurements on various *Sepia* muscles. Their values for diffusion distances in tentacle muscle (61 μm), and for the muscles of the collar flap (32 μm), are close to those measured for the central and the outer zones of the mantle respectively in *Alloteuthis*. As Abbott and Bundgaard point out, the tentacle muscles are used only occasionally (the tentacles are normally coiled up in a pouch) whereas the collar flap is rhythmically active (see p. 316). It seems clear from the morphological data on mitochondrial content and vascularization that the inner and outer zones of the mantle operate aerobically, as do those of the collar flap, while the central zone operates anaerobically, as do the tentacle muscles. This conclusion is supported by studies of enzyme activity levels in the two fibre type zones of the mantle (Mommsen *et al.* 1981). The radial muscle fibres of the chromatophores, whose contraction expands the chromatophore sac and darkens the skin, have a relatively large mitochondrial core (Cloney and Florey 1968; Florey and Kriebel 1969), and in view of their capability for sustained contraction, may be assumed to operate aerobically.

Cross-striated fibres

Although the longitudinal and circular fibres of the tentacles in squid are oblique striated, Kier (1985) found that almost all the transversely oriented fibres are cross-striated [Fig. 20.6(a) and (b)]. Among the mass of cross-

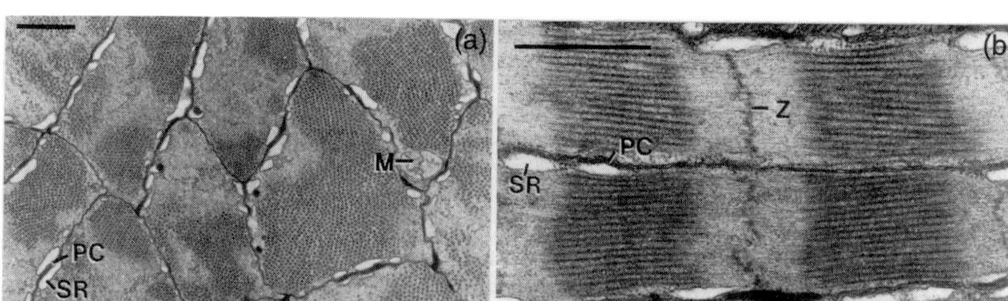

Fig. 20.6. Transverse and longitudinally sectioned cross-striated tentacle muscle of *Loligo pealei*. Peripheral couplings (PC) of the sarcoplasmic reticulum (SR) with the sarcolemma are conspicuous and may be artefactually swollen. Mitochondria (M) are sparse. The Z-discs (Z) are diffuse. Scale bars 1 μm. (From Kier 1985.)

Table 20.1 Morphology of cross-striated transverse tentacle fibres

Species	A-band length	Resting sarcomere length	Thick filament diameter
Illex illecebrosus	0.52 ± 0.03 μm	0.86 ± 0.6 μm	23.7 ± 1.0 nm
Loligo pealei	0.94 ± 0.08 μm	1.64 ± 0.2 μm	31.4 ± 2.2 nm

striated transverse fibres, however, Kier observed a few bundles of obliquely striated transverse fibres. In the arms, the fibres in all three planes are obliquely striated. Table 20.1 shows measurements obtained on the cross-striated fibres in two species.

These are small fibres, maximum diameter only being up to 3 μm. M-lines are absent, and the Z-lines are more diffuse than in vertebrate-striated fibres. Transverse sections of the myofilaments array show up to 11 thin filaments surrounding the thick filaments (which have an electron-lucent core), but the array is less regular than in vertebrate fibres. In contrast to the obliquely striated fibres, there are no SR tubules among the myofilament array, and the SR consists of sub-sarcolemmal cisternae coupled to the sarcolemma with dense junctional feet, as in other cephalopod fibres.

Kier (1985) suggested that these unique cross-striated fibres, whose rôle when they contract is to shoot out the tentacles to capture the prey, differ from those of other fibres in cephalopods (the transverse fibres of the arms are obliquely striated) because they are more rapidly contracting. Calculations from kinematic observations of the strike indicated a shortening velocity of around 8 lengths/s (temperature not given). However, some obliquely striated muscle fibres (in the locomotor muscle bands of the small tunicate *Doliolum*; Bone and Trueman 1984) contract equally rapidly, normalized contraction velocity being 8.5 lengths/s at 20°C, so that there appears no intrinsic reason why obliquely striated fibres should be slower than cross-striated fibres, and it remains unclear why the tentacle transverse fibres should be cross-striated.

Mechanical responses

Mantle muscle

Marceau's (1905, 1906) early work on muscle twitches in *Sepia* and *Loligo*, (where he stimulated mantle and tentacle muscle strips directly via two platinum wire electrodes connected to a Du Bois Reymond coil, and recorded contractions on a Marey myograph), showed rapid twitches followed by slower recovery of original tension (Fig. 20.7). As he demonstrated, the twitches were almost as fast as those of frog sartorius. Tetani occurred at stimulus frequencies at about 20 Hz. This pioneering work was followed later

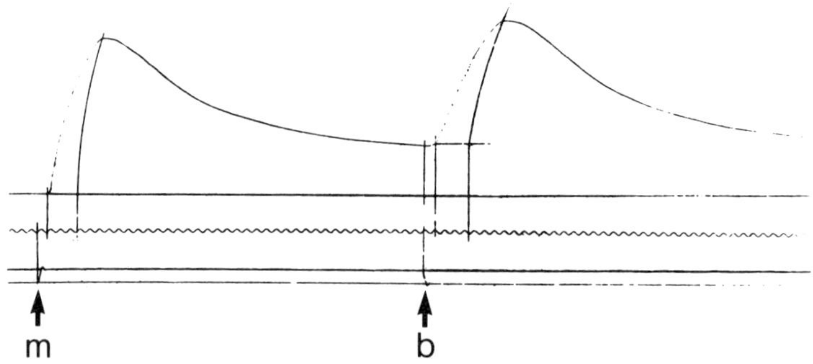

Fig. 20.7. Mechanical response of *Sepia* mantle muscle. An early record from a strip of circular mantle fibres showing twitches on closing (make =m) and then opening (break =b) stimulating coil. Time base: 0.1 Hz. (From Marceau 1905.)

by Prosser and Young (1937) who examined the responses of squid mantle muscle stimulated via the stellar nerves containing the third-order giant axons as well as smaller fibres. They observed rapid contractions of similar amplitude on repetitive stimulation at 4 Hz; tetani occurred at stimulation frequencies at about 35 Hz. Since both the giant axon and the smaller fibres of the stellar nerve were stimulated, some uncertainty remained, and their observation that in fatigued preparations although tensions were now much reduced, increasing frequency of stimulation evoked increased tension indicated that some change had occurred as the preparation aged. As Wilson (1960) observed, these two types of mechanical response are the consequence of failure of the giant axon-evoked contraction of the mantle musculature as the preparation aged, although it could still be stimulated via the small fibres of the stellar nerves. Wilson examined mantle muscle responses in squid when the stellar nerve was stimulated before and after the giant axon was inactivated (by fatigue or by damage) so that the small fibres alone were active. Young (1938) had previously stimulated the giant axon and small nerve fibres of the stellar nerves separately, finding that small fibre stimulation resulted in slower and weaker responses from the mantle than those evoked by stimulation of the giant axon alone. Wilson's observations of the two kinds of mechanical responses were similar to those figured by Prosser and Young (1937). Lowy and Millman (1962) examined the twitches of the funnel retractor muscles of *Octopus* and *Sepia* in more detail, finding that in *Octopus*, tetani and twitches at 18°C were similar in form and time course to those of frog sartorius at 0°C, while those of *Sepia* were about twice as fast. Maximum tensions observed were 49 kN/m^2 in *Octopus* and 19.6 kN/m^2 in *Sepia*. O'Dor (1988) has estimated maximum stresses of mantle muscle during peak jet propulsion as 110 kN/m^2 in *Nautilus* and 250 kN/m^2 in the squid *Illex*.

More recently Usher (unpublished) has examined the isometric mechanical responses of small bundles of mantle circular muscle fibres in the small squid

Alloteuthis and in *Sepia*, with similar results [Fig. 20.8(a) and (b)]. The most interesting feature of these results is the increase of force as stimulus frequency increases, leading to a short summation when three stimuli are given at a frequency of 5 Hz, peak force (2.8 mN) then being approximately the sum of three individual twitches (1.133 mN). As yet, all records have been from the central zone of circular muscle fibres, and thus the interesting question whether there are differences in mechanical properties between the two types of circular fibre has not been examined.

These results from direct stimulation of the muscle are paralleled by observations on circular muscle responses in the intact half mantle when a single giant axon in a stellar nerve is stimulated (after the small fibres of the nerve have been removed). In contrast to the previous work by Prosser and Young (1937) and Wilson (1960), however, succeeding stimuli at short intervals

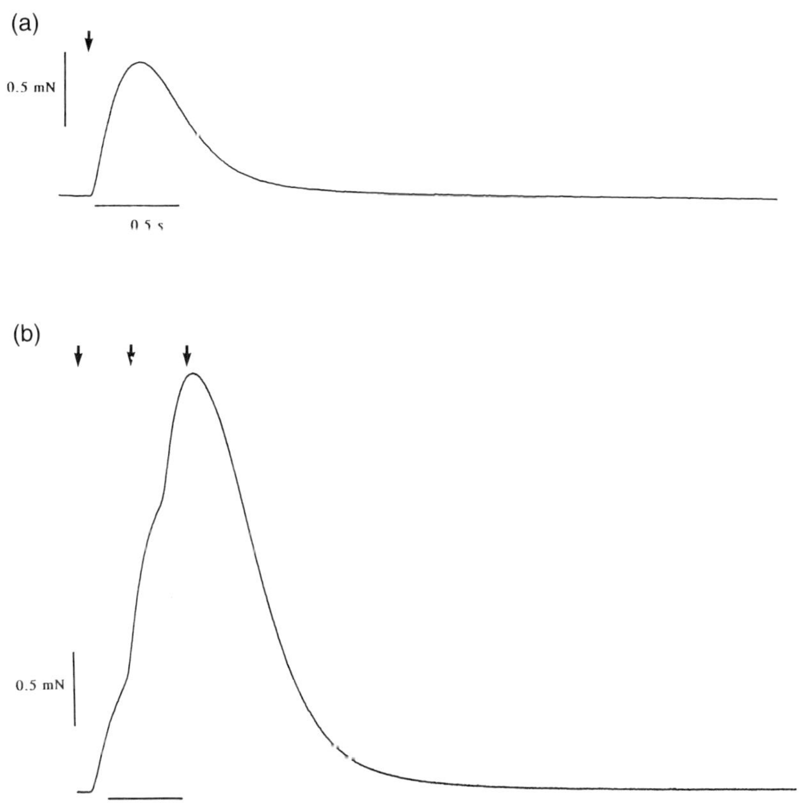

Fig. 20.8. Recent records of mechanical responses of small bundles of *Sepia* mantle circular muscle fibres. In (a), a single twitch response, in (b) successive stimuli (arrows) evoke summated responses. (From Usher, unpublished.)

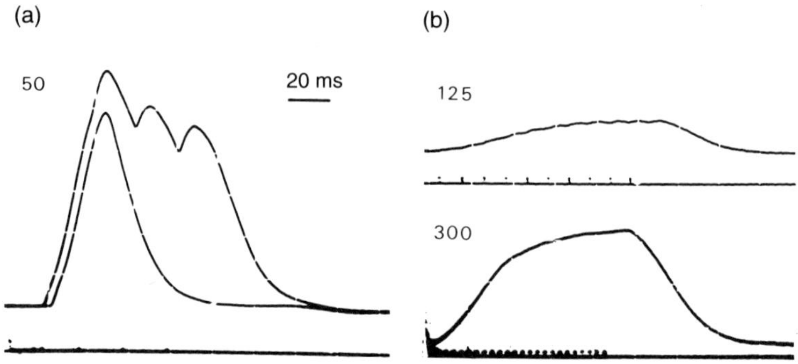

Fig. 20.9. Mechanical responses of circular muscles in intact half mantle of *Alloteuthis* following stimulation of (a) the giant axon of a stellar nerve, and (b) the small fibres of the same nerve after the giant axon response had failed. Numbers: stimulus frequency in Hz. The responses in (a) were recorded at half the gain of those in (b).

augment the initial contraction [Fig. 20.9(a)]; in intact squid the giant axon can fire at similar intervals, as seen in Fig. 20.14. Repetitive small fibre stimulation evokes less rapid responses, increasing markedly in amplitude as stimulus frequency increases [Fig. 20.9(b)]. The simplest explanation for this difference between the two types of neural stimulation would be that the small fibres innervate a different muscle fibre type to the giant axons, but this is improbable (see p. 317). More probably, the neuromuscular junctions of the small fibres upon the central zone of circular fibres soon fail owing to anoxia in the isolated mantle (as do those of the giant axons), and only those on the outer and inner 'aerobic' zones then remain viable, so that effectively, only one of the mantle fibre types is stimulated by the small fibre system, even though the small fibres in fact innervate both types.

O'Dor and Webber (1991) have tabulated data on jet-propelled swimming performance from different cephalopods, which show that in the fast squid *Illex*, peak mantle cavity pressures are about five times those of *Nautilus*, and peak power some 14 times greater. As Alexander (1992) points out, at maximum swimming speeds, *Illex* mantle muscles must put out more than 50 J/kg of work. This compares favourably with human and avian maximum work outputs.

Chromatophore radial muscle

Since it is easy to obtain thin flat preparations of cephalopod skin, contractions of the radial muscles around a single chromatophore can simply be recorded under a microscope by monitoring changes in light transmitted through the preparation, as the muscles expand the chromatophore sac. This technique was employed by Bozler (1928, 1931), and later by Florey (1966),

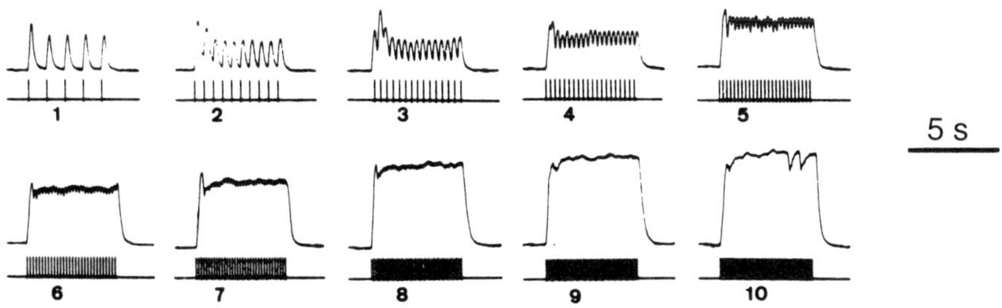

Fig. 20.10. Twitch and summated responses of chromatophore muscle fibres from the squid *Lolliguncula brevis* following repetitive nerve stimulation. Numbers: stimulus frequency in Hz. (From Florey *et al.* 1985.)

Florey and Kriebel (1969), Florey *et al.* (1985), and Dubas and Boyle (1985). Florey and Kriebel stimulated small bundles of skin nerves, and showed that repetitive stimulation evoked twitch responses, summating to produce fused responses at higher stimulus frequency (Fig. 20.10). In *Loligo* (Florey *et al.* 1985) and in *Eledone* (Dubas 1987), the chromatophore fibres followed stimuli at 20 Hz or higher, but in *Lolliguncula*, repetitive stimuli were usually only followed up to 3 Hz, although occasionally one-to-one responses occurred up to 40 Hz (Florey *et al.* 1985). Curiously, in aged preparations [even in 'dead' skin the following day, or in preparations where nerve conduction is blocked by tetrodotoxin (TTX)], the radial muscles can still contract even after the chromatophore nerve fibres are no longer active; the chromatophores pulsate or remain expanded as the radial fibres twitch spontaneously or are tonically contracted. Although the radial muscles of individual chromatophores are usually independent, Froesch-Gaetzi and Froesch (1977) have shown that they may be interconnected and suggest that in 'aged' preparations, this may account for the waves of colour which may pass across the skin.

Functional role of the different mantle muscle fibres

The discovery of zones of two circular fibre types with very different mitochondrial content in the mantles of *Sepia* and squid, naturally suggested that they performed different functions. Those in the central zone, with few mitochondria and a relatively poor vascular bed, appear adapted to intermittent anaerobic activity, while those of the inner and outer zones appear adapted to more sustained aerobic activity. Both Mommsen *et al.* (1981) and Bone *et al.* (1981) proposed that the thin inner and outer zones of mitochondria-rich fibres were employed during the shallow respiratory contractions of the mantle, whereas the much thicker central zone of mitochondria-poor fibres was used during escape jetting. Mommsen *et al.* (1981) examined different squid, and found that the relative proportions of the inner and outer zones differed between species, and that in the sluggish benthic *Berryteuthis* both were reduced as compared with the active *Ommastrephes* and *Symplectoteuthis*.

The constant back and fore movements of *Loligo* and *Alloteuthis* as they school, are not the result of jetting, but are due to fin movements, and in *Loligo, Sepia*, and *Eledone* it has recently been found that the resting expiratory respiratory flow of water over the gills is *not* driven by contractions of the mantle, rather, it is produced by the movements of the collar flaps and associated muscles; the mantle circular fibres are inactive during respiration (Bone *et al.* 1994). The inner and outer zones of mantle circular muscle in these cephalopods are apparently used during slow jetting, the central zone during more powerful jetting.

Electromyographic (EMG) records from tethered *Loligo opalescens* (Gosline *et al.* 1983) mantle showed two patterns of mantle muscle activity during respiration. In both, radial muscle activity (which expands the mantle cavity) was seen during inspiration, but in the first no circular muscle activity occurred during expiration. In the other respiratory pattern (regarded as probably the most natural), circular fibres were active during expiration. Our more recent observations on untethered *Loligo forbesi* suggest that in fact, the more natural pattern does not involve circular muscle fibre activity. Unexpectedly, in view of their regular respiratory activity, the radial fibres are all of the mitochondria-poor type (Bone *et al.* 1981). The mantle radial muscle fibres also act to expand (hyperinflate) the mantle cavity prior to escape jetting both in *L. opalescens* and in *Sepia* (Fig. 20.11).

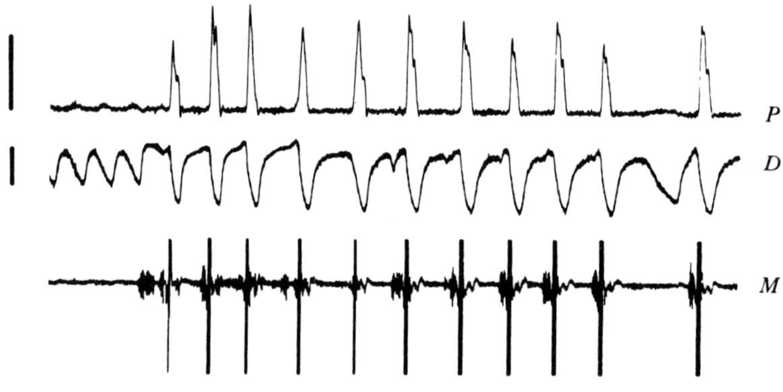

Fig. 20.11. Simultaneous records of mantle cavity pressure (*P*), mantle diameter (*D*), and mantle muscle electrical activity (*M*) in the squid *Loligo opalescens* during a series of jets. Note radial muscle activity (small potentials) during expansion of mantle cavity, prior to circular muscle activity (large potentials) during expulsion of jet. Scale bars: 3 s; 12.5 kPa; 0.75 cm. (From Gosline *et al.* 1983.)

Innervation

General pattern

Both reduced silver and supravital methylene blue techniques work well on cephalopod material, but the small size of the terminal motor axon branches, as well as the small size of the muscle fibres, make light microscopic observations unsuitable for determining innervation pattern except in a gross manner. For example, in squid or *Sepia* mantle, thick silver-impregnated slices show that the giant axons branch together with the small axons in the stellar nerves, but the innervation of single muscle fibres cannot be determined. Nerve terminals have been observed at the ultrastructural level on mantle, gut, sucker, and chromatophore muscle fibres; in each case, the nerve terminals contain electron-lucent vesicles some 50 nm in diameter [Fig. 20.2(c)]. In the mantle of *Alloteuthis*, the terminals are similar, whether they are upon the 'cholinergic' longitudinal fibres at the anterior region of the mantle, or on the 'glutamatergic' circular fibres. While fibres containing much larger (100 nm) dense cored vesicles are seen in the mantle, such vesicles have not been seen at nerve terminals on the muscle fibres, and perhaps these nerve fibres pass to the skin. The chromatophore muscles of *Eledone cirrhosa* appear to have terminals of two types, one with the usual electron-lucent vesicles, the other with many larger dense-cored vesicles (Dubas 1987). Subjunctional folds are lacking in all cases, although the subsynaptic membrane is denser than the remainder of the sarcolemma. Graziadei (1966) found in the lip muscles of *Sepia* that the axons were often embedded within infoldings of the sarcolemma, so that the junctions actually lay 'inside' the muscle fibres, as Dubas (1987) observed on occasion in the chromatophore muscles of other cephalopods.

It is not possible to say whether more than one nerve terminal may occur on a single muscle fibre. In the mantle, which is supplied both by giant axons and by the small fibres in the stellar nerves, it is not yet clear whether both types of axon supply individual muscle fibres. Although some authors (e.g. Moon and Hulbert 1975) have claimed that there is a true double innervation, 'analogous to the arthropod situation', there is no direct evidence for this assertion (see p. 322). Indeed, the evidence for dye-coupling in the mantle might indicate that if the bundles of fibres in the circular muscle bands of the mantle (separated by the lines of radial fibres) are electrically coupled, only some fibres within the bundle may be directly innervated, others being activated via gap junctions. The circular fibres are innervated by terminals which probably contain L-glutamate, while the longitudinal fibres at the anterior border of the mantle and the radial fibres are apparently supplied by cholinergic nerves; in the latter case (if acetylcholine indeed acts postsynaptically on the muscle fibre membrane (see next section) it should be a relatively simple matter to stain for acetylcholinesterase to determine the number of terminals on each fibre after maceration. This study is yet to be performed.

Possible neurotransmitters: acetylcholine, 5-hydroxytryptamine, and L-glutamate

Since Bozler (1931) demonstrated that acetylcholine caused chromatophore expansion, confirmed by Kahr (1959) and by Florey (1966), most work on cephalopod neuromuscular transmitters has been upon chromatophore muscle fibres, where the situation is complex. Although as Kahr showed, acetylcholine evokes chromatophore muscle contraction, relaxed by serotonin (5-hydroxytryptamine, 5HT), careful work by Florey and Kriebel (1969) showed that acetylcholine acted presynaptically by augmenting release of the actual transmitter (as evidenced by increased miniature end plate potentials; Fig. 20.12). The relaxation by 5HT of the tonic contractions evoked by acetylcholine (which are not seen in normal untreated fresh preparations) was attributed by Florey and Kriebel to an intracellular action of 5HT on Ca^{2+} binding by the SR, and possibly also, to a similar decrease of free Ca^{2+} in the nerve terminals but modern understanding of 5HT receptors suggests that a plasmalemmal site of action is more likely. The identity of the actual transmitter, probably L-glutamate, was established by Florey *et al.* (1985) who reinvestigated these rather complex pharmacological responses of the chromatophore muscles in the light of the discovery of L-glutamate as the putative neurotransmitter for the circular muscle fibres of the mantle.

Simple experiments upon circular muscle strips from the mantles of *Sepia*, *Alloteuthis*, and *Loligo* (Bone and Howarth 1980) showed that these were insensitive to acetylcholine, but contracted when L-glutamate was applied (at bath concentrations of 2×10^{-4} M). This response was completely abolished by the glutamate blocking agent 2-amino-4-phosphono butyric acid (2APB or AP4). Topical application of L-glutamate solutions to the skin caused immediate chromatophore expansion, and caused contractions of the muscles of the branchial heart, ink-sac, and gut. Florey *et al.* (1985) extended the observations on chromatophore expansion, concluding that L-glutamate was a strong candidate for the transmitter released by the motoneurones innervating the

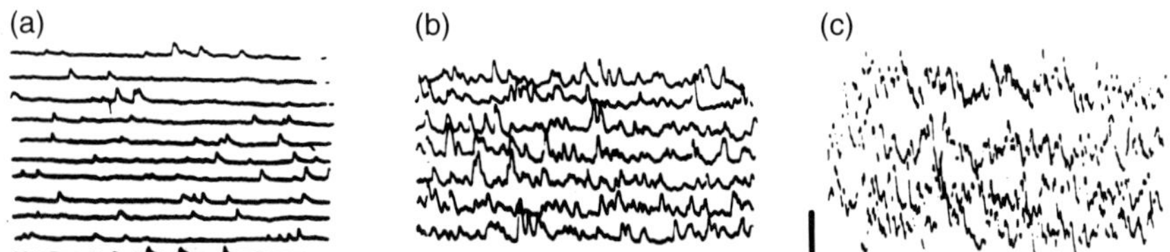

Fig. 20.12. The effects of acetylcholine on miniature potentials in chromatophore muscle fibres of the squid *Loligo opalescens*. In each intracellular record, successive sweeps are vertically displaced. (a) During moderate and (b) during strong tonic contraction; (c) during maximal contractions evoked by application of acetylcholine at 10^{-5} g/ml. Scale bars: 5 mV; 200 ms. (From Florey and Kriebel 1969.)

chromatophore muscles, although they cautiously did not exclude the release of other possible transmitters or modulators. More recent immunocytochemical observations by Cornwell and Messenger (see Chapter 22) showed the presence of L-glutamate-like material in the nerve fibres upon the chromatophore muscles. They also found fibres containing 5HT-like material associated with the chromatophore muscles (it is as yet unknown whether this is in the same fibres as those containing L-glutamate-like material), hence 5HT (unlike acetylcholine) is apparently involved in the normal chromatophore control mechanism.

In *Alloteuthis* and *Loligo*, Bone *et al.* (1982), showed that the longitudinal fibres on the inner margin of the mantle, the radial fibres throughout the mantle, the head retractor, and siphonal muscles, and the longitudinal fibres of the arms and tentacles, all contracted in response to acetylcholine and were insensitive to glutamate. In the mantle, the acetylcholine contractions were largely blocked by D-tubocurarine.

Evidently, many other criteria require to be satisfied before L-glutamate and acetylcholine can be accepted as the actual neurotransmitters of these different cephalopod muscles, and the observations on the effects of acetylcholine by Bone *et al.* (1982) mentioned above, require to be extended by examination of the effects of acetylcholine on muscle membrane potential. However, there is good evidence that L-glutamate and acetylcholine are likely to be widespread transmitter substances in cephalopods. Recent work by De Santis and Messenger (1989; see also Chapter 19) on the giant synapse in *Alloteuthis* and the most recent study using 'caged' glutamate (Corrie *et al.* 1993) has shown that L-glutamate is very probably the transmitter, despite doubts raised earlier by reversal potential values (Miledi 1969). Tansey (1972) gives references to studies showing the presence of L-glutamate and acetylcholine (and other possible transmitters) in the cephalopod CNS, while Andrews *et al.* (1981) showed that a number of neurotransmitters (including acetylcholine and L-glutamate) exerted central effects on the chromatophore lobes, manifested by chromatophore expansion or contraction. Auerbach and Budelmann (1986) have shown that acetylcholine is a likely transmitter in the *Octopus* statocyst.

Electrical activity

The difficulty of obtaining suitable preparations has inhibited electrophysiological work on cephalopod muscle fibres; even in the chromatophore muscle fibres, which have been closely studied by Florey and his colleagues, stable intracellular recording has not proven an easy matter, although they were able to impale single fibres with two electrodes. Florey and Kriebel (1969) impaled chromatophore muscle fibres and stimulated small bundles of chromatophore nerve fibres. Their results showed that in fresh preparations, propagated spikes were not evoked, rather, graded depolarizing postsynaptic potentials were seen, varying widely in latency and rise time. Since they showed that a degree of electrical coupling occurred between radial fibres (with coupling coefficients between 0.1 and 0.16), as expected from the presence of gap junctions between the fibres, some postsynaptic potentials apparently were electrotonic currents from postsynaptic potentials in neighbouring muscle fibres. However, since in fresh skin individual muscle fibres of a single chromatophore can be activated independently (Bozler 1928; Florey 1966), spread of excitation from one radial fibre to those adjacent is not likely to be of functional significance in the normal animal. Repetitive stimulation showed minimal facilitation, and spikes were never observed, even when current was injected via a second electrode.

In contrast, however, in aged preparations, where rhythmic contractions and relaxations of the chromatophore muscles at about 1 Hz are often seen, spike potentials followed by hyperpolarizing after potentials now appear, which may overshoot the 20–50 mV resting potential. Muscle contractions increase with increased amplitude of the spikes (as Florey 1966, had previously observed), presumably a consequence of greater Ca^{2+} influx during the action potential (see below). In aged skin, the simultaneous activation of all the radial fibres of a single chromatophore suggests that the spikes which occur in such material propagate from one fibre to the next, and that one of the fibres acts as a pacemaker for the rhythmic pulsations of the chromatophore. Florey and Kriebel (1969) suggested that the changes in the behaviour of the chromatophore fibres on ageing involved not only changes in membrane properties but also in Ca^{2+} binding within the fibre.

In the mantle, muscle action potentials apparently are a normal feature of fresh preparations. Brown (1991) has found that in *Alloteuthis* central zone circular mantle muscle fibres, rapid overshooting spike potentials are evoked by giant axon stimulation, from resting potentials of about 70 mV (Fig. 20.13). The ionic basis of these action potentials is yet to be determined, but muscle contractions are reversibly abolished in nominally zero-Ca^{2+} solutions, and it is possible that part (if not all) of the action potential is carried by Ca^{2+}.

As yet, intracellular records have not been obtained from mantle circular fibres when the small axons of the stellar nerves are stimulated.

Dual control of mantle circular muscles

Both the giant axon and small nerve fibres of the stellar nerves supply the mantle circular fibres in squid and *Sepia*, and it long seemed obvious that the powerful escape jetting was driven by the giant axons, whereas the small fibres were involved in the much weaker jets of slow swimming, and also innervated the radial fibres to inflate the mantle cavity. However, very unexpectedly, when Otis and Gilly (1990) recorded mantle pressure pulses and nerve activity during escape jetting of restrained *L. opalescens*, they found escape jets could be driven by the small fibre nerves alone.

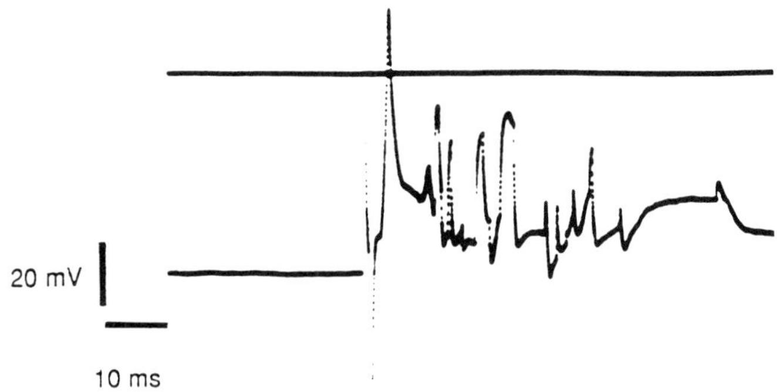

Fig. 20.13. Spike potential evoked from circular muscle fibre of central zone in *Alloteuthis* mantle muscle following stimulation of third-order giant axon in stellar nerve. (From Brown 1991.)

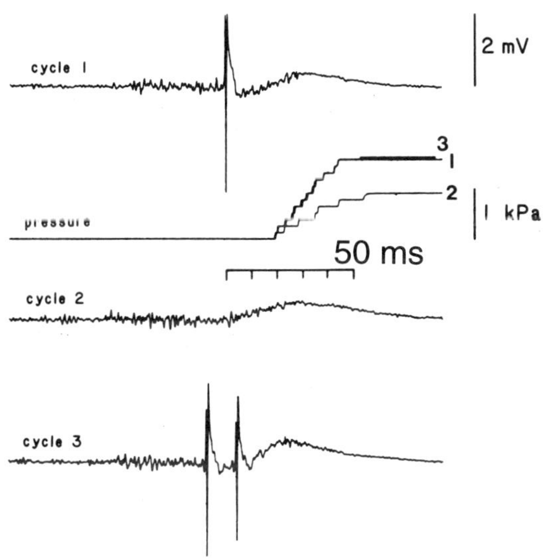

Fig. 20.14. Simultaneous extracellular records of nerve activity and of pressure pulses during three jet cycles in the squid *Loligo opalescens* showing pulses evoked by small fibre activity with and without one or more giant axon spikes (see text). (From Otis and Gilly 1990.)

While the giant axons can drive a short latency escape response, they can also be employed in a longer latency more complex response, where they fire after the jet has been begun by the small axon system, thus boosting the jet (Fig. 20.14). Since Young (1936, 1939) has described two separate second-order giant fibres in the pallial nerve (one smaller than the other), synapsing with the third-order giant fibres of the stellar nerves, it is possible that the longer latency response may be driven by the smaller second order giant fibres but as yet, the function of this axon has not been investigated experimentally. Mantle pressure pulses when the circular fibres were activated by the small axon system, were about half those resulting from the short latency giant axon activity (Fig. 20.14, cycle 2) and could be 'topped up' by appropriately timed later giant axon spikes. In view of the observations on the mechanical responses of intact half mantles, where it is clear that closely spaced stimuli to the giant axon produce summating and augmented tension responses (see p. 314) it is interesting that two giant axon spikes in rapid succession (Fig. 20.14, cycle 3) produce a slightly larger pressure pulse than a single giant axon spike (Fig. 20.14, cycle 1).

Since the thin inner and outer layers of mitochondria-rich fibres only make up a small part of the circular mantle muscle, it is evident that the escape jets driven by the small fibre system, as well as those driven by the giant axons, must involve the large mass of central zone muscle fibres. The central fibres then, must presumably receive a dual innervation. One possibility accounting for the lower pressure pulses when the system is driven by the small nerve fibres alone is that all regions of the mantle do not then contract simultaneously to give a maximum pressure pulse, as they do when the giant axons fire. Indeed, it is difficult to see how different conduction delays of the small axons in the different stellar nerves could be avoided. Simultaneous EMG records from front and rear of the mantle have yet to be obtained when the small axon system alone is activated.

Studies on the development of the control of jetting in embryos and hatchlings of *Loligo opalescens* and *Sepioteuthis* (Gilly *et al.* 1991) have shown that in hatchlings of both species, strong escape jets can be driven by the small axon system alone, with no giant axon activity, but as in adults, similar patterns of small axon and giant fibre control of jetting are seen (Fig. 20.15). Since small axon control of jetting is seen at the earliest stage that Gilly *et al.* examined (stage 26), but giant axon responses appear later (at stage 28), the two motor systems do not develop in perfect synchrony. Hence, as these authors point out, developing embryos and hatchlings would provide a valuable system in which to investigate the dual innervation of the mantle circular muscles.

Mechanoreceptors associated with muscle

There is good physiological evidence for mechanoreceptors in the mantle of *Octopus* (Gray 1960; Boyle 1976), responding to mantle stretch [Fig. 20.16(a)]. Multipolar stellate cells with centripetal axons have been described among the muscular layers of the mantles of *Octopus* and *Eledone* (Sereni and Young 1932; Alexandrowicz 1960; Boyle 1976). Alexandrowicz found 20 or more of these cells associated with a special muscular plexus underneath

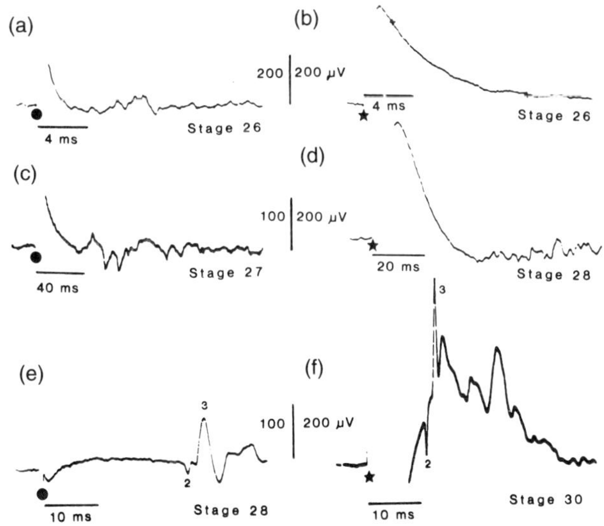

Fig. 20.15. Maturation of jet pulse system in *Loligo opalescens*. Indirect electrical (●) or light flash (★) stimuli were used to evoke responses recorded by suction electrodes on stellate ganglion. (a) Small fibre activity only, at stage 26. (b) A light flash produces no responses at this stage. (c) Small fibre activity only, at stage 27. (d) At stage 28 a light flash only evokes small fibre activity. (e) Electrical stimuli at the same stage evoke giant axon activity (2: second-order giant axon; 3: third-order giant axon). (f) At stage 30, a light flash evokes a giant axon response. (From Gilly *et al.* 1991.)

the stellate ganglion, the substellate plexus. He mentioned that despite a special search he had not observed them (or other possible proprioceptors) elsewhere in cephalopod muscles.

More recently, Kier *et al.* (1985) have demonstrated both phasic and tonic units in *Sepia* fins, responding to fin beat and to mechanical stimulation of the fins [Fig. 20.16(b)]. These authors found curious muscular structures within the fin muscle around the borders of the fin, but were unable to show that these were innervated, so that it is still unclear if these are the mechanoreceptors they recorded from. Further investigations of cephalopod mechanoreceptors which might have a proprioceptive function in muscle control would clearly be rewarding. It is interesting that Gray (1960) found no receptor unit activity in *Octopus* stellar nerves during respiratory movements of the mantle.

Biochemical studies

Lipid levels are low in cephalopod muscle, and although direct mobilization of muscle protein may take place during starvation (O'Dor *et al.* 1984), normal muscle metabolism is carbohydrate-based, and the muscle fibres

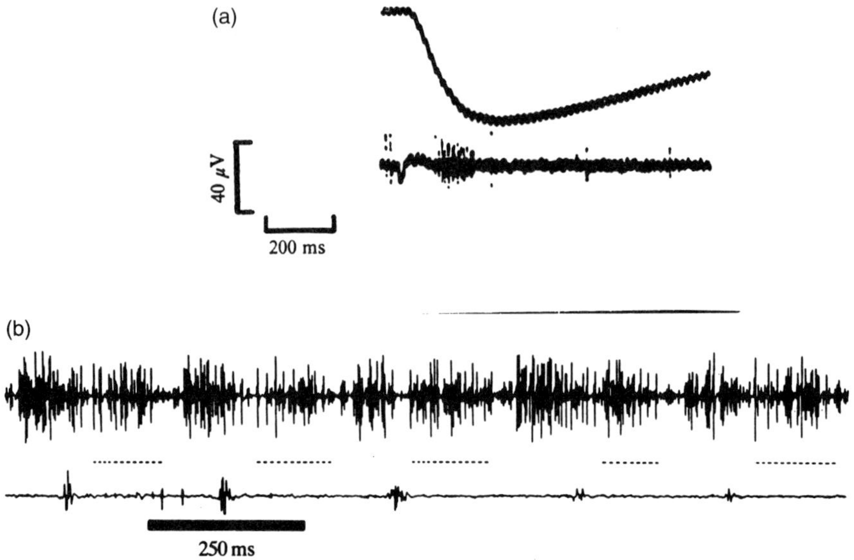

Fig. 20.16. Proprioceptive activity. In (a) mantle movement (upper trace) elicits a short burst from the distal stump of a severed stellar nerve in *Octopus*. (From Boyle 1976.) In (B) from *Sepia* fin, the upper record is from the proximal and the lower record from the distal cut ends of a fin nerve, showing afferent activity near the bottom of the downstroke. The dotted lines represent upward movement of the fin. (From Kier *et al.* 1985.)

contain glycogen. Storey and Storey (1983) have given an excellent review of carbohydrate metabolism in cephalopod muscle, which will therefore not be dealt with in any detail here. Both aerobic and anaerobic muscle contractions are fuelled by glycolysis, aerobic metabolism being similar to that of insect flight muscle in the operation of the α-glycerophosphate cycle and the coupling of proline oxidation with carbohydrate catabolism. Anaerobic metabolism does not involve the production of lactate, and lactate dehydrogenase (LDH) levels are low; instead, anaerobic glycolysis gives rise to octopine; octopine dehydrogenase (ODH) occurs in tissue-specific forms and some (possibly most) of the octopine produced in the mantle muscle is catabolized in other tissues. The ratios of the two key enzymes involved in aerobic and anaerobic metabolism, namely α-glycerophosphate dehydrogenase (α-GPDH) and ODH offer a useful measure of the aerobic/anaerobic capacity of different tissues; low ODH/α-GPDH ratios indicating aerobic metabolism, high ratios anaerobic metabolism. As Abbott and Bundgaard (1987) point out, in aerobic muscles such as those of the fins in squid, the ratio is around about one-sixth of that in tentacle muscle, according well with the evidence for metabolic type from vascularity.

The parallels between the lactate-LDH system of vertebrates and the octopine-ODH system of cephalopods are emphasized by Storey and Storey (1983), who also consider possible advantages of octopine as compared with lactate for the glycolytic end-product in molluscan muscles.

Conclusions

There are evidently many gaps in our knowledge of cephalopod muscle, partly no doubt, because the small diameter of cephalopod fibres makes them difficult 'preparations' compared with vertebrate muscle fibres which are much larger, ranging up to 300 μm, but also because physiologists have naturally devoted most attention to the uniquely large giant axons, rather than to the muscles. It is striking that it is only very recently that the functional role of the giant axons in escape responses has been reconsidered, and that it is still unclear how they and the small axons innervate the mantle muscle. Obvious gaps also remain in knowledge of the mechanical responses of cephalopod muscle; the kinds of studies that have been done on vertebrate single fibres to investigate the work done during contraction have yet to be begun on cephalopods. Nothing is known of the growth of cephalopod muscles; in *Sepia* and squid mantles, although it seems probable that growth zones are found along the insertions of the mantle circular fibres on to the cuttlebone and pen, this remains to be demonstrated; 'myosatellite' cells have not been seen so far in the mantle.

Acknowledgements

This work has been supported by an Emeritus Fellowship from The Leverhulme Trust to Q.B., a Wellcome Trust post-doctoral Research Assistantship to E.R.B., and an MBA Bursary to M.U.

References

Abbott, B. C. and Lowy, J. (1956). A new muscle preparation for the study of optical changes during contraction. *Nature*, **177**, 788–9.

Abbott, N. J. and Bundgaard, M. (1987). Microvessel surface area, density and dimensions in brain and muscle of the cephalopod *Sepia officinalis. Proceedings of the Royal Society of London, Series B*, **230**, 459–82.

Alexander, R. McN. (1992). The work that muscles can do. *Nature*, **357**, 360–1.

Alexandrowicz, J. S. (1960). A muscle receptor in *Eledone cirrhosa. Journal of the Marine Biological Association of the United Kingdom*, **39**, 419–31.

Alexandrowicz, J. S. (1963). A pulsating ganglion in the Octopoda. *Proceedings of the Royal Society of London, Series B*, **157**, 562–73.

Amsellem, J. and Nicaise, G. (1980). Ultrastructural study of muscle cells and their connections in the digestive tract of *Sepia officinalis*. *Journal of Submicroscopic Cytology*, **12**, 219–31.

Andrews, P. L. R., Messenger, J. B., and Tansey, E. M. (1981). Colour changes in cephalopods after neurotransmitter injection into the cephalic aorta. *Proceedings of the Royal Society, London, Series B*, **213**, 93–9.

Auerbach, B. and Budelmann, B. U. (1986). Evidence for acetylocholine as a neurotransmitter in the statocyst of *Octopus vulgaris*. *Cell and Tissue Research*, **243**, 429–36.

Ballowitz, E. (1892). Ueber der feineren Bau der Muskelsubstanzen. 1. Muskelfaser der Cephalopoden. *Archiv für Mikroskopische Anatomie*, **29**, 291–324.

Boll, F. (1869). Beiträge zur vergleichenden Histologie des Molluskentypus. *Archiv für Mikroskopische Anatomie*, **5** (Suppl.), 1–111.

Bone, Q. and Howarth, J. V. (1980). The role of L-glutamate in neuromuscular transmission in some molluscs. *Journal of the Marine Biological Association of the United Kingdom*, **60**, 619–26.

Bone, Q. and Ryan, K. P. (1974). The structure and innervation of the muscle bands of *Doliolum* (Tunicata: Cyclomyaria). *Proceedings of the Royal Society of London, Series B*, **187**, 315–27.

Bone, Q. and Trueman, E. R. (1984). Jet propulsion in *Doliolum* (Tunicata: Thaliacea). *Journal of Experimental Marine Biology and Ecology*, **76**, 105–18.

Bone, Q., Pulsford, A., and Chubb, A. D. (1981). Squid mantle muscle. *Journal of the Marine Biological Association of the United Kingdom*, **61**, 327–42.

Bone, Q., Packard, A. P., and Pulsford, A. L. (1982). Cholinergic innervation of muscle fibres in squid. *Journal of the Marine Biological Association of the United Kingdom*, **62**, 193–9.

Bone, Q., Brown, E. R., Nicholson, D., and Travers, G. (1994). On the respiratory flow in *Sepia* and other cephalopods. *Journal of Experimental Biology*, **192** (in press).

Boyle, P. R. (1976). Receptor units responding to movement in the *Octopus* mantle. *Journal of Experimental Biology*, **65**, 1–9.

Bozler, E. (1928). Uber die Tätigkeit der einzelnenglatten Muskelfaser bei der Kontraktion. II. Mitteilung: Die Chromatophorenmuskeln der Cephalopoden. *Zeitschrift für vergleichenden Physiologie*, **5**, 371–4.

Bozler, E. (1931). Ueber die Tätigkeit de einzelnen glatten Muskelfaser bei der Kontraktion. III Mitteilung: Registrierung der kontraktionen der Chromatophero-enmuskelzellen von Cephalopoden. *Zeitschrift für vergleichenden Physiologie*, **13**, 763–72.

Brown, E. R. (1991). Axon–Schwann cell interaction in the squid. PhD Thesis, University of London.

Clark, R. B. (1964). *Dynamics in Metazoan evolution. The origin of the coelom and segments*. Clarendon Press, Oxford.

Cloney, R. A. and Florey, E. (1968). Ultrastructure of cephalopod chromatophore organs. *Zeitschrift für Zellforschung*, **89**, 250–80.

Corrie, J. E. T., DeSantis, A., Katayama, Y., Khodakhah, K., Messenger, J. B., Ogden, D. C., and Trentham, D. R. (1993). Postsynaptic activation at the squid giant synapse by photolytic release of L-glutamate from a 'caged' L-glutamate. *Journal of Physiology*, **465**, 1–8.

DeSantis, A. and Messenger, J. B. (1989). New evidence that L-glutamate is a transmitter at the squid giant synapse. *Quarterly Journal of Experimental Physiology*, **74**, 219–22.

Dubas, F. (1987). Innervation of chromatophore muscle fibres in the octopus *Eledone cirrhosa*. *Cell and Tissue Research*, **248**, 675–82.

Dubas, F. and Boyle, P. R. (1985). Chromatophore motor units in the octopus *Eledone cirrhosa* (Cephalopoda: Octopoda). *Journal of Experimental Biology*, **117**, 415–31.

Florey, E. (1966). Nervous control and spontaneous activity of the chromatophores of a cephalopod, *Loligo opalescens*. *Comparative Biochemistry and Physiology*, **18**, 305–24.

Florey, E. and Kriebel, M. E. (1969). Electrical and mechanical responses of chromatophore muscle fibres of the squid, *Loligo opalescens*, to nerve stimulation and drugs. *Zeitschrift für Vergleichenden Physiologie*, **65**, 98–130.

Florey, E., Dubas, F., and Hanlon, R. T. (1985). Evidence for L-glutamate as a transmitter substance of motoneurons innervating squid chromatophore muscles. *Comparative Biochemistry and Physiology*, **82C**, 259–68.

Froesch-Gaetzi, V. and Froesch, D. (1977). Evidence that chromatophores of cephalopods are linked by their muscles. *Experientia*, **33**, 1448–50.

Gilly, W. F., Hopkins, B., and Mackie, G. O. M. (1991). Development of giant motor axons and neural control of escape responses in squid embryos and hatchlings. *The Biological Bulletin, Woods Hole*, **180**, 209–20.

Gosline, J. M. and Shadwick, R. E. (1983). The role of elastic energy storage mechanisms in swimming: an analysis of mantle elasticity in escape jetting in the squid *Loligo opalescens*. *Canadian Journal of Zoology*, **61**, 1421–31.

Gosline, J. M., Sleeves, J. D., Harman, A. D., and Demont, M. D. (1983). Patterns of circular and radial muscle activation in respiration and jetting of the squid *Loligo opalescens*. *Journal of Experimental Biology*, **104**, 97–109.

Gray, J. A. B. (1960). Mechanically excitable receptor units in the mantle of *Octopus* and their connections. *Journal of Physiology*, **153**, 573–82.

Graziadei, P. (1966). The ultrastructure of the motor nerve endings in the muscle of cephalopods. *Journal of Ultrastructural Research*, **15**, 1–13.

Hanson, J. and Lowy, J. (1957). Structure of smooth muscles. *Nature*, **180**, 906–9.

Hanson, J. and Lowy, J. (1960). Structure and function of the contractile apparatus in the muscles of invertebrate animals. In *Structure and function of muscle*, I (ed. G. H. Bourne), pp. 265–35. Academic Press, New York and London.

Hochachka, P. W., French, C. J., and Meredith, J. (1978). Metabolic and ultrastructural organization in *Nautilus* muscles. *Journal of Experimental Zoology*, **205**, 51–62.

Jolyet, F. and Sellier, J. (1899). Contribution à l'étude de la physiologie comparée de la contraction musculaire chez les invertébrés. *Travaux des Laboratoires de la Societé Scientifique d'Arcachon*, pp. 73–8.

Kahr, H. (1959). Zur endokrinen Steuerung der Melanophorenreaktion bei *Octopus vulgaris*. *Zeitschrift für vergelichenden Physiologie*, **41**, 435–48.

Kawaguti, S. and Ikemoto, N. (1957). Electron microscopy of the smooth muscle of a cuttlefish, *Sepia esculenta*. *Biological Journal of Okayama University*, **3**, 196–209.

Kier, W. M. (1985). The musculature of squid arms and tentacles: ultrastructural evidence for functional differences. *Journal of Morphology*, **185**, 223–9.

Kier, W. M. and Smith, K. K. (1985). Tongues, tentacles and trunks: the biomechanics of movement in muscular-hydrostats. *Zoological Journal of the Linnean Society*, **83**, 307–24.

Kier, W. M. and Smith, K. K. (1990). The morphology and mechanics of octopus suckers. *The Biological Bulletin, Woods Hole*, **178**, 126–36.

Kier, W. M., Messenger, J. B., and Miyan, J. A. (1985). Mechanoreceptors in the fins of the cuttlefish, *Sepia officinalis*. *Journal of Experimental Biology*, **119**, 369–73.

Lanzavecchia, G. (1981). Morphofunctional and phylogenetic relations in helical muscles. *Bolletino di Zoologia*, **48**, 29–40.

Lowy, J. and Millman, B. M. (1962). Mechanical properties of smooth muscles of cephalopod molluscs. *Journal of Physiology*, **160**, 353–63.

Marceau, F. (1905). Recherches sur la structure des muscles du manteau des Céphalopodes en rapport avec leur mode de contraction. *Bulletin de la Societé Biologique d'Arcachon*, Huitième année, 1904–5, 1–18.

Marceau, F. (1906). Note complémentaire sur la structure du manteau des céphalopodes en rapport avec leur mode de contraction. *Bulletin de la Societé Biologique d'Arcachon*, Neuvième année, 1906, 1–4.

Margo, T. (1860). Ueber die Muskelfasern der Mollusken. *Wiener Sitszungber. Math. naturwiss. Cl.*, **XVXIX**.

Miledi, R. (1969). Transmitter action in the giant synapse of the squid. *Nature*, **223**, 1284–6.

Mommsen, T. P., Ballantyne, J., MacDonald, D., Gosline, J., and Hochachka, P. W. (1981). Analogues of red and white muscle in squid mantle. *Proceedings of the National Academy of Sciences, USA*, **78**, 3274–8.

Moon, T. W. and Hulbert, W. C. (1975). The ultrastructure of the mantle musculature of the squid *Symplectoteuthis oualariensis*. *Comp. biochem. Physiol.*, **52B**, 145–9.

O'Dor, R. K. (1988). Limitations on locomotor performance in squid. *Journal of Applied Physiology*, **64**, 128–34.

O'Dor, R. K. and Webber, D. M. 1991. Invertebrate athletes: trade-offs between transport efficiency and power density in cephalopod evolution. *Journal of Experimental Biology*, **160**, 93–112.

O'Dor, R. K., Mangold, K., Boucher-Roderic, R., Wells, M. J., and Wells, J. (1984). Nutrient absorption, storage and remobilization in *Octopus vulgaris*. *Marine Behaviour and Physiology*, **11**, 239–58.

Otis, T. S. and Gilly, W. F. (1990). Jet-propelled escape in the squid *Loligo opalescens*: concerted control by giant and non-giant motor axon pathways. *Proceedings of the National Academy of Sciences, USA*, **87**, 2911–15.

Packard, A. (1972). Cephalopods and fish: the limits of convergence. *Biological Reviews*, **47**, 241–307.

Prosser, C. L. and Young, J. Z. (1937). Responses of muscles of the squid to repetitive stimulation of the giant nerve fibres. *Biological Bulletin, Woods Hole*, **73**, 237–41.

Sereni, E. and Young, J. Z. (1932). Nervous degeneration and regeneration in cephalopods. *Pubblicazione della Stazione Zoologica di Napoli*, **12**, 173–208.

Storey, K. B. and Storey, J. M. (1983). Carbohydrate metabolism in cephalopod molluscs. In *The Mollusca*, 1 (ed. P. W. Hochachka), pp. 91–136. Academic Press, New York and London.

Tansey, E. M. (1979). Neurotransmitters in the cephalopod brain. *Comparative Biochemistry and Physiology*, **64C**, 173–82.

Wagenknecht, T., Grassuci, R., Frank, J., Saito, A., Inui, M., and Fleischer, S. (1989). Three-dimensional architecture of the calcium channel-foot structure of sarcoplasmic reticulum. *Nature*, **338**, 167–70.

Ward, D. V. and Wainwright, S. A. (1972). Locomotory aspects of squid mantle structure. *Journal of Zoology*, **167**, 437–49.

Wilson, D. M. (1960). Nervous control of movement in cephalopods. *Journal of Experimental Biology*, **37**, 57–72.

Young, J. Z. (1936). Structure of nerve fibres and synapses in some invertebrates. *Cold Spring Harbor Symposia on Quantitative Biology*, **4**, 1–6. The Biological Laboratory, Cold Spring Harbor, New York.
Young, J. Z. (1938). The functioning of the giant nerve fibres of the squid. *Journal of Experimental Biology*, **15**, 170–85.
Young, J. Z. (1939). Fused neurons and synaptic contacts in the giant nerve fibres of Cephalopods. *Philosophical Transactions of the Royal Society of London, Series B*, **229**, 465–503.

21

Organization of cephalopod chromatophore systems: a neuromuscular image-generator

ANDREW PACKARD

Summary

The skin of some common cephalopods contains several hundred chromatophores per mm^2. Modulation of chiaroscuro (brightness contrast) information through differential expansion of these during natural patterning employs combinations of distinct components (up to 40 in some species). The physiological mechanisms and spatial characteristics of the motor units involve principles used in the half-tone process and tuned to the visual perception processes of animals. They can be traced (a) to epigenesis of the skin and facultative coupling between chromatophores, controlled in part by firing frequency, and (b) to central coupling between motor units of a given class and reciprocal inhibition between spatially complementary units. Myogenic coordination within populations of chromatophores that behave in many ways like visceral muscle is reversibly suppressed by nerves (denervation studies). A revised view of the role of higher and lower brain centres in the selection, integration and modulation process is presented in the light of recent evidence.

The physiology of cephalopod chromatophores has had a curious history. Eminent researchers, no doubt drawn by the pyrotechnics offered by the skin—even after the death of its owner—made discoveries that were at the forefront of neuromuscular investigations at the time (Harless 1846; von Uexkuell 1892; Bozler 1931; Florey 1966)—and then returned to more familiar ground.

Likely reasons for the relative neglect are that a chromatophore system of the cephalopod type occurs nowhere else in the animal kingdom; it lacks the simplicity that promotes some other invertebrate models; it is not well suited to conventional recording methods; and optical recording, to which it is suited, presents such wealth of data that neurophysiologists accustomed to looking down the barrel of a microelectrode may be merely baffled.

The main biological interest of the subject is an evolutionary one: the close mapping that exists between patterns generated in the skin and mechanisms of pattern recognition by the eye and brain (Packard 1988). A common language links the two. The appearances generated are often complex, but underlying physiology is probably no less complex than that of other well-studied muscle systems. Moreover, the demand for rigorous interpretation of optically recorded chromatophore activity—stored either on videocasette and with a digital clock or as a library of commercially printed colour photographs—can now be met by image analysing techniques (La Gioia 1992).

Chromatophores always act in concert with other chromatophores (see Box 21.1). Therefore the parallel-processing capacities of a camera, so good at capturing the displays and patterns produced by cephalopods as part of their behavioural repertoire (see for instance Hanlon and Messenger 1988, and for references), also makes it the instrument of choice for recording neuromuscular details of these visual effectors: both in whole animals (Packard 1990*a*), with which much of this account is concerned, and in the isolated skin. Conversely, the fact that in life chromatophores always act at the population level limits the usefulness of interpretations based on single chromatophore recordings.

Where possible, in what follows, data recorded photographically or by video-camera are related to findings arrived at by more traditional means. A number of phenomena—for example effects of certain anaesthetics—were first encountered during routine procedures connected with other kinds of experiments.

While concentrating on general principles, the chapter illustrates the special contribution that cephalopod chromatophore studies can make to several fields of neuroscience—neuromuscular and visceral muscle physiology, motor control, nerve regeneration, developmental neurobiology—and to pattern generation theory.

Anatomical arrangements linking the levels discussed are shown later in Fig. 21.16.

Box 21.1. Results of electrical stimulation (von Uexkull 1892) *(Eledone)*
1. Thresholds for chromatophore activity also produce skin movements.
2. Chromatophores never respond singly, even the large well isolated ones.
3. 'In areas of high chromatophore density I was never able to stimulate a discrete number which would correspond to the lines of current'; that is, chromatophores not stimulated directly [but through nerves].
4. Activity is sometimes restricted to grooves of the patch-and-groove system of the skin (see Fig. 21.9).
5. Chromatophores which could not be stimulated directly, can respond to stimuli at another point in the skin.

Chromatophore populations and the 'half-tone' process

Chromatophores are pigmented brightness contrast elements that generate visual information by differential expansion against a pale background. The principle is the same as in the half-tone process used by the printing industry (Fig. 21.1). Modulation of luminance may be either temporal (i.e. phasic changes in the sizes of chromatophores) or spatial (i.e. expansion of some chromatophores relative to others elsewhere in the population, or in other populations, generally as sustained, or tonic, activity). In parts of the skin generating the highest brightness contrast, light areas are underlain by white pigment (leucophores).

Modulation is of three kinds: colour (quality) control, regional modulation, modulation of gradients (Fig. 21.2). An idea of the wide range of spatial frequencies available—still wider potentially than the half-tone process used to print them—can be gained from inspection of Fig. 21.3 and other photographs illustrating this article.

In photoelectric (single channel) recording, either of single chromatophores

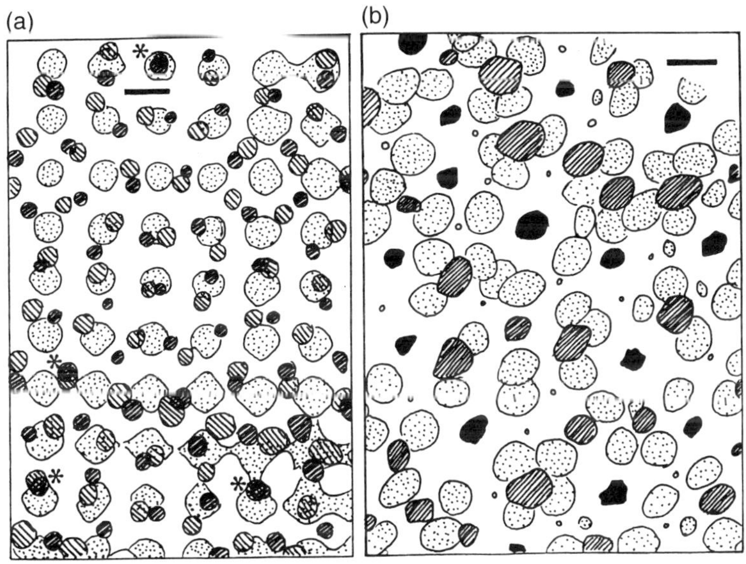

Fig. 21.1. Image generation by the half-tone process: (a) process engraving using three superimposed 150-line screens of coloured inks (blue, thick hatching; red, thin hatching and yellow, stippled); (b) squid mantle (*Loligo vulgaris*; 'brown', black, 'red', hatched and 'yellow', stippled chromatophores). Note that the serially regular screens of (a) have different orientations and produce higher order irregularities—spaces, clusters and overlapping centroids (∗)—that are avoided in the autopoietic screens of (b) (see text). Scale bars: (a) 0.2 mm, (b) 0.5 mm.

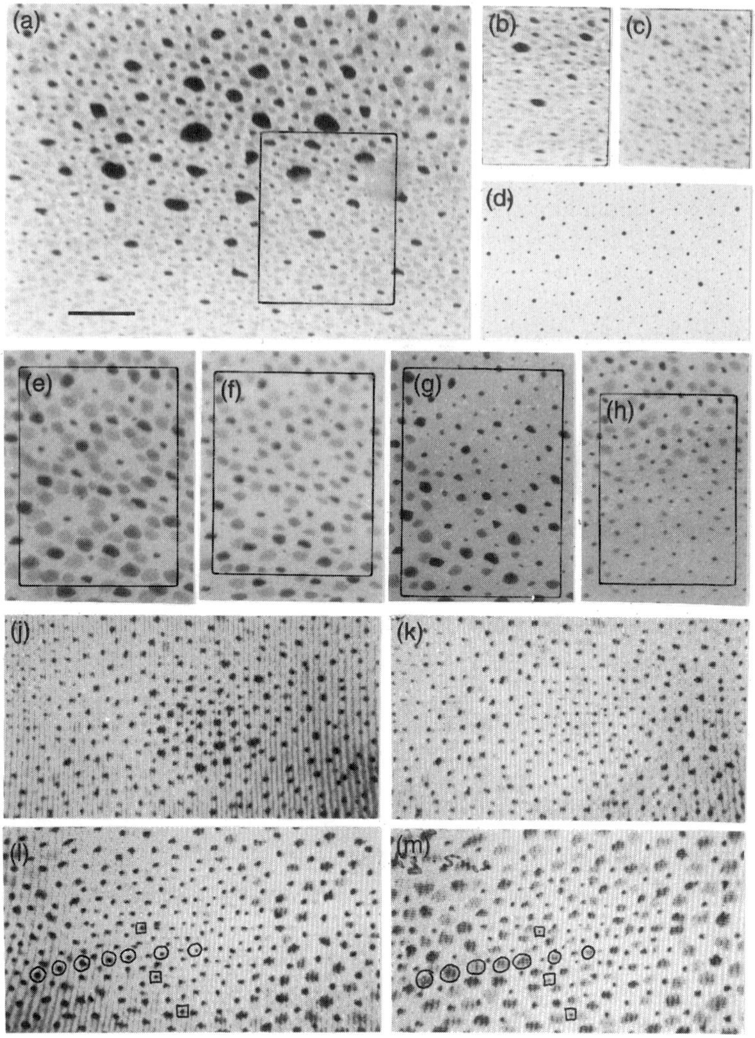

or of groups of chromatophores, brightness contrast information is converted into grey level (luminance) information and positional and spatial frequency information is lost. At some spatial frequencies and at low levels of illumination, similar integration occurs within the eye. Most (and in the natural environment perhaps all) chromatophores are too small to be resolved individually, even when fully expanded, but they contribute to the average grey level of the skin, and to modulation in lower spatial frequency bands, by expanding in combination with others in their neighbourhood. Figure 21.2(a) shows a cluster of chromatophores, each generating its own contrast, that the underwater eye would normally perceive either as one dark spot or as part of

Fig. 21.2. Details of image generation in two squids, (a)–(d) and (e)–(h), during natural patterning showing regional, colour, and gradient modulations. (*Loligo vulgaris*, hand-held camera, telephoto lens, dorsal upwards. Scale bar: 2 mm, same for all photographs ± 10 per cent.) (a) Typical mantle 'dark spot' feature generated by dark 'nuclear' and brown and red chromatophores on a yellow ground (grey); (b) and (c) modulation of activity in framed area of (a); (d) same region of skin post-mortem, dots (in descending size order) are 'nuclear', brown, red, and yellow chromatophores, all in the retracted state. (e)–(h) Modulation of activity (differential expansion) of red and brown chromatophores only. ('Nuclear', black, chromatophores retracted, yellow retracted, invisible in these photographs.) (j)–(m) Dark units responding to pallial nerve stimulation on the mantle of a juvenile octopus (*Octopus vulgaris*). (Photographs from video-monitor. Identical areas in all photographs. Scale bar: 1 mm.) (j) Localized 'patch' response ('off' to d.c.); (k) same unstimulated; (l) frequency-sensitive 'groove' response (5 Hz) which at 20 Hz (m) invades the central 'patch'. (Paired shocks, 0.5 ms; sample chromatophores circled.) Note that at high frequency there is further retraction (muscle relaxation) of some melanophore classes (squares) within the patch.

---

a larger feature. Some additional repertoires available at this location—involving regional suppression, grey level changes, and gradations in grey level—are shown alongside.

The colour (i.e. 'hue') of the chromatophores is probably of secondary importance to their chiaroscuro and grey level effects. (Spectral hues and changes in spectral hue are generated by iridocytes, but these structures are not considered further in this chapter.)

Spatial distribution and other characters of chromatophores: a question of development

To generate this kind of information, individual brightness contrast elements need to be appropriately spaced in the image plane: for instance at the interstices of a grid (the half-tone screen) or series of grids in the case of colour printing.

Cephalopods succeed in this rather better than the printing industry does, because the grid is an autocatalytic one (Packard 1982, 1985). Chromatophores —pale yellow or orange when they first appear (Table 21.1)—are added through interdigitation into spaces between already established ones [Fig. 21.4(a)] and into slightly different layers of the dermis (which means that the positional information supplied by the pre-existing state of the array transcends the separate layers). They increase in diameter and become darker as they mature (Froesch and Packard 1979) but never grow as large as their neighbours belonging to older classes. The process repeats, and old chromatophores do not die or disappear.

These simple epigenetic rules have been confirmed for all three major

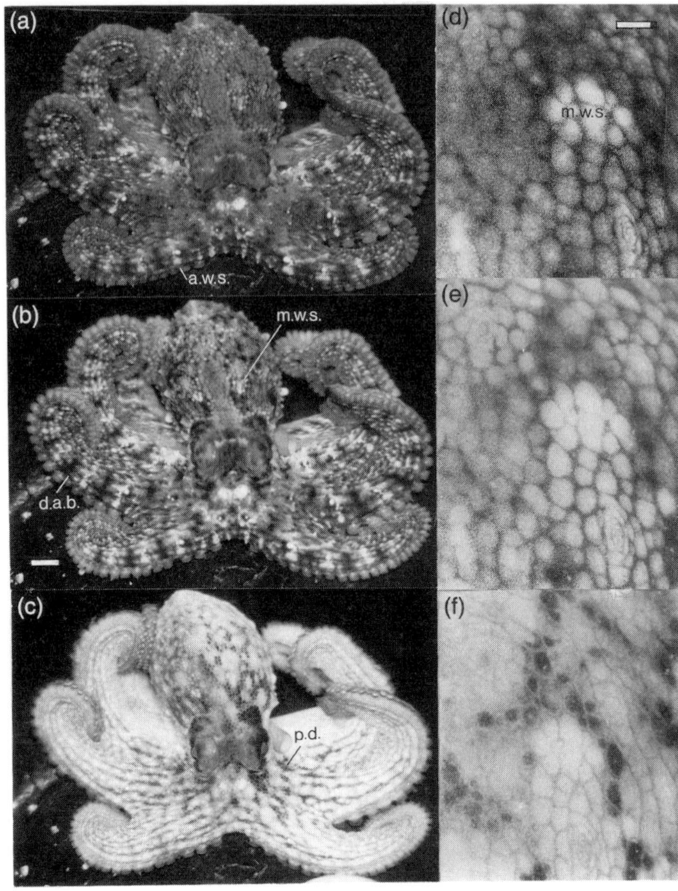

Fig. 21.3. Pattern generation by *Octopus vulgaris*: from camouflage to warning (conspicuous) in three steps without change of posture. In (a), a dark screen covering most of the skin produces low contrast amplitude in areas containing dark components (e.g. d.a.b., dark arm bands). (b) This screen is partially suppressed (step 1, grey-level reduced) heightening local contrast in the medium- and high-frequency bands of the 'acute mottle'. (c) The octopus is responding to a threatening stimulus approaching on the animal's left front, and all the round and orthogonally arranged dark features on the mantle and arms have been suppressed (step 2). This creates brightness contrast in the low-frequency band (octopus stands out against background). Simultaneously (step 3) a dark component of unique patch-only units—'polka dots' (p.d.)—replaces the previous dark units on and around the head. (Note the proximodistal gradient in expression of this component.) (d)–(f) details of these same changes in the right mantle white spot (m.w.s.) area of another animal [orientation of photographs inverted with respect to (a)–(c)] showing the contribution of patches and grooves. Individual dark chromatophores (melanophores) just visible. Scale bars 1 cm (a)–(c) and 2 mm (d)–(f). a.w.s., arm white spot.

Table 21.1 Chromatophore life histories (octopus)

Yellow (concentrated—just visible)
|
Yellow (expanded)
|
Yellow (responsive)
↘
Orange → dark (brown)
↘ ↘
Red 'Black'
(melanophores)

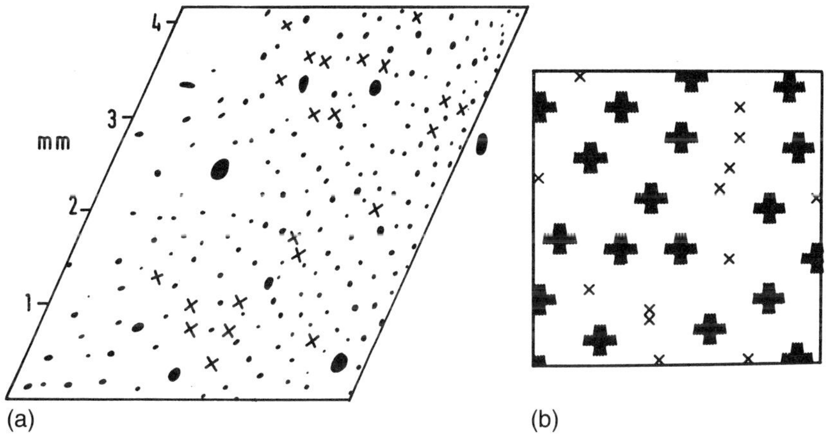

Fig. 21.4. Autopoietic half-tone screen. (a) Fall of new spots (X) in the ventral skin of a juvenile *Octopus vulgaris* over a 6-day interval and (b) computer model of the same, based on the principle of lateral inhibition, that has inserted the latest spots into an array of pre-existing spots (+) (see text). Both illustrations from Meinhardt (1982), with permission.

groups of cephalopods. Differences between squids, cuttlefish, and octopuses—especially in the adult—are due to differences in the values and number of repetitions of the controlling algorithms [Fig. 21.4(b); Meinhardt and Gierer 1974; Meinhardt 1982]. There is a fourfold difference in scale between the chromatophore patterns in Fig. 21.5(a) and (b); nevertheless the arrangement in the squid and the octopus is similar, if somewhat less regular in the latter. As development proceeds, squid skin largely conserves the arrangement seen here, but after several further iterations of the controlling algorithm the skin of octopus, dorsally, becomes very different in appearance.

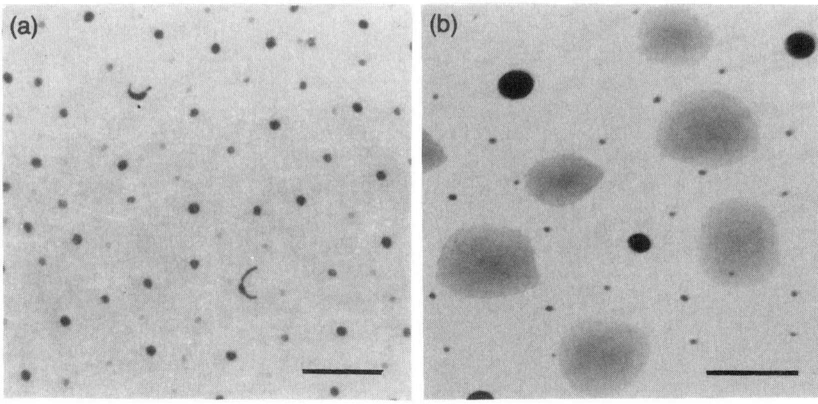

Fig. 21.5. Arrangement of chromatophores in an octopus and a decapod. (a) Ventral skin of baby (4 g body weight) *Octopus vulgaris*. All chromatophores in the relaxed state (crescents: broken founder and early chromatophores). (b) Ventral skin of *Loligo vulgaris* showing typical co-ordinated spontaneous activity (48 h post-mortem) of a single size/colour (red) class of squid chromatophores (14 per cent of all chromatophores in the field). Scale bars: (a) 0.2 mm: (b) 0.5 mm.

Relationship between chromatophore genesis and feature generation

One result of the repeating process of growth by interdigitation, and hierarchical control over the sizes attained by each generation of chromatophore, is that however small the area of skin under consideration, it will contain several size/colour/age classes. The history of the skin can be read in the hierarchy of resting sizes (Fig. 21.6). As each generation has slightly different physiological characteristics from its predecessors—and is likely to belong to a separate motor unit (see below)—the spectrum of chromatophore sizes in any one part of the skin is also a spectrum of age and of physiological class. The importance of this for the neurophysiologist can hardly be overstated.

The effects of developmental history on the appearances of the skin—discerned by applying the age/size rule—are illustrated by three well-known features: one from the octopus, one from the cuttlefish, and one from the squid.

The patch of the 'patch and groove' system [so conspicuous on the dorsal surface of octopuses; Figs 21.3(d)–(f), 21.8, and 21.9] first appears as a small area (~ 0.1 mm^2) in which rates of chromatophore production are relatively higher than elsewhere. The process of production and growth pushes aside some of the older chromatophores so that they come to be stretched (oval in shape rather than circular) and to lie in 'grooves', forming the 'trellis' units. Consequently younger units tend to lie within the patches, especially towards their centres, producing 'dots' when active [Fig. 21.2(j)]. In keeping with the

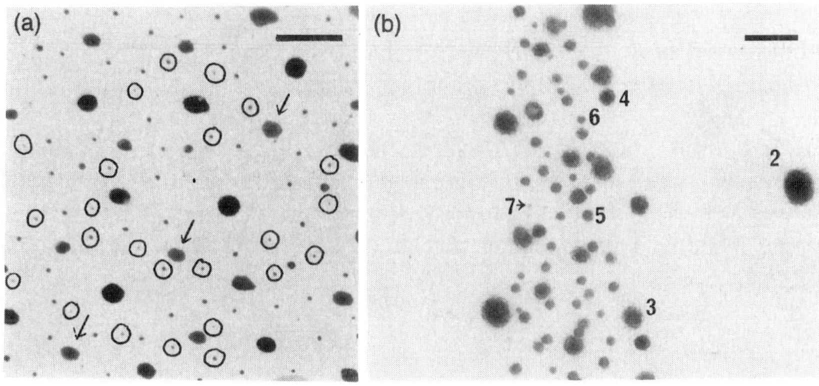

Fig. 21.6. (a) Part of a position-sensitive yellow motor unit (circled chromatophores, all in the relaxed state) on ventrolateral skin of the isolated mantle of *Loligo vulgaris* revealed by a stimulating electrode located outside the field. The unit represents 15 per cent of chromatophores shown. Most other chromatophores also relaxed: one class of red chromatophores (arrows) incompletely relaxed. Scale bar: 0.5 mm. (b) Detail of 'lateral flame' on an adult squid, *Loligo vulgaris*. All chromatophores relaxed. Numbers indicate gross size classes. The higher chromatophore concentrations along the flame compared with the rest of the skin can be explained as the result of increased activator levels and/or reduction in the distances over which lateral inhibition is effective during the birth of generations 3–7, while in the empty flanking zones chromatophore production has been completely suppressed during generations 3–7. (See p. 362.) Scale bar: 0.2 mm.

rules, the third, punctate unit of the three patch-centred 'visual units' described in Packard (1983) has the smallest chromatophore resting sizes of the three, and is found only fairly late in ontogeny.

The 'zebra stripes' [Figs 21.17(a) and 21.19], particularly well developed in male *Sepia*, have a somewhat similar history. The white segments of the 'zebra' (underlain, like the patches, by leucophores) only make their appearance and become progressively wider when the cuttlefish approaches sexual maturity. They are regions of high chromatophore production and are colonized by young motor units. The chromatophores responsible for the intervening dark of the 'zebra' are of slightly larger size class and belong to older motor units that have been pushed aside by the higher rate of chromatophore genesis over the 'white' areas (see below).

The 'lateral flame' (Hanlon 1982) of *Loligo*—which arises during sexual development on top of the uniform array of the juvenile—is composed of a population of very small (originally yellow) chromatophores bordered by 'stretch marks' [Fig. 21.6(b)]. They have independent innervation from the classes of larger chromatophores. In the 'stretch marks' only a very few, large, early, chromatophores are present. Chromatophore genesis has been subsequently suppressed in these zones.

Co-ordinated and differential expansion and retraction of chromatophores as coupled ensembles

A further requirement of the half-tone process is that the spots of the half-tone grid, the chromatophores, should be expanded or retracted in groups—as ensembles—for only then can they produce visual effects. This is achieved by both peripheral and central mechanisms, without local feedback (von Uexkull 1894).

Peripheral mechanisms

Peripheral mechanisms account for co-ordinated contraction and relaxation of muscle fibres within a local population of chromatophores and for the topographical details of patterns generated (high spatial frequencies and colour separations—see the developmental considerations above—and spatial gradients within a unit). Co-ordination within the skin is of two kinds: (1) neurogenic, many chromatophores being innervated by a single nerve fibre and forming a motor unit; and (2) myogenic, through a network linking the muscles of separate chromatophores.

Single chromatophore studies

Fundamental to the understanding of these peripheral mechanisms, particularly the relationships between nerve and muscle, is the work on squids during the 1960s by Ernst Florey and his collaborators. The essentials (Florey and Kriebel 1969) are summarized in Box 21.2 and Fig. 21.7; their examination of the role of acetylcholine (ACh) in squid chromatophore activity is one of the first descriptions of neuromodulation.

Motor unit and motor field studies

The shapes of the fields innervated by nerves to the skin and of motor units within those fields, vary from one species to another and in relation to their ecology.

Besides *Loligo opalescens* (Florey 1966) they have been examined in another loligonid, the Bay squid *Lolliguncula brevis* (Dubas *et al.* 1986*a*,*b*; Ferguson *et al.* 1988), which lives in murky waters and has a simpler patterning system, also in the cuttlefish *Sepia officinalis* (Maynard 1967) and *Octopus vulgaris* (Froesch 1973*a*; Packard 1974, 1983; Andrews *et al.* 1982)—both with complex patterns, and in the deeper water North Sea octopus *Eledone cirrhosa*—with its less complex patterns (Dubas and Boyle 1985). Methods include nerve lesions, fine dissection of skin nerves to separate out single fibres, stimulation of distal branches evoking axon reflexes, and central stimulation of motor neurones (Dubas *et al.* 1986*a*).

Stimulation of whole nerves or their branches can activate all the units in a

Box 21.2. Results of stimulation and recording studies in squid *Loligo opalescens* (Florey 1966; Florey and Kriebel 1969)
(A) Motor unit analysis:
1. Yellow chromatophores are separately innervated: no common innervation with other colour classes. This provides the basis of natural responses such as Fig. 21.2.
2. Dark chromatophores (red and brown) are multiply innervated, their muscle fibres belonging to one of a number of motor units (as also reported for the dark screening chromatophores of the cuttlefish (Maynard 1967)).
3. Red and brown chromatophores (different colour classes) may belong to the same motor unit.
4. The responses of single chromatophores to stimulation of nerve bundles on the dorsal surface increase with increases both in frequency and voltage (Fig. 21.7), but the responses of chromatophores in poorly innervated ventral skin do not increase with voltage.
5. Chromatophore muscles do not normally respond to direct stimulation.
(B) Electrophysiology (intracellular and photoelectric recording) and pharmacology of chromatophore muscle fibres:
1. Phasic and tonic contractions are not normally accompanied by muscle action potentials and are not suppressed by tetrodotoxin.
2. Evoked postsynaptic potentials and their associated contractures are of several sizes indicating polyneuronal innervation.
3. Both repetitive stimulation and applied serotonin (5HT) induce relaxation of tonically contracted chromatophore muscles, and 5HT shortens the duration of phasic contractions (Fig. 21.7).
4. Acetylcholine induces tonic contraction and miniature postsynaptic potentials but does not alter muscle resting potentials.
5. Spikes and pacemaker activity can be recorded from the radial muscle fibres of spontaneously pulsating chromatophores.
6. Adjacent fibres of such chromatophores are electrically coupled (Fig. 21.13).
7. End plate potentials generated in other muscle fibres can be recorded at a distance in neighbouring fibres.

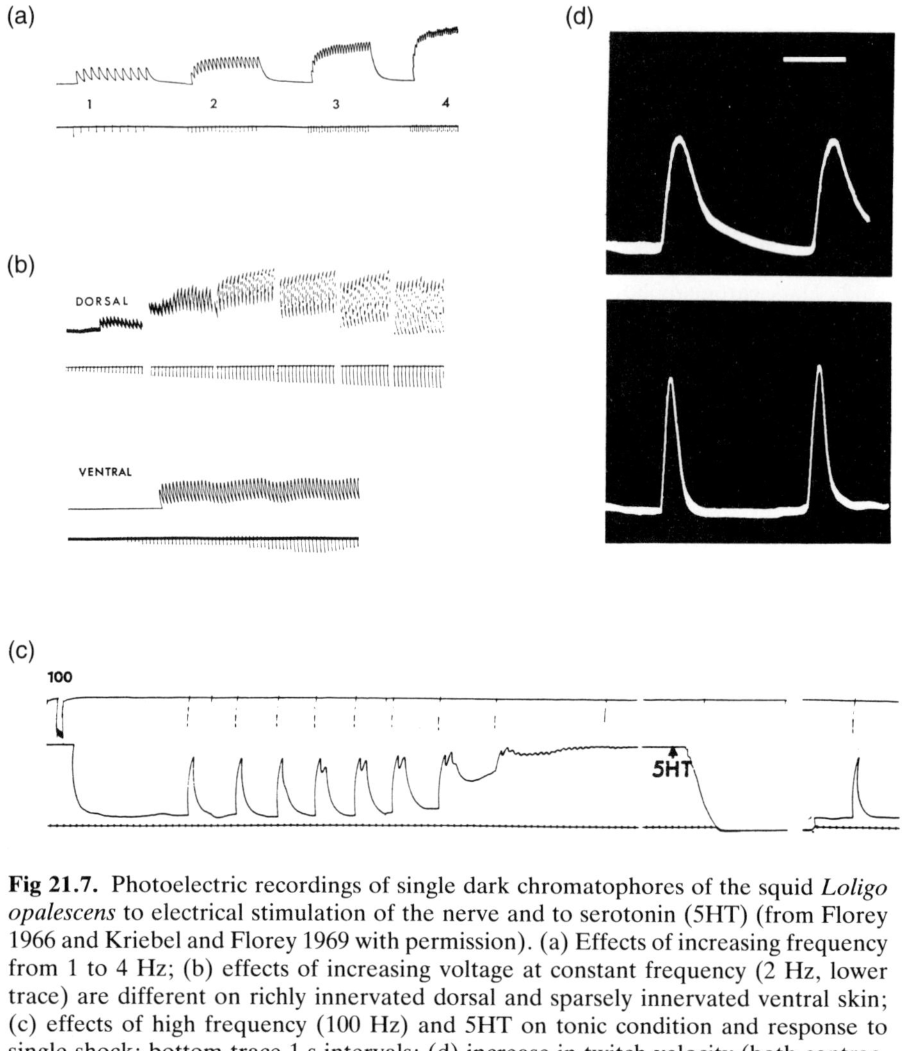

Fig 21.7. Photoelectric recordings of single dark chromatophores of the squid *Loligo opalescens* to electrical stimulation of the nerve and to serotonin (5HT) (from Florey 1966 and Kriebel and Florey 1969 with permission). (a) Effects of increasing frequency from 1 to 4 Hz; (b) effects of increasing voltage at constant frequency (2 Hz, lower trace) are different on richly innervated dorsal and sparsely innervated ventral skin; (c) effects of high frequency (100 Hz) and 5HT on tonic condition and response to single shock; bottom trace 1 s intervals; (d) increase in twitch velocity (both contraction and relaxation phases) of the muscles of a chromatophore in response to single shocks following 10^{-6} g/ml 5HT (lower trace). Upper trace, same chromatophore before 5HT application. Scale bar: 1 s.

motor field (Fig. 21.8), but, near to threshold, also allows individual components to be dissected out of the response (see below).

While responses vary from one part of the skin to another—notably as between dorsal and ventral skin—and from one type of unit to another: (1) the shapes of responses depend on the position of the stimulus (either central or peripheral [Figs 21.6(a), 21.9, and 21.11]; (2) the areas covered by motor units and motor fields interdigitate and/or overlap with those of

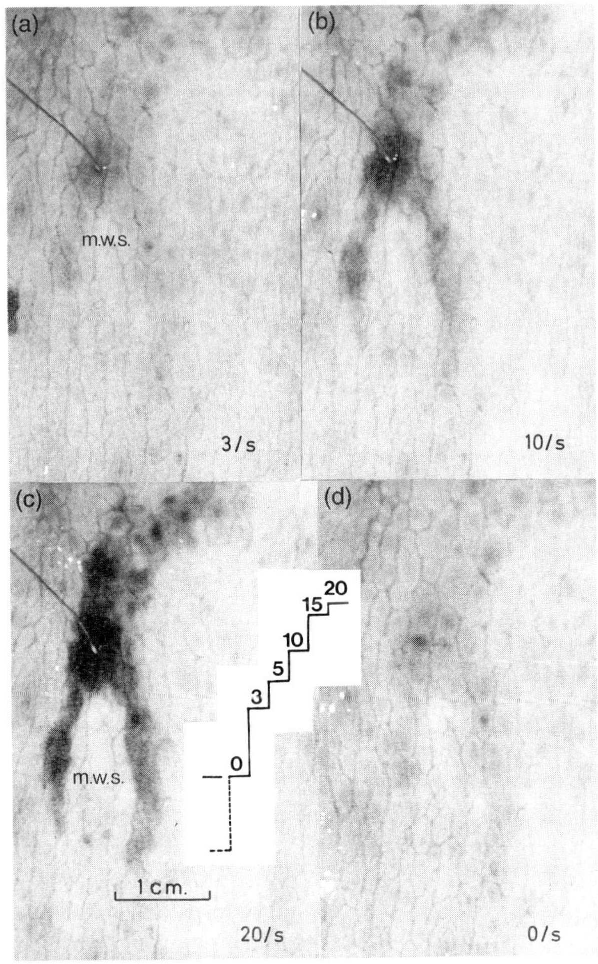

Fig 21.8. (a)–(c) Frequency-sensitive responses of a motor field (right 'horseshoe' area) in the isolated mantle of an adult *Octopus vulgaris* to shocks from a monopolar electrode near the centre of the field. Frequency in Hz. Inset, grey levels at these and intermediate frequencies relative to the unstimulated condition (d) and the adjacent mantle white spot (m.w.s.), broken line (from peak readings of microdensitometer scansions of the photographic negatives in the region of the electrode tip). Note that 50 per cent grey level increase is achieved at about 4 Hz.

neighbouring units and fields; (3) different motor units may share one and the same red or brown chromatophore (i.e. such chromatophores are multiply innervated); and (4) responses enlarge (spatial summation; Figs 21.8 and 21.10) with an increase in stimulus frequency and/or voltage (e.g. see fig. 9C in Dubas *et al.* 1986*a*) much as the responses of single chromatophores do (see Box 21.2, A4).

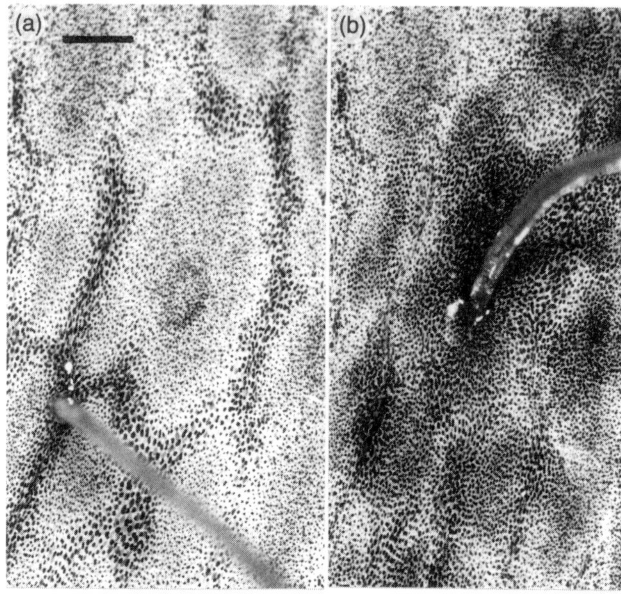

Fig. 21.9. Spatially complementary responses of 'groove' and 'patch' melanophores on the mantle of an adult *Octopus vulgaris* to a stimulating electrode, (a) on a groove, (b) on the patch it encircles. Identical skin areas. Scale bar: 1 mm.

Typical of frequency-sensitive responses is the 'mostly groove' units of the patch-and-groove system of the octopus skin which, at low frequencies, emphasize the grooves by darkening, and at higher frequencies spread on to the slopes of the patches [Fig. 21.2(l) and (m)]. Graded responses during natural patterning (see Fig. 21.2) are referred to below.

Typical of position-sensitive units, evoked by selective placing of the stimulating electrode at threshold intensities, are the patch-centred (centre/surround organized) 'visual units' of the octopus skin (Packard 1983). Such units do not enlarge much with increase in frequency. In the lateral and ventral skin of squids where motor unit organization is relatively simple, small shifts in the position of the stimulating electrode give pure colour responses, confined to discrete size/age/colour classes [Fig. 21.6(a)].

Details of dually innervated melanophores (black chromatophores) (Fig. 21.11) were published by Maynard (1967). In the overlapping 'visual units' of octopus just referred to, the extent of participation of a particular chromatophore (measurable as percentage expansion of the chromatophore) when a particular unit is active is a function of the difference between its resting size (i.e. age) and that of the rest of the chromatophores in the unit. The further apart they are in age, the less the participation (see p. 364, Fig. 21.21), but the subject of multiple innervation has been little explored.

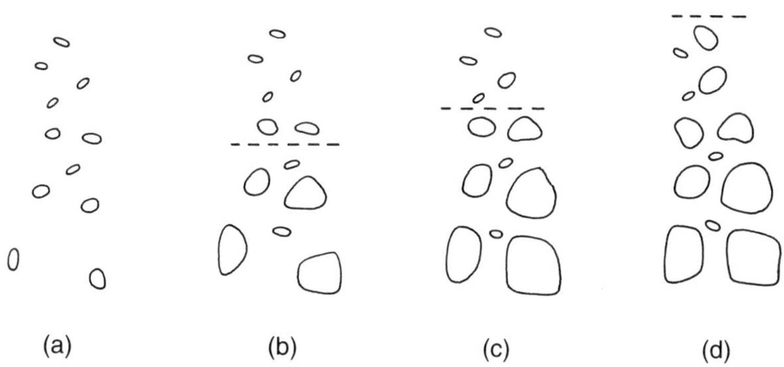

Fig. 21.10. Spatial summation. The diagram shows the enlargement in the response of chromatophores to stepwise increases of stimulus frequency [(b)–(d), values arbitrary] in a frequency-sensitive unit. Note that summation involves increases both in the number of responding chromatophores and in the degree of expansion of those already responding. Dashed line: shifting boundary of responsiveness (40 per cent increase in diameter) above resting level (a).

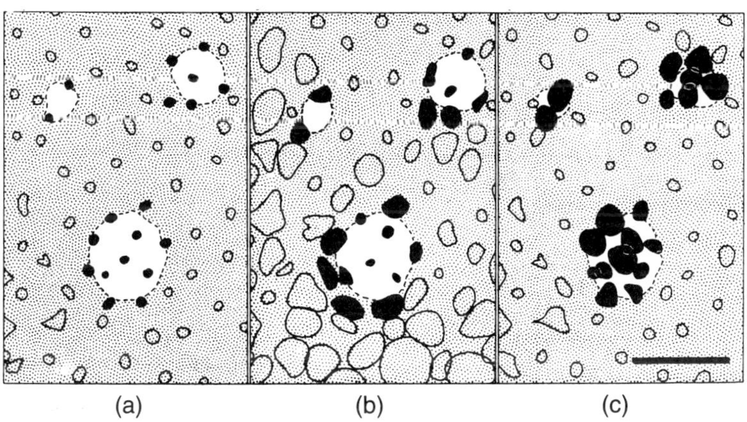

Fig. 21.11. Responses of 'Maynard units' (cuttlefish fin) to stimulation of adjacent nerves. (a) Melanophores (small open and black circles) at rest; (b) 'white spots' enhanced by melanophore expansion in the 'surround'; (c) 'white spots' screened, melanophores in the 'surround' relaxed. (See also Hanlon and Messenger 1988, figs 18 and 19.) Note dual innervation of some of the melanophores (black for emphasis) involved in both responses. Scale bar: 0.5 mm. (From Maynard 1967 with permission.)

Anatomical details of chromatophore innervation within a given population are poorly known. The question of particular interest is whether or not all chromatophores or all muscle fibres receive innervation. All squid chromatophore muscle fibres are said to be innervated. The picture in octopus skin is different. Instead of a web of nerve fibres, the single strands of which run along a muscle wherever they encounter one, as in the classic picture of red and brown squid chromatophores (see Chapter 22), innervation may be limited to rare *en passant* contacts (Froesch 1973*b*; Dubas 1987), often made without the contacting fibre leaving the nerve bundle. The contacts contain two types of synaptic vesicle. At the same level as the '*en passant*' synapses, there may be a presynaptic modifier synapse—reminiscent of crustaceans—from a second fibre which does not degenerate when the nerve to the chromatophores is severed (Froesch 1973*b*). Innervation on the ventral surface is very sparse.

In the skin of *Eledone* the northern octopus, Dubas estimated that less than 1 per cent of chromatophores are innervated, but if the cobalt used in this study did not enter the finest (0.2 μm diameter) branches of the motor neurones this will prove to be a considerable underestimate.

Little is known of the size of motor units. Obviously they are at least as large as the area responding to stimulation of a particular nerve fibre in one particular nerve branch, but they may be larger depending on whether the fibre divides into other branches proximal to the point of stimulation. The only central stimulations of single motor units are those of Dubas *et al.* (1986*a*), on a squid species (*Lolliguncula brevis*) possessing rather few chromatophores (some 40 000 in the adult). The number of chromatophores recruited at threshold was small (two to 60). Their definition of a motor unit is 'the smallest number of chromatophores that always respond synchronously to a threshold stimulus' but in view of the phenomenon of spatial summation, it is doubtful whether this definition is particularly useful.

No doubt, as in their other characteristics, qualitative differences in innervation exist between different size/colour/age classes of chromatophores, but no systematic comparisons have been conducted. There have been no electrical recordings of chromatophore activity at the population level, and nothing is known of the electrical behaviour of the peculiarly thin terminations of octopus motor nerves.

Motor units change their characteristics during development (Packard 1990*b*): the spatial and the pharmacological characteristics of younger motor units are different from those of older units (pp. 363–4).

Myogenic co-ordination

Most motor unit studies have interpreted the results of nerve stimulation as due to direct activation of chromatophore muscle fibres contacted by nerves. The 'wandering clouds' examined by Hofmann (1910), which I have recently reinvestigated, provide physiological evidence for the existence of a muscle network linking chromatophores of the same size/colour class (as well as of different classes) through which chromatophore muscle fibres communicate and can be activated other than through nerves. Waves of colour, much like a spreading depression, also pass across the skin under some natural conditions—for instance down the arms of a hunting cuttlefish. They can best be studied in denervated preparations (see Box 21.3) under conditions of supersensitivity

Box 21.3. Short- and long-term effects of interrupting nerve supply to skin (pallial nerve section) in *Octopus vulgaris* (Fig. 21.12).
At operation. Palor. (NB if operation is performed without anaesthetic, palor is preceded by injury mottle lasting 5–10 min after operation).
0–2 days (22°C). Gradual appearance of dark blotches, especially at boundary with intact skin, spreading to whole of operated side as peripheral nerve fibres die. Development of denervation supersensitivity.[1]
3–30 days (22°C). Tonic darkening. Bouts of spontaneous activity (waves). Denervation supersensitivity maximal (5 days).
Paramedian area (pma) appears.[2] *Paling* in pma and adjacent skin triggered by activity on intact side.[3] From *ca.* 25 days, partial reconnection of denervated skin (regeneration) produces paling (relaxation) of tonically darkened skin.
30 days plus. Gradual return of function culminating in chromatic and skin textural patterns mirroring intact side;[4] some details of normal patterning (e.g. small dark spots) never return, however.
1. Packard (1991*a*); 2. Packard (1991*b*); 3. Packard (1992); 4. Sanders and Young (1974).

once the nerve has died (Fig. 21.13). Denervation also gives rise to spontaneous propagated movements of the skin like the peristalsis of gut, bladder, and blood vessels, etc.; indeed the 'wandering clouds' are a kind of colour peristalsis. As recognized earlier (Bozler 1948), the physiology of chromatophores is in many ways like that of smooth muscle (Packard 1990*a*).

Waves are of two kinds, fast and slow (cf. Dahms *et al.* 1986). They may be either dark waves (of contraction) or pale waves (of relaxation in tonically contracted, dark, skin).

Dark waves follow routes that reflect the territories of nerves, but do not coincide with them. In octopus, waves in A zones (characterized by black melanophores) may fail to cross into B zones (characterized by brown melanophores). They can often be confined to a single patch or delay in passing from one patch to another; others may be restricted to the grooves. Even within a given area of skin, waves may be confined to a single class of chromatophore—for instance to the black (nuclear) chromatophores of squids—and then be substituted by a wave in another class—for instance the red chromatophores [Fig. 21.13 (b)]. In octopuses waves appear to be more prevalent in dark chromatophores (melanophores) than in other classes.

Fast dark waves (travelling at a few millimetres per second) are propagated pulsations of colour in pale or partially darkened skin which often originate,

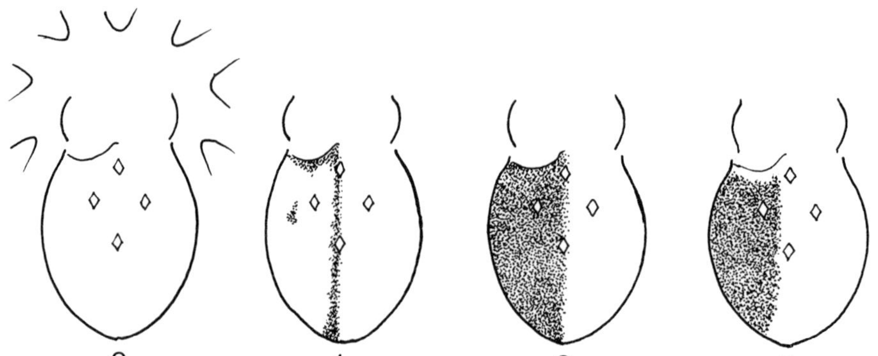

Fig. 21.12. Course of development of denervation supersensitivity (dark due to tonic expansion of chromatophores) on the left side of the mantle of an octopus following section of the pallial nerve on that side. Before operation (0) and at 1, 3, and 5 days after operation at summer temperatures. Diamond landmarks are the two mid-line and left and right long mantle papillae.

repeatedly, from the same point in the skin as if driven by a pacemaker [Fig. 21.13 (a)]. [NB The pacemaker activity of squid chromatophore muscles that produce synchronous, all-or-none, phasic contractions in all the fibres of that chromatophore also trigger pulsations in neighbouring chromatophores of the same class (Florey and Kriebel 1969).]

Slow waves (travelling at a few millimetres per minute) may be: (1) propagations of tonic chromatophore expansion (dark); or (2) propagated relaxation (palor). They are responsible for the sustained or slow-moving blotchiness of the skin evident in denervated skin and in the skin post-mortem.

The structural nature of the network through which activity propagates from one chromatophore to another has received little attention, but it has long been known that the distal ends of chromatophore muscle split up into several fine branches in the dermis criss-crossing with those of other chromatophores lying in the same and adjacent layers. Froesch-Gaetzi and Froesch (1977) traced the patterns of muscle connection between synchronously contracting chromatophore muscles of *Loligo, Sepia, Eledone* (and *Octopus*) in the living state. They estimated that in *Eledone* one chromatophore may contact from eight to 14 other chromatophores. Adjacent chromatophores sometimes share muscle fibres (Fig. 21.14). Although no electron micrographs of the inter-chromatophore junctions have been published, myomuscular gap junctions occur in other molluscs (Nicaise and Ansellem 1983), and may be expected here. See also Chapter 20.

Neurogenic/myogenic interactions

Spontaneous chromatophore activity and the 'wandering clouds' (waves of contraction) to which it gives rise are normally suppressed by the presence of

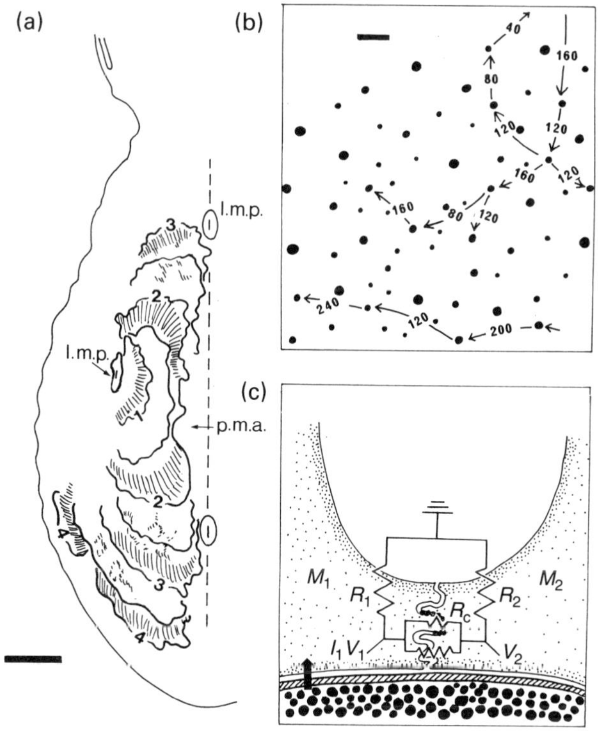

Fig. 21.13. (a) and (b) Myogenic propagation. (a) Fast dark wave spreading across the long-term denervated dorsal half of the mantle of an octopus (*Octopus vulgaris* in alcoholic sea water, 3 per cent) from a partially innervated focus, the left long mantle papilla (left l.m.p.). Wavefront followed at 0.5 s intervals (numbering in seconds), state of darkening at 1 s intervals. The wave peters out laterally and posteriorly and near the mid-line (dashed line) where it meets the paramedian area (p.m.a. see text) and the innervated skin of the intact (right) half of the mantle, and is slowed where it crosses from one motor field to another. Scale bar: 5 mm. (b) Frame by frame analysis of video recording to show path of excitation of single chromatophores on the edge of a passing wave (out of the picture) in the denervated skin of a squid (*Loligo vulgaris*), and the transmission times (in ms to the nearest 40 ms) between one all-or-none expansion and the next. Note that the wave is confined to red chromatophores of the same size/colour class. All chromatophores shown in relaxed state. Scale bar: 0.5 mm. (c) Electrical circuit and relevant resistances, R_1, R_2, and R_c (coupling resistance), at the point where the membranes of two chromatophore muscle fibres (M_1 and M_2) meet. R_c is composed of two junctional resistances, the regular and a variable close (or gap) junction (shown black). V_1 and V_2 are the voltage changes when current (I_1) is injected into cell 1. Below, membrane of the pigment body, arrow indicates potential path of photogenic excitation. (After Florey and Kriebel 1969, modified with permission.)

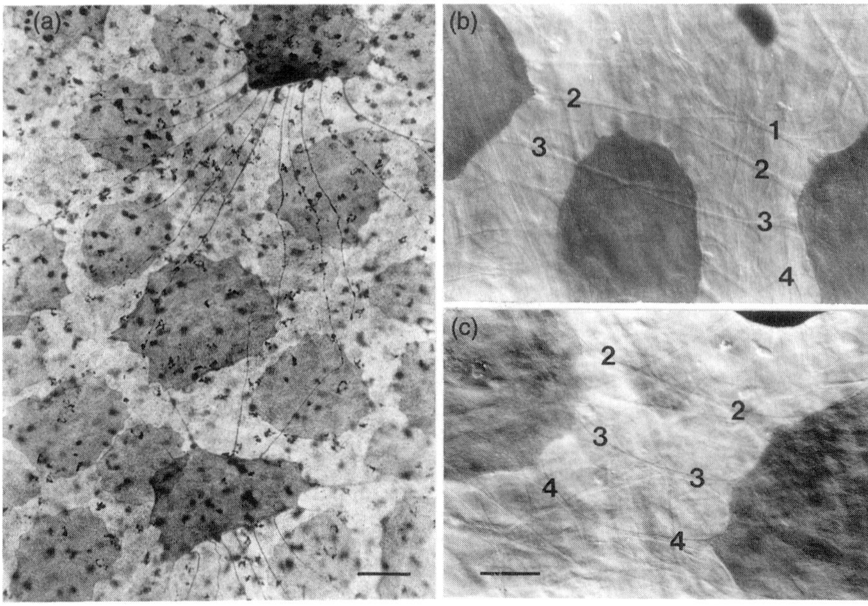

Fig. 21.14. Isolated fresh skin of an octopus (*Eledone cirrhosa*). (a) Two dye-coupled chromatophores showing joined muscle fibres among a population of unstained chromatophores. (Methylene blue; contents of gland cells also dyed. Photograph supplied by D. Froesch.) (b) and (c) Interference contrast photographs of living chromatophores (same individuals in both photographs) sharing common (2, 3) and interconnected muscle fibres (4). Note the extensive changes in shape and relative positions that accompany spontaneous activity. (Scale bars: 20 μm.) (From Froesch-Gaetzi and Froesch 1977, and Froesch unpublished, with permission.)

nerves: a property in common with visceral muscle (see Bozler 1948). In addition, an intact nerve supply induces active relaxation of chromatophore muscles (Packard 1992), recalling the neurogenic relaxation of Gillespie (1972) now reported for a great many smooth muscles. Conversely, agents which abolish nerve activity (including anaesthetics) also abolish the ability of nerves to suppress spontaneous activity and can give rise to waves.

Nevertheless some nervous activities—such as high firing rates in frequency-sensitive units exhibiting spatial summation—facilitate the spread of muscle contractions, and this ceases when the nerve stops firing. Under ordinary circumstances, myogenic communication of this kind cannot be readily distinguished from responses which might be due to progressive invasion of the fine branches of a motor unit with increase in firing rate, and the necessary experiments have still to be done. However, close inspection of the frequency-sensitive responses of groove chromatophores which invade bordering patches, reveals that the sequence of chromatophore recruitment with increase in stimulus frequency is the same as the path followed by a passing wave in denervated skin although the nerve has long since died.

During their examination of intrachromatophore junctions, Florey and Kriebel (1969) found stepwise increases of local depolarizations of muscle with increase in stimulus strength which they thought to be excitatory post-synaptic potentials (EPSPs) in neighbouring fibres recorded through low resistance junctions. They raised the possibility that EPSPs in one muscle fibre may summate to produce a spike that could propagate to other muscles of the same chromatophore. If interchromatophore junctions have the same properties, then the basis of neurogenic and myogenic interactions in normal skin could be partly explained (see also Muneoka and Twarog 1983). The variations in number of steps recorded (up to six) from one fibre could be examples of differences in the number of junctions or routes through which the increasing stimulus was able to invade the fibre and account for the graded responses seen throughout the skin. Frequency-dependent graded responses and facultative dye-coupling (dependent on extracellular Ca^{2+}) have now been found in squid mantle muscle bundles (see Chapter 20).

The distinction between neurogenic and myogenic mechanisms can best be seen on the ventral surface. Although chromatophores are found throughout the ventral surface, innervation is sparse and even at high stimulus intensities, there are only a few positions in which an electrode exploring the under-surface of the mantle encounters a nerve fibre and produces a response. When it does, the response covers a large area, including points which had been unresponsive to the electrode. Apparently the co-ordination of this response is mostly through muscle and not through nerve, the nerve only having access to the majority of responding chromatophores indirectly through their muscles. Motor units on the ventral surface must thus be regarded as having boundaries which extend well beyond the ramifications of the motor neurone itself. (The other possibility is that the finest branches of nerves are simply refractory to electrical stimulation even at the highest intensities.)

The paramedian area (PMA) of octopus skin is a special case of neurogenic/myogenic interactions (Packard 1991*b*) revealed some time after unilateral denervation (Fig. 21.12). Within the PMA—a kind of two-man's land wider on the ventral surface than the dorsal and roughly parallel to the mid-line—chromatophores can be controlled by nerves either from the left or the right side. When nerves on the denervated side have degenerated, the chromatophores on that side of the PMA continue to behave like chromatophores on the intact side. Apparently nerves are able to trigger normal responses myogenically in areas of skin from which at least half of the nerves have been lost and can keep the musculature of these areas in a relaxed state—though not necessarily with the same signals—and free from signs of supersensitivity.

Figure 21.15 summarizes some of the neurogenic/myogenic phenomena described here, but as excitation–contraction coupling processes vary from

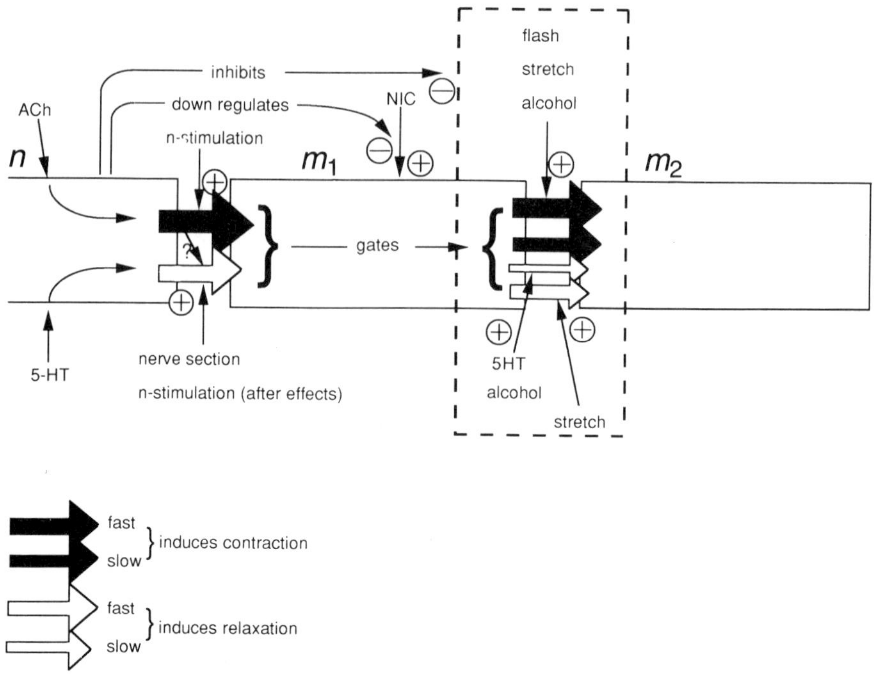

Fig. 21.15. Scheme of the relationship between neurogenic and myogenic mechanisms in chromatophore muscle activity discussed in this essay. m_1, m_2, muscle fibres; *n*, nerve fibre; NIC, nicotine. Broad arrows indicate both transmission and activation.

muscle to muscle (see Muneoka and Twarog 1983) possibly even within the same tissue, they should not be generalized to all chromatophore systems.

Denervated chromatophores respond directly to light energy [Packard, 1993; Fig. 21.13(c)].

Putative transmitters and the effects of serotonin

Evidence that L-glutamate—a force-inducing agonist promoting slow contractures and occasional twitches dependent on external calcium—is also a transmitter of some chromatophore classes is presented elsewhere in this volume (see Chapter 22).

The relaxation caused by 1 μM serotonin (5-hydroxytryptamine, 5HT) is immediate, produces greater palor than that of normal resting skin and spreads from the site of injection at the same slow rate as the spontaneous relaxation of partially innervated skin (Packard 1992) (see above). 5HT should not be considered an inhibitor, however. It alters the mechanical properties of some chromatophore muscle fibres (see Fig. 21.7) but does not impair their excitability: for example, 'flashing brown spot' (FBS) units [Fig. 21.17(d)] continue to be expressed in treated skin (see plate III in Andrews *et al*. 1982). It raises the excitability of dark squid and cuttlefish chromatophores to previously applied L-glutamate while abolishing tonus (Packard, unpublished).

Central mechanisms

The brain (Fig. 21.16) is responsible for co-ordination and modulation of motor unit activity including: amplitude control, selection of components and

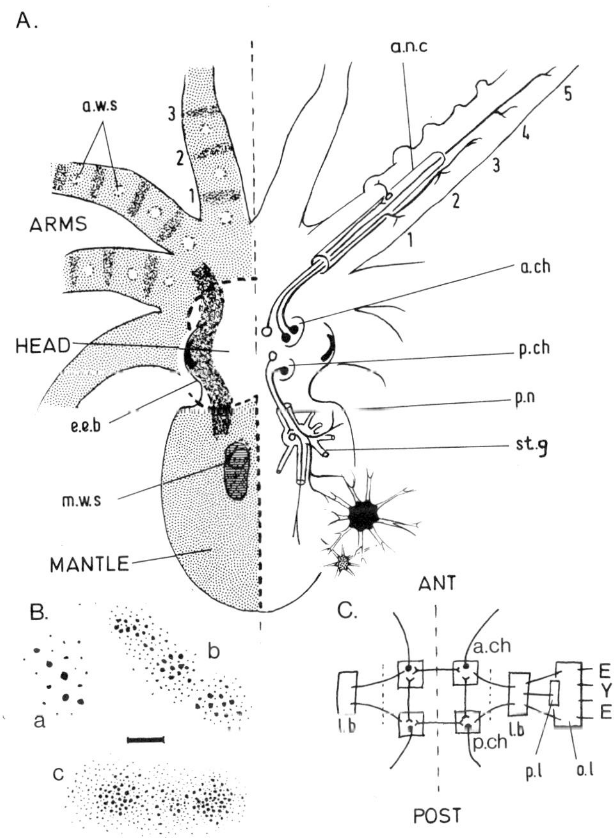

Fig. 21.16. (A) Relationship between the anatomy of an octopus, right, and some transient pattern features, left, (a.ch, anterior chromatophore lobes; a.n.c, arm nerve cord; l.b, lateral basal lobe; o.l, optic lobe; p.ch, posterior chromatophore lobe, p.l, peduncle lobe; p.n, pallial nerve; st.g, stellate ganglion; a.w.s, arm white spots; e.e.b, extended eye-bar; m.w.s, mantle white spot). Note that chromatophore motor neurones (filled cell bodies) proceed without synapse to their ramifications in the skin. 1–5, positions of dark arm bars. (B) responses of a) *Lolliguncula*, b) *Sepia*, c) *Octopus* chromatophore units for rough comparisons of scale; scale bar 1 mm. (C) central connections.

of spatial frequency bands, colour mixtures, generation of gradients within components, generation of low spatial frequencies, and of whole patterns including integration of the chromatophore system with other systems (notably with the dermal musculature responsible for skin texture, and with the body musculature responsible for shape and postures).

Components (classes of motor units)

The name 'components'—originally adopted in order to classify the various permutations of colour, shape, and texture observable in cephalopods (Packard and Sanders 1971)—applies to recognizably like activity occurring in different places of the skin and varying independently of other activity (Fig. 21.17). [More or less incomplete lists of components have been compiled for octopus (Packard and Sanders 1971; Packard and Hochberg 1977); cuttlefish (Hanlon and Messenger 1988), and squid (Hanlon 1982; Moynihan and Rodaniche 1982; Porteiro *et al.* 1990.] Components and their subdivisions can be classified into: (1) diffuse screens responsible for altering ground colour and grey-level; and (2) *feature generators*, of a range of spatial frequencies, which tend to be round or centre/surround organized, or elongated bars orientated with respect to the main axes of the body.

The neurophysiological interest of 'components' is that ideally they are the co-ordinated output of single classes of motor units (see above); that is synergic groups of muscle fibres belonging to motor neurones of the same class situated in the chromatophore lobes.

Whether one is in fact dealing with a single motor unit or many synchronously firing motor units, is rarely possible to decide, even though the patterns offer a great deal of information on the question. (In practice, the name 'component' is often used for one part of the component only—such as a single arm bar.) All components are subdivisable according to the primary divisions of the body (head, arms, mantle; right, left) into single or multi-unit subcomponents more or less strongly coupled in the brain with their companions responsible for like activity in the other divisions. Those on the head tend to be one-feature components.

During 'acute' patterning of the kind seen in Fig. 21.3, the various discontinuous components [such as the dark arm bars (d.a.b.), the arm white spots (a.w.s.) and the groups of expanded orange chromatophotes ('coral tips') screening the tips of all papillae in *Octopus vulgaris*], as well as the various continuously arranged components (such as the 'all dark' and the countershading of squids, or, in octopus, the fine 'trellis' and the centre/surround features of the 'conflict mottle'), appear across the whole of their spatial extent (Fig. 21.3) and with only little variation in contrast amplitude. In other words in acute patterns there is strong coupling between the units comprising a component. But when not maximally expressed, the contrast amplitude of the component falls off as a gradient (for instance d.a.b. may only be expressed

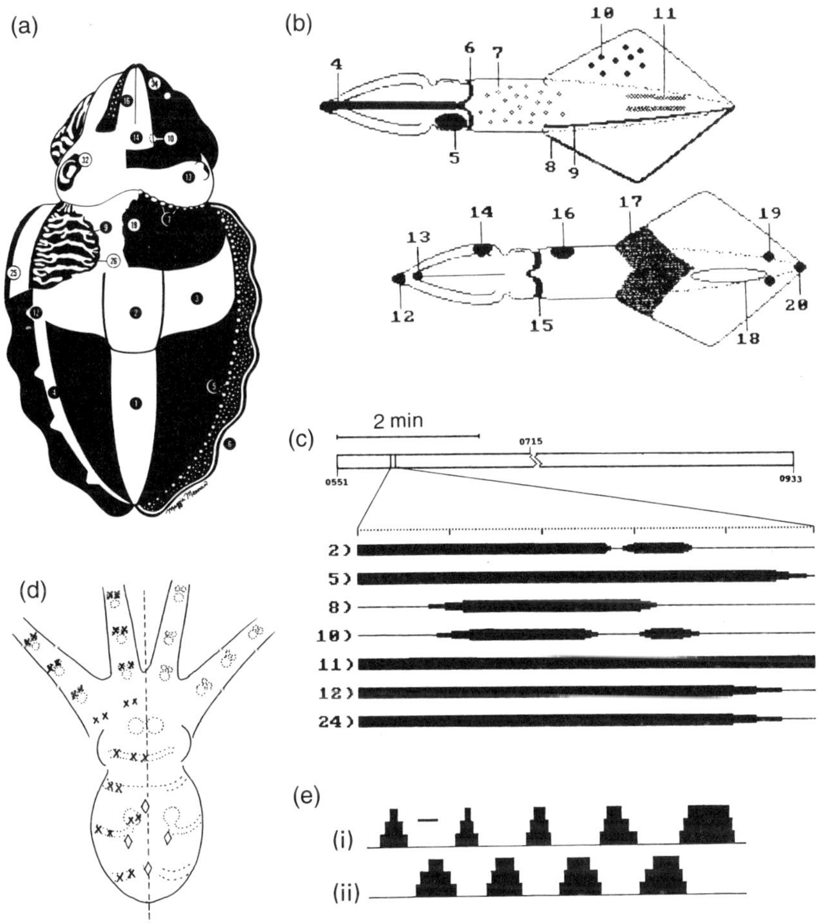

Fig. 21.17. Composite diagrams showing numbered components (brightness-contrast features)—which can vary independently of each other—in (a), the cuttlefish *Sepia officinalis* (from Hanlon and Messenger 1988, modified with permission); (b) the squid *Loligo vulgaris*. (c) Time course (over a 5-s stretch of video-recording) of seven such numbered components in a female squid interacting with a second female. (b) and (c) from Brancato *et al.*, in preparation. (d) Distribution of synchronous 'flashing brown spot', units (x)—a spatially discontinuous component revealed by cold anaesthesia (see text)—on the dorsal surface of *Octopus vulgaris*, and its relation to 'white spot' areas (dotted) (from Andrews *et al.* 1982, with permission). (e) Spatially and temporally complementary components (Maynard units, see Fig. 21.11) revealed by weak alcohol anaesthesia and followed over four and a half cycles, (i) screen over fin 'white spot', (ii) dark surround. From video-recording. Scale bar 1 s.

on the proximal parts of the first and second arms), revealing that there is also partial coupling within a multi-unit component (see below).

During natural patterning, spatial gradation within a component is the rule rather than the exception. All of the responses of Fig. 21.2 are so graded and recall the graded responses obtained with electrical stimulation of motor fields and single units. But as these are multi-unit responses the graded nature cannot be due to peripheral mechanisms alone. There must also be gradients in the strength of coupling between the central neurones (motor neurones or their co-ordinating interneurones) comprising a component. Fluorescent dye-coupling between motor neurones in the chromatophore lobes of cuttlefish, squid, and octopus, recently demonstrated by Ferguson *et al.* (unpublished; see also Chapter 25) provides an anatomical basis for multi-unit responses, and presumably also for electrical modulation of coupling.

Evidence for central coupling of another kind, which lines up motor units according to one or other of the main axes of the body, can be seen in the mutually exclusive transversely and longitudinally organized patterns of some species (see figs 32 and 33 in Packard and Hochberg 1977).

Pharmacological or other manipulations of whole animals have confirmed the physiological discreteness of a number of components in octopus (Table 21.2; see also Andrews *et al.* 1983). Those treatments acting centrally have the advantage that they reveal the full spatial extent of a component.

However, the commonest category of colour effect—the umber or neutral density shading supplied by screens of differentially expanded chromatophores—is far from discrete and has so far defied analysis (see below).

Reciprocal innervation of brightness contrast units (Fig. 21.18)

Even a casual observer will notice that cephalopods are able to modulate the efferent (chiaroscuro) information produced by a single feature generator

Table 21.2 Separate components (classes of motor units) revealed experimentally in *Octopus vulgaris* (see text)

Component	System	Revealed by:	Peripheral/ central	Present in Decapods
Black features	A	Urethane	C	Squid, cuttlefish
Orange-brown screen	B	Ethyl alcohol	C	
Papilla coral tips	B	Nicotine	C	
Flashing brown spots	B	Cold, Flash	C	
Injury mottle (groove system only)		Nerve trauma	P	
Polka dots	A	Carbachol, ACh (local effect)	P	
All-dark		Optic stimulation	C	Yes
Reverse countershading		Postural reflex	C	Cuttlefish
Acute mottle	A/B	Serotonin	C	

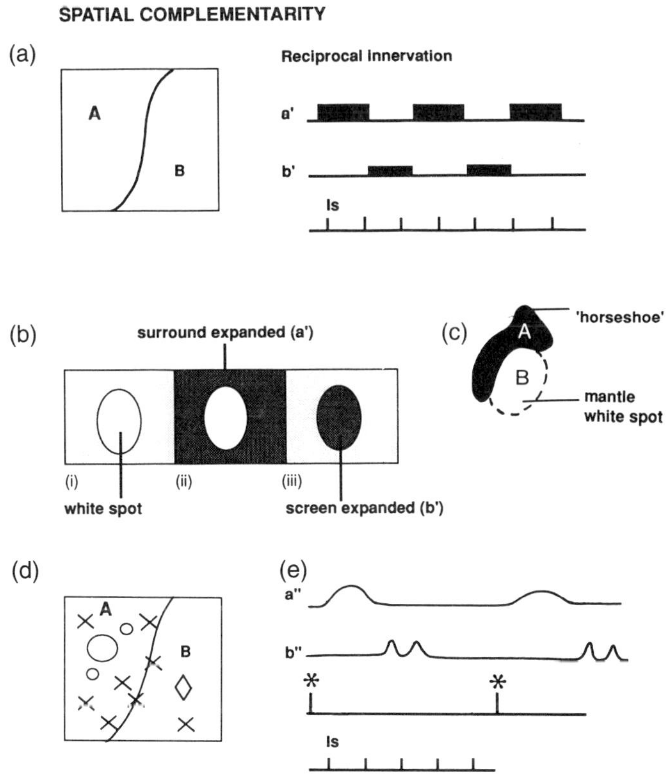

Fig. 21.18. Reciprocal innervation. (a) Spatially complementary areas A and B, laid down early in development (B being subordinate to A), receive reciprocal innervation, which expresses itself as inhibition b′ of units during excitation of a′—for instance between 'Maynard units' on the cuttlefish (a) and (b), and between the 'horseshoe' and enclosed mantle 'white spot' of the octopus (c). (e) A phasic unit in the latter (b″ 'flashing brown spot') is released from inhibition following excitation of a″ by a strong stimulus (∗) (flash to the eyes). (d) A and B areas later become morphogenetically further subdivided (open symbols): (X) hypothetical units originating before division of the field into A and B. The time markers in (a) and (e): 1 s.

such as the dark components of the mottle of octopus, the 'zebra' stripes which come and go on the flank of a cuttlefish or the 'spots' on the mantle of a squid [Figs 21.2(a) and 21.17(b)]. The change in contrast amplitude is brought about in two ways: (1) only the amplitude of the dark component alters, through variations in the activity of the excited units, without change in the baseline grey level of adjacent areas (Fig. 21.8); and (2) reciprocal inhibition of the spatially complementary non-excited units notably of the screens over the 'white' components by their adjacent dark components (i.e. lateral inhibition).

Excitation (or suppression) of one set of units producing suppression (or excitation) of another—much as in the reciprocal innervation between flexors and extensors generating patterns of movement and posture—is seen in the small complementary units on the fin of *Sepia*, here called 'Maynard units' (Fig. 21.11), and in some of the components of Table 21.2. Indeed the alcohol-sensitive and cold-sensitive components shown in Table 21.2 only came to light through the release of central inhibition resulting from the treatment. With ethyl alcohol a plane of anaesthesia is reached which suppresses the dark components of A areas in octopus and their equivalent, the dark 'rings', in the squid, but evokes reciprocal expression of the screens over the normally suppressed areas (i.e. B areas in octopus, white fin spots and white of the 'zebra' in *Sepia* and the intervals between 'rings' in *Loligo*). The effect—a kind of negative of the normal—is never seen in untreated animals and does not occur in denervated skin. Cold has a somewhat similar effect in releasing from inhibition the command neurones of a phasic component the FBS (Andrews *et al.* 1983) located within the 'white spots' of B areas of *Octopus vulgaris* [Fig. 21.17(d)]. When a temperature of 4–5 °C is reached at which the dark components of A areas ('arm bars' and 'horseshoe') are suppressed, the brown spots flash twice (*cf.* Williamson and Roberts 1982). The reciprocal relationship of these phasic units with neurones controlling darkening is confirmed by the fact that FBS also appear as a rebound ('off') effect of the darkening response to electronic flash [Fig. 21.18(e)]. The central neurones responsible for the spatially complementary (lateral) inhibition appear to be in the lateral basal lobes (see below).

Functional levels in the brain (Table 21.3)

This subject (Boycott, 1953) was first investigated by German authors in the latter part of the last century (see von Uexkuell 1894). Four brain centres or levels are known to be involved in the control of colour patterns: (1) optic lobes; (2) peduncle lobes; (3) lateral basal lobes; (4) chromatophore lobes (see Fig. 21.16). But it is not known precisely at what level in the brain reciprocal inhibition and co-ordination of banks of motor units (components) take place.

Because the eyes, and therefore the optic lobes, clearly play an important part in selecting patterns and modifying them in association with events taking place in the visual environment there has been a tendency to think of patterns as residing in the optic lobes. Stimulation of these lobes produces them (Boycott 1961; Chichery and Chanelet 1976). But recent findings from long-term brain-lesioned octopuses (Packard and Young, in preparation) suggest that their role is limited to selection, modulation (especially of high spatial frequencies), and gain control.

Days after the optic tracts have been severed, eliminating levels 1 and 2, octopuses exhibit very dramatic and sustained patterns including some so-called 'acute' ones from the total repertoire, that are rarely and only transiently seen in intact animals and then usually in juveniles (Packard and

Table 21.3 Central nervous integration of chromatophore system: (1) and (2) Optic and peduncle lobe(s); (3) lateral basal lobe(s); (4) chromatophore lobe(s)

PATTERN CONTROL	EXCITATORY
Selects patterns Modulates —high spatial frequencies —low spatial frequencies —spatial gradation Controls amplitude (screens) L/R co-ordination and unilateral projection	Pattern generator Integrates components, (colours, textures, and postures) Generates contrast Reciprocal inhibition Sustains motor tone (all-or-none)
(1) and (2)	(3)
	(4) Generates components Co-ordinates anterior/posterior Synchronizes units within a component

Sanders 1971). They are characterized by exaggerated posture (hypertonia) and lack of modulation—stiffness, maximum contrast amplitude, and maximum extension of papillae, etc.—recalling the decerebrate rigidity of mammals after forebrain removal. Sometimes the activity is expressed only unilaterally.

Evidently removal of levels 1 and 2 does not remove the ability to produce the static combinations of components called 'patterns', the circuitry and excitatory output for which must therefore be located in levels 3 and 4. Much as in the mammalian basal ganglia, patterns generated at level 3 are normally inhibited from expression (or damped) by levels 1 and 2. Removal of 1 and 2 removes amplitude control (regulation of tone) and the ability to switch patterns off and on. Bilateral co-ordination is impaired.

The relative roles of 3 and 4 have been studied by removing or undercutting the supra-oesophageal brain. The octopus becomes inert and pale. Motor tonus can become so low that slow waves (patches of spontaneously expanded chromatophores) develop in the skin as if it were denervated. But occasionally—either autonomously or if vigorously stimulated—these animals express complete components, unilaterally or bilaterally, both on the arms and the mantle, and sometimes in exaggerated form. Textural components are separated from their usual combinations with other components. Evoked responses from a 'suboesophageal' animal include the dark component of the acute mottle (but not the 'white spots' produced by lateral inhibition), and an expression of primary and secondary papillae called 'bossed'. When the brain is challenged pharmacologically only some of the responses seen in normal animals

(Table 21.2) can be evinced (including the 5HT-mottle followed by deep red, Andrews *et al.* 1983) and only at high thresholds.

The chromatophore lobes must therefore contain the coupled banks of motor units of a single class called components, both within and between subdivisions of the body (left/right, anterior/posterior), although not apparently the circuitry for combining components into normal 'patterns' (nor for reciprocal inhibition). In the absence of higher centres, the banks of motor units at this level are normally switched off and tonic activity is minimal, either because they are not activated from above or because of inhibitory activity within the lobes. So long as some motor tone is present, however, chromatophore nerves issuing from the chromatophore lobes retain their power to keep the skin of a 'suboesophageal' animal relaxed (see above) and electrical stimulation post-mortem gives entirely normal responses even though the skin has been pale for several weeks.

Apparently, level 3 is an excitatory centre which also co-ordinates single components (located in 4) into a whole pattern and activates the lateral inhibition between reciprocally innervated components that heightens brightness contrast.

One of the functions of levels 1 and 2 is evidently to modulate the extent of coupling between banks of motor units that produce a given component. The proximodistal and mediolateral gradients in the expression of 'chronic' patterns typical of normal animals suggest there are fixed mechanisms for producing regular gradients in the extent of coupling within a bank. This too appears to be a function of the higher lobes, altering the strength of coupling within the chromatophore lobes. Table 21.3 summarizes these findings.

Details of the organization of the chromatophore lobes are becoming better known. The finding that cell bodies of motor neurones innervating the mantle and the arms, respectively, are located in the anterior, posterior, and controlateral lobes (Dubas *et al.* 1986*b*) helps to account for the strong coupling—in the engineering sense—between like components in the mantle and arms. Miyan and Messenger (see Chapter 25) have described interneurones in the octopus chromatophore lobe, but it is not known whether these are inhibitory. The chromatophore lobe undergoes considerable differentiation during ontogeny and its cortex comes to be organized in columns (Young 1971) separated by bundles of fibres from and to the neuropil (Froesch 1972) which have been likened to the columns of the mammalian cortex (Young 1971). Ever since von Uexkuell (1894) it has been a source of interest that the motor patterns of cephalopod colour proceed without internal feedback from the periphery, and there may be a similar lack of internal feedback centrally. Although ascending fibres have been found between levels 3 and 1 (*Lolliguncula*; Novicki *et al.* 1990) none have been found between levels 4 and 3.

The role of the eyes in darkening the skin—raising the grey-level by increasing the amplitude of expression of dark components already active—when stimulated by electronic flash has been mentioned above. It is likely that the reflectance matching between object (i.e. animal) and background seen in camouflaged animals (Messenger 1974)—requiring measurement of the albedo (ratio of reflected to downwelling light)—is obtained by the same mechanism (ratio of ventral to dorsal retinal stimulation) as in flatfish. Partial denervation of optic nerves—removing the contribution from dorsal retina by cutting above the chiasma—produces an animal which remains darker than the background during the days following operation (Packard, unpublished).

Visual elements and visual units

In view of the claims of this chapter it is worth summarizing the elementary properties that the skin, and some of its feature-generating units (driven by motor neurones and their switches at higher levels in the brain), have in common with retinal receptors and units of the vertebrate visual system (feature-detecting ganglion cells or cells of the visual cortex or optic tectum).

1. Colour selectivity (and probably also separate colour and monochrome channels).
2. Single units connecting with a large number of primary peripheral elements of a given class.
3. Units characterized by their patterns of connection at the periphery (i.e. by their shapes, often circular).
4. Separate units that share one and the same element at the periphery.
5. Brightness contrast effects/sensitivities of units sinusoidal rather than square wave.
6. Some transient units characterized by 'off' responses.
7. Units of similar type linked at higher levels to generate features with characteristic orientation, etc.—for example bars running transversely or horizontally.
8. Feature-generators/feature-detectors, that fall into one of several separate spatial frequency channels.
9. Stereo-units (raising papillae, etc.).

Although lateral inhibition (in the electrophysiological sense) between units innervating adjacent fields undoubtedly occurs centrally (see below)—recalling suppression in the striate cortex between complementary receptive fields (Kim *et al.* 1993)—there is no clear evidence that it occurs peripherally or within a unit.

Theoretical considerations: chronomers

The idea that a form owes its characteristics to the manner in which it grows must be as old as biology itself. The finding that components, or classes of motor unit, are successions of developmental classes which have pushed up

UNIVERSITY
COLLEGE LONDON
LIBRARY

through each other acquiring their positions, shapes, sizes, and other characteristics in large part from local conditions established by those that have gone before (and, in turn, modifying the latter) is the essential key for understanding the surface complexities of the patterning systems of cephalopods. (In a sense the spatial complexities are trivial.) Some years ago I coined the word 'chronomer' (Packard 1982) for a unit of form whose shape is a result of the point in time (as well as in space) at which it arises. The succession of rings of a fish scale are chronomers. The 'flame' on the flank of an adult squid (Fig. 21.6) is a chronomer and so is the vertical white 'zebra' stripe (Fig. 21.19) on the mantle and arms of a male cuttlefish.

Although we do not know the developmental sequence in the brain, the chromatophore lobes grow considerably in thickness and complexity in the juvenile adding new motor neurones and other cells. The original sketch illustrating the chronomer concept (Fig. 21.20) shows motor neurones growing out and establishing connections with each new generation of chromatophore. It is the conditions that the nerve fibres encounter at the periphery which determine the shapes of motor units.

One should not expect a topological map of cell bodies in the brain corresponding to locations in the periphery. Even though some kind of isomorphism presumably exists, one area of skin houses many generations of chromatophore and many components one on top of another, and most components

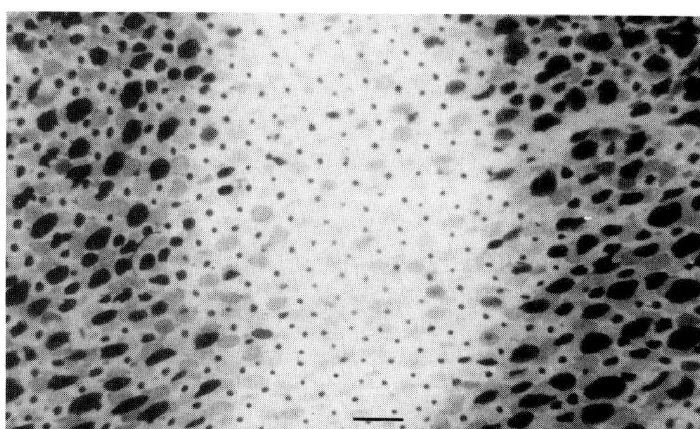

Fig. 21.19. Separate classes of chromatophore muscle fibre. Detail of 'zebra' component on the mantle of adult cuttlefish (*Sepia officinalis*) several hours post-mortem generated by tonic contraction of the muscles of some melanophore (black) and orange (grey) chromatophore classes (expanded) to right and left, and by complete relaxation of the muscle fibres of the melanophores and most orange chromatophores (retracted) over the central strip. (Brightness contrast enhanced by presence of leucophore material in the central strip.) Retracted orange and yellow chromatophores not visible in this photograph. Scale bar: 0.2 mm.

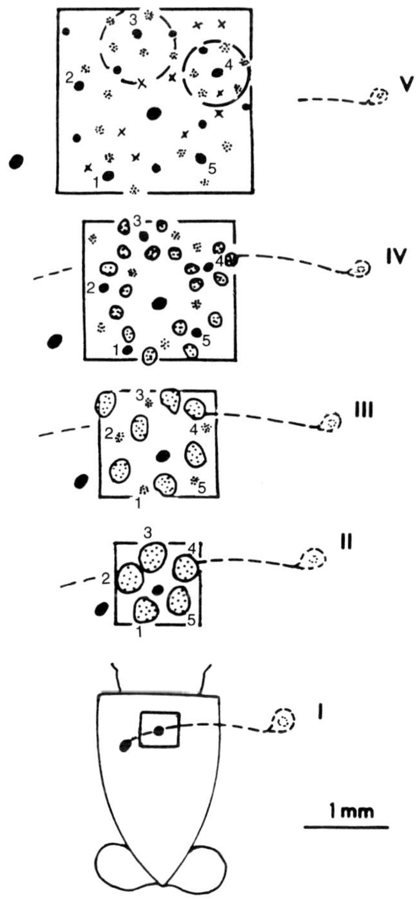

Fig. 21.20. Sketch of the developing skin of a squid to illustrate the chronomer concept. The placings of each new generation of chromatophore and the shapes of the motor units innervating them (*I–V*) are determined by the positions of preceding generations. As recruitment is by interdigitation, these chronomers are also intercalary. (From Packard 1982, with permission.)

(whether feature or screen) are a distributed global category tightly co-ordinated centrally; which is why dye taken up from a discrete site in the periphery and transported centrally (Dubas *et al.* 1986*b*) reveals cell bodies not in one but in several locations in the chromatophore lobes.

What of the physiological and pharmacological characteristics of chronomers? It turns out that these too may be the result of arrangements at the periphery rather than the centre. The 'zebra' stripes of the adult cuttlefish are responsible for some of the highest brightness contrast amplitude observable in cephalopods and their fleeting expression by the male during courtship

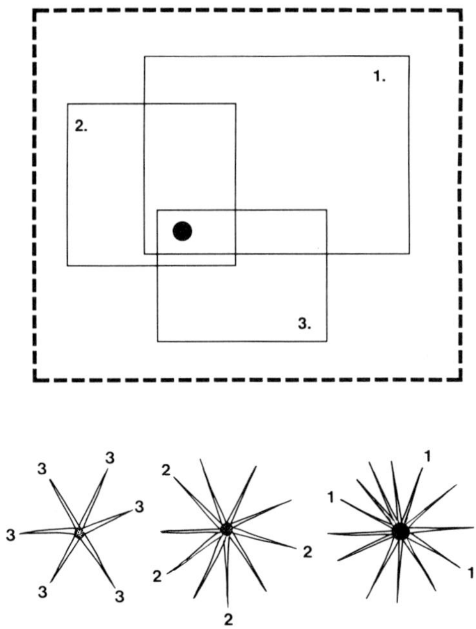

Fig. 21.21. The spot in this diagram is a chromatophore belonging to three motor units (1–3) of decreasing size and age. A theoretical developmental history of this chromatophore which starts yellow and acquires innervation from the latest motor unit (3) and further muscles, as it matures, innervated by units 2 and 1, is shown below. The position and rate of development of the chromatophore are assumed to be determined by morphogens in its immediate surroundings, and by field effects (dashed line) operating before it was born.

involves centrally controlled reciprocal innervation (see above). When 'zebra' is expressed most chromatophores in the 'dark' segments expand, and all size and colour classes of chromatophores lying in the 'white' stripe retract. Yet, in conditions promoting tonus, this same effect can be obtained when the influence of the nervous system has been abolished—for instance if the skin is denervated or is removed from the body and left to dry (Fig. 21.19). This must mean that the biophysical characteristics of the tonically contracted muscle fibres contributing the 'black' part of 'zebra' are distinct from those of the relaxed ones over the 'white' zones. (Local modulatory influences can probably be ruled out because there are also many retracted chromatophores in the 'black' zones.) The tonus is difficult to abolish even in the muscle relaxant magnesium chloride. In keeping with their developmental history, the mean resting sizes of the melanophore population over the 'white' is slightly smaller than over the 'black' (see above).

A theoretical scheme of how a given chromatophore, which starts yellow and with its own innervation, may acquire, at a decreasing rate, innervation in common with older (slightly larger and darker) chromatophores as it in turn becomes older (and darker) is shown in Fig. 21.21. It takes into account the information mentioned above under 'multiple innervation' and implies some kind of competition between units.

The dynamics of coupling that enable this spot to contribute to the whole have been barely touched on. As the subject is ripe for investigation, squid workers might in future think twice before discarding that skin.

References

Andrews, P. L. R., Packard, A., and Tansey, E. M. (1982). A physiologically discrete population of chromatophores in *Octopus vulgaris* (Mollusca). *Journal of the Zoological Society of London*, **198**, 131–40.

Andrews, P. L. R., Messenger, J. B., and Tansey, E. M. (1983). The chromatic and motor effects of neurotransmitter injection in intact and brain-lesioned *Octopus*. *Journal of the Marine Biological Association of the United Kingdom*, **63**, 355–70.

Boycott, B. B. (1953). The chromatophore system of cephalopods. *Proceedings of the Linnaean Society of London*, **164**, 235–40.

Boycott, B. B. (1961). The functional organisation of the brain of the cuttlefish *Sepia officinalis*. *Proceedings of the Royal Society of London, Series, B*, **153**, 503–34.

Bozler, E. (1931). Ueber die Taetigkeit der einzelnen glatten Muskelfaser bei der Kontraktion, 3. *Zeitschrift für vergleichende Physiologie*, **13**, 762–72.

Bozler, E. (1948). Conduction, automaticity, and tonus of visceral muscles. *Experientia*, **4**, 213–18.

Chichery, R. and Chanelet, J. (1976). Motor and behavioural responses obtained by stimulation with chronic electrodes of the optic lobes of *Sepia officinalis*. *Brain Research*, **105**, 525–32.

Dahms, V., Prosser, C. L., and Suzuki, N. (1986). Two types of 'slow waves' in intestinal smooth muscle of cat. *Journal of Physiology*, **392**, 51–69.

Dubas, F. (1987). Innervation of chromatophore muscle fibres in the octopus *Eledone cirrhosa*. *Cell and Tissue Research*, **248**, 675–82.

Dubas, F. and Boyle, P. R. (1985). Chromatophore motor units in *Eledone cirrhosa* (Cephalopoda: Octopoda). *Journal of Experimental Biology*, **117**, 415–31.

Dubas, F., Hanlon, R. T., Ferguson, G. H., and Pinsker, H. M. (1986*a*). Localization and stimulation of chromatophore motoneurones in the brain of the squid, *Lolliguncula brevis*. *Journal of Experimental Biology*, **121**, 1–25.

Dubas, F., Leonard, R. B., and Hanlon, R. T. (1986*b*). Chromatophore motoneurons in the brain of the squid *Lolliguncula brevis*. *Brain Research*, **374**, 21–9.

Ferguson, G. P., Martini, F. M., and Pinsker, H. M. (1988). Chromatophore motor fields in the squid, *Lolliguncula brevis*. *Journal of Experimental Biology*, **134**, 281–95.

Florey, E. (1966). Nervous control and spontaneous activity of the chromatophores of a cephalopod *Loligo opalescens*. *Comparative Biochemistry and Physiology*, **18**, 305–24.

Florey, E. and Kriebel, M. E. (1969). Electrical and mechanical responses of the chromatophore muscle fibres of the squid, *Loligo opalescens*, to nerve stimulation and drugs. *Zeitschrift für vergleichende Physiologie*, **65**, 98–130.

Froesch, D. (1972). The synaptic profiles in the posterior chromatophore lobe of the *Octopus* brain: a study in normal and degenerated tissue. *Brain Research*, **48**, 1–9.

Froesch, D. (1973*a*). Projection of chromatophore nerves on the body surface of *Octopus vulgaris*. *Marine Biology*, **19**, 153–5.
Froesch, D. (1973*b*). On the fine structure of the *Octopus* iris. *Zeitschrift für Zellforschung*, **145**, 119–29.
Froesch, D. and Packard, A. (1979). Octopus chromatophores accumulate nickel. *Experientia*, **35**, 828–9.
Froesch-Gaetzi, V. and Froesch, D. (1977). Evidence that chromatophores of cephalopods are linked by their muscles. *Experientia*, **33**, 1448–50.
Gillespie, J. S. (1972). The rat anococcygeus muscle and its response to nerve stimulation and some drugs. *British Journal of Pharmacology*, **45**, 409–16.
Hanlon, R. T. (1982). The functional organisation of chromatophores and iridescent cells in the body patterning of *Loligo plei* (Cephalopoda: Myopsida). *Malacologia*, **23**, 83–119.
Hanlon, R. T. and Messenger, J. B. (1988). Adaptive coloration in young cuttlefish (*Sepia officinalis* L.): the morphology and development of body patterns and their relation to behaviour. *Philosophical Transactions of the Royal Society of London*, **320**, 437–87.
Harless, E. (1846). Untersuchung über die Chromatophoren bei *Loligo*. *Archiv für Naturgeschichte*, **1**, 34–44.
Hofmann, F. B. (1910) Gibt es in der Muskulatur der Molluske periphere, kontinuerlich leitende Nervennetze bei Abwesenheit von Ganglienzellen? *Archiv für die Gesamte Physiologie*, **132**, 43–81.
Kim, J.-N., Forthergill, L. K., and Hammond, P. (1993). Excitatory and suppressive receptive fields of striate cortical neurones in the anaesthetized cat. *Journal of Physiology*, **459**, 133P.
La Gioia, E. (1992). Un sistema per la valutazione quantitativa di caratteristiche morfologiche e densitometriche in pattern naturali. Tesi di Laurea in Ingegneria Elettronica, Università di Napoli Federico II, gennaio 1992. (Unpublished thesis, available on request).
Maynard, D. M. (1967). Organization of central ganglia. In *Invertebrate nervous systems. Their significance for mammalian physiology* (ed. C. A. G. Wiersma), pp. 231–55. University of Chicago Press.
Meinhardt, H. (1982). *Models of biological pattern formation*. Academic Press, London.
Meinhardt, H. and Gierer, A. (1974). Applications of a theory of biological pattern formation based on lateral inhibition. *Journal of Cell Science*, **15**, 321–46.
Messenger, J. B. (1974). Reflecting elements in cephalopod skin and their importance for camouflage. *Journal of Zoology*, **174**, 387–95.
Moynihan, M. and Rodaniche, A. F. (1982). The behavior and natural history of the Caribbean reef squid *Sepioteuthis sepioidea*. *Advances in Ethology*, **25**, 1–125.
Muneoka, Y. and Twarog, B. M. (1983). Neuromuscular transmission and excitation–contraction coupling in molluscan muscle. In *The Mollusca*, Vol. 4, *Physiology, Part 1* (ed. A. S. M. Saleuddin and K. M. Wilbur), pp. 35–76. Academic Press, London and New York
Nicaise, G. and Amsellem, J. (1983). Cytology of muscle and neuromuscular junction. In *The Mollusca*, Vol. 4, *Physiology, Part 1* (ed. A. S. M. Saleuddin and K. M. Wilbur), pp. 1–33. Academic Press, London and New York
Novicki, A., Budelmann, B. U., and Hanlon, R. T. (1990). Brain pathways of the chromatophore system in the squid *Lolliguncula brevis*. *Brain Research*, **519**, 315–23.
Packard, A. (1974). Chromatophore fields in the skin of the octopus. *Journal of Physiology*, **238**, 38–40P.

Packard, A. (1982). Morphogenesis of chromatophore patterns: are morphological and physiological units the same? *Malacologia*, **23**, 193–201.
Packard, A. (1983). 'Visual units' in the skin of *Octopus*. *Journal of Physiology*, **346**, 34P.
Packard, A. (1985). Sizes and distribution of chromatophores during post-embryonic development in cephalopods. *Vie et Milieu*, **35**, 285–98.
Packard, A. (1988). Visual tactics and evolutionary strategies. In *Cephalopods—present and past*, J. Wiedmann and J. Kullmann, (ed.) pp. 89–103. Schweizerbart'sche Verlagsbuchhandlung, Stuttgart.
Packard, A. (1990*a*). Advantages (and disadvantages) of octopus skin for smooth muscle physiology: a protocol for visual recording *in vivo*. *Journal of Physiology*, **427**, 57P.
Packard, A. (1990*b*). Four-dimensional analysis of patterned activity in a complex smooth muscle system: octopus skin. *Journal of Physiology*, **427**, 58P.
Packard, A. (1991*a*). Uses of nicotine to follow denervation supersensitivity in unilaterally denervated octopus *in vivo*. *Journal of Physiology*, **438**, 325P.
Packard, A. (1991*b*). Spatial entrainment of denervated by innervated muscle fibres in a paramedian area of overlapping innervation (octopus skin). *Journal of Physiology*, **438**, 326P.
Packard, A. (1992). Nerve-induced and myogenically propagated relaxation of skin muscle (*Octopus vulgaris*). *Journal of Physiology*, **452**, 130P.
Packard, A. (1993). Some responses of *Octopus* chromatophores to light. *Journal of Physiology*, **459**, 429P.
Packard, A. and Hochberg, F. G. (1977). Skin patterning in *Octopus* and other genera. In *The biology of cephalopods*. *Symposium of the Zoological Society of London*, no. 38 (ed. M. Nixon and J. B. Messenger.), pp. 191–231. Academic Press, London and New York.
Packard, A. and Sanders, G. D. (1971). Body patterns of *Octopus vulgaris* and maturation of the response to disturbance. *Animal Behaviour*, **19**, 780–90.
Porteiro, F. M., Martins, H. R., and Hanlon, R. T. (1990). Some observations on the behaviour of adult squids, *Loligo forbesi*, in captivity. *Journal of the Marine Biological Association of the United Kingdom*, **79**, 459–72.
Sanders, G. D. and Young, J. Z. (1974). Reappearance of specific colour patterns after nerve regeneration in *Octopus*. *Proceedings of the Royal Society of London, Series B*, **186**, 1–11.
von Uexkuell, J. (1892). Physiologische Untersuchungen an *Eledone moschata*. *Zeitschrift fur Biologie*, **28**, 550–66.
von Uexkuell, J. (1894). Physiologische Untersuchungen an *Eledone moschata*, IV Zur Analyse der Functionen des Centralnervensystems. *Zeitschrift für Biologie*, **31**, 584–609.
Williamson, R. M. and Roberts, B. L. (1982). Body cooling as a supplement to anaesthesia for fishes. *Journal of the Marine Biological Association of the United Kingdom*, **69**, 129–31.
Young, J. Z. (1971). *The anatomy of the nervous system of Octopus vulgaris*. Clarendon Press, Oxford.

22

Neurotransmitters of squid chromatophores

C. J. CORNWELL and J. B. MESSENGER

Summary

Squid chromatophores expand as a result of the contraction of a set of 15–25 radial muscles set around a sac containing pigment granules. Running along the surface of each radial muscle are excitatory nerve fibres that arise directly from cells lying in the brain. There appear to be no inhibitory fibres. The hypothesis that L-glutamate may be an excitatory transmitter at the neuromuscular junctions on the radial muscles has been confirmed by the application of a series of newly developed specific glutamate receptor agonists to the skin of the small loliginid squid, *Alloteuthis subulata*. All the non-NMDA receptor agonists tested on *Alloteuthis* expanded the chromatophores. More importantly, immunohistochemical staining of the skin with antibodies raised against L-glutamate shows that this putative transmitter is endogenous in the radial muscle nerves. Serotonin (5-hydroxytryptamine, 5HT) has long been known to affect the responses of chromatophores to nerve stimulation, and to cause paling of the skin (radial muscle relaxation) when applied topically. Similar effects have now been obtained with 5HT receptor agonists and immunohistochemical staining reveals that 5HT is also endogenous in the radial nerves. It remains to be established whether these two substances co-exist in the same nerve fibres. The mode of action of 5HT is not clear.

Surely nobody who has ever seen a living cephalopod can have failed to marvel at the beautiful skin, with its abrupt changes of colours and patterns. Such behaviour is possible because the skin contains thousands, sometimes many tens of thousands of chromatophores, under the direct control of the brain. Our understanding of their structure and function effectively dates from the late 1960s, when Florey and his collaborators, using *Loligo opalescens*, described their ultrastructure for the first time and explored their physiology using modern techniques (Florey 1966, 1969; Cloney and Florey 1968; Florey and Kriebel 1969). Each chromatophore comprises several different cell types: the chromatophore proper, containing pigment granules in an elastic sac; a set of 15–25 separate radial muscle fibres; nerves associated with these fibres; glial cells; and sheath cells.

Squids change colour by expanding and retracting the chromatophores. Axons from cell bodies within the brain directly innervate the chromatophore

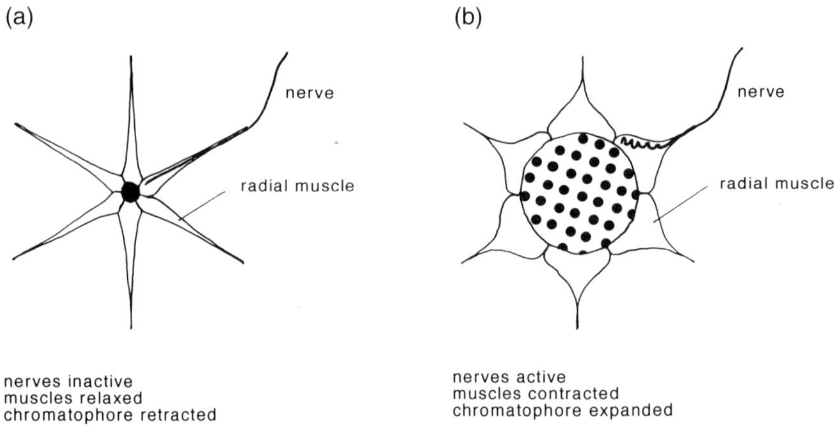

Fig. 22.1. Diagram of a squid chromatophore organ retracted (a) and expanded (b). For simplicity only six radial muscles are shown and only a single nerve fibre on one of them (see text). Note how the nerve folds when the muscle is contracted.

radial muscles (Sereni and Young 1932). When the nerves are active, the radial muscles contract and the chromatophore expands; when they are inactive, energy stored in the elastic sac causes the chromatophores to retract as the muscles relax (Fig. 22.1). Expansion of the chromatophores reveals more pigment and thus alters the quality of light reflected from the skin: the change in area of pigment exposed, from fully retracted to fully expanded, can be 80-fold or more. Squids use their chromatophores primarily for concealment, in particular for countershading (Cott 1940; Ferguson and Messenger 1991), and it is important to recall that in order to provide effective camouflage the chromatophores must be continuously active (see Discussion).

One of the questions left unanswered by Florey's original study was the nature of the transmitter at the neuromuscular junction of the chromatophore. Subsequently, however, Florey *et al.* (1985), acting on a suggestion by Bone and Howarth (1980), provided compelling evidence that it could be L-glutamate. This chapter presents two kinds of evidence to support this hypothesis: pharmacological and immunohistochemical.

We studied the chromatophores of the loliginid squid, *Alloteuthis subulata*, a small, coastal species (Rodhouse *et al.* 1988) with comparatively large chromatophores (Figs 22.2 and 22.3), whose skin is peculiar in that the dermis can be easily removed to allow access of drugs, antibodies and stains.

We first used supravital methylene blue and silver nitrate staining (Bielschowsky method; see Drury and Wallington 1967) to confirm that the morphology and innervation of the chromatophores was like that of *Loligo*. The results of this staining were in agreement with the descriptions of Hofmann (1907), Weber (1968), and Cloney and Florey (1968). Like these authors we

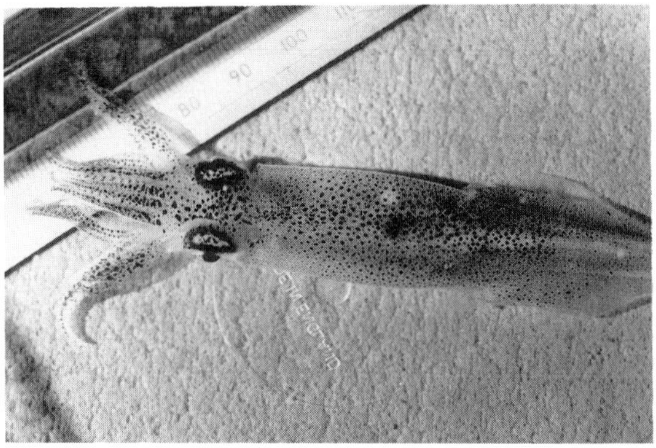

Fig. 22.2. Living *Alloteuthis* showing the chromatophores all over the arms, head, and mantle. Scale is in millimetres.

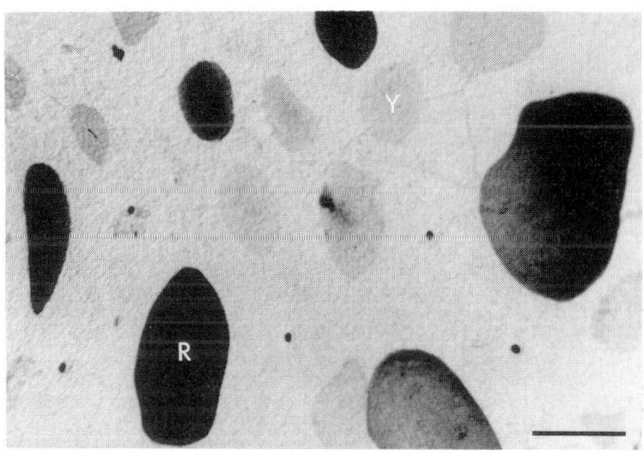

Fig. 22.3. Chromatophores at higher magnification. R, red; Y, yellow chromatophore. Scale bar: 1 mm.

saw clear evidence of multiple innervation: there were two to six nerve fibres running along each radial muscle (Figs 22.4 and 22.5).

Pharmacology

Florey's early physiological investigations of the chromatophores (Florey 1966; Florey and Kriebel 1969) included a study of the effects of two putative neurotransmitters, acetylcholine (ACh) and serotonin (5-hydroxytryptamine, 5HT), which at that time had been implicated in the action of another molluscan muscle, the *Mytilus* anterior byssus retractor muscle (ABRM; Twarog 1960). Florey (1966) showed that, in *Loligo opalescens*, ACh caused

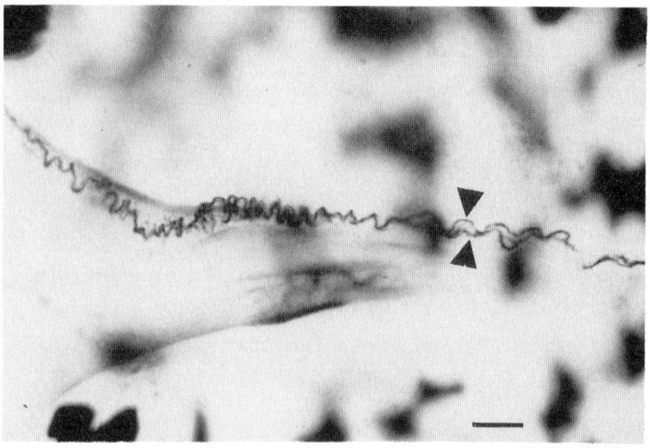

Fig. 22.4. Chromatophore muscle with two nerves (arrows). Silver staining. Scale bar: 1 μm.

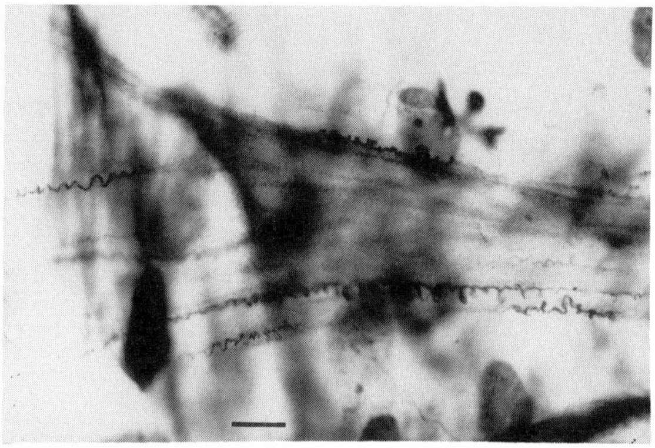

Fig. 22.5. Multiple innervation in a radial muscle (some fibres are out of the plane of focus). Silver staining. Scale bar: 1 μm.

the radial muscles to contract and the chromatophores to expand, although there were often some pulsations as well. 5HT facilitated both the contraction and relaxation of the radial muscles induced by nerve stimulation. Tetrodotoxin abolished the effects of nerve stimulation but did not interfere with the action of ACh and 5HT when they were applied topically. By further careful experiments Florey and Kriebel (1969) also established: (1) that ACh is not itself the natural transmitter but acts by causing the release of some other transmitter, at that time unknown; and (2) that there was no inhibitory innervation of the radial muscles. They suggested that 5HT was probably acting intracellularly within the muscle fibre.

In 1980 Bone and Howarth showed that L-glutamate expands chromatophores of *Alloteuthis* and other cephalopods and established this amino acid as a prime candidate for being an endogenous transmitter at the neuromuscular junction on the radial muscles. Florey *et al.* (1985) extended these observations using the squid, *Lolliguncula brevis*. They found that glutamate and the glutamate receptor agonist kainate caused the radial muscles to contract (and hence the chromatophores to expand), and that several glutamate receptor antagonists blocked nerve-induced and glutamate-induced contractions.

Since 1985 many new drugs active at glutamate receptors have been developed (see Chapter 19 and Messenger 1991) and we took advantage of this to study *Alloteuthis* chromatophores. By making a hole in the dermis we could expose some chromatophores to solutions of drugs flooded on to the underside of the skin. Two such experiments are shown in Figs 22.6 and 22.7 and the overall results are listed in Table 22.1. In summary, they show that glutamate and non-*N*-methyl-D-aspartic acic receptor agonists elicit chromatophore expansion when topically applied to *Alloteuthis* skin, and that the receptors would be classified 'kainate' or 'quisqualate' if this were a mammal (Monaghan *et al.* 1989; see also Discussion and Chapter 19).

Table 22.1 Effect of L-glutamate agonists on the chromatophores

Agonist	Active?	Threshold	Inactive (maximum concentration)
Alloteuthis subulata			
L-glutamate	yes	2×10^{-4} M	
Kainate	yes	2×10^{-4} M	
AMPA	yes	5×10^{-4} M	
Bromowillardiine	yes	5×10^{-5} M	
Quisqualate	yes	2×10^{-5} M	
L-ODAP	yes	2×10^{-5} M	
Domoate	yes	5×10^{-7} M	
NMDA	no		1×10^{-2} M
D-glutamate	no		1×10^{-2} M
L-, D-aspartate	no		1×10^{-2} M
Lolliguncula brevis			
L-glutamate	yes	1×10^{-4} M	
Kainate	yes	1×10^{-4} M	
D-glutamate	no		1×10^{-2} M
Loligo vulgaris			
L-glutamate	yes	$1–2 \times 10^{-4}$ M	
Kainate	yes	$2–5 \times 10^{-4}$ M	

AMPA, (RS)-α-amino-3-hydroxy-5-methyl-4-isoxazolepropionic acid.
L-ODAP, β-*N*-oxalyl-D-α,β-diaminopropionic acid.
NMDA, *N*-methyl-D-aspartic acid.

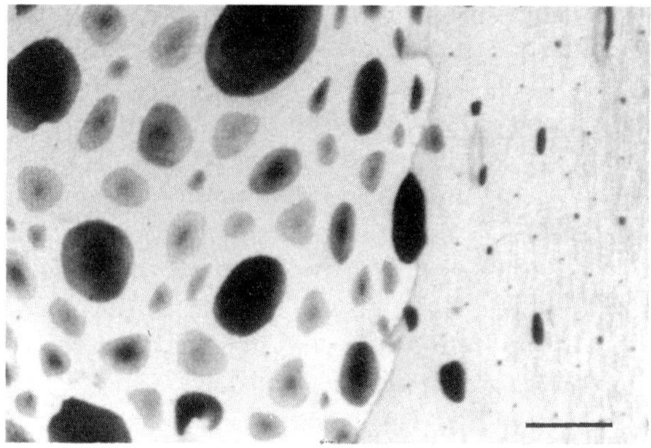

Fig. 22.6. Exposed chromatophores expand after the skin is flooded with L-glutamate (left, 5×10^{-4} M); controls (right) remain retracted. Scale bar: 1 mm.

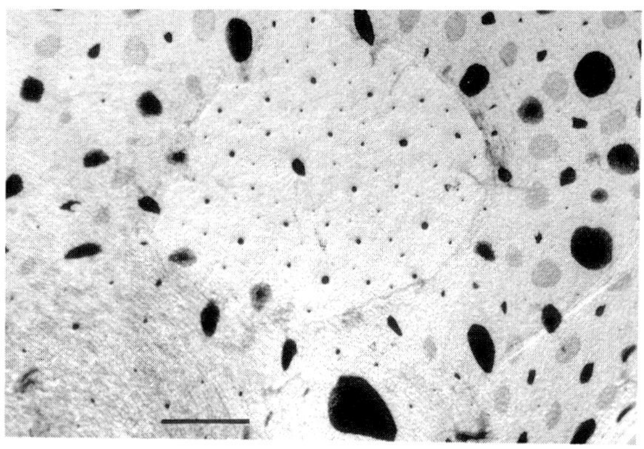

Fig. 22.7. Tonically expanded chromatophores exposed to 5HT (centre, 1×10^{-6} M) retract; controls (surround) remain expanded. Scale bar: 1 mm.

Florey (1966) and Florey and Kriebel (1969) also investigated the action of 5HT on the chromatophores of *Loligo opalescens*. They found that it relaxed tonically contracted radial muscles but proposed that 5HT was acting intracellularly (see Discussion). *Alloteuthis* and *Sepia* chromatophores also respond to 5HT; tonically expanded chromatophores in ageing preparations retract instantly when it is applied. Retraction also follows applications of 8-OH-DPAT (DPAT = (±)-8-hydroxy-dipropylaminotetralin HBr (Research Biochemicals International, Natick, MA, USA)] and α-methyl-5HT. These are $5HT_1$ and $5HT_2$ receptor agonists, respectively, according to the mammalian classification (Peroutka 1990; Bonate 1991). Although this classification will almost certainly not apply to invertebrates, 5HT receptors have now been described in molluscan neurones (e.g. *Helix*; Vehovszky and Walker 1991) and these findings strengthen our claim that there are 5HT receptors on the radial muscles, and imply that 5HT may play an important part in regulating the chromatophores of living squid.

Immunohistochemistry

In view of this pharmacological background we decided to employ immunohistochemistry to investigate whether L-glutamate and 5HT are endogenous in the nerves innervating the chromatophore muscles. An antibody to L-glutamate was obtained from Nottingham University (Budd and Duce 1988) and an antibody to 5HT purchased from Immunonuclear Corporation, USA. Both primary antibodies were rabbit polyclonals. We used the peroxidase–antiperoxidase (PAP) staining method of Sternberger (1979), with a swine anti-rabbit second antibody and rabbit PAP from Dakopatts, Copenhagen, Denmark Ltd. Diaminobenzidine tetrachloride (DAB) from Sigma, UK, which yields a dark brown deposit, was used as a chromogen. Skin was fixed for 1 h in a solution consisting of equal parts of a fresh 2 per cent paraformaldehyde solution in distilled water and a 2 per cent glutaraldehyde solution in chilled filtered sea water. (For a detailed description of the methods see Messenger *et al.*, in preparation.)

Staining with L-glutamate antibodies (Fig. 22.8) and with 5HT antibodies (Fig. 22.9) produced lines of dark brown 'blobs' of DAB. Staining was intense against a low background and was confined to the regions of radial muscles along which the axons have been shown to run. Exhaustive controls with both primary antisera revealed that the staining was the result of specific antibodies to L-glutamate and 5HT (Pool *et al.* 1983; Sternberger 1979). We conclude that L-glutamate and 5HT are endogenous in the radial nerve fibres of squid chromatophores.

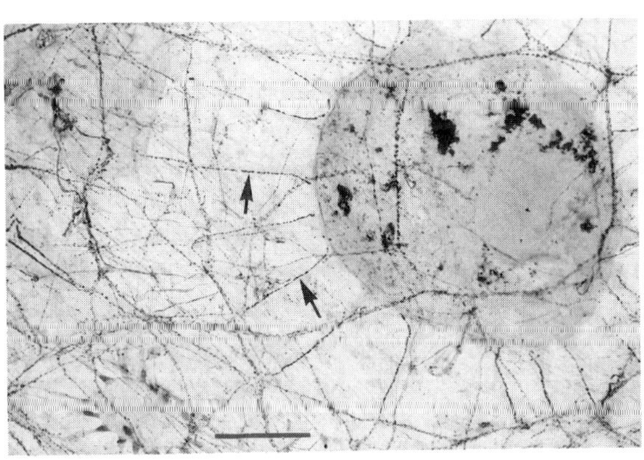

Fig. 22.8. Nerves (arrows) stained with L-glutamate antibodies and DAB running along the radial muscles of a chromatophore. Scale bar: 500 μm.

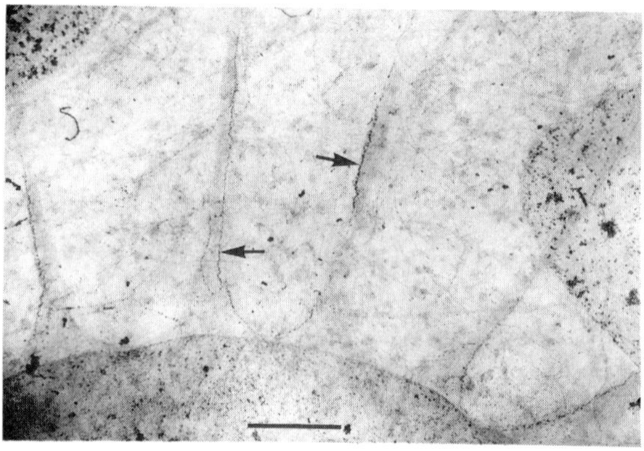

Fig. 22.9. Nerves (arrows) stained with 5HT antibodies and DAB. Scale bar: 200 μm.

Discussion

Previous studies at the light- and electron-microscopic levels (Hofmann 1907; Weber 1968; Cloney and Florey 1968) have shown that the radial muscles of *Loligo* chromatophores are innervated by nerves that run along them. A similar picture emerges in *Alloteuthis* and our silver staining shows beyond doubt that radial muscles are multiply innervated by as many as six axons. This could permit stepwise gradation of muscle contraction (Florey 1966, 1969), and also allow individual chromatophore muscles to participate in several motor units (Florey 1966; Hanlon and Messenger 1988).

The immunohistochemical and pharmacological data presented here for L-glutamate are consistent and confirm the findings of Bone and Howarth (1980) and of Florey *et al.* (1985). They show that glutamate is endogenous in the chromatophore motor neurones and suggest that it is a naturally occurring transmitter substance used by squids to regulate their skin tone. We also have complementary data for other cephalopods. In *Loligo vulgaris* positive staining was obtained with anti-glutamate antibodies. In the cuttlefish, *Sepia*, the chromatophores are sensitive to 'caged' glutamate (Messenger *et al.* 1991). In *Lolliguncula* it is possible to denervate the mantle skin and show dramatically increased sensitivity to topically applied L-glutamate (Messenger, unpublished data). We cannot say yet whether L-glutamate is the only excitatory transmitter. In *Lolliguncula* the experiments of Florey *et al.* (1985), have shown that there are at least three effects of applied glutamate, suggesting that the story might not be a simple one.

As far as 5HT is concerned, the entirely novel and unexpected finding that it is endogenous in (some of?) the chromatophore nerve fibres poses several questions and opens up new avenues for research. To begin with it makes

Florey's original findings (Florey 1966; Florey and Kriebel 1969) doubly interesting, for the pharmacological effects he reported may well be mimicking the physiological action of 5HT in subtly regulating the contraction of the radial muscles and hence the degree of expansion of the chromatophores and the overall skin tone. Our preliminary experiments with 5HT agonists certainly suggest that there are specific receptor sites for 5HT on the radial muscles.

Since the physiological evidence for the absence of inhibitory nerves is strong (Florey 1969) it seems unlikely that 5HT is restricted to some nerve fibres only. Instead, it seems probable that 5HT and glutamate coexist in all of the chromatophore nerve fibres and that 5HT may act routinely in some way to modify the response of the muscle to the excitatory transmitter, L-glutamate, which is presumably being released continuously to maintain the chromatophores in the state of tonic expansion needed to achieve camouflage. 5HT is known to act as a modulator in several invertebrate phyla (Walker and Holden-Dye 1991), including the molluscs, where it has been shown to relax the ABRM of *Anodonta* and *Mytilus*, the penis retractor muscle of *Helix* and *Strophocheilos*, and the pharyngeal retractor muscle of *Helix*, where the excitatory transmitter is L-glutamate (for references see Leake and Walker 1980).

Florey and Kriebel (1969) suggested that 5HT might be acting intracellularly but now that there is evidence for receptors on the radial muscle membrane it seems more likely that it is operating via a G-protein system (Hille 1989). Obviously more experiments are required.

However 5HT acts, it is clearly essential to establish whether or not it coexists with 5HT in the same axons. Preliminary attempts to do this by using immunofluorescence double staining with two differently coloured fluorochromes (Wessendorf and Elde 1985) were not successful, however, so that this important question remains unresolved. It is interesting to recall that although many examples are now known of co-localized transmitters and neuroactive substances, especially in the molluscs (Walker and Holden-Dye 1991), there seems to no recorded case of co-localization of L-glutamate and 5HT.

Acknowledgements

We would like to thank the Director and staff of the Marine Biological Association, Plymouth, for providing facilities and animals and also for a research bursary enabling us to carry out some of the experimental work. It is also a pleasure to thank Dr Ian Duce and Dr Tim Budd from Nottingham University for their technical assistance and generosity in supplying L-glutamate antibodies. We are grateful to David Hollingworth for photographic assistance. C.J.C. was supported by an SERC studentship.

References

Bonate, P. L. (1991). Serotonin receptor subtypes: functional, physiological and clinical correlates. *Clinical Neuropharmacology*, **14**, 1–16.

Bone, Q. and Howarth, J. V. (1980). The role of L-glutamate in neuromuscular transmission in some molluscs. *Journal of the Marine Biological Association of the United Kingdom*, **60**, 619–26.

Budd, T. C. and Duce, I. R. (1992). Immunocytochemical localisation of L-glutamate in motor innervation of locust (*Schistocerca gregaria*) skeletal muscle. *Pesticide Science*, **24**, 177–8.

Cloney, R. A. and Florey, E. (1968). Ultrastructure of cephalopod chromatophore organs. *Zeitschrift für Zellforschung*, **89**, 250–80.

Cott, H. B. (1940). *Adaptive coloration in animals*. Methuen, London.

Drury, R. A. B. and Wallington, E. A. (ed.) (1967). *Carleton's histological technique*, 4th edn. University Press, Oxford.

Ferguson, G. P. and Messenger, J. B. (1991). A countershading reflex in cephalopods. *Proceedings of the Royal Society of London, Series B*, **243**, 63–7.

Florey, E. (1966). Nervous control and spontaneous activity of the chromatophores of the cephalopod, *Loligo opalescens. Comparative Biochemistry and Physiology*, **18**, 305–24.

Florey, E. (1969). Ultrastructure and function of cephalopod chromatophores. *American Zoologist*, **9**, 429–42.

Florey, E. and Kriebel, M. E. (1969). Electrical and mechanical responses of chromatophore muscle fibres of the squid *Loligo opalescens* to nerve stimulation and drugs. *Zeitschrift für vergleichende Physiologie*, **65**, 98–130.

Florey, E., Dubas, F., and Hanlon, R. T. (1985). Evidence for glutamate as a transmitter of motoneurons innervating squid chromatophore muscles. *Comparative Biochemistry and Physiology*, **82C**, 259–68.

Hanlon, R. T. and Messenger, J. B. (1988). Adaptive coloration in young cuttlefish (*Sepia officinalis* L.): the morphology and development of body patterns and their relation to behaviour. *Philosophical Transactions of the Royal Society of London, Series B*, **320**, 437–87.

Hille, B. (1989). Ionic channels: evolutionary origins and modern roles. *Quarterly Journal of Experimental Physiology*, **74**, 785–804.

Hofmann, F. B. (1907). Histologische Unterschungen über die Innervation der glatten und der ihr verwandten Muskulatur der Wirbeltiere und Mollusken. *Archiv für Mikroskopische Anatomie*, **70**, 361–413.

Leake, L. D. and Walker, R. J. (1980). *Invertebrate neuropharmacology*. Blackie, Glasgow.

Messenger, J. B. (1991). Transmitters, toxins and phylogeny. *Journal of Zoology*, **223**, 687–94.

Messenger, J. B., Katayama, Y., Ogden, D. C., Corrie, J. E. T., and Trentham, D. R. (1991). Photolytic release of glutamate from 'caged' glutamate expands squid chromatophores. *Journal of Physiology*, **438**, 293*P*.

Monaghan, D. T., Bridges, R. J., and Cotman, C. W. (1989). The excitatory amino acid receptors. *Annual Revue of Pharmacology and Toxicology*, **29**, 365–402.

Pool, C. W., Buijs, R. M., Swaab, D. F., Boer, G. J., and Van Leeuwen, F. W. (1983). On the way to a specific immunocytochemical localization. In *Immunohistochemistry* (ed. A. C. Cuello), IBRO handbook series: Methods in the neurosciences, Vol. 3, pp. 1–46. Wiley, New York.

Peroutka, S. J. (1990). 5-hydroxytryptamine receptor subtypes. *Pharmacology and Toxicology*, **67**, 373–83.

Rodhouse, P. G., Swinfen, R. C., and Murray, W. A. (1988). Life cycle, demography and reproductive investment in the myopsid squid *Alloteuthis subulata. Marine Ecology Progress Series*, **45**, 245–53.

Sereni, E. and Young, J. Z. (1932). Nervous degeneration and regeneration in cephalopods. *Pubblicazioni della Stazione Zoologica di Napoli*, **12**, 173–208.

Sternberger, L. A. (1979). *Immunocytochemistry*. Wiley, New York.

Twarog, B. M. (1960). Effects of acetylcholine and 5-HT on the contraction of a molluscan smooth muscle. *Journal of Physiology*, **152**, 236–42.

Vehovszky, A. and Walker, R. J. (1991). An analysis of the 5-hydroxytryptamine (serotonin) receptor subtypes of central neurones of *Helix aspersa. Comparative Biochemistry and Physiology*, **100C**, 463–76.

Walker, R. J. and Holden-Dye, L. (1991). Evolutionary aspects of transmitter molecules, their receptors and channels. *Parasitology*, **102**, S7–29.

Weber, W. (1968). Multiple Innervation der Chromatophorenmuskelzellen von *Loligo vulgaris. Zeitschrift für Zellforschung*, **92**, 367–76.

Wessendorf, M. W. and Elde, R. P. (1985). Characterization of an immunofluorescence technique for the demonstration of coexisting neurotransmitters within nerve fibers and terminals. *Journal of Histochemistry and Cytochemistry*, **33**, 984–94.

PART VI

Central nervous system

23

Diffusion properties of the microenvironment of cephalopod brain

CHARLES NICHOLSON, JALEEL A. MIYAN, KATHERINE T. POTTER, RODDY WILLIAMSON, and N. JOAN ABBOTT

Summary

The diffusion characteristics of brain extracellular space can be summarized by two parameters, the tortuosity (λ) and the volume fraction (α). Volume fraction is a measure of the volume of extracellular space relative to whole tissue volume while tortuosity measures the extent to which diffusion is hindered by cellular obstruction in comparison to a free medium. By using local iontophoresis of tetramethylammonium from a micropipette in conjunction with suitable ion-selective microelectrodes, λ and α can be determined.

This study sought to measure λ and α in an invertebrate neural mass to determine if the parameters had similar values to those previously found in vertebrates. The cephalopod vertical lobe and optic lobes were chosen because of their size and accessibility. We found that the average value of the diffusion parameters in the vertical lobe of intact anaesthetized *Sepia* (cuttlefish), intact anaesthetized *Eledone* (octopus) and isolated slices of *Sepia* were $\lambda = 1.65$ and $\alpha = 0.10$. In isolated slices from the optic lobes of *Sepia* the average values were $\lambda = 1.86$ and $\alpha = 0.29$. These data suggest that while there may be definite structural differences between the two neural lobes of these cephalopods, the diffusion characteristics are quite similar to those in vertebrates (typically, $\lambda = 1.60$ and $\alpha = 0.20$). This implies that the connectivity and size of the extracellular space is conserved in neuronal aggregates and that this has functional consequences for the movement of metabolic substrates and chemical signals in the extracellular microenvironment.

Analysis of the diffusion properties of brain tissue in several vertebrates has revealed similarities between regions and species (Nicholson 1992). In particular the fraction of tissue that comprises the extracellular space, the extracellular volume fraction, designated by α, is about 0.2 or 20 per cent of the tissue while the tortuosity, λ, a measure of the extent to which diffusion is hindered by obstructions, is generally 1.6. To be sure, deviations from these values exist (McBain *et al.* 1990) but the generality of the results is striking and prompted us to question whether this was a feature peculiar to the vertebrates or was a necessary property of all brain cell aggregates. To explore this we analysed the diffusion properties of an invertebrate. The cephalopod brain was chosen because we needed a certain mass of relatively homogeneous tissue for the intended method and because one of the parameters, α, had already been measured in *Sepia* by a different method (Abbott 1985; Abbott *et al.* 1985).

Our *in vivo* experiments were carried out over four brief summer research projects in various preparations of *Sepia officinalis* and *Eledone cirrhosa* with measurements predominantly in the vertical lobe. We found that the tortuosity was strikingly similar to that obtained in vertebrates but that the volume fraction was smaller than that seen in typical vertebrates and smaller than the value previously obtained from *Sepia*. Moreover the statistical distribution of the volume fraction data had a significant tail of very small values suggesting that it was not entirely accounted for by a Gaussian distribution of random errors of measurement. We believed that these data might be a reflection of poor perfusion of the brain vasculature (Magnum 1990; O'Dor *et al.* 1990). To investigate this further we devoted an additional summer to studies on *Sepia* brain slices in both the vertical and optic lobes on the premise that this would circumvent any perfusion problem. The results on slices showed significant differences in the volume fractions of the vertical lobe and the optic lobes, suggesting that underlying structural factors, rather than functional state, might play a primary role in the small values obtained in the intact vertical lobe.

Because each experimental season only lasted a few weeks, and involved a fairly complex quantitative methodology, our data show more scatter than the vertebrate studies but generally confirm that the architecture of brain cell aggregates in these invertebrates is similar to that of vertebrates.

Methods

Intact preparations

During the first three summers, we used intact *Sepia* exclusively. Initially, several experiments were made with a perfused preparation using filtered sea water or artificial sea water (ASW) as the perfusate. This was abandoned in favour of a preparation anaesthetized by immersion in circulating sea water containing a small percentage of ethanol in a special holding tank, developed by Miyan and Messenger (see Chapter 25), at about 10°C. Finally a 'decerebrate' preparation was developed, as described below, using the approach developed by Miyan and Messenger. Brain structures were identified on the basis of the monograph by Young (1971).

The animal was placed in a bucket containing 1.2–1.5 per cent ethanol until there was a loss of response to pinching the skin over the eyes, but continued ventilation. The animal was then transferred to a cloth soaked in sea water. The cornea over each eye was cut open and the lens removed to expose the retina to sea water which immediately blinded the preparation, thereby removing the major source of sensory stimulation. The front projection of the cuttlebone was cut sufficiently far back to remove any possibility that subsequent jetting would thrust the cuttlebone into the recording electrodes. The skin between the eyes was cut back so that the cartilage enclosing the brain could be seen and then the cartilage was cut through from the front until a thin sheet was left above the vertical lobes, now clearly visible. The lobes could now be exposed by making a rectangular slot with fine scissors or a fragment of a razor blade. Finally the preparation was placed in the holding tank beneath a plexiglass plate with a suitable cut-out and anchored with no. 16 barbless fish hooks and monofilament line inserted into the cartilage at the edge of the orbit on either side of the muscle bands running forward from the cranium. The animal was bathed in circulating, oxygenated, sea water cooled to not more than 2–3°C below the holding tank temperature, which was usually 10°C.

In the fourth summer, the methods described above were adapted to *Eledone*. The preparation was initially anaesthetized as described for *Sepia*. The animal was then placed on a cloth soaked in sea water and the arms amputated with a new scalpel blade. So long as the anaesthesia was not too deep, no bleeding occurred when the animal revived. The skin overlying the cranium was opened by a mid-line incision, which was extended over each orbit enabling the optic tract to be cut and the optic lobes removed. The eyes were left in place since this seemed to aid in the recovery of blood circulation. The cranial cartilage was opened by a dorsal mid-line longitudinal incision to expose the entire remaining central nervous system. Six barbless no. 16 fish hooks were attached to the sides of the cartilage and the cranium secured to the plexiglass holder using monofilament line. The jelly surrounding the brain was removed using a Pasteur pipette attached to a vacuum pump and fine scissors. After that procedure any region of the superficial neural lobes could be exposed by cutting an appropriately located small window in the sheath. As with *Sepia*, the animal was bathed in circulating, oxygenated, sea water cooled to not more than 2–3°C below the holding tank temperature of about 10°C. More extensive cooling resulted in intermittent respiration. We found it best to change the sea water every 2–3 h to avoid mucus build-up. The state of the preparation could be judged by the chromatophore colour and respiration. A light reddish-brown with regular breathing indicated an animal in good condition; paling indicated deterioration which was often accompanied by irregular breathing and excessive bleeding from the cut stumps of the arms, presumably due to the relaxation of the muscles.

Slice preparation

As outlined earlier (see p. 384), to assess the possible contribution of inadequate vascular perfusion to our results, in the fifth summer we used slice preparations developed from the work of Williamson and Budelmann (1991). The brain of *Sepia* was dissected out under ice-cold ASW with the following composition: 470 mM NaCl, 10 mM KCl, 10 mM $CaCl_2$, 50 mM $MgCl_2$, 10 mM HEPES, and 20 mM glucose. The ASW was initially at pH 7.8 and this was reduced to 7.6 by bubbling with 95 per cent O_2 and 5 per cent CO_2 which also added HCO_3 to the solution. Either the optic lobe or vertical lobe, appropriately oriented, was attached to a circular piece of filter paper by dripping over it a warm solution of 4 per cent agar dissolved in 0.5 M NaCl and 10 mM KCl. The filter paper and brain were then placed on a McIlwain Tissue Chopper and sections cut at 400 μm. Slices were put into a submersion chamber with a Peltier cooling device based on the design used by Rice and Nicholson (1991). The chamber was perfused with the ASW described above at 2–3 ml/min cooled to about 14°C. In two experiments (not described further here) the optic nerve of the optic lobe was stimulated with a local bipolar electrode and field potentials recorded with a saline-filled micropipette from the optic lobe. These tests confirmed the viability of the preparation after the slicing procedure.

Biophysical techniques

The geometry of the extracellular space can be measured by allowing a substance to diffuse through it and analysing this process. Some of the advantages of this method, compared with previous approaches based on radiotracers or quantitative histology, are that changes in the extracellular space can be followed in real time and the method can be applied to very small regions.

Diffusion in brain tissue obeys Fick's law with two important modifications. First, the diffusion coefficient, D, is reduced by the square of the tortuosity, λ, to an apparent diffusion coefficient $D^* = D/\lambda^2$. This is because a diffusing particle encounters membranous obstructions as it executes a random walk in the intercellular clefts. Second, the strength of an imposed source of diffusing substance is modified by the volume fraction, α, of the extracellular space which means that a given quantity of released substance becomes more concentrated than in a free medium.

In brain tissue a third factor, non-specific uptake, which has a linear dependence on extracellular concentration, enters the diffusion process as a term, k', (s^{-1}) for loss of material across the cell membranes due to the concentration gradient and leakiness. For a more complete discussion see Nicholson and Phillips (1981) or Nicholson (1992).

An experiment consisted of releasing an ion that remains predominantly in the extracellular space, namely tetramethylammonium (TMA^+), from a micropipette by iontophoresis and recording its arrival 100–200 μm away with a suitable ion-selective microelectrode (ISM). Using the appropriate solution to the relevant diffusion equation and curve fitting procedures, α, λ, and k' were determined.

If the iontophoresis is applied for a finite duration, then there are separate expressions for the rising phase of the curve, during the iontophoresis, and for the falling phase of the curve, after cessation of the iontophoresis. These expressions were derived by Nicholson and Phillips (1981, expression A14) and can be expressed as (Nicholson 1992):

$$C(r, t \leqq d) = G(r, t);\ C(r, t > d) = G(r, t) - G(r, t - d);$$
$$G(r, u) = (Q/(8\pi D^*\alpha r))\ \{\exp[r(k'/D^*)^{1/2}]\ \mathrm{erfc}(r/(2(D^*u)^{1/2}) + (k'u)^{1/2}] + \exp[-r(k'D^*)^{1/2}]\ \mathrm{erfc}[r/(2(D^*u)^{1/2}) - (k'u)^{1/2}]\}. \quad (23.1)$$

The term 'erfc' is the complementary error function. In this expression C is concentration, r is distance, t is time, and d is the duration of the iontophoretic pulse. The function G is defined as shown and the variable u is replaced by t or $t-d$ as required. $Q = In/zF$ and I is the iontophoresis current applied to the iontophoresis electrode, n is the transport number of the iontophoresis pipette for the iontophoretic substance, z is the charge number of the ion, and F is Faraday's electrochemical equivalent.

ISMs were made by conventional techniques (e.g. Nicholson and Rice 1988), except that the majority of ISMs were single barrelled, since there was no necessity to subtract a signal from the reference barrel in these measurements. Corning 477317 was used as ion-exchanger and the backfill solution was typically 100 mM TMA-Cl. The signals from the ISM were recorded using the circuits described by Nicholson and Rice (1988). The slope and interference of the ISM were measured using the fixed interference method with doubling steps of [TMA^+] from 1 to 16 mM in a solution containing 0.5 M NaCl and 10 mM KCl, to simulate the major interferents found in sea water. Typical slopes were about 60 mV/decade with an interference of 0.1 mM (see Nicholson and Rice 1988, for further details of calibration methods).

The iontophoresis micropipette usually had a tip diameter of about 1 μm and was filled with either 100 mM or 1 M TMA-Cl. To ensure precise spacing between ISM and iontophoresis electrode they were glued together with dental cement or Rapid Araldite™ under a compound microscope equipped with a measuring eyepiece. For iontophoresis, the simple circuit described by Nicholson and Rice (1988) was used with a forward bias current of 20–50 nA throughout the experiment to maintain the tip concentration at a level where the transport number of the iontophoresis electrode was not subject to 'warm-up'. To make diffusion measurements the current was stepped to 100–200 nA for 100 s. The micropipette arrays were advanced into the preparation, and measurements were usually made at about 150–400 μm deep in the vertical and optic lobes in both *in vivo* and slice preparations. Because penetrations of the *in vivo* preparations were always from the dorsal surface they were made in the more peripheral zones while the slice orientation permitted measurements from all areas.

Control measurements were made in a 0.3 per cent agarose (NuSieve, FMC BioProducts, Rockland, Maine, USA) or 0.3 per cent agar gel (Special Agar

Noble, Difco, Detroit, Michigan, USA) made up in 0.5 M NaCl, 10 mM KCl, and 1 mM TMA-Cl to determine *n* and *D*.

The signals from the ISM amplifier were recorded on a chart recorder and the baseline determined from the chart record. In order to have a stable calibration during the experiment the ISM was withdrawn into a saline solution, over the top of the brain, that contained 1 mM TMA^+.

For analysis the ISM signals were acquired by an SPI-500 analogue-to-digital converter (Thornton Associates Inc., Waltham, Massachusetts, USA) with serial interface and appropriate calibration and curve fitting procedures carried out using the program VOLTORO on a Sharp portable PC computer equipped with a mathematics co-processor. The diffusion analysis programme fitted the theoretical curve (23.1) to the experimental data using a non-linear simplex algorithm to obtain α, λ, and k'.

Results were summarized as quantile plots (Chambers *et al.* 1983; Rice and Nicholson 1991) for λ and α using the scientific graphing program, Sigmaplot (Jandel Scientific, Corte Madera, CA, USA). These were made by ordering the variable in question and plotting it as a function of its rank, normalized to 1. This provided an overview of the entire data distribution. To test whether the data had a Gaussian distribution, synthetic data were generated that had the same mean, standard deviation, and same number of values as the experimental data using an appropriate pseudorandom number generator to provide a Gaussian distributed data set. To simulate our experimental procedures, synthetic data which were outside the range of physical possibility ($\alpha < 0$; $\lambda < 1$) were discarded. A quantile–quantile plot of the experimental and synthetic data then enabled the conformity of the experimental data to a Gaussian distribution to be assessed.

Results

The data from studies during the first summer with intact *Sepia* showed great variability associated with the many changes in experimental paradigm. We decided that this was not representative of the more stable experimental conditions that we achieved in subsequent years, so these data have been omitted. The data from the second and third summers with intact *Sepia* have been combined and data with *Eledone* obtained in the fourth summer and the data obtained with the two types of slice in the fifth year are all presented individually (Table 23.1).

Intact preparations

Typical diffusion records for the vertical lobes of intact *Sepia* and intact *Eledone* are shown in Fig. 23.1(a) and (b), respectively. A period of approximately 100 s of iontophoretic ejection of TMA^+ resulted in the signals shown

Table 23.1 Summary of diffusion characteristics

Preparation	Region	λ	α	k' (s^{-1})	N
Sepia intact	VL	1.68 ± 0.40	0.087 ± 0.090	0.015 ± 0.012	101
Eledone intact	VL	1.75 ± 0.40	0.126 ± 0.129	0.016 ± 0.008	45
Sepia slice	VL	1.54 ± 0.24	0.117 ± 0.049	0.012 ± 0.008	36
Sepia slice	OL	1.86 ± 0.47	0.289 ± 0.175	0.004 ± 0.004	54
Sepia inact + slice	VL	1.65 ± 0.37	0.095 ± 0.082	0.014 ± 0.011	137
Rat slice*	ST	1.54 ± 0.10	0.21 ± 0.05	0.01 ± 0.006	173

VL, vertical lobe; OL, optic lobes; ST, striatum.
Numbers are mean ± SD.
* Rice and Nicholson (1991).

on the ISM located 150–300 μm away. The exact parameters are detailed in the figure legend. The fitted theoretical curve, using expression (23.1) and the values of α, λ, and k' shown in each panel, are depicted with dashed lines together with the theoretical curve for agar (this constitutes a free medium with $\alpha = 1$, $\lambda = 1$ and $k' = 0$).

In general the data were more noisy than those typically obtained in vertebrates. This could be largely attributed to the respiratory movements of the preparation as well as the general motion of the circulation water in the bath containing the animal. The statistical distribution of the extracted parameters is illustrated in the next two figures.

Quantile plots of λ and α for *Sepia* are shown in Fig. 23.2(a) and (c) and the data are summarized in Table 23.1. To test whether the experimental distribution was Gaussian it was plotted against its corresponding synthetic distribution [Fig. 23.2(b) and (d)]. If the experimental data did have a Gaussian distribution then this type of plot would conform to a straight line at 45° to the axes. It is evident that λ is quite well described by a Gaussian distribution but α shows significant deviation. It is of course not possible to assign a cause to these deviations, simply from the statistics.

Figure 23.3 shows a set of data for the intact *Eledone* similar to those shown for *Sepia* in Fig. 23.2. Here again the tortuosity data conform to a Gaussian distribution better than the volume fraction results.

Slice preparations

A diffusion curve obtained from a slice of the vertical lobe of *Sepia* is depicted in Fig. 23.1(c) and the results of curve fitting to a set of such curves is shown in Fig. 23.4. In Fig. 23.4(a), the quantile plot of the tortuosity is shown while Fig. 23.4(b) illustrates a quantile–quantile plot, constructed as detailed for Fig. 23.2. There is significant deviation from a Gaussian curve, but the total

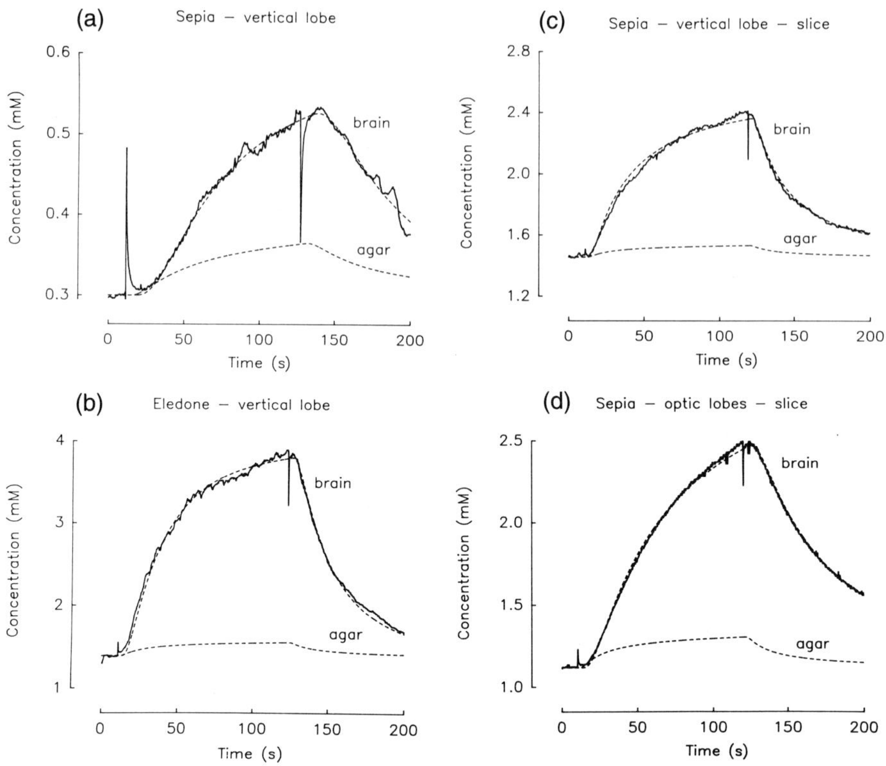

Fig. 23.1. Examples of diffusion curves from both intact and slice preparations of *Sepia* and *Eledone*. In all cases the iontophoresis of TMA^+ took place during the period of approximately 100 s delineated by the on/off current artefacts in the records. Each record shows the best-fit theoretical curve representing expression (23.1). The dotted 'agar' curve is the theoretical curve obtained by setting $\alpha = 1$, $\lambda = 1$ and $k' = 0$. (**a**) Vertical lobe, *Sepia*. Electrode spacing 269 μm, main current 100 nA, bias current 25 nA. Iontophoresis electrode transport number (n) was 0.4, temperature 10°C. $\alpha = 0.26$; $\lambda = 1.57$; $k' = 0.009$/s. (**b**) Vertical lobe, *Eledone*. Electrode spacing 159 μm, main current 99 nA, bias current 43 nA, n was 0.7, temperature 16°C. $\alpha = 0.09$; $\lambda = 1.60$; $k' = 0.012$/s. (**c**) Vertical lobe, *Sepia*, slice preparation. Electrode spacing 151 μm, main current 84 nA, bias current 26 nA, n was 0.3, temperature 13°C. $\alpha = 0.30$; $\lambda = 1.84$; $k' = 0.002$/s. (**d**) Optic lobe, *Sepia*, slice preparation. Electrode spacing 161 μm, main current 200 nA, bias current 26 nA, n was 0.2, temperature 13°C. $\alpha = 0.12$; $\lambda = 1.36$; $k' = 0.004$/s.

number of observations is quite low. Figure 23.4(c) shows volume fraction data and Fig. 23.4(d) the quantile–quantile plot, here the conformity to a Gaussian distribution is better.

Figure 23.1(d) shows a diffusion curve obtained from a slice of an optic lobe of *Sepia*. The results of curve fitting to a set of such curves is shown in

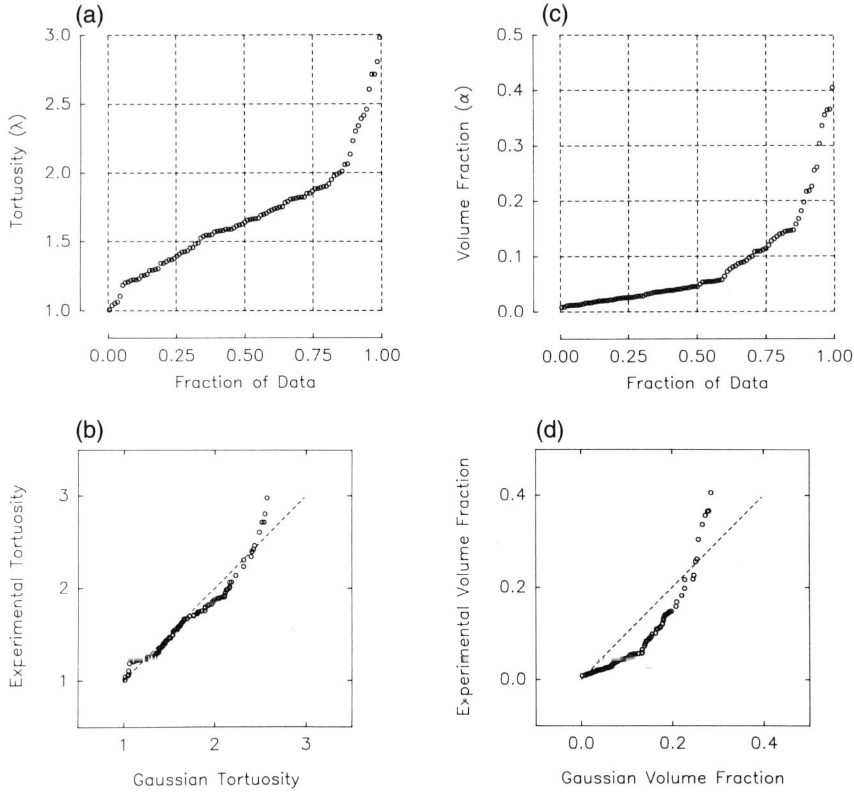

Fig. 23.2. Quantile plots of tortuosity and volume fraction for the intact *Sepia* vertical lobe. (a) Shows the tortuosity with mean of 1.68, standard deviation 0.40, and number of observations 101. Based on these statistics, a Gaussian distribution was generated and the quantile for these synthetic data used in the quantile–quantile plot shown in (b). In a similar manner (c) shows the volume fraction with mean of 0.087, standard deviation 0.090, and number of observations 101. Again, based on these statistics a Gaussian distribution was generated and the quantile for these synthetic data used in the quantile–quantile plot shown in (d). See Methods for description of procedure used to generate the Gaussian distributions. The broken lines at 45° in panels (b) and (d) indicate where the points would lie if the distributions conformed to a Gaussian distribution. The tortuosity conforms quite well for much of the data but the volume fraction shows significant deviations.

Fig. 23.5. Figure 23.5(a) displays the quantile plot of the tortuosity, while Fig. 23.5(b) illustrates a quantile–quantile plot. The tortuosity fits well to a Gaussian curve in this case. Figure 23.5(c) shows volume fraction data and Fig. 23.5(d) the quantile–quantile plot. The conformity to a Gaussian distribution is somewhat erratic.

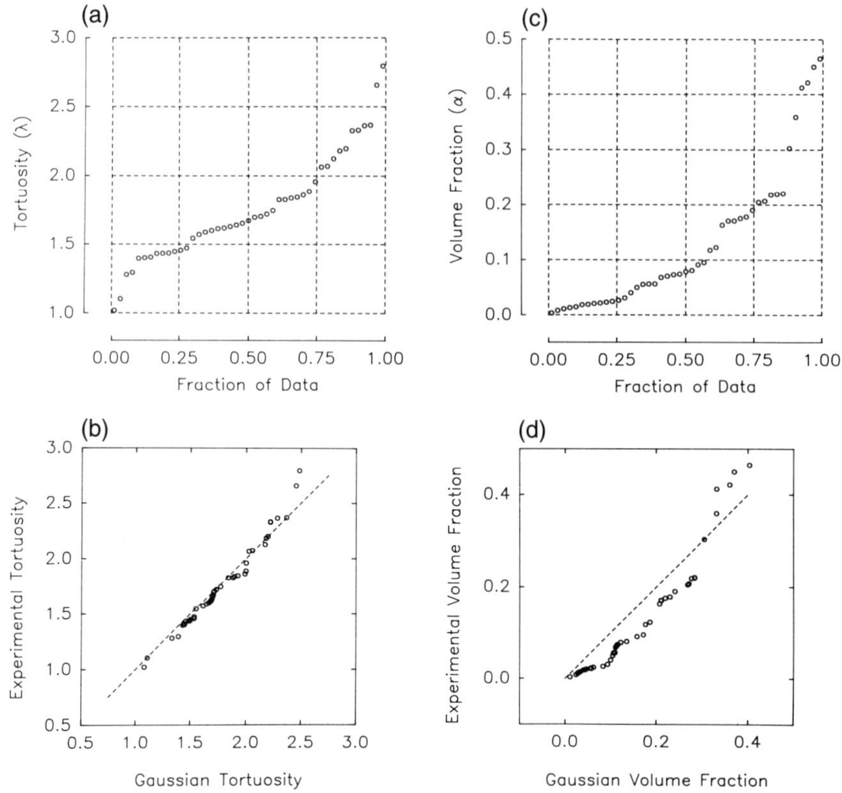

Fig. 23.3. Quantile plots of tortuosity and volume fraction for the intact *Eledone* vertical lobe. The plots are similar to those shown in Fig. 23.2. (a) Shows the tortuosity with mean of 1.76, standard deviation 0.39, and number of observations 45. (c) Shows the volume fraction with mean of 0.131, standard deviation 0.128, and number of observations 45. (b) and (d) Show the respective quantile–quantile plots. Here the tortuosity conforms well to a Gaussian curve but the volume fraction again shows deviation.

Comparison of data

Overall the tortuosities of all the data are quite similar and the volume fractions of all the vertical lobes resemble each other but are less than half of the value obtained in the optic lobe (Table 23.1). In order to further study these features a *t*-test was performed on the various data sets. First it was determined that the *Sepia* vertical lobe data from the intact preparation were not different from the slice data in the same region and species at the $p = 0.05$ level for either λ or α. Given this result the two sets of data were combined

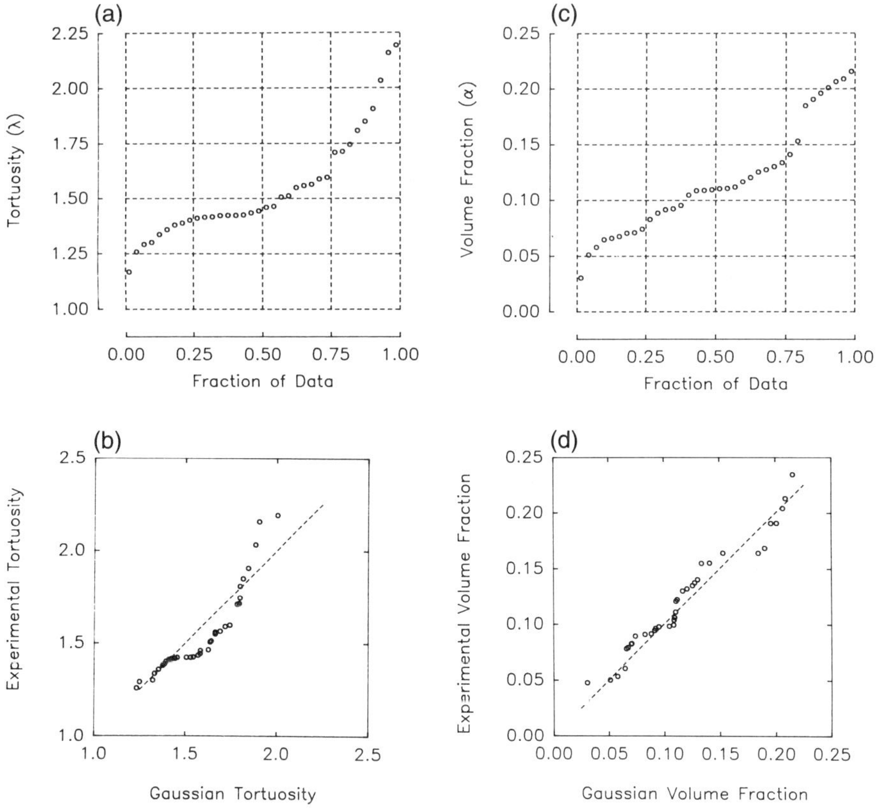

Fig. 23.4. Quantile plots of tortuosity and volume fraction for slices from *Sepia* vertical lobe. The plots are similar to those shown in Fig. 23.2. (a) Shows the tortuosity with mean of 1.54, standard deviation 0.24, and number of observations 36. (c) Shows the volume fraction with mean of 0.117, standard deviation 0.049, and number of observations 36. (b) and (d) Show the respective quantile–quantile plots. Here the tortuosity deviates from a Gaussian curve but the volume fraction conforms quite well.

for further comparisons. The combined data are shown in Table 23.1 in the penultimate row. The combined *Sepia* vertical lobe data did not differ at the $p = 0.05$ level from the intact *Eledone* vertical lobe data for either λ or α. Finally, however, when the *Sepia* vertical lobe combined data were tested against the optic lobe data obtained in slices from the same species both the λ and α data differed at the $p = 0.05$ level.

The uptake data were not tested, since they are presented primarily for comparison with other species but it is evident that the uptake in the *Sepia* optic lobes was only about one third as large as that in the vertical lobes.

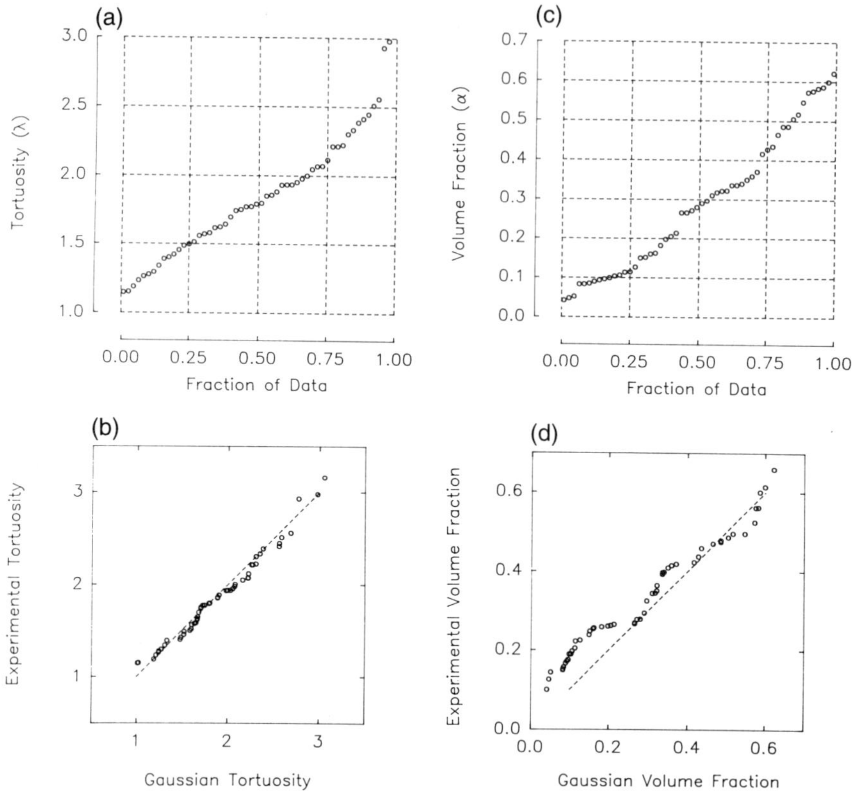

Fig. 23.5. Quantile plots of tortuosity and volume fraction for slices from *Sepia* optic lobe. The plots are similar to those shown in Fig. 23.2. (a) Shows the tortuosity with mean of 1.86, standard deviation 0.47, and number of observations 54. (c) Shows the volume fraction with mean of 0.289, standard deviation 0.017, and number of observations 54. (b) and (d) Show the respective quantile–quantile plots. The tortuosity conforms well to a Gaussian curve but the volume fraction shows deviation.

Discussion

From Table 23.1 it is apparent that the tortuosities are quite consistent among the four preparations and they range from 1.54 to 1.86. Since the diffusion coefficient is reduced by λ^2 this means that the apparent diffusion coefficient will be between 2.4 and 3.5 times less than the aqueous, or unhindered, coefficient.

Volume fraction shows a more interesting pattern. For the vertical lobe in both the intact preparations, *Sepia* and *Eledone*, and in the vertical lobe slices from *Sepia* the average value for α is about 0.1 or 10 per cent. This value is in sharp contrast to that found in the slices of *Sepia* optic lobes where the average value of α was 0.29 or 29 per cent. These differences suggest that the structures of the two regions may differ.

One curious fact about all the volume fraction data was the occurrence of many very small values, of only a few per cent. We were never able to establish if this was some form of unrecognized artefact or a real measurement. Given the persistence and the occurrence of more normal values it seems likely that these small values were real and we have wondered if they represent something akin to an anoxic state of the tissue due to the failure of local vascular circulation, since it is recognized that this is not robust in cephalopods (Magnum 1990; O'Dor 1990). It is also to be noted that determination of volume fraction is more prone to error than determination of tortuosity because it is based on combining more measurements, including accurate knowledge of the baseline TMA^+ concentration. The wide range of data values [see Figs 23.2(c)–23.5(c)] attests to the errors present, due in part to the difficult nature of the preparation and in part due to the relatively brief time available to set up and make measurements each summer. Consequently we urge caution that the volume fraction data not be overinterpreted.

We were fortunate that one prior set of measurements of the total volume fraction of *Sepia* brain existed (Abbott 1985; Abbott *et al.* 1985). Those measurements were made using a radiotracer and revealed a volume fraction of about 18 per cent for optic lobes and 20 per cent for 'vertical lobes', which actually included parts of the frontal and subvertical lobes. As it happens, 20 per cent is close to the value we obtained by averaging the two regions studied with the TMA^+ method. This suggests the two sets of data may be consistent because, although different neural lobes may have quite large variations in volume fraction, the mean overall lobes may be consistent with the radiotracer result.

The underlying purpose of this study was to compare the diffusion characteristics of an invertebrate brain tissue mass with previous data from vertebrates. To this end we have included recent values obtained with the TMA^+ method from the rat striatal slice as the final row of Table 23.1. These mammalian data are quite representative of other studies. They include measurements on the brains of rats, turtles, and elasmobranchs using the TMA^+ method and earlier work with radiotracers (for a summary and references see: Nicholson and Rice 1986, 1991; Nicholson 1992). We note that the tortuosity values in the cephalopods are slightly higher than the mammalian example and the volume fractions straddle the mammalian value of 21 per cent, which, as noted above, is close to the value obtained in the radiotracer study for the whole cephalopod nervous system. Uptake values for TMA^+ were roughly comparable with those seen in the mammalian slice.

While the diffusion data obtained in the cephalopods do show some deviations from those seen in the vertebrate, the most striking fact is that the diffusion properties are fairly comparable in both the invertebrates and vertebrates. These results strengthen the argument that these two anatomical parameters, λ and α, have similar magnitudes in the majority of nerve cell aggregates. This suggests that cells of the brain pack in similar ways regardless of location, and since both λ and α are related to the size and connectivity of extracellular space, these data suggest that there is a consistent size associated with these spaces. This is indeed likely to be true because of the need for nutrients and waste products to diffuse between the vascular system or ventricles of the brain and the cell surfaces. The existence of such a channel also offers the opportunity for 'informational traffic' (Schmitt 1984; Nicholson and Rice 1986), for example via neuromodulator substances circulating through the interstitium, and lends credence to the possibility of chemical communication as a universal feature of the brain cell microenvironment.

Acknowledgements

Supported by NIH Grant NS-13742 to C.N. and Royal Society Marine Science Grants to J.A.M. and N.J.A. K.T.P. received an MBA Bursary. R.W. was a Wellcome Trust Senior Research Fellow.

References

Abbott, N. J. (1985). Size of extracellular space in isolated brain of cuttlefish *Sepia* and squid *Alloteuthis*. *Journal of Physiology*, **369**, 57P.

Abbott, N. J., Bundgaard, M., and Cserr, H. F. (1985). Brain vascular volume, electrolytes and blood–brain interface in the cuttlefish *Sepia officinalis* (Cephalopoda). *Journal of Physiology*, **368**, 197–212.

Chambers, J. M., Cleveland, W. S., Kleiner, B., and Tukey, P. A. (1983). *Graphical methods for data analysis*. Wadsworth & Brooks/Cole Publishing Company, Pacific Grove, CA.

Magnum, C. P. (1990). Gas transport in the blood. In *Squid as experimental animals* (ed. D. L. Gilbert, W. J. Adelman Jr, and J. M. Arnold), pp. 443–68. Plenum Press, New York.

McBain, C. J., Traynelis, S. F., and Dingledine, R. (1990). Regional variation of extracellular space in the hippocampus. *Science*, **249**, 674–7.

Nicholson, C. (1992). Quantitative analysis of extracellular space using the method of TMA^+ iontophoresis and the issue of TMA^+ uptake. *Canadian Journal of Physiology and Pharmacology*, **70**, Suppl., S314–S322.

Nicholson, C. and Phillips, J. M. (1981). Ion diffusion modified by tortuosity and volume fraction in the extracellular microenvironment of the rat cerebellum. *Journal of Physiology*, **321**, 225–57.

Nicholson, C. and Rice, M. E. (1986). The migration of substances in the neuronal microenvironment. *Annals of the New York Academy of Sciences*, **481**, 55–71.

Nicholson, C. and Rice, M. E. (1988). Use of ion-selective microelectrodes and voltammetric microsensors to study brain cell microenvironment. In *Neuromethods, Vol 9: The neuronal microenvironment* (ed. A. A. Boulton, G. B. Baker, and W. Walz), pp. 247–361. Humana, Clifton, NJ.

Nicholson, C. and Rice, M. E. (1991). Diffusion of ions and transmitters in the brain cell microenvironment. In *Volume transmission in the brain* (ed. K. Fuxe and L. F. Agnati), pp. 279–94. Raven Press, New York.

O'Dor, R., Pörtner, H. O., and Shadwick, R. E. (1990). Squid as elite athletes: Locomotory, respiratory, and circulatory integration. In *Squid as experimental animals* (ed. D. L. Gilbert, W. J. Adelman, Jr. and J. M. Arnold), pp. 481–503. Plenum Press, New York.

Rice, M. E. and Nicholson, C. (1991). Diffusion characteristics and extracellular volume fraction during normoxia and hypoxia in slices of rat neostriatum. *Journal of Neurophysiology*, **65**, 264–72.

Schmitt, F. O. (1984). Molecular regulators of brain function: a new view. *Neuroscience*, **13**, 991–1001.

Williamson, R. and Budelmann, B. U. (1991). Convergent inputs to *Octopus* oculomotor neurones demonstrated in a brain slice preparation. *Neuroscience Letters*, **121**, 215–18.

Young, J. Z. (1971). *The Anatomy of the Nervous System of* Octopus vulgaris. Clarendon Press, Oxford.

24

Cephalopod brains: promising preparations for brain physiology

BERND U. BUDELMANN, THEODORE H. BULLOCK, and RODDY WILLIAMSON

Summary

The brains of cephalopods are the most sophisticated brains of all invertebrates and their gross anatomy and neuronal pathways are well known. Also much is known in cephalopods about learning and memory functions. Yet physiological recordings from cephalopod brains are scanty. Recently, however, three preparations have been developed and are now available for experiments on cephalopod brain physiology: (i) a brain slice preparation that allows intracellular recordings from identified brain neurones, (ii) an intact animal preparation that permits multiple electrode recordings of spikes and compound field potentials from unanaesthetized and unrestrained cuttlefish, and (iii) mapping of metabolic brain activity with $[^{14}C]$deoxyglucose. These preparations are complementary and allow a variety of physiological experiments to be done. With further improvements of the techniques and in combination with the morphological information that already exists on pathways in the cephalopod brain, these new preparations are promising tools for cephalopod brain physiology. They may even serve as supplementary or alternative invertebrate preparations for vertebrate research.

For more than 50 years the cephalopod brain and nervous system have played an important role in neurobiological research. Beyond question, cephalopods have the most complex of any invertebrate brains and a tremendous body of precise information already exists about the embryology, the gross anatomy, and the neuronal pathways of their brains (e.g. Boycott 1961; Young 1965, 1971, 1974, 1976, 1977*a*,*b*, 1979; Woodhams 1977; Messenger 1979; Budelmann and Young 1985, 1987; Plän 1987; Marquis 1989; Gleadall 1990). In addition to these morphological data, much is known in cephalopods about motor behaviour, learning, and memory functions (e.g. Sanders and Young 1940; Boycott 1961; Muntz 1963; Messenger 1973; Chichery 1983; Young 1983, 1991; Budelmann and Young 1984; Gilly *et al.* 1991; for further references, see Wells 1978; Mangold 1989).

This wealth of information, however, contrasts with the fact that electrophysiological recordings from cephalopod brains are scanty. Two reasons may account for this. One is the difficulty in adequately immobilizing cephalopods for brain recordings, and the other is the poor survival rate of the animals when anaesthetized or perfused. Because of these difficulties only two papers (Laverack 1980; Bullock 1984) and several short notes (Mislin 1955; Bullock and Uter 1976; Robertson *et al.* 1991; Bleckmann *et al.* 1991; Miyan and Messenger, Chapter 25 of this volume) have been published that describe electrophysiological recordings from cephalopod brains, either in isolated brain preparations or in heavily restrained animals. Several recordings, however, have been successfully made from the optic lobes (MacNichol and Love 1960*a,b*; Lettvin and Maturana 1961, 1962; Boycott *et al.* 1965; Corner and Schadé 1967; Hartline and Lange 1974; Lange and Hartline 1974; Stephen 1974; Patterson and Silver 1983; Saidel *et al.* 1983).

Recently, substantial progress has been made. Three new preparations have been described in detail: (1) a brain slice preparation that allows intracellular recordings from cells of any area of the brain (Williamson and Budelmann 1991); (2) an intact animal preparation that permits multiple electrode recordings of spike and compound field potentials in different brain areas of unanaesthetized and unrestrained cuttlefish that are capable of behaving (Bullock and Budelmann 1991); and (3) mapping of metabolic brain activity with [^{14}C]deoxyglucose ([^{14}C]DG; Novicki *et al.* 1992).

These three preparations are complementary and allow a variety of physiological experiments to be done. This chapter will briefly describe the preparations and discuss the impact they may have on future research on cephalopod and vertebrate brain functions.

Brain slice preparation

For many years thin brain slices have been successfully used in many types of research on the physiology of the vertebrate brain. In contrast, a slice preparation of the cephalopod brain has only recently been developed (Williamson and Budelmann 1991). Although the preparations used an octopod species (*Octopus bimaculoides*), slices from any decapod species should function just as well.

To obtain the brain slices, the animals were killed by decapitation (without prior anaesthesia) and the brain quickly exposed under artificial sea water (ASW) and removed from its cartilaginous capsule. The ASW contained 10 mM KCl, 470 mM NaCl, 55 mM $MgCl_2$, 11 mM $CaCl_2$, 10 mM glucose, and was buffered to pH 7.6 with MOPS. It was chilled to about 4°C. The low temperature during all processing of the brain tissue is probably an essential factor for good survival of the brain. Removal of the left and right optic lobes permits brain slices to be cut in any desired direction (transverse, sagittal, horizontal, or oblique). Williamson and Budelmann (1991) used a tangential sagittal slice, up to 1.5 mm thick, cut by hand with a razor blade,

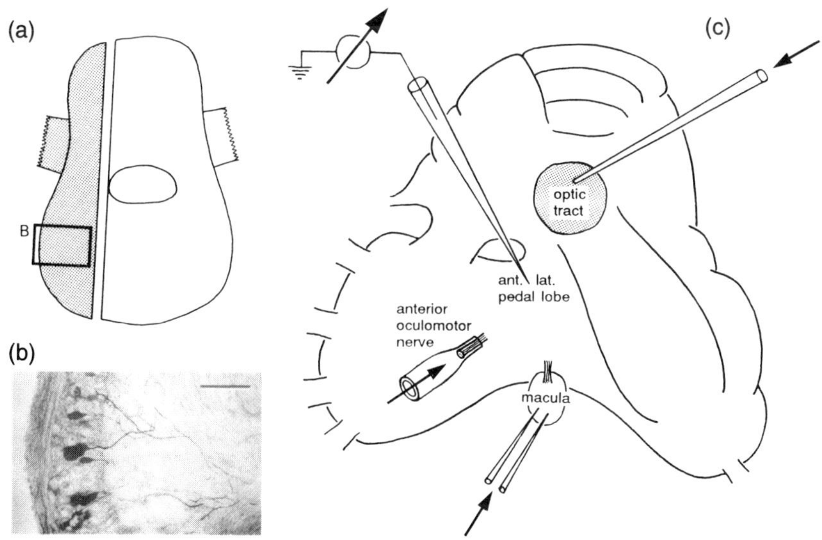

Fig. 24.1. Brain slice preparation of *Octopus vulgaris*. (a) Posterior view of the whole brain (with the two optic lobes removed), showing the direction of the cut and the brain slice (stippled). (b) Transverse section through the anterior lateral pedal lobe [as outlined in (a)], showing cobalt-labelled oculomotor neurones stained via the anterior oculomotor nerve. Scale bar: 200 μm. (c) Brain slice seen laterally. Intracellular recordings were made with glass microelectrodes from oculomotor neurones in the anterior lateral pedal lobe [cf. (b)]. Stimulation of the anterior oculomotor nerve was with a polyethylene suction electrode, stimulation of the sensory cells in the macula epithelium was via a bipolar pair electrode, and stimulation of the optic tract was via a concentric needle electrode. (From Williamson and Budelmann 1991; reprinted with the permission of Elsevier Scientific Publishers.)

and containing the most lateral ~one-quarter (by volume) of the brain of a 100–150 g octopus [Fig. 24.1(a)]. The beauty of such a relatively thick slice preparation is that it leaves intact many of the axons that run from the somata of the neurones to other parts of the brain [Fig. 24.1(b)]. Depending upon the angle of sectioning, slices can even be obtained to which one, or more, nerves are still attached [Fig. 24.1(c)]. This permits intracellular recordings from brain neurones while stimulating a nerve, or recordings from one or several nerves while stimulating brain neurones.

For recordings, the brain slice was transferred to a small dish and sandwiched between two nylon grids. The dish was continuously perfused with cooled (12–14°C) oxygenated ASW, which had access to both sides of the slice. Under such conditions the preparation remained viable and stable for 3–4 h, and intracellular recordings could be held for up to 20 min.

Using such a slice preparation, Williamson and Budelmann (1991) obtained the first intracellular recordings from identified neurones in a cephalopod

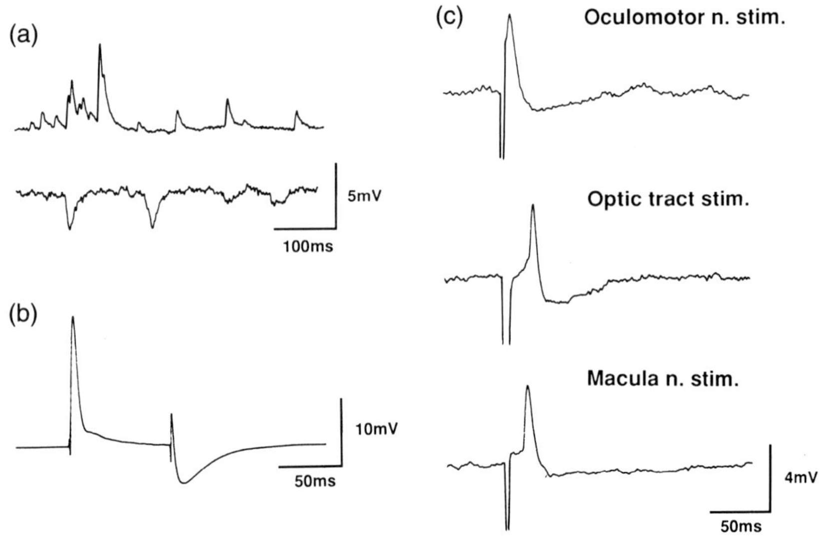

Fig. 24.2. Intracellular recordings from neurones in the perikaryal layer of the anterior lateral pedal lobe of the *Octopus* brain. (a) Recordings from two neurones showing membrane resting potential fluctuations similar to junctional evoked postsynaptic potentials (upper trace) and inhibitory postsynaptic potentials (lower trace). (b) Recording showing an action potential evoked by injecting a depolarizing pulse into the neurone. (c) Recordings from an oculomotor neurone to electrical stimulation of the anterior oculomotor nerve to give an antidromic spike (response latency 8 ms), of the optic tract to represent a visual input (response latency 27 ms), and of sensory cells in the statocyst macula epithelium to represent an equilibrium receptor input (response latency 20 ms). (From Williamson and Budelmann 1991; reprinted with the permission of Elsevier Scientific Publishers.)

brain. The neurones investigated were from the 'oculomotor centre' (anterior lateral pedal lobe). They showed membrane resting potentials of about −40 mV and spontaneous action potentials of up to 20 mV in amplitude. Some neurones showed membrane potential fluctuations similar to junctional excitatory and inhibitory postsynaptic potentials [Fig 24.2(a)].

Williamson and Budelmann (1991) used a tangential sagittal section of the brain to verify physiologically the convergence of a visual and an equilibrium receptor input on to oculomotor neurones in the anterior lateral pedal lobe that had earlier been described morphologically (Budelmann and Young 1984). Figure 24.2(c) shows intracellular recordings from oculomotor neurones (identified as such by antidromic stimulation of one of the oculomotor nerves) that could be depolarized—with some latency—by stimulating the optic tract (latency between 10 and 32 ms) as well as sensory cells in the statocyst macula epithelium (latency up to 20 ms).

Thick brain slice preparations thus are suitable tools to examine the properties of identified neurones in the cephalopod brain, and their connections with other parts of the brain, as well as with sense organs and effector systems.

Intact animal preparation

Boycott (1961), Chichery and Chanelet (1976), and Chichery (1983) were the first to describe an intact unrestrained cephalopod preparation for experiments on brain function, although they used the implanted electrodes for stimulation, not for recording. Bullock and Budelmann (1991) recently developed a preparation that allows such recordings in animals that are unanaesthetized, unrestrained, and capable of behaving. They used large specimens (15–18 cm mantle length) of the European cuttlefish, *Sepia officinalis*. Only these large specimens have a layer of cartilage above the vertical lobe of the brain that is thick enough (about 5 mm) for a mechanically stable electrode implantation.

Several days prior to the experiment, 10–15 mm of the rostrally protruding tip of the cuttlebone was cut off. This exposes the anterior part of the nuchal cartilage (which overlies the neck and the posterior end of the head) and avoids damage to the implanted electrodes if the animal retracts its head underneath the cuttlebone.

On the day of the experiment, the animal was gently transferred (by scooping up, not netting) from the holding tank to the laboratory and, after a resting period of about 20 min, was eased into a small experimental tank. Such gentle handling is essential to avoid any stress prior to the experiment, and thus to avoid any startle or other nervous reactions. The experimental tank contained a 15 mm layer of gravel and 70 mm depth of sea water and was continuously aerated with a mixture of 95 per cent oxygen and 5 per cent carbon dioxide. The total volume of sea water in the tank was three to four times the volume of the animal. A fenced partition of the tank was just large enough to allow the animal some forward and backward, and some lateral movements, but no turns.

With the animal in the experimental tank the electrodes were implanted into the brain in two steps, involving no surgery or anaesthesia. In a first step, a 15 mm length of a 25 gauge hypodermic needle was pushed vertically down, or in any desired direction, through the anterior end of the nuchal cartilage, or the skin slightly anterior to it; it was then pushed through the muscle layer and the dorsal cranial cartilage above the vertical lobe to a depth of 10 mm so that 5 mm of the needle still protruded above the nuchal cartilage, or above the skin. Because the muscle layer and the cranial cartilage are together 10 mm thick, the 15 mm needle did not penetrate into the brain (Fig. 24.3). Four such hypodermic needles were usually placed, aiming at selected parts of the brain. This aiming needs some experience and a good knowledge of the

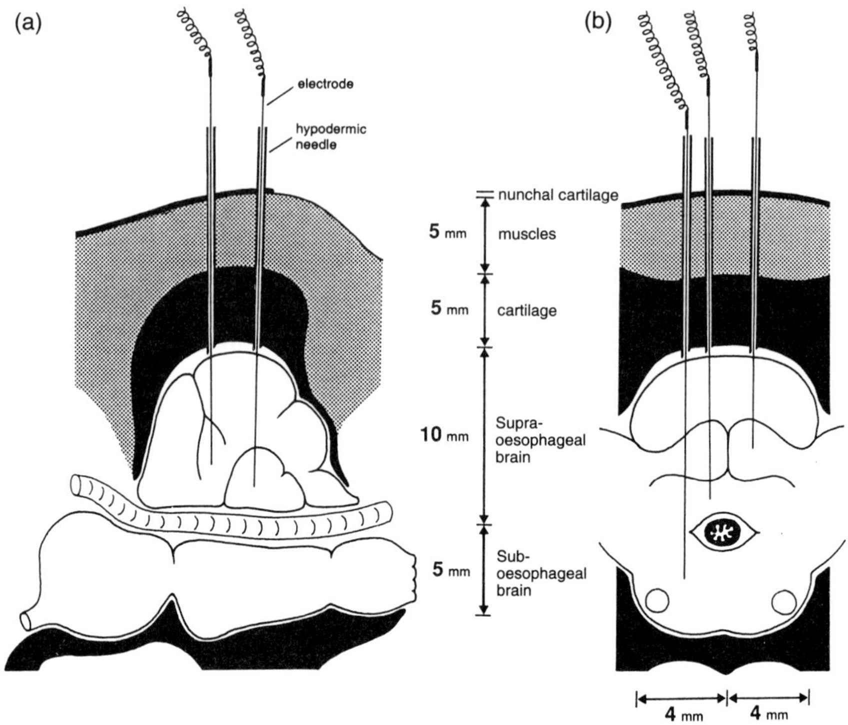

Fig. 24.3. Diagrams of sections through the brain of an adult cuttlefish (*Sepia officinalis*, mantle length 17–18 cm), showing the dimensions of the supra- and suboesophageal brain, and the cartilage and muscle layers above the brain. (a) Lateral view; (b) anterior view. For technique of electrode positioning, see text.

gross anatomy of the cuttlefish brain. In a second step, commercial platinum/iridium recording electrodes of straight, thin, sharpened wire were inserted through the hypodermic needles and beyond into the brain to a predetermined depth, according to the dimensions of the brain (Fig. 24.3). These dimensions, of course, vary according to the animal's size. The depth (D) of penetration of the electrode into the brain can easily be calculated as $D = L_E - (L_H + L_{EO})$, with L_E being the total length of the electrode, L_H the length of the hypodermic needle, and L_{EO} the length of the electrode outside the needle (Fig. 24.3). Surprisingly, throughout the insertion of the hypodermic needles and of the electrodes, and afterwards for up to 12 hours of recordings, the animals lay quietly in their confined space. They moved forwards from time to time, but always slowly without strong jetting; inking never occurred. With the electrodes implanted, the animals could then be exposed to various stimuli, such as visual, acoustic, electrical, and mechanical.

To localize exactly where the tips of the electrodes had been during the recordings, the animals were anaesthetized in the experimental tank at the end of the experiment by slowly adding an overdose of ethanol. Subsequently, the electrodes were replaced by needles of the same diameter and length, pushed through the hypodermic needles down to the depth at which the tips of the electrodes had been during the recordings. With all these needles in place, the head of the animal was then carefully removed and fixed in formalin. After fixation the brain was dissected out, with the needles in place, and cut parallel to the needles, sagittally or transversely, into about 1 mm thick slices. These slices are thin enough to become transparent and show the borderlines between the various lobes when illuminated from below. Stereomicroscopical analysis of the set of sections allows an accurate localization of the tip of the needles (= tip of the electrodes) within the brain, and even within a given brain lobe.

Bullock and Budelmann (1991) recorded compound field potentials and spikes during ongoing, spontaneous activity, and after sensory stimulation from seven different supra- and suboesophageal areas of the brain. The overall aspect of the compound field potentials has been found to be highly variable from locus to locus and from moment to moment (Fig. 24.4); thus, cephalopods seem to have more steeply differentiated electric fields than mammals. Slow waves have been the major feature, as in the vertebrate electrocorticogram, but they are more prominent than in gastropod or arthropod brain activity (Bullock 1945; Bullock and Basar 1988). Spikes, if present, were mostly large and occurred with a frequency of a few, to a few dozen times per second.

In cephalopods, potentials evoked by sensory stimuli, in general, were remarkably similar to those known in mammals and other vertebrates. Visual evoked potentials to a single flash were of long duration (100–125 ms, or more; Fig. 24.5). Potentials to trains of flashes showed an upper following frequency of usually more than 20 Hz; i.e. they had a fusion frequency of about 50 ms (Fig. 24.6). At the end of a train, sometimes a slow wave occurred, representing a special form of central 'off' potential. It resembled the 'omitted stimulus potential' of fishes and belongs by definition to the 'Event related potentials' (Fig. 24.7; Bullock *et al.* 1990).

Mapping of metabolic brain activity with [^{14}C]DG

For several years, the [^{14}C]DG technique has been successfully used for mapping metabolic activity in vertebrate brains (e.g. Sokoloff *et al.* 1977; Sokoloff 1987), and has been applied to a few invertebrate preparations as well (e.g. arthropods: Buchner *et al.* 1984, Bausenwein 1990; gastropod molluscs: Sejnowski *et al.* 1980; Chase 1985). Since the cephalopod brain metabolizes glucose (Cory and Rose 1969), Novicki *et al.* (1992) successfully adapted the [^{14}C]DG technique to cephalopod preparations. They used animals, unilaterally visually deprived by lens or eye removal, or by cutting

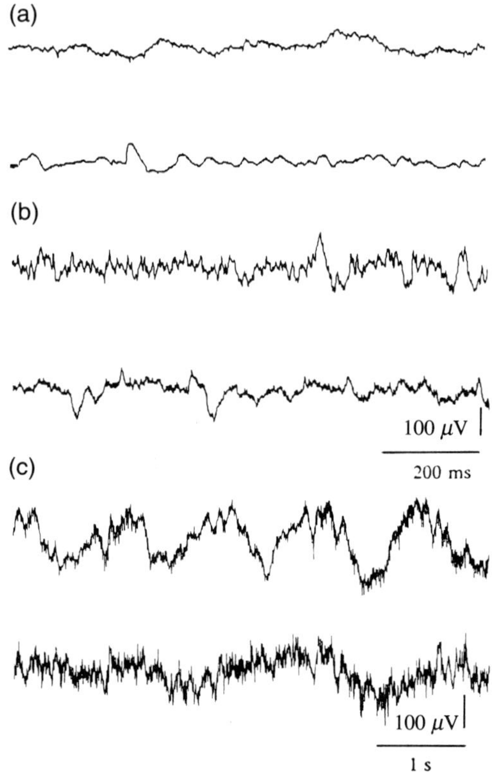

Fig. 24.4. Ongoing background activity (EEG) in the brain of the cuttlefish *Sepia officinalis* without intentional stimulation, recorded at two time scales in different loci of the brain. (a) Two simultaneous recordings, one dorsally in the posterior part of the anterior basal lobe, close to the subvertical lobe (upper trace), and one in the middle of the anterior basal lobe (lower trace). (b) Two simultaneous recordings in the anterior basal lobe, one dorsally in its posterior part (upper trace) and one ventrally (lower trace), digitized at 100 Hz, filtered from 10 to 500 Hz. (c) Two simultaneous recordings, one in the dorsal precommissural lobe (upper trace), and one from the border between the anterior part of the median basal lobe and the precommissural lobe, near the mid-line, possibly in the optic commissure (lower trace), digitized at 300 Hz, filtered from 1 to 100 Hz. (From Bullock and Budelmann 1991; reprinted with the permission of Springer-Verlag.)

the optic nerves. In such deprived octopods the dorsal aorta was cannulated with a fine polyethylene tubing (Andrews and Tansey 1981) to inject [^{14}C]DG, dissolved in sterile ASW. Cuttlefish were injected directly into the blood sinus on the ventral side of the head. Shortly after injection of [^{14}C]DG, the animals received a variety of visual stimuli for 45 min, before they were killed by decapitation. The brain was then quickly processed for cryostat sectioning. The sections were exposed for about 9 days to a calibrated X-ray film and the

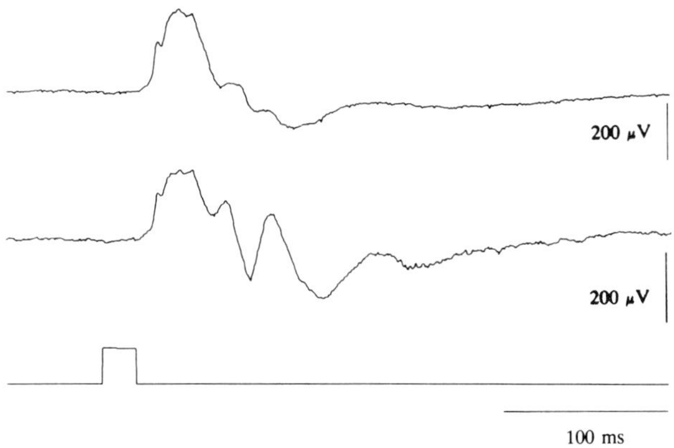

Fig. 24.5. Visual evoked potentials in the brain of the cuttlefish *Sepia*, recorded in two loci to a 20 ms light flash (lower trace), one in the dorsal precommissural lobe (upper trace) and one from the border between the anterior part of the median basal lobe and precommissural lobe, near the mid-line, possibly in the optic commissure (middle trace). Single sweeps, filtered from 1 to 100 Hz. (From Bullock and Budelmann 1991; reprinted with the permission of Springer-Verlag.)

^{14}C concentrations were measured by means of a charge-coupled-device camera and an image analysis system.

Novicki *et al.* (1992) showed that the autoradiographs of the optic lobe sections of the animals injected with $[^{14}C]DG$ had a significantly lower density in the lobe of the visually deprived side, indicating a lower $[^{14}C]DG$ uptake and thus a lower metabolic activity in the deprived lobe. Differences between the non-deprived and the deprived side were also seen at the level of the superior frontal–vertical lobe system, which is involved in visual memory (Young 1965). Thus, Novicki *et al.* (1992) were able to demonstrate that the $[^{14}C]DG$ autoradiographic technique can be profitably applied to octopod and decapod cephalopods.

Concluding remarks

For many years the three preparations described above have been very valuable and powerful tools for physiological experiments on vertebrate brain function. They have now been successfully adapted to physiological research on cephalopod brains, which at the level of gross anatomy and behaviour are undoubtedly the best understood of all the invertebrate brains. Thus, these new preparations finally allow us to make use of the wealth of morphological information we have from the outstanding anatomical and behavioural work of J. Z. Young and others over the past 50 years on the pathways, connections, and functions of the various lobes of the octopus, cuttlefish, and squid brains.

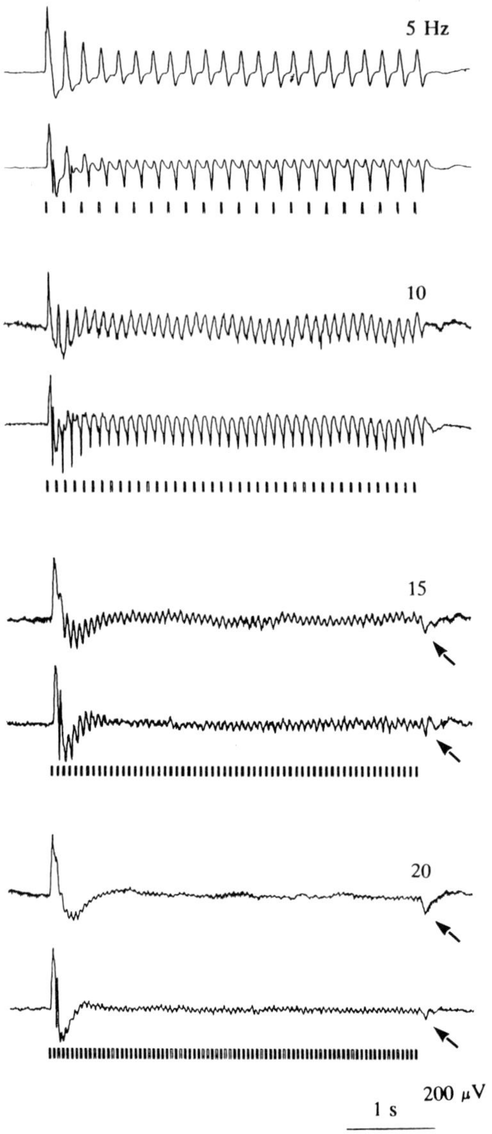

Fig. 24.6. Visual evoked potentials to trains of flashes of various frequencies, recorded in the brain of the cuttlefish *Sepia*. Every 10 s a 4.2 s train of 20 ms flashes of 2100 lux was directed from the end of a light pipe on to the animal's right eye from a distance of 80 mm. The frequencies of the flashes are shown on the right and by the photodiode ticks. Records from two brain loci, filtered from 1 to 1000 Hz, are shown for each frequency (same preparation and loci as in Fig. 24.5). Records at 5 Hz are averages of 10 sweeps, the others are single sweeps showing small spikes. At 15 and 20 Hz, there is a slow wave after the last flash, the 'omitted stimulus potential (OSP)' (arrows). Vertical calibration bar = 200 μV for both channels. (From Bullock and Budelmann 1991; reprinted with the permission of Springer-Verlag.)

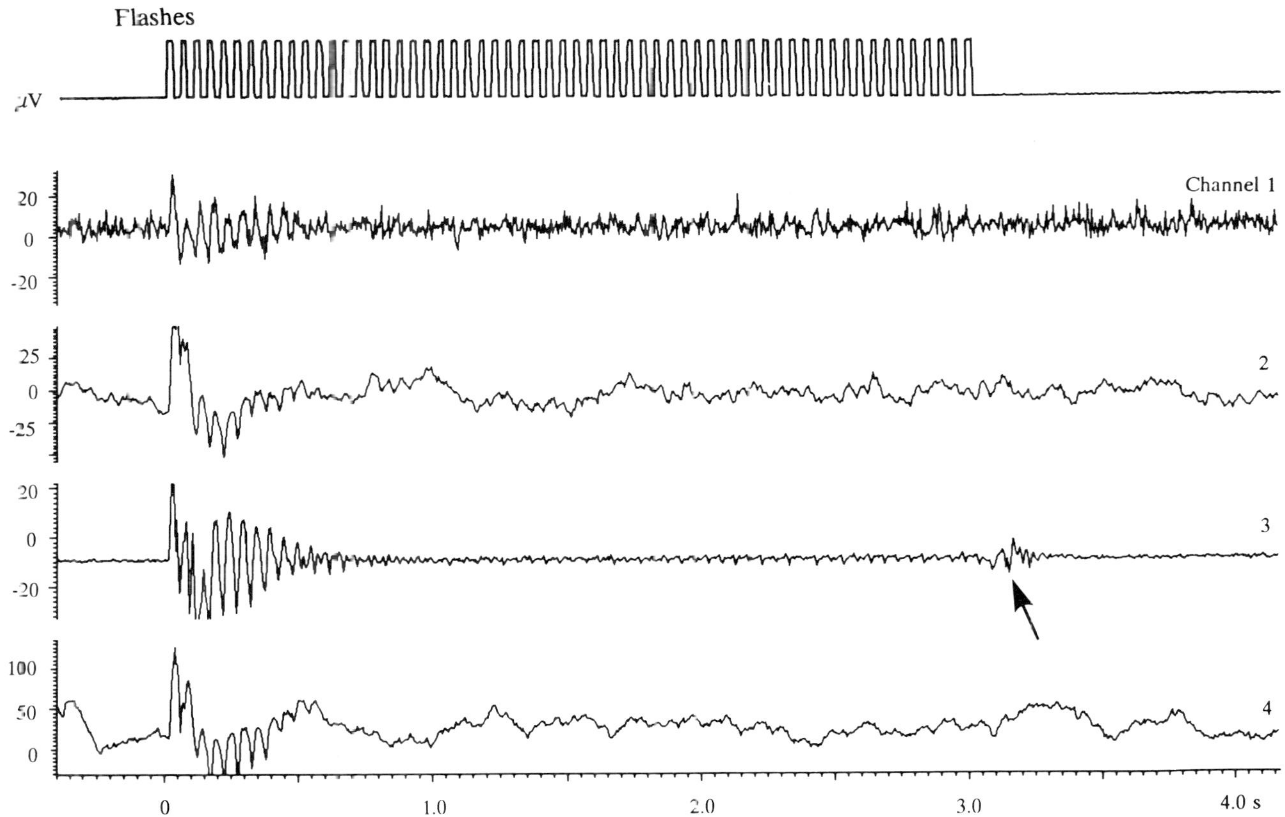

Fig. 24.7. Visual evoked potentials and the omitted stimulus potential recorded in the brain of the cuttlefish *Sepia* in four loci: *channel 1*, between superior and inferior frontal lobe; *channels 2 and 3*, optic lobe; *channel 4*, precommissural lobe. Light flashes at 20 Hz every 15 s. Recordings are averages of 17 sweeps, showing that all loci have visual evoked activity for a few flash cycles before following fails. Only channel 3 shows the omitted stimulus potential after the end of the flash train (arrow). (From Bullock and Budelmann 1991; reprinted with the permission of Springer-Verlag.)

All three preparations still offer several opportunities for further improvements. Once these have been achieved, it might become possible—at the functional level—to do such diverse experiments as comparison of different ontogenetic stages and behavioural states, or of results of lesions, drug applications, regeneration, stimulation, and training. The cephalopod preparations have the potential of becoming extremely interesting and powerful alternative (not substitute!) invertebrate preparations for the study of brain function, perhaps including the evolution of learning, memory, motivation, and cognition. Certainly they are able to open a new decade of fascinating physiological research on cephalopod brains.

References

Andrews, P. L. R. and Tansey, E. M. (1981). A technique for central drug administration in *Octopus vulgaris. Journal of Neuroscience Methods*, **4**, 243–7.

Bausenwein, B. (1990). 2-Deoxyglucose labelling of central brain compartments in *Drosophila. Biological Chemistry*, **371**, 8.

Bleckmann, H., Budelmann, B. U., and Bullock, T. H. (1991). Peripheral and central nervous responses evoked by small water movements in a cephalopod. *Journal of Comparative Physiology A*, **168**, 247–57.

Boycott, B. B. (1961). The functional organization of the brain of the cuttlefish *Sepia officinalis. Proceedings of the Royal Society of London, Series B*, **153**, 503–34.

Boycott, B. B., Lettvin, J. Y., Maturana, H. R., and Wall, P. D. (1965). Octopus optic responses. *Experimental Neurology*, **12**, 247–56.

Buchner, E., Buchner, S., and Bülthoff, I. (1984). Deoxyglucose mapping of nervous activity induced in *Drosophila* brain by visual movement. I. Wildtype. *Journal of Comparative Physiology A*, **155**, 471–83.

Budelmann, B. U. and Young, J. Z. (1984). The statocyst–oculomotor system of *Octopus vulgaris:* Eye muscles, eye muscle nerves, statocyst nerves, and the oculomotor centre in the central nervous system. *Philosophical Transactions of the Royal Society of London, Series B*, **306**, 159–89.

Budelmann, B. U. and Young, J. Z. (1985). Central pathways of the nerves of the arms and mantle of *Octopus. Philosophical Transactions of the Royal Society of London, Series B*, **310**, 109–22.

Budelmann, B. U. and Young, J. Z. (1987). Brain pathways of the brachial nerves of *Sepia* and *Loligo. Philosophical Transactions of the Royal Society of London, Series B*, **315**, 345–52.

Bullock, T. H. (1945). Problems in the comparative study of brain waves. *Yale Journal of Biological Medicine*, **17**, 657–79.

Bullock, T. H. (1984). Ongoing compound field potentials from octopus brain are labile and vertebrate-like. *Electroencephalography and Clinical Neurophysiology*, **57**, 473–83.

Bullock, T. H. and Basar, E. (1988). Comparison of ongoing compound field potentials in the brains of invertebrates and vertebrates. *Brain Research Revision*, **13**, 57–75.

Bullock, T. H. and Budelmann, B. U. (1991). Sensory evoked potentials in unanesthetized unrestrained cuttlefish: a new preparation for brain physiology in cephalopods. *Journal of Comparative Physiology A*, **168**, 141–50.

Bullock, T. H. and Uter, T. G. (1976). Brain waves of octopus. *Society for Neuroscience Abstracts*, **2**, 236.

Bullock, T. H., Hofmann, M. H., Nahm, F. K., New, J. G., and Prechtl, J. P. (1990). Event-related potentials in the retina and optic tectum of fish. *Journal of Neurophysiology*, **64**, 903–14.

Chase, R. (1985). Response to odors mapped in snail tentacle and brain by ^{14}C-2-deoxyglucose autoradiography. *Journal of Neuroscience*, **5**, 2930–9.

Chichery, R. (1983). Motor and behavioral responses obtained by electrical stimulation of peduncle and basal lobes: the control of visual-static centres on oculomotor reactions and locomotion in the cuttlefish, *Sepia officinalis*. *Fortschritte der Zoologie*, **28**, 231–40.

Chichery, R. and Chanelet, J. (1976). Motor and behavioral responses obtained by stimulation with chronic electrodes of the optic lobe of *Sepia officinalis*. *Brain Research*, **105**, 525–32.

Corner, M. A. and Schadé, J. P. (1967). Neurobiological studies on cephalopods. VI. The electrical response of the eye and retinal nerves in *Sepia officinalis*. *Pubblicazoni della Stazione Zoologica di Napoli*, **35**, 353–73.

Cory, H. Y. and Rose, S. P. R. (1969). Glucose and amino acid metabolism in *Octopus* optic and vertical lobes *in vitro*. *Journal of Neurochemistry*, **16**, 979–88.

Gilly, W. F., Hopkins, B., and Mackie, G. O. (1991). Development of giant motor axons and neural control of escape responses in squid embryos and hatchlings. *Biological Bulletin*, **180**, 209–20.

Gleadall, I. G. (1990) Higher motor function in the brain of *Octopus*: the anterior basal lobe and its analogies with the vertebrate basal ganglia. *Annals of Applied Information Sciences*, **16**, 1–30.

Hartline, P. H. and Lange, G. D. (1974). Optic nerve responses to visual stimuli in squid. *Journal of Comparative Physiology*, **93**, 37–54.

Lange, G. D. and Hartline, P. H. (1974). Retinal responses in squid and octopus. *Journal of Comparative Physiology*, **93**, 19–36.

Laverack, M. S. (1980). Electrophysiology of the isolated central nervous system of the northern octopus *Eledone cirrhosa*. *Marine Behavioural Physiology* **7**, 155–69.

Lettvin, J. Y. and Maturana, H. R. (1961). Octopus vision. *Massachusetts Institute of Technology Quarterly Progress Report*, **61**, 194–209.

Lettvin, J. Y. and Maturana, H. R. (1962). Neapolitan studies. *Massachusetts Institute of Technology Quarterly Progress Report*, **64**, 288–92.

MacNichol, E. F. and Love, W. E. (1960*a*). Electrical responses of the retinal nerve and optic ganglion of the squid. *Science*, **132**, 737–8.

MacNichol, E. F. and Love, W. E. (1960*b*). Impulse discharges from retinal nerve and optic ganglion of the squid. In *Symposium on the visual system* (ed. R. Jung and J. Kornhuber), pp. 97–103. Springer-Verlag, Berlin.

Mangold, K. (1989). Le système nerveuz. In *Traitè de Zoologie, Tome V, Fascicule 4, Céphalopodes* (ed. K. Mangold), pp. 163–240. Masson, Paris.

Marquis, F. (1989). Die Embryonalentwicklung des Nervensystems von *Octopus vulgaris* Lam. (Cephalopoda, Octopoda), eine histologische Analyse. *Verhandlungen der Naturforschenden Gesellschaft Basel*, **99**, 23–76.

Messenger, J. B. (1973). Learning in the cuttlefish, *Sepia*. *Animal Behaviour*, **21**, 801–26.

Messenger, J. B. (1979). The nervous system of *Loligo*. IV. The peduncle and olfactory lobes. *Philosophical Transactions of the Royal Society of London, Series B*, **285**, 275–309.

Mislin, H. (1955). Die rhytmischen Spontanentladungen im Zentralnervensystem der Tintenfische. *Acta Medica Scandinavica Supplementum*, **307**, 58.
Muntz, W. R. A. (1963). Interocular transfer and the function of the optic lobes in *Octopus*. *Quarterly Journal of Experimental Psychology*, **15**, 116–24.
Miyan, J. A. and Messenger, J. B. (1995). Intracellular recordings from the chromatophore lobes of *Octopus*. In *Cephalopod Neurobiology* (ed. N. J. Abbott, R. Williamson, and L. Maddock), pp. 415–29. Oxford University Press, Oxford.
Novicki, A., Messenger, J. B., Budelmann, B. U., Terrell, M. L., and Kadekaro, M. (1992). [^{14}C]Deoxyglucose labelling of functional activity in the cephalopod central nervous system. *Proceedings of the Royal Society of London, Series B*, **249**, 77–82.
Patterson, J. A. and Silver, S. C. (1983). Afferent and efferent components of *Octopus* retina. *Journal of Comparative Physiology*, **151**, 381–7.
Plän, T. (1987). Functional neuroanatomy of sensory-motor lobes of the brain of *Octopus vulgaris*. Dissertation, University of Regensburg, Federal Republic of Germany.
Robertson, J. D., Gillette, R., Lee, P., Meadows, S., and Zitz, J. (1990). Neuronal somata in the *Octopus* central nervous system are inexcitable and label retrogradely with DiI and DiO. *Society for Neuroscience Abstracts*, **16**, 605.
Saidel, W. M., Lettvin J. Y., and MacNichol, E. F. (1983). Processing of polarized light by squid photoreceptors. *Nature*, **304**, 534–6.
Sanders, F. K. and Young, J. Z. (1940). Learning and other functions of the higher nervous centres of *Sepia*. *Journal of Neurophysiology*, **3**, 501–26.
Sejnowski, T. J., Reingold, S. C., Kelley, D. B., and Gelperin, A. (1980). Localization of ^{3}H-2-deoxyglucose in single molluscan neurones. *Nature*, **287**, 449–51.
Sokoloff, L. (1987). Mapping cerebral functional activity with radioactive deoxyglucose. In *Encyclopedia de neuroscience* (ed. G. Adelman), pp. 604–9. Birkhäuser, Boston.
Sokoloff, L., Reivich, M., Kennedy, C., Des Rosiers, M. H., Patlak, C. S., Pettigrew, K. D. *et al.* (1977). The [^{14}C]deoxyglucose method for the measurement of local cerebral glucose utilization: theory, procedure, and normal values in the conscious and anesthetized albino rat. *Journal of Neurochemistry*, **28**, 897–916.
Stephen, R. O. (1974). Electrophysiological studies of the brain of *Octopus vulgaris*. *Journal of Physiology*, **240**, 19–20P.
Wells, M. J. (1978). *Octopus. Physiology and behaviour of an advanced invertebrate*. Chapman & Hall, London.
Williamson, R. and Budelmann, B. U. (1991). Convergent inputs to *Octopus* oculomotor neurones demonstrated in a brain slice preparation. *Neuroscience Letters*, **121**, 215–18.
Woodhams, P. L. (1977). The ultrastructure of a cerebellar analogue in *Octopus*. *Journal of Comparative Neurology*, **174**, 329–46.
Young, J. Z. (1965). The organization of a memory system. *Proceedings of the Royal Society of London, Series B*, **163**, 285–320.
Young, J. Z. (1971). *The anatomy of the nervous system of* Octopus vulgaris. Clarendon Press, Oxford.
Young, J. Z. (1974). The central nervous system of *Loligo*. I. The optic lobe. *Philosophical Transactions of the Royal Society of London, Series B*, **267**, 263–302.
Young, J. Z. (1976). The nervous system of *Loligo*. II. Suboesophageal centres. *Philosophical Transactions of the Royal Society of London, Series B*, **274**, 101–67.
Young, J. Z. (1977*a*). The nervous system of *Loligo*. III. Higher motor centres: the basal supraoesophageal lobes. *Philosophical Transactions of the Royal Society of London, Series B*, **276**, 351–98.

Young, J. Z. (1977*b*). Brain, behaviour and evolution in cephalopods. *Symposia of the Zoological Society of London*, **38**, 377–434.
Young, J. Z. (1979). The nervous system of *Loligo*. V. The vertical lobe complex. *Philosophical Transactions of the Royal Society of London, Series B*, **285**, 311–54.
Young, J. Z. (1983). The distributed tactile memory system of *Octopus*. *Proceedings of the Royal Society of London, Series B*, **218**, 135–76.
Young, J. Z. (1991). Computation in the learning system of cephalopods. *Biological Bulletin*, **180**, 200–8.

25

Intracellular recordings from the chromatophore lobes of *Octopus*

J. A. MIYAN and J. B. MESSENGER

Summary

This is an account of a whole brain preparation that has enabled us for the first time to dye-fill, and record intracellularly from, cells in the brain of *Octopus*. Over a hundred cells in the posterior chromatophore lobes of nearly 50 octopuses were penetrated. Most cells were spontaneously active, firing at 1–12 Hz, some were silent and others showed regular bursting activity. Lucifer Yellow fills confirmed that many cells in this lobe are unusual by invertebrate standards in having cell body 'dendrites' as well as axonal processes with branches. Many cells are also dye-coupled. This new preparation could be modified for longlasting intracellular recording in many lobes of the brain of semi-intact cephalopods.

Many physiological preparations have been developed in cephalopods to study fundamental features of nerve and glial physiology. In contrast, there have been few attempts to apply modern neurobiological techniques to the understanding of neural function in the intact, whole brain of these animals. The detailed mapping of the brain of cephalopods by Young and his colleagues (see Chapter 26) has not been followed by any serious attempt to investigate it physiologically. This failure has largely been due to the lack of a suitable preparation in which single cell electrophysiology can be correlated with behavioural output, although one recent technique seems to offer great possibilities for electrophysiological studies in totally intact, unrestrained animals (Bullock and Budelmann 1991; see also Chapter 24). Perfused, isolated brains have yielded some information about the electrical properties of large cells in the sub-oesophageal lobes (Laverack 1980) and activity in some of these cells has been correlated with ventilatory activity of the mantle in semi-intact, perfused preparations (Boardman and Laverack, personal communication), but these preparations are short-lived and the supra-oesophageal lobes rapidly lose vascular flow (Miyan, unpublished observation). Brain slice techniques are now being applied with great success to the cephalopod CNS (Robertson *et al.* 1990; see also Chapter 24) but these are not suitable for neuroethological studies.

As an extension to the preparation developed for recordings from peripheral nerves innervating the skin of living cuttlefish (Messenger and Miyan 1986), we sought to develop a preparation in which glass micropipettes could be inserted into individual neurones for recording and subsequent dye injection. For this study we concentrated our attention on the chromatophore system of semi-intact octopuses as this has been well characterized and the posterior chromatophore lobes are readily accessible.

The chromatophore system

It has been known since antiquity that octopuses and squids can change colour; and over the last 150 years a number of workers have addressed themselves to the natural history of colour change in these animals (e.g. Cowdry 1911; Holmes 1940; Packard and Sanders 1969, 1972; Packard and Hochberg 1977; Hanlon and Messenger 1988). There is also an extensive literature on the physiology of the chromatophore organs, admirably summarized in the important papers of Florey (1966, 1969; Florey and Kriebel 1968; Cloney and Florey 1968; Florey *et al.* 1985).

The nerves to the chromatophores originate in the brain itself (Sereni and Young 1932; but see Rowell 1963), specifically in the paired anterior and posterior chromatophore lobes (Fig. 25.1). These form part of the hierarchically arranged chromatophore control system (Boycott 1953, 1961; Young 1971), which is driven principally by visual cues, although open to other

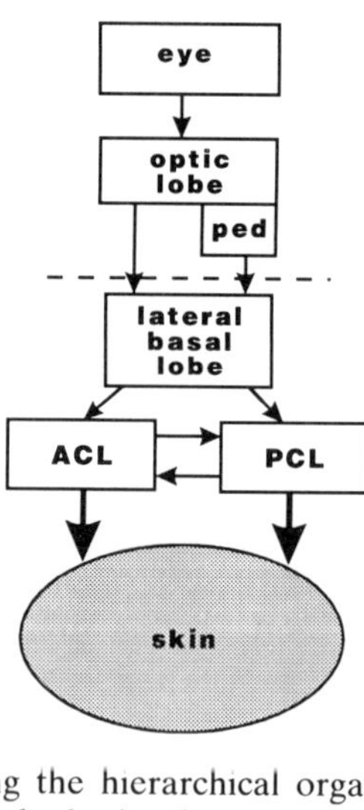

Fig. 25.1. Diagram illustrating the hierarchical organization of the chromatophore control system for one side of the brain. Commissures link the left and right anterior (ACL) and posterior (PCL) chromatophore lobes. In the experiments described here, everything above the dotted line was removed. Ped = peduncle lobe.

influences (Budelmann and Young 1985; Ferguson and Messenger 1991). Motor neurones in the chromatophore lobes innervate the skin of the arms, head, and mantle ipsilaterally. In *Octopus* the anterior chromatophore lobes are small and inaccessible and we made no attempt to penetrate these lobes with microelectrodes. The posterior chromatophore lobes (PCL) are relatively large and conspicuous, however, lying as hemispheres on the lateral, dorsal surface of the palliovisceral lobe. And because these lobes extend more posteriorly than the supra-oesophageal brain, they are accessible to electrodes, as in *Lolliguncula* (Dubas *et al.* 1986).

In *Octopus vulgaris* each PCL comprises about 150 000 cells (Young 1963, 1971) lying in a rind, or cell layer, around a central neuropil [Fig. 25.2(a)]. There are small cells (8–12 μm) lying close to the neuropil and larger cells (15–30 μm) lying peripherally. The cells appear to form vertical columns, and clusters of columns are grouped together leaving spaces between them into which the neuropil extends as a series of 'canals' [Fig. 25.2(b)]. The columnar arrangement is especially evident in thin sections prepared for electron microscopy [EM; Fig. 25.2(c)] and EM studies have shown that many of the cells lie very close to one another (Camm 1986). Many of the cells are unusual for an invertebrate in that their perikarya bear moderately long processes or 'dendrites' that appear to receive synaptic input (Young 1971). In the light microscope, degeneration has been observed around these processes after lesions to optic and lateral basal lobes (Camm and Messenger, unpublished). Degeneration is also seen in the canals between clusters of columns, and it is assumed that they must provide a route for incoming fibres to influence cells with dendrites in the cell layer. Most cells in the PCL have a main process or axon that runs down into the neuropil, where it gives off fine branches before running either to the skin or to another chromatophore lobe. The neuropil appears as a regularly arranged meshwork of fibres, in between which are the efferent bundles. There are also some cells whose processes are confined to the PCL. Efferent pathways from the PCL run to: (1) the ipsilateral mantle and arm skin; (2) the contralateral PCL; and (3) the anterior chromatophore lobes.

Materials and methods

Octopus vulgaris (Cuvier) of both sexes and weighing between 600 and 1200 g were caught in the Bay of Naples and kept in holding tanks with locking lids and circulating sea water. They were fed daily with live crabs. All experiments were carried out between May and September 1986.

The preparation

Individuals selected for experiments were anaesthetized in a 1.5 per cent solution of ethanol in sea water. Anaesthesia was deemed to be complete

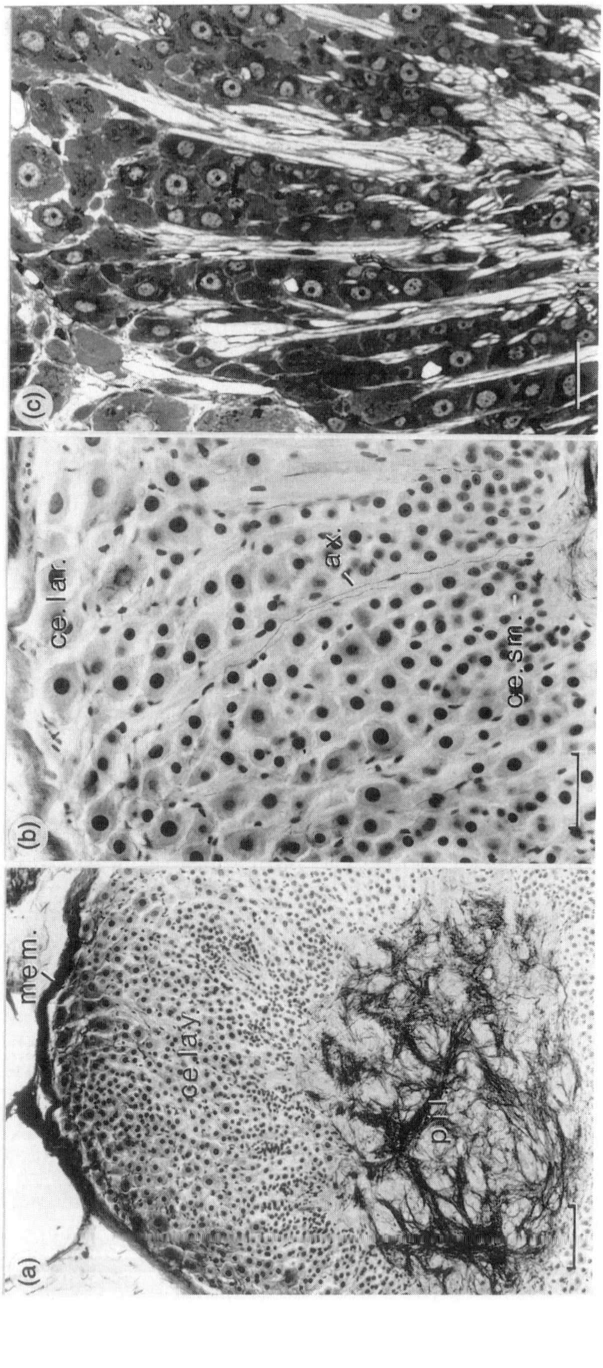

Fig. 25.2 (a) *Octopus* posterior chromatophore lobe (PCL) in transverse section, Cajal silver stain. Note the characteristic dark meshwork in the neuropil formed by afferent fibres projecting into the lobe. ce.lay., cell layer; mem., membrane over lobe; pil., neuropil. Scale bar: 500 μm. (b) Higher power view of previous section showing large and small cells in the cell layer (ce.lar., ce.sm.) and a 'canal' with nerve axons (ax.). Scale bar: 60 μm. (c) Columnar arrangement of cells in the PCL seen in a semi-thin section stained with toluidine blue (by kind permission of Dr J.-P. Camm). Scale bar: 100 μm.

when animals ceased to respond to a pinch applied to the skin over the eyes (Andrews and Tansey 1981). This generally occurred before cessation of respiration. The animals were weighed and sexed before surgery. In all our experiments we opened the orbits, cut the optic tracts central to the peduncle complex and removed the optic lobes bilaterally. Removal of these highest centres of the brain resulted in a very quiet preparation (Messenger 1967); the eyes were also removed to ensure rigid mounting of the cranium to the holder (see below).

During anaesthesia, all eight arms were rapidly amputated close to the head. The contraction of the arm muscles prevented any serious blood loss, unless the level of anaesthesia was too deep, when massive haemorrhage occurred. Experiments proved impossible when the arms were intact (see below).

After removal of the optic lobes, eyes, and arms, a mid-dorsal incision was made in the cranium to expose the brain. The incision was continued posteriorly to reveal the PCLs, branched cephalic aorta, and salivary gland ducts. Six fish hooks (no.16, barbless, bronze) attached to tungsten wire were inserted into the cranial cartilage. The preparation was then positioned under a 2 cm hole drilled in an 8 mm thick Perspex plate, the wires pulled taught and clamped under six washers held by screws radially arranged around the hole (Fig. 25.3). The screws provided the additional tension required to hold the cranium rigid within the Perspex holder. This was then clamped on to the top of a tank of circulating, cooled and filtered sea water in which the suspended animal rapidly recovered from anaesthesia to show slow, strong, rhythmical respiration. The tank had an integral cooling jacket that was connected to a Grant chiller unit and pump. The system utilized an open reservoir circuit for pressure compensation and secondary cooling of a Fluval, sealed filter pump that fed the preparation bath with sea water. Bath temperature was held between 10 and 15°C at ambient temperatures of about 27°C. A stainless steel plate covering the floor of the tank formed the ground reference and earthing point for the preparation. A copper pipe integrated into the plastic pipework of the cooling system provided a convenient earthing point for that system. The whole tank was bolted to the steel plate of the experimental table.

The preparation was viewed through a Wild stereo zoom microscope. Half an hour was allowed for recovery and then the cranial jelly surrounding the brain was quickly removed, using a Pasteur pipette attached to a vacuum pump. Very little damage resulted from this procedure. A window was then cut in the membrane overlying one of the PCLs, with care being taken to avoid damaging the branches of the cephalic aorta. Puncturing of the membrane was essential for electrode penetration as the lobes are surrounded by a tough collagenous sheath (see Fig. 25.2).

Electrophysiology

Standard electrophysiological techniques were used throughout. We found that a hydraulic microdrive (Narashige MO-15N) was essential for successful

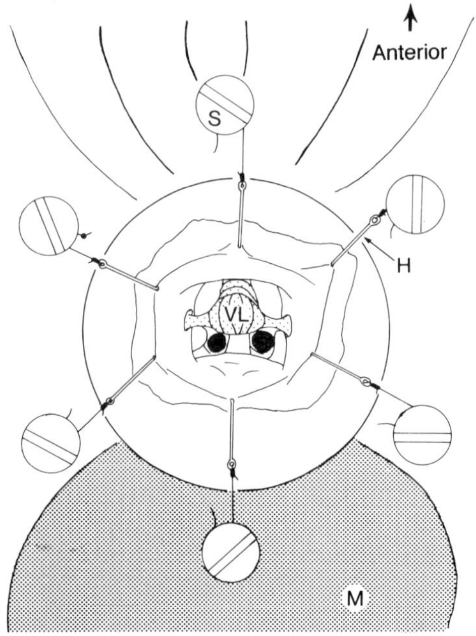

Fig. 25.3. Drawing of the *in situ Octopus* brain preparation (dorsal view) prior to insertion of electrodes. Fish hooks (H) are used to anchor the cranium on to the perspex plate under a hole. The posterior chromatophore lobes (dark stipple) are visible at the posterior end of the brain with easy access behind the vertical lobes (VL). S, screw; M, mantle.

penetrations, which were best achieved with brief capacity overcompensation. Records were photographed from the screen of a storage oscilloscope. For injection of Lucifer Yellow, we found that penetrations of less than 3 min were insufficient to fill the fine processes of cells, although the cell body and axon were filled.

Our criteria for interpretation of both electrophysiological and morphological data were that a single penetration was defined as that in which no sudden changes in membrane potentials nor spike amplitude occurred during a recording and dye-filling experiment. Data from single clean recordings in which the first cell impaled was maintained, together with the fact that most of our observations were of superficial cells led us to conclude that we had maintained a single cell penetration. Extracellular, multi-unit recordings were made, but these were clearly different having varying amplitude spikes. Strong respiration in the preparation led to movement artefacts in our recordings but in most cases we did not lose our penetration unless there was also jetting, a rare occurrence.

Histology

Immediately after dye injection, the preparation was re-anaesthetized and the entire brain (together with a short length of the ipsilateral pallial nerve) was removed. Fixation was in 10 per cent formalin in sea water overnight. Initially, thick transverse sections were cut by hand with a razor blade. These were dehydrated, cleared in methyl salicylate and viewed either unmounted in a Petri dish, or mounted in Canada balsam on a slide. Later preparations were embedded in celloidin for thick sectioning (100–200 μm) using a Reichert sliding microtome. Sections were made in the sagittal plane in an attempt to trace the destination of axons from filled cells. Cells were visualized using a Leitz Orthoplan, or Zeiss Standard 16 microscope equipped for epifluorescence. All photographs were taken on Kodak Ektachrome 400 film. Cells were drawn by hand without the aid of a drawing tube.

Results

The behaviour of the preparation

Octopuses survived the extensive surgery necessary for these experiments surprisingly well. Although they are well known to be very resilient animals that can withstand rough handling by fishermen as well as the insults of surgery (see Boycott 1954; Wells 1962, 1978; Young 1971) the stability of this preparation was remarkable.

Virtually every preparation recovered rapidly from anaesthesia and resumed regular, 'quiet' respiratory movements within minutes of being attached to the head plate. No spasmodic or violent 'jetting' movements were made, even initially, and there was no spontaneous inking in undisturbed preparations. Nor were there abrupt colour changes (see below). To observers familiar with the behaviour of healthy intact octopuses these preparations appeared quiescent and relatively untraumatized. The duration of most experiments was less than 6 h but early on we found that undisturbed preparations could survive overnight.

It needs emphasizing that we had recourse to arm amputation only because we could devise no other way of immobilizing the preparation. A pilot experiment showed that with the arms intact movements could not be eliminated even if the entire supra-oesophageal brain was removed in addition to the optic lobes (cf. Boycott and Young 1950), nor did it prove possible to eliminate movements by tying the arms up in a bag (Budelmann 1970). Although sectioning the brachial nerves produces atonia initially, spontaneous arm movements soon arise of sufficient magnitude to move the head in its holder.

The behaviour of the skin

Opening the membrane over the PCL produced a remarkable change in the appearance of the mantle. Immediately there was strong expansion of the chromatophores, ipsilaterally, and, even more striking, the erection of long papillae all over the ipsilateral half of the mantle, ceasing abruptly at the mid-line. This effect, a consistent feature of cutting the membrane under sea water, confirms that the so-called chromatophore lobes control the muscles of the skin as well as the chromatophore muscles (Boycott 1961). This dark spiky condition persisted for 10–15 min. Subsequent insertion of micro-electrodes caused no perceptible change in the appearance of the skin.

Neurone morphology

From 47 preparations, 120 cells were successfully filled with Lucifer Yellow. The majority were cells with axons that passed out of the lobe. Most of these were probably motor neurones but some may have been contributing to the inter-chromatophore lobe connectives and commissures. These cells could be subdivided into a number of classes: (1) those with no arborizations; (2) those with arborizations restricted to the neuropil [Fig. 25.4(a)]; (3) those with arborizations in the neuropil and also within the cell layer [Fig. 25.4(b)]; (4) those with major axon collaterals that return along the efferent path to end in the cell layer [Fig. 25.4(c)]; (5) those with 'dendrites' on the soma [Fig. 25.4(d)]; and (6) those that, in addition to other characteristics, are dye-coupled to one or more adjacent cells [Fig. 25.4(e) and (f), (g) and (h)].

Another category of cells was filled that had no fibre leaving the lobe (Fig. 25.5). These, which we took to be interneurones, had similar additional characteristics to those listed for the motor neurone/commissural cells above. One of these interneurones was an intracortical neurone [Fig. 25.5(b)], having no processes that passed out of the cell layer.

Details of morphology

Most of the cell penetrations were made at, or near, the dorsal surface of the lobe. Thus the cells filled were among the largest found in the lobe (Young 1971; and see Fig. 25.2 above) with soma diameters between 35 and 83 μm.

Young (1971) has commented on the small size of cell processes in the chromatophore lobes and one of the most striking characteristics of many of the cells we filled with Lucifer Yellow is the way the axon tapers sharply as it leaves the soma and passes through the cell layer (Figs 25.4–25.6). This was not an artefact as other cells filled in the same preparations under the same conditions do not possess such tapered processes.

Many of the cells filled had terminal swellings of their finer arborizations. These were of three size categories [Fig. 25.6(d)–(f); see also Fig. 25.4 (i) and

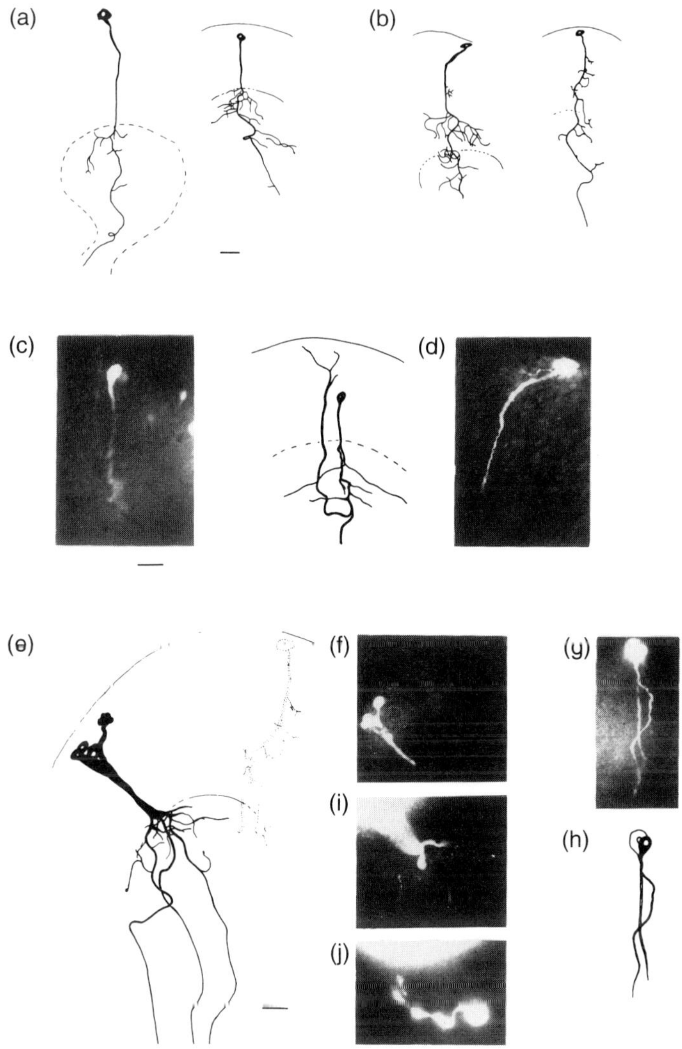

Fig. 25.4. Lucifer Yellow filled cells in the posterior chromatophore lobe. Dorsal is to the top in all cases. The solid line indicates the surface of the lobe/brain, while the dotted line indicates the boundary between the cell layer and neuropil. (a) Cells with arborizations restricted to the neuropil. (b) Cells with arborizations in both cell layer and neuropil. (c) Cell with a major axon collateral returning through the cell layer. (d) Another cell with a short 'dendritic' process running in the cell layer. (e) and (f), (g) and (h) Dye-coupled cells visualized with Lucifer Yellow. (i) and (j) High power views of terminal swellings within the neuropil. Scale bar: 50 μm (a,b), 30 μm (c), 50 μm (c), (f), (g), and (h), 10 μm (i), 1μm (j).

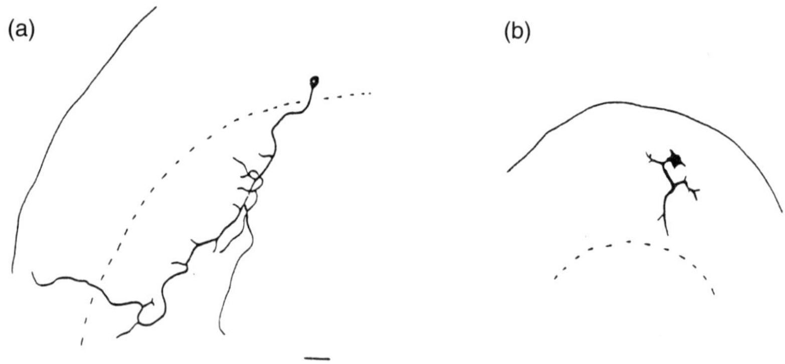

Fig. 25.5. Drawings of interneurones within the posterior chromatophore lobe. (a) A cell with a process extending through the neuropil and into the cell layer at some distance from the soma. (b) An intracortical cell with no process leaving the cell layer. Scale bar: 20 μm.

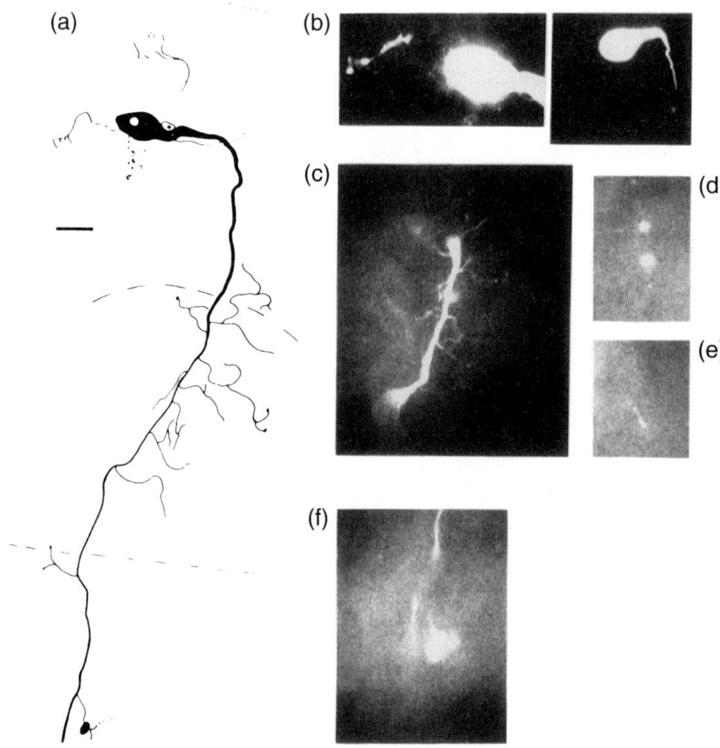

Fig. 25.6. Lucifer Yellow filled cell in the posterior chromatophore lobe with all three sizes of terminal swelling. The most striking of these is the large 30 μm swelling on a small branch off the main axon leaving the lobe (a) and (f). The function of such large structures is unknown. (b) Views of the soma and tapering axon; (c) view of the axon and branches within the neuropil; (d)-(f) the three sizes of terminal swellings present on this cell. Scale bar: 50 μm.

(j)]: (1) 1–6 μm diameter; (2) 7–13 μm, the most common size; and (3) 30 μm diameter (with only a single example).

Dye-coupling

In approximately one-quarter of the preparations in which electrophysiological evidence suggested we had maintained a single cell recording for the duration of the dye injection (see later), two or more closely apposed cells were filled with Lucifer Yellow [Fig. 25.4(e) and (f), (g) and (h)]. These were, therefore, probably dye-coupled across electrotonic gap junctions (Stewart 1978). A second type of coupling was observed between a filled neurone and what appeared to be a sheath of small cells running along its axon (Fig. 25.7). It is possible that these are glial cells, which are numerous in the PCL (Young 1971), coupled to each other and/or to the axon of the impaled neurone.

Physiology

Intracellular recording revealed three different classes of cell in the PCL. The majority of the cells exhibited continuous tonic spontaneous activity, with firing frequencies between 1 and 12 Hz [Fig. 25.8(a) and (b)]. Assuming that most of the cells sampled were motor neurones, this finding is consistent with: (1) the fact that in isolated squid skin the chromatophores show maximum expansion when stimulated at 12–15 Hz (Florey 1966); and (2) the observation that in healthy intact octopuses, and in our preparations, a tonic level of chromatophore expansion is always present (cf. Young 1971). Cells that did

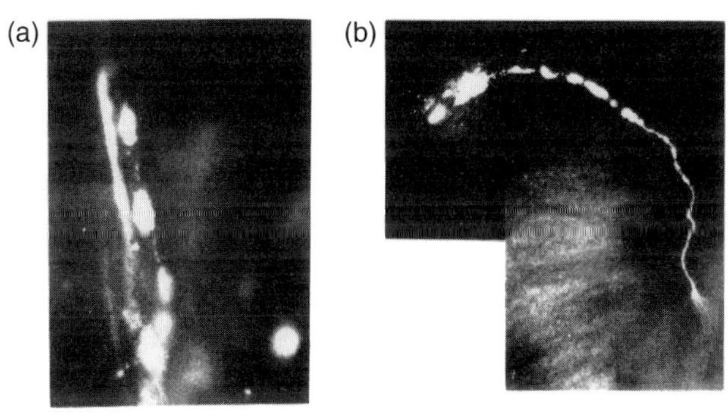

Fig. 25.7. Lucifer Yellow filled cells in the posterior chromatophore lobe demonstrating dye-coupling to cellular elements associated with the major axonal process. In (a) what appear to be nuclei have stained within otherwise unstained cells. It is likely that these are glial cells. (b) In this fill no axon is visible but a line of cell nuclei, similar to those in (a), are stained along the characteristic course of an axonal process. This remarkable fill suggests there may be dye-coupling between (probable) glial cells and particular neurones. Scale bars: 20 μm.

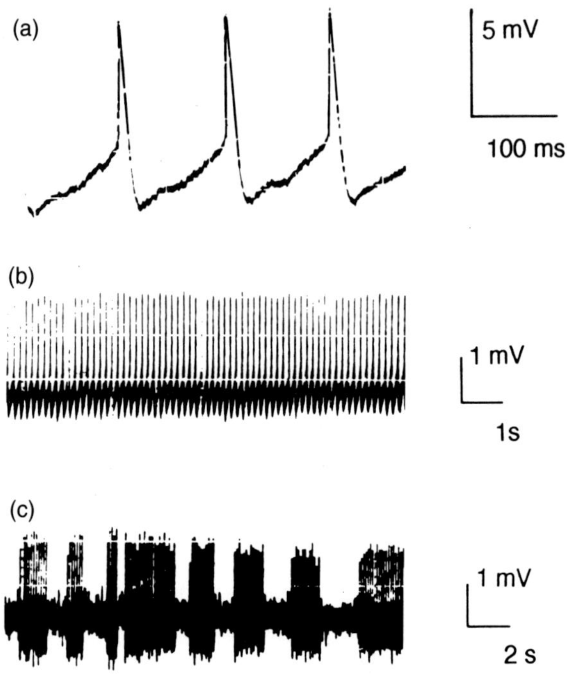

Fig. 25.8. Intracellular recordings from cells in the posterior chromatophore lobe (a) on a fast time base to illustrate the time course of action potentials which was similar in all cells examined. (b) Example of tonically firing cell, commonly encountered. (c) Example of a bursting cell, rarely encountered.

not fire tonically were either silent or showed regular bursting activity [Fig. 25.8(c)]. The resting potential of the cells was between −20 and −40 mV. Our best and later preparations had resting potentials between −35 and −45 mV. The small size of recorded spikes may reflect the antidromic nature of somal activity together with the distance to the site of spike activity in these cells, which is unknown.

Discussion

A major outcome of this study has been the development of a preparation that allows the investigation of a functioning cephalopod brain with modern neurophysiological techniques. This preparation should be ideal for studies of lower motor centres located in the sub-oesophageal lobes, such as those controlling respiration. An important limitation of the method, however, is that it cannot be applied to the highest centres of the cephalopod brain, the optic lobes. These comprise the visual memory store, systems for initiating and controlling whole sets of complex motor programmes, and the visual analysing machinery in the outer layers of the lobe (Cajal 1917; Young 1971, 1974). The preparation, adapted slightly for cuttlefish, is already being utilized in studies of the brain interstitium (see Chapters 23 and 28).

Using this preparation with *O. vulgaris*, we have successfully recorded from, and dye-filled, single cells in the PCL. Lucifer Yellow fills confirm the observations of Young (1971) that some PCL cells bear dendrites and that in the cell layer there are processes wandering among the cell bodies, features that are unusual in an invertebrate. However, since we have only succeeded in sampling less than 0.1 per cent of the 150 000 cells that form a single PCL, it would be premature here to make any statement about this or about the neural organization of the chromatophore system. The variety of cell types we have found suggests the chromatophore control system is very complex. Nevertheless, the demonstration of tonically firing cells, confirming Mislin *et al.* (1954), and agreeing with the behaviour of *Octopus* skin, is encouraging and we believe this preparation may prove invaluable for a neuroethological study of colour patterning and other behaviours in cephalopods.

Acknowledgements

We would like to record our thanks to the Director and staff of the Stazione Zoologica, Naples for the use of facilities, provision of animals and stimulating discussions. This work was supported by the SERC (grant GR/D/73478 to J.B.M.) and the Royal Society (to J.B.M. and J.A.M.).

References

Andrews, P. L. R. and Tansey, E. M. (1981). The effects of some anaesthetic agents in *Octopus vulgaris*. *Comparative Biochemistry and Physiology*, **70C**, 241–7.

Boycott, B. B. (1953). The chromatophore system of cephalopods. *Proceedings of the Linnean Society of London*, **164**, 235–40.

Boycott, B. B. (1954). Learning in *Octopus vulgaris* and other cephalopods. *Pubblicazioni della Stazione Zoologica di Napoli*, **25**, 67–93.

Boycott, B. B. (1961). The functional organisation of the brain of the cuttlefish *Sepia officinalis*. *Proceedings of the Royal Society of London, Series B*, **153**, 503–34.

Boycott, B. B. and Young, J. Z. (1950). The comparative study of learning. *Symposia of the Society for Experimental Biology*, **4**, 432–53.

Budelmann, B. U. (1970). Die Arbeitsweise der Statolithenorgane von *Octopus vulgaris*. *Zeitschrift für vergleichende Physiologie*, **70**, 278–312.

Budelmann, B. U. and Young, J. Z. (1985). Central pathways of the nerves of the arms and mantle of *Octopus*. *Philosophical Transactions of the Royal Society of London, Series B*, **310**, 109–22.

Bullock, T. H. and Budelmann, B. U. (1991). Sensory evoked potentials in unanesthetized unrestrained cuttlefish: a new preparation for brain physiology in cephalopods. *Journal of Comparative Physiology A*, **168**, 141–50.

Cajal, S. Ramon y (1917). Contribución al conociento de la retina y centros ópticos de los cefalópodos. *Trabajos del Laboratorio de investigaciones biológicas de la Universidad de Madrid*, **15**, 1–82.

Camm, J.-P. (1986). The chromatophore lobes and their connections in *Octopus*. PhD Thesis, University of Sheffield.

Cloney, R. A. and Florey, E. (1968). Ultrastructure of cephalopod chromatophore organs. *Zeitschrift für Zellforschung*, **89**, 250–80.

Cowdry, E. V. (1911). The colour changes of *Octopus vulgaris. University of Toronto Studies, Biology* Series No. 10, 1–53.

Dubas, F., Hanlon, R. T., Ferguson, G. P., and Pinsker, H. M. (1986). Localization and stimulation of chromatophore motoneurones in the brain of the squid, *Lolliguncula brevis. Journal of Experimental Biology*, **121**, 1–25.

Ferguson, G. P. and Messenger, J. B. (1991). A countershading reflex in cephalopods. *Proceedings of the Royal Society, London, Series B*, **243**, 63–7.

Florey, E. (1966). Nervous control and spontaneous activity of the chromatophores of a cephalopod, *Loligo opalescens. Comparative Biochemistry and Physiology*, **18**, 305–24.

Florey, E. (1969). Ultrastructure and function of cephalopod chromatophores. *American Zoologist*, **9**, 429–42.

Florey, E. and Kriebel, M. E. (1969). Electrical and mechanical responses of chromatophore muscle fibres of the squid, *Loligo opalescens*, to nerve stimulation and drugs. *Zeitschrift für vergleichende Physiologie*, **65**, 98–130.

Florey, E., Dubas, F., and Hanlon, R. T. (1985). Evidence for L-glutamate as a transmitter substance of motoneurons innervating squid chromatophore muscles. *Comparative Biochemistry and Physiology*, **82C**, 259–68.

Hanlon, R. T. and Messenger, J. B. (1988). Adaptive coloration in young cuttlefish (*Sepia officinalis* L.): the morphology and development of body patterns and their relation to behaviour. *Philosophical Transactions of the Royal Society of London, Series B*, **320**, 437–87.

Holmes, W. (1940). The colour changes and colour patterns of *Sepia officinalis* L. *Proceedings of the Zoological Society of London*, **110A**, 17–35.

Laverack, M. S. (1980). Electrophysiology of the isolated central nervous system of the Northern Octopus *Eledone cirrhosa. Marine Behaviour and Physiology*, **7**, 155–69.

Messenger, J. B. (1967). The effects on locomotion of lesions to the visuo-motor system in *Octopus. Proceedings of the Royal Society, London, Series B*, **167**, 252–81.

Messenger, J. B. and Miyan, J. A. (1986). Neural correlates of colour change in cuttlefish. *Journal of Experimental Biology*, **125**, 395–400.

Mislin, H., Riesterer, L., and Schroeder, D. (1954). Die Aktionspotentiale (EEG) der visceralen Chromatophorenzentren bei Cephalopoden (*Octopus vulgaris*). *Helvetica Physiologica et Pharmacologia Acta*, **12**, C85.

Packard, A. and Hochberg, F. G. (1977). Skin patterning in *Octopus* and other genera. In *The biology of cephalopods* (ed. M. Nixon and J. B. Messenger), pp. 191–231. Academic Press, London.

Packard, A. and Sanders, G. D. (1969). What the octopus shows to the world. *Endeavour*, **28**, 92–9.

Packard, A. and Sanders, G. D. (1971). Body patterns of *Octopus vulgaris* and maturation of the response to disturbance. *Animal Behaviour*, **19**, 780–90.

Robertson, J. D., Gillette R., Lee P., Meadows S., and Zitz, J. (1990). Neuronal somata in *Octopus* central nervous system are inexcitable and label retrogradely with DiI and DiO. *Society for Neuroscience Abstracts*, **19**, 605.

Rowell, C. H. F. (1963). Excitatory and inhibitory pathways in the arm of *Octopus*. *Journal of Experimental Biology*, **40**, 257–70.

Sereni, E. and Young, J. Z. (1932). Nervous degeneration and regeneration in cephalopods. *Pubblicazioni della Stazione Zoologica di Napoli*, **12**, 173–208.

Stewart, W. W. (1978). Functional connections between cells as revealed by dye-coupling with a highly fluorescent naphthalimide tracer. *Cell*, **14**, 741–59.

Wells, M. J. (1962). *Brain and behaviour in cephalopods*. Heinemann, London.

Wells, M. J. (1978). *Octopus. Physiology and behaviour of an advanced invertebrate*. Chapman & Hall, London.

Young, J. Z. (1963). The number and sizes of nerve cells in *Octopus*. *Proceedings of the Zoological Society of London*, **140**, 229–54.

Young, J. Z. (1971). *The anatomy of the nervous system of* Octopus vulgaris. Clarendon Press, Oxford.

Young, J. Z. (1974). The central nervous system of *Loligo*. I. The optic lobe. *Philosophical Transactions of the Royal Society of London, Series B*, **267**, 263–302.

26

Multiple matrices in the memory system of *Octopus*

J. Z. YOUNG

Summary

The systems in *Octopus* for visual and touch learning are distinct but have some overlap. In the visual system the impulses from groups of retinal receptors are passed through the optic lobes and then a series of four further matrices of intersecting axons. With appropriate synaptic changes these form associations between conjunctions of signals of visual events and their consequences, producing attack or retreat. The touch memory consists primarily of four matrices but the tactile impulses from the arms also have access to the visual centres so that no less than eight centres are involved. Injury to any of them reduces the accuracy of learned discriminations of rough and smooth objects. The arrangement of the matrices recalls that of the mammalian hippocampus and associated neocortical centres; there are recursive pathways that would re-inforce the conjunctions. These are therefore excellent examples of how memories of complex situations can be set up by passing impulses from numerous channels through a series of matrices. Such distributed degenerate and redundant systems provide the efficient memories allowing for generalization that are characteristic of complex nervous systems.

The capacity to learn by experience is probably a large factor in the success of modern cephalopods. More than half of the volume of the brain is given up to recognition of objects and memory of the outcome of situations. Unfortunately little is known about learning in the wild but the memory capacity of *Octopus vulgaris* has been extensively studied, and distinct centres for visual and tactile memory identified. Anatomical investigation has made it possible to begin to understand the pattern of organization of the large number of channels that are involved, and to suggest an hypothesis as to their functions. The system is organized as a series of matrices of crossing fibres, somewhat like the mammalian hippocampus (see Rolls 1990). This constitutes the most fully investigated example, outside the mammals, of a distributed memory system based on a large population of parallel channels. It is striking that memory mechanisms based upon a series of matrices should have evolved in such different animals.

The centres involved in cephalopod learning are developments of the feeding system, perhaps serving originally to bridge the delays involved as the distance receptors became more efficient. They provide the capacity to find out which things that are seen or touched are likely to be good to eat or should be avoided. To do this the system must categorize the properties of complex objects and situations and retain records of their probable consequences. This is achieved by statistical selection of invariant combinations among numerous parallel input channels arising from varied feature detectors.

The two memory systems

In an octopus there are distinct memory systems for vision and touch, with some overlap (Figs 26.1 and 26.2). For vision there are 120 million receptors in each retina, whose signals are analysed in each optic lobe by some 60 million cells, including presumed feature detectors. These provide information about the orientation of changing outlines and other characteristics of the visual scene. In the chemotactile system the receptors of the edges of the suckers provide signals indicating the degree of roughness of objects and their chemical properties.

The hypothesis is that the process of learning consists in finding invariant conjunctions between sets of inputs by passing the signals through a series of matrices of intersecting axes (Young 1991). There are no less than eight anatomical matrices for touch, and four, plus the optic lobe, for vision (Figs 26.1

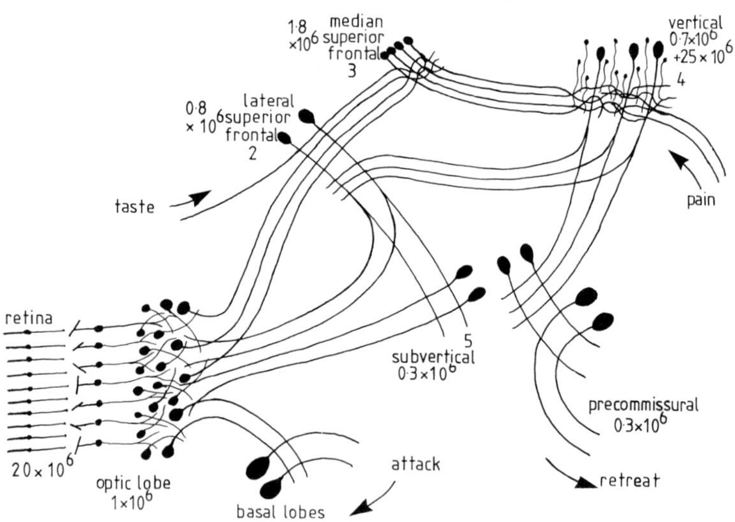

Fig. 26.1. A diagram of the matrices in the visual memory system of *Octopus*. The successive matrices are numbered 1–5 and the number of cells in each lobe is given. (Reproduced from Young 1991, with permission.)

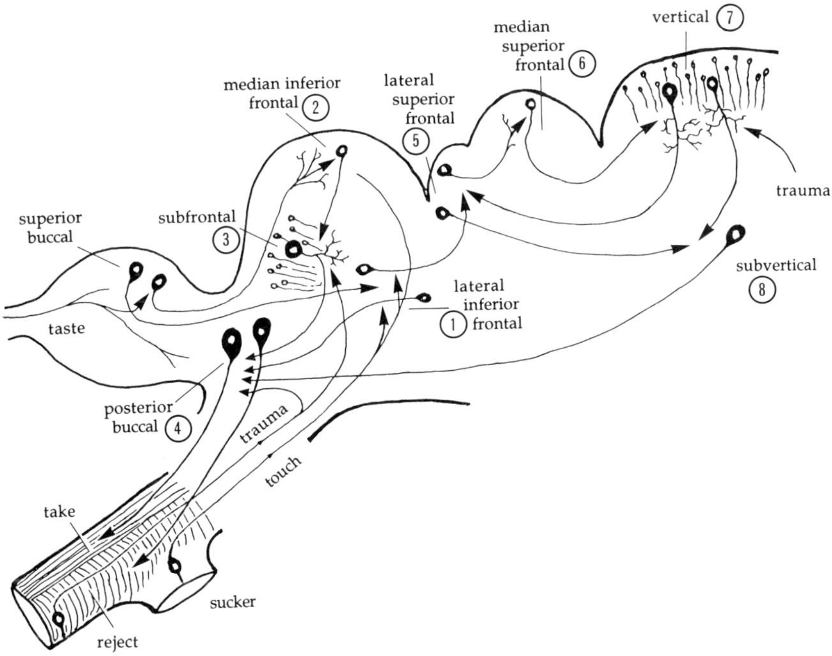

Fig. 26.2. A diagram of the matrices in the tactile memory system of *Octopus*. The matrices are labelled 1–8, the first four being purely tactile while 5–8 are also parts of the visual system.

and 26.2). These sets of intersecting fibres serve to find and store associations between the combinations of the signals of input events and their consequences. The outputs provide for effective behaviour in the light of previous experience. Thus the systems enable identification of categories of input, say the sight of a crab or the feel of a clam, and the storage of records of the probable value of each. The records have the form of bias to particular directions of action, whether to come forward or to retreat.

The principle of the matrices is to provide for selection of paths that are used and inhibition of those that are not. The fibres coming in to each matrix make synapses with many cells of the lobe. Each cell will thus receive signals from several inputs. Synaptic changes will ensure that paths that are used are facilitated, and differences will be enhanced by lateral inhibition. The preferred pathways may also be reinforced by recurrent fibres. The octopus system thus has some similarity with the mammalian hippocampus. With appropriate synaptic changes 'the response of the system as a categoriser climbs over repeated iteration' (Rolls 1990).

The essence of the system is the presence of a sequence of matrices in each of which further interactions can occur, forming a series of conjunctions of conjunctions. The population of cells in the later stages of the sequence then ensures appropriate behaviour. The parallel with the series of centres in the mammalian cortex is striking.

The visual memory system

The axons of the retinal cells end in the outer plexiform zone of the optic lobes, in contact with the dendrites of a large number of cells that probably act as feature detectors (Young 1971). These dendritic fields are of various shapes and lengths, some extending for great distances. This variety provides for the recognition of the relevant features of objects and scenes that an octopus meets in its life. Physiological investigation of the responses of these cells is urgently needed and should be possible by the use of slices of the optic lobe (see Daw and Pearlman 1969). The axons of these second-order visual cells form columns, with progressively wider interconnections proceeding centrally. The detailed arrangements of the fibres within this system are little understood and are shown only diagrammatically in Fig. 26.1. They provide the first opportunities for categorization by the selection of sets of inputs that occur together. Some synaptic changes that result in memory formation can probably take place within the optic lobes but learning also involves four supra-oesophageal centres (Young 1965*b*).

There are three relevant output pathways from third or higher order visual neurones (Fig. 26.3; Young 1991). A direct path proceeds to the magnocellular lobe and is probably concerned with avoidance reactions. A second pathway leads to the system of basal lobes, which direct precise movements. This is probably the pathway used for attack in situations that have become familiar during learning.

The fibres that lead to the higher centres from all parts of the optic lobes proceed first across the dendrites of the cells of the lateral superior frontal lobes, forming a typical lattice-like matrix. Signals also enter this lobe indirectly from the lips (see Fig. 26.2), so that here particular patterns of visual input can be combined with signals of taste. The axons of the cells of the lobe pass either back to the optic lobe or down to the subvertical lobe. Their action is to increase the tendency to attack when a visual input is accompanied by signals of taste. Accompanied by impulses in the recurrent tract from the vertical lobe these cells probably set up memory records tending to promote attack. In an unfamiliar situation an octopus will not make rapid attacks, even at crabs. It first advances and puts out an arm, after a delay of a minute or more, to test the object. If it proves favourable later visual attacks quickly come to be made in as little as 2 s. This learning probably occurs in the lateral superior frontal lobe; if this lobe is removed an octopus will no longer make attacks at an object seen at a distance, even a crab. The octopus is not blind and will put out an arm to a crab nearby.

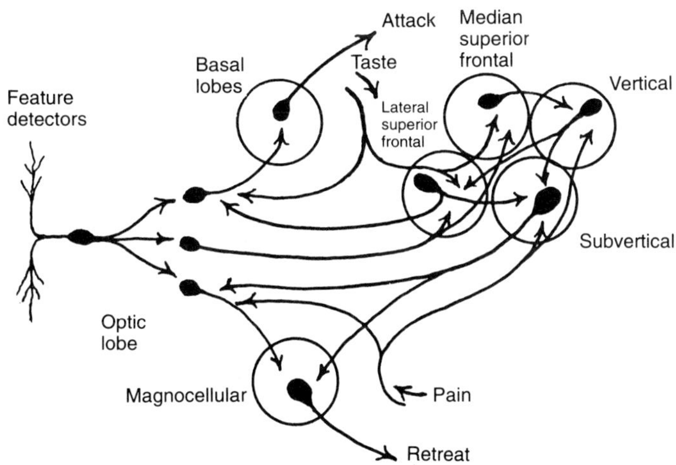

Fig. 26.3. Diagram of the three main pathways from the feature detectors of the optic lobe of an octopus. The output of the magnocellular lobe is probably for retreat, that to the basal lobes may produce attack or retreat. The third output leads to the matrices of the vertical lobe system. Here particular patterns of visual signals are combined with those of taste to produce records tending to increase attacks or, with signals of pain, to produce avoidance. (Reproduced from Young 1991, with permission.)

The fibres from the optic lobe pass on across the trunks of the cells of the second lobe, the median superior frontal. Here they and the fibres of taste interweave in a complex plexus. This provides for the maximum possibility that each of the cells may be activated by a particular combination of the inputs.

The cells of the median superior frontal lobe are all alike and their axons all pass through the plexus to the third member of the series, the vertical lobe (Young 1971). This also has a matrix structure but its cells are very different from those of the previous two lobes. Some 65 000 of the cells are large and their trunks take irregular courses through the neuropil (Young 1971). They carry large lateral branching bushy dendrite twigs (Fig. 26.2). Each of these cells is thus in a position to receive signals from many of the incoming fibres. Some of their axons proceed back to the lateral superior frontal lobe, others down to the fourth lobe of the system, the subvertical. They are probably of two sorts, some promoting attack, others retreat. Possibly those promoting attack are the ones that pass to the lateral superior frontal, but there is no evidence for this. The large cells of the vertical lobe are embedded among 25 million minute amacrine cells whose axons are restricted to the neuropil of the lobe, mixing with the dendrites of the large cells. To accommodate this vast number of cells the upper surface of the brain is folded so that the vertical lobe consists of six cylinders, with cells around the outside and neuropil at the centre.

This great set of cells seems to be concerned with monitoring the tendency to attack and in particular ensuring that the actions of the octopus are suitably restrained. Besides the input from the median superior frontal there is a large set of fine fibres entering the vertical lobe from below. The origin of these has never been exactly determined but they are probably fibres signalling the occurrences of trauma. Their precise method of connection within the lobe is also uncertain (see Young 1983).

Animals from which the vertical lobe has been removed show no abnormalities of posture or locomotion (Boycott and Young 1955). They can be distinguished from normal only in situations where some aspect of learned behaviour is involved. The most striking of these is the lack of capacity to refrain from attack after receipt of shocks. If a normal octopus is given a 12 V shock when it attacks a crab it will not attack again for many hours, whereas animals without vertical lobes attack within a few minutes (Boycott and Young 1955). In an octopus without a vertical lobe the process of learning to discriminate between visual figures is greatly impaired. The errors that are made are nearly all attacks at the unwanted figure, which had been the subject of shocks.

The simplest view of the function of the vertical lobes is that they serve to increase the tendency to attack objects associated with food but that they act especially to prevent attack at certain specific inputs that have been associated with trauma; however, the lobes also have further functions. Without them an octopus is unable to learn with delay of reward, which can be tolerated up to 30 s in a normal animal (Wells and Young 1968). It has also been suggested that the lobes may act to carry short-term memory or as a read-in device, or for generalization or transfer. Being part of a complex system the matrices may indeed participate in these and other activities, such as control of the level of motivation.

In terms of the present treatment of the system as an association memory, we may speculate that the vertical and subfrontal lobes serve to increase the total ensemble and so the storage capacity. Inhibition by the amacrine cells would then serve to reduce the proportion of the output fibres that are excited by each input pattern. This would be a function similar to that suggested by Marr for the Golgi and basket cells of the cerebellum (see Tyrrell and Willshaw 1992).

The tactile memory system

Sensory cells in the rims of the suckers provide signals that serve to identify objects by their degree of roughness. This was first studied by M. J. Wells and J. Wells, who also showed that the suckers are very sensitive to chemical changes (see Sanders 1974; Wells 1978). An octopus is not able to identify the shape of objects by touch, probably because with the mobility of the arms it is not possible to judge the relative position of the receptors.

The nerve impulses from the arms are passed through a system of four matrices precisely analogous to those for vision (Fig. 26.2; Young 1965*a*). In addition the tactile impulses are also passed through the four visual lobes, which contribute further reliability to learned reactions to touch. Investigations of touch learning have made use of a series of plastic balls, either smooth or engraved with between one and 13 depressions (Wells 1978; Young 1983). The octopus quickly learn to take either the rougher or smoother of a pair when takes of one are rewarded with a piece of fish and the other is either unrewarded or followed by a small electric shock. The animals are able to make quite fine discriminations, for instance between balls with four or seven rings. After training, the accuracy of discrimination is tested by a series of 24 rapid extinction tests, using balls of various roughness; habituation proceeds faster according to similarity to the balls used for positive and negative training (Fig. 26.4). With this technique the effect of removing the various lobes can be tested.

The four central lobes of the touch learning system are matrices with a structure strikingly similar to those of the visual system (Fig. 26.2). The touch fibres from the arms, together with taste fibres from the lips pass across the

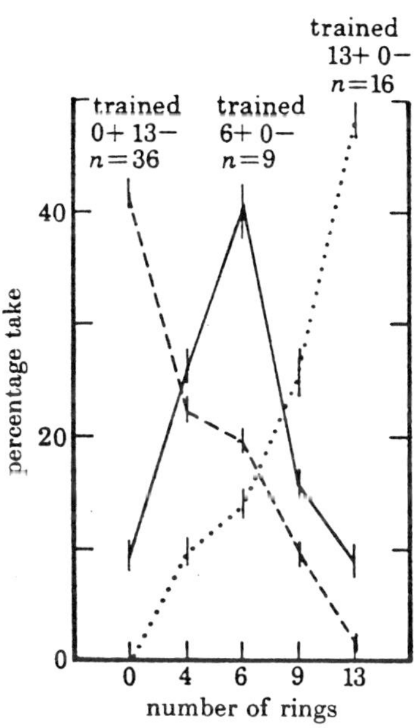

Fig. 26.4. Results of unrewarded tests after training in different directions. All trained for 20 sessions of four trials daily with each ball, one with food the other with shock. The figure shows the means and standard errors of the takes of each ball as a proportion of all takes. (Note that after training with smooth negative and six rings positive the balls with nine and 13 rings were seldom taken; the 'positive' training was specific to six rings.) (Reproduced from Young 1983, with permission.)

trunks and synapses of the cells of the first lobe, the lateral inferior frontal. The axons of these cells pass directly to the fourth lobe, the posterior buccal. Here some of the cells send axons direct to the arms and control the action of taking or rejecting objects touched. This path from the lateral inferior frontal to the posterior buccal is comparable with that from the lateral superior frontal and probably increases the tendency to take.

The second lobe of the touch system, the median inferior frontal, is a matrix organized precisely like the median superior frontal. It probably serves to make further and more refined combinations of particular sets of signals of touch and to combine them with taste, if appropriate. Incoming fibres of touch from the arms and taste from the lips mingle in a series of interweaving bundles. Each of the million small cells of the lobe has the possibility of receiving a combination of inputs. Unfortunately there is no information as to the coding systems involved. Possibly they consist of patterns of frequency of impulses.

All the fibres of the cells of the median inferior frontal proceed through the plexus to the third matrix, the subfrontal lobe. This is organized precisely like the vertical lobe. There are about 1000 large cells and no less than 5 million minute amacrine cells whose axons all remain within the lobe (Young 1965*a*). The large cells have twisted trunks running through the neuropil, with many collaterals, some ending in bushy tufts. These cells can thus be influenced by many incoming fibres and they provide for the setting up of a record of the common occurrence of a large number of sets of signals as recorded by the cells of the median inferior frontal. These cells of the subfrontal are presumably of two sets, like those of the vertical lobe; some tend to promote taking by the arms, others to promote rejection (there is, however, no evidence of the two sorts). The fact that there are so few of the large cells (about 1000) is puzzling. Perhaps an octopus can only store touch records of rather few objects. The axons of the large cells run to the fourth tactile lobe, the posterior buccal. Many small nerve fibres enter the subfrontal from below and are presumed to signal pain, like those of the vertical lobe.

Using these facts about the connections it is possible to make an hypothesis as to the sequence of events during touch learning. When an octopus touches an unfamiliar object it is drawn in only after a delay, and then finally tasted. If it proves favourable the appropriate synapses on cells of the first matrix in the lateral inferior frontal are modified and these cells send impulses to those cells of the posterior buccal that promote taking. On later presentations the object is drawn in more readily. In the second matrix, in the median inferior frontal, there are further and more detailed conjunctions between patterns of signals of touch. If these are also accompanied by satisfactory signals of taste then the appropriate output fibres will further activate those cells of the posterior buccal lobe that enhance the tendency to take, already primed by stimulation from the cells of the lateral inferior frontal. However, if the result of taking the object was a shock then those large cells of the subfrontal that operate rejection are activated by the combined operation of the fibres signalling pain and the signals of touch from the fibres from the median inferior frontal.

Each of the actions of taking or rejecting follows a different pattern of signals of touch. The synapses associated with each pattern become reinforced by use. On subsequent occasions each then produces the appropriate result. The synaptic changes may occur in all the lobes but are probably especially important in the subfrontal, without which no learning is possible. The huge number of amacrine cells must be somehow active in producing the synaptic changes. They end among the branches of the larger cells but there is no evidence as to how they influence them.

Evidence for this interpretation comes from experiments in which parts of the system are removed (Young 1965*a*, 1983; Wells 1978). To do this we have often made use of octopuses in which the whole supra-oesophageal lobe has been split in the sagittal plane. Each side can then be trained separately, if necessary in opposite directions. It is convenient to leave one side as a control. Using this technique it was shown that tactile learning is possible only if some part of the region between the subfrontal and posterior buccal lobes is left intact. A few tens of thousands out of the 5 million amacrine neurones were enough to allow some learning. It was accordingly suggested that this is the seat of the chemotactile memory record (Wells and Young 1975). This is much too simple a view. In a distributed memory system such as this, synaptic changes may occur at many places. We have already suggested them in the lateral and median inferior frontals. The failure to learn after lesions of the subfrontal–posterior buccal region may be due not only to the absence of amacrine cells but also to the fact that this is a key point of access to the main output cells of the whole system.

The changes that take place in tactile learning are indeed even more widely distributed (Young 1983). In particular the whole vertical lobe system is involved. The tactile input to it passes from the lateral inferior frontal to the lateral superior frontal and thence round the whole visual circuit, except of course the optic lobes. The output comes from the cells of the subvertical and median basal lobes, which proceed to the posterior buccal lobe (Robertson *et al.* 1993). There are therefore eight distinct matrices involved in the tactile learning system.

Evidence for the involvement of each of the various lobes in the memory is provided by comparison of the responses given during extinction tests by animals with various lesions (Fig. 26.5; Young 1983). The difference between the takes of rough and smooth balls in the tests provides a measure of the accuracy of the memory after the lesions (Table 26.1). Using the difference in takes by unoperated controls as 100 per cent we find that the animals without vertical lobes are rather less than half as efficient and those without median inferior frontal lobes only one-third. After the subfrontal lesions, performance was perverse, presumably due to some continuing effect of the inherited response system that remains after this lesion.

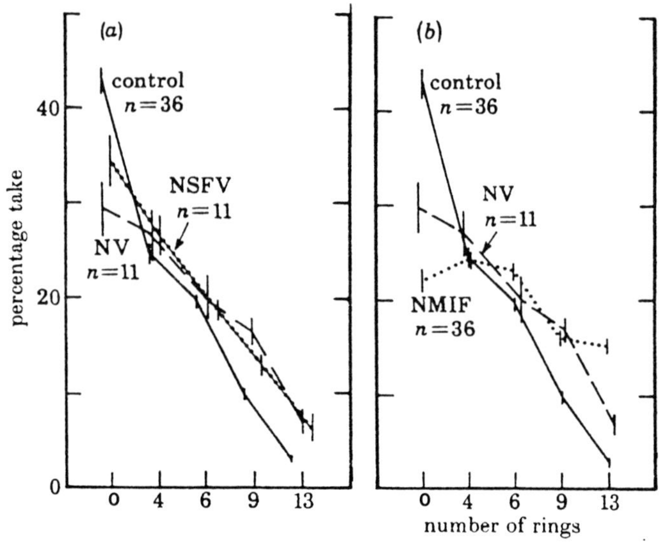

Fig. 26.5. Tests after training animals with various parts removed: (a) 11 octopuses with the vertical lobe removed on one side (NV) and the tract from the median superior frontal to vertical lobe cut in the other (NSFV); 36 control half brains for comparison. (b) 36 half brains with the median inferior frontal lobe removed (NMIF), compared with 11 without vertical lobes (NV) and 36 controls; all were trained with smooth positive and 13 rings negative. Tests as for Fig. 26.4. (Reproduced from Young 1983, with permission.)

Table 26.1 Mean takes in 24 extinction tests after various lesions

Lobe removed	Smooth ball positive	Rough ball negative	Difference	% accuracy
Controls	20.22	2.44	17.76	100
Vertical	17.68	10.55	7.13	40
Median inferior frontal	16.22	10.69	5.63	32
Cerebrobrachial tract cut	13.78	10.94	2.84	16
Subfrontal	7.89	8.42	−0.53	—

The last column shows the capacity for learned discrimination that remains after each lesion, expressed as a percentage of difference in takes of the two balls compared with the controls. (Reproduced from Young 1983, with permission.)

A further interesting complication is that a small capacity for learning is seen after lesions that prevent all tactile impulses from reaching the supra-oesophageal lobes (Fig. 26.6; Young 1983). The changes involved are perhaps in the prebrachial lobes, where there is a large number of small cells. This emphasizes that the capacity to make appropriate adaptive changes may be widespread in nervous tissues and is especially developed in those devoted to

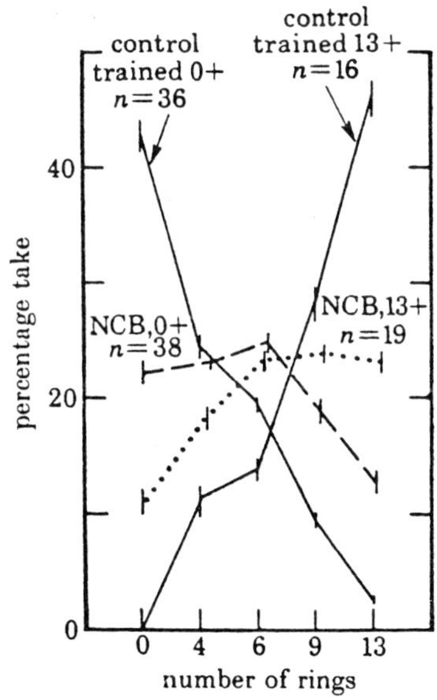

Fig. 26.6. Suboesophageal half brains. The cerebrobrachial tracts have been cut (NCB) in 38 half brains trained with smooth positive and 13 rings negative, and 19 half brains the reverse. Controls as in Fig. 26.4 for comparison. (Reproduced from Young 1983, with permission.)

learning. In this connection it is relevant that the two sets of memory centres in the octopus brain have evidently become developed during evolution as backward extensions of the superior buccal lobe, which is concerned with feeding. This lobe contains both large and small cells and the latter make a layer directly continuous with the amacrine cells of the inferior frontal and vertical lobe systems. The memory systems that we see today have been evolved by development of the original simple feeding system (see Young 1965, 1983).

Conclusions

Knowledge about memory in cephalopods is in a tantalizing state. Great capacities for learning have been demonstrated. The anatomy of the centres concerned has been investigated at the light microscope level, but only a little with the electron microscope. An hypothesis for the mode of functioning is suggested and can be partly tested and confirmed by the study of lesions. Proper validation of the hypothesis requires further detailed knowledge of the connections, and especially study of their physiology and pharmacology. These have been little investigated for want of suitable preparations, but this should soon be remedied. It has been shown that reflexes can readily be studied physiologically in slices of octopus brain (Williamson and Budelmann 1991). This technique should be applicable to the learning centres, using, for instance, the large cells of the posterior buccal lobes. A search for long-term potentiation should also be possible in slices. It is also urgently necessary to find ways to study the small amacrine cells that are such a prominent feature of the memory systems. Identification of their pharmacology would be a useful beginning and should not be too hard. Further electron microscope studies of them would also be very helpful.

The information already available shows that the octopus memory involves selection during learning among large numbers of parallel pathways. The capacity to set up and hold the memory is distributed among many distinct matrices. This seems to be the case also in the complex memories of mammals, birds, and insects (Rolls 1990; Horn 1991; Strausfeld and Lee 1991). This distribution implies that each area contributes a part of the whole and the record is also distributed within each part. The results of removing individual lobes or parts of them in an octopus have shown that the memory is indeed distributed in this way. Each matrix of the sequence plays a different part in the whole process but none is uniquely essential (except at the input and output).

Such a system provides the characteristics that are necessary for a useful memory. It allows for fault tolerance, for generalization and for completion; that is, recognition of situations in spite of their variations. Organization of numerous channels into a series of matrices is a method of achieving those results given the properties of nervous tissues.

Acknowledgements

I am grateful for the assistance over many years of Dr M. Nixon, also of the President, Director, and Staff of the Stazione Zoologica at Naples. Work has been assisted by a grant from the Wellcome Trust and by the secretarial help of my wife, Raye Young.

References

Boycott, B. B. and Young, J. Z. (1955). A memory system in *Octopus vulgaris* Lamarck. *Proceedings of the Royal Society of London, Series B*, **143**, 449–80.

Daw, N. W. and Pearlman, A. L. (1969). Receptive field studies of single units in the optic lobe of the squid *Loligo pealei*. *Biological Bulletin*, **137**, 398.

Horn, G. (1991). Neural bases of recognition memory investigated through an analysis of imprinting. *Philosophical Transactions of the Royal Society, Series B*, **329**, 133–42.

Robertson, J. D. Schwartz, O. M., and Lee, P. (1993). Carbocyanine dye labeling

reveals a new motor nucleus in *Octopus* brain. *Journal of Comparative Neurology*, **328**, 485–500.
Rolls, E. T. (1990). The representation and storage of information in neural networks in the primate cerebral cortex and hippocampus. In *The computing neuron* (ed. R. Durbin, C. Miall, and G. Mitchison), pp. 125–9. Addison-Wesley, Wokingham.
Sanders, G. D. (1974). The cephalopods. In *Invertebrate learning*, Vol. 3 (ed. W. C. Corning, J. A. Dyal, and A. O. D. Willows), pp. 1–101. Plenum Press, New York.
Strausfeld, N. J. and Lee, J.-K. (1991). Neuronal bases for parallel processing in the fly. *Visual Neuroscience*, **7**, 13–34.
Tyrrell, T. and Willshaw, D. (1992). Cerebellar cortex; its simulation and the relevance of Marr's theory. *Philosophical Transactions of the Royal Society of London, Series B*, **336**, 239–57.
Wells, M. J. (1978). *Octopus. Physiology and behaviour of an advanced invertebrate.* Chapman & Hall, London.
Wells, M. J. and Young M. J. (1968). Learning with delayed rewards in *Octopus. Zeitschrift für vergleichende Physiologie*, **61**, 103–28.
Wells, M. J. and Young, J. Z. (1975). The subfrontal lobe and touch learning in the *Octopus* brain. *Brain Research*, **92**, 103–21.
Williamson, R. and Budelmann, B. U. (1991). Convergent inputs to *Octopus* oculomotor neurones demonstrated in a brain slice preparation. *Neuroscience Letters*, **121**, 215–18.
Young, J. Z. (1965*a*). The centres for touch discrimination in *Octopus. Philosophical Transactions*, **249**, 45–67.
Young, J. Z. (1965*b*). The organisation of a memory system. Croonian Lecture. *Proceedings of the Royal Society of London, Series B*, **163**, 285–320.
Young, J. Z. (1971). *The anatomy of the nervous system of* Octopus vulgaris. Clarendon Press, Oxford.
Young, J. Z. (1983). The distributed tactile memory system of *Octopus. Proceedings of the Royal Society of London, Series B*, **218**, 135–76.
Young, J. Z. (1991). Computation in the learning system of cephalopods. *Biological Bulletin*, **180**, 200–8.

27

A novel occluding junction forms the blood–brain barrier in cephalopod molluscs

MAGNUS BUNDGAARD, N. JOAN ABBOTT, and NANCY J. LANE

Summary

Earlier studies using radioisotopic tracer techniques showed that the cuttlefish *Sepia officinalis* has a blood–brain barrier (BBB) as tight as that of mammals. This chapter reviews recent high resolution electron microscopic studies of the blood–brain interface in *Sepia* which demonstrate that the BBB is formed by a novel restricting junction, between vascular pericytes in arterial vessels, and between perivascular glial cells in capillaries and venous vessels. The filtering properties of the junction appear to be determined by the extracellular matrix within the intercellular cleft, and a model is proposed by which adsorbed plasma proteins cross-link the matrix and reduce its effective mesh size. Electron microscopic examination of electron-dense tracer distribution in octopus and squid brain confirms that a BBB to protein is also present in these groups. The similarity between parts of the *Sepia* junctional 'skeleton' and junctions in other animal groups suggests an evolutionary sequence in which a relatively wide 'ancestral' junctional structure could have developed into the range of modern occluding junctional types.

Introduction

In vertebrates, the tightest cellular barriers are formed by intercellular tight junctions or *zonulae occludentes* (Farquhar and Palade 1963; Staehelin 1974; Gumbiner 1987). This is true of tight epithelia such as frog skin, and in the tight endothelium lining brain microvessels and forming the blood–brain brain barrier (Reese and Karnovsky 1967). Among invertebrates, several junctional types have been described, but again tight junction variants have been found at most sites of low permeability, such as the arthropod blood–brain and blood–testis barriers (Lane and Chandler 1980; Lane 1984; Baldwin *et al.* 1987). It has been proposed that a BBB is necessary for the fine control of the brain microenvironment (Cserr and Bundgaard 1984), which in turn is required for higher integrative functions of the central nervous system (Abbott *et al.* 1986*a*). A BBB is lacking in many invertebrates with simple nervous systems, including bivalve and gastropod molluscs (reviewed in Abbott *et al.* 1986*b*).

Cephalopod molluscs such as octopus, squid, and cuttlefish, are among the most advanced invertebrates known, with large brains and highly developed sensory and motor capabilities (Young 1971; Nixon and Messenger 1977; Wells 1978). In experiments on the cuttlefish *Sepia officinalis* using radio-labelled non-electrolytes (EDTA and polyethylene glycol) we showed that the blood–brain barrier (BBB) is as tight as that of mammals (Abbott *et al.* 1985*a,b*). This chapter reviews recent evidence that the BBB of *Sepia* is formed by a novel restricting junction, located between vascular pericytes in arterial vessels, and between perivascular glial cells in small and venous vessels (Abbott and Bundgaard 1992; Bundgaard and Abbott 1992; Lane and Abbott 1992). The filtering properties of the junction appear to be determined by the extracellular matrix within the intercellular cleft (Abbott *et al.* 1992). The similarity between parts of the *Sepia* junctional 'skeleton' and junctions in other animal groups may provide information about junction evolution.

Studies in brain of *Sepia officinalis*

Methods

Sepia of mass 10–400 g were trawled off the coast of Plymouth, kept in circulating sea water tanks, and anaesthetized by immersion in 0.2 per cent chloral hydrate in sea water (Abbott and Bundgaard 1987). Fixation for electron microscopy used glutaraldehyde/paraformaldehyde fixatives and immersion or perfusion fixation (Abbott and Bundgaard 1987; Bundgaard and Abbott 1992). In tracer studies with horse-radish peroxidase (HRP), the tracer was injected intravenously and allowed to circulate for 30 min before fixation; reaction product was developed by standard methods (Abbott and Bundgaard 1987, 1992). In experiments with ionic lanthanum as tracer, an initial aortic perfusion with saline to wash out haemocytes was followed by perfusion with lanthanum chloride in saline for 10 min, then with fixative containing phosphate buffer to precipitate lanthanum ions (Abbot and Bundgaard 1992). For freeze-fracture examination, tissue was fixed by perfusion in glutaraldehyde fixatives, mounted in yeast paste in metal holders and frozen in liquid Freon 22 cooled by liquid nitrogen. Replicas were prepared using a Balzers freeze-fracturing device, followed by shadowing with tungsten–tantalum, backing with carbon, and coating with celloidin. Replicas were cleaned and mounted on grids for electron microscopic examination (Lane and Abbott 1992).

Fine structure of the vessel wall

Bundgaard and Abbott (1992) have described the fine structure of the vessels in *Sepia brain*, including the smallest microvessels (=capillaries), arterioles,

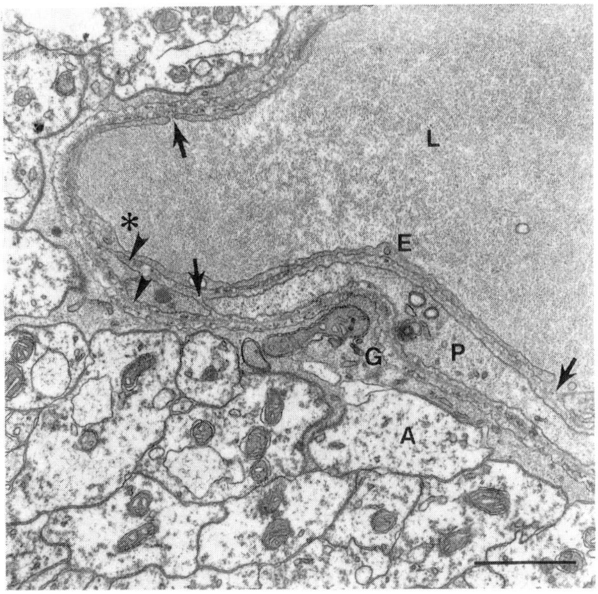

Fig. 27.1. Transverse section showing the wall of a microvessel in vertical lobe of *Sepia* brain, lumen (L) top right. There are patent gaps or clefts (arrows) between both endothelial cells (E) and pericytes (P), but not across the layer of perivascular glial cells (G). Small axon profiles (A) are seen deeper in the parenchyma. The inner and outer basal laminae (arrowheads) are prominent landmarks. Above the asterisk, only a single cell type (endothelial/pericyte) overlies the perivascular glial layer. Terminology as in Barber and Graziadei (1965) and Gray (1969). Scale bar: 1.0 μm.

arteries, and venules/veins. Figure 27.1 shows part of the wall of a brain microvessel, equivalent to a capillary. The lumen is lined by a discontinuous endothelium with an associated fine basal lamina. This is surrounded by a layer of pericytes, also incomplete, and a second, thicker basal lamina. In the smallest vessels, the distinction between endothelial cells and pericytes breaks down, with the inner basal lamina sometimes missing, and parts of the outer basal lamina occasionally devoid of cells on the luminal side. Beyond the outer basal lamina is a complete layer of perivascular glial cells. In the smallest vessels this is formed by a single cell following the whole circumference, and in 90 per cent of sample vessel profiles where the glial membranes were in focus around the entire periphery (27 of 30), no intercellular cleft interrupted the glial layer. In the remaining profiles, where an intercellular cleft was seen, there was an extensive region of overlap between glial processes either side of the cleft. The perivascular glial sheath of larger venous channels, and of the brain surface facing the surrounding venous sinus, was multilayered, with a lamellated arrangement of overlapping glial processes.

In arterial vessels, there was a complete layer of thickened pericytes containing myofilaments.

Electron-dense tracer studies

Electron-dense tracers introduced intravascularly were used to assess the permeability of the blood–brain interface; these were HRP (40 000 Da molecular weight, *ca.* 2.5 nm radius) and ionic lanthanum (La^{3+}, hydrated radius 0.4 nm) (Abbott and Bundgaard 1992). In small vessels and venous

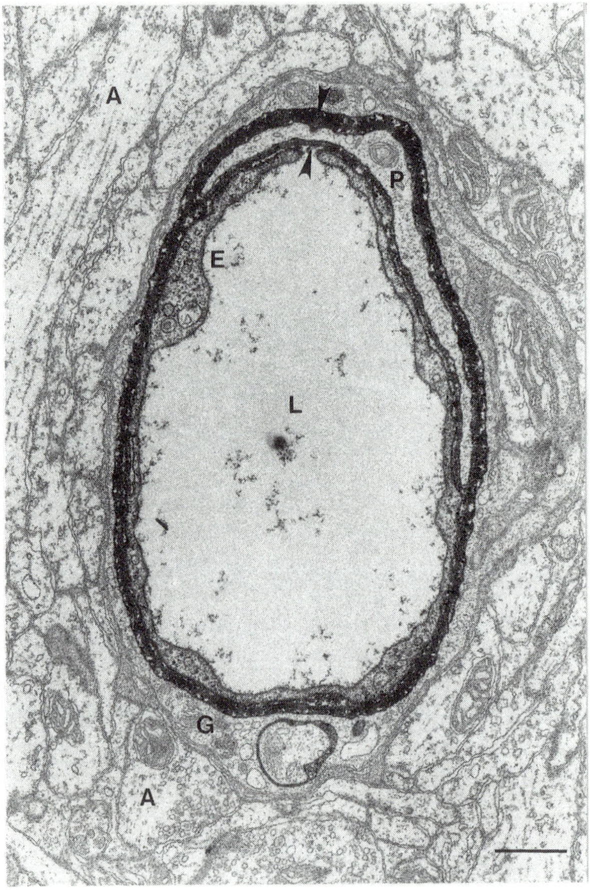

Fig. 27.2. Transverse section of a microvessel in *Sepia* vertical lobe, immersion fixed. Horse-radish peroxidase was injected intravenously and allowed to circulate for 30 min before fixation. Electron-dense reaction product is seen in the vessel lumen (L), and in both basal laminae (arrowheads), but is not seen beyond the layer of perivascular glial cells (G). E, endothelium; P, pericyte; A, axon. Scale bar: 0.5 μm.

channels, HRP reaction product and La^{3+} penetrated up to the outer basal lamina and the perivascular glial layer, but could not be detected in the brain parenchyma beyond (Fig. 27.2). In small arteries and arterioles, the restriction occurred at the level of the pericytes. Examination of the site of tracer block revealed a relatively wide (20 nm) intercellular cleft, generally without focal narrowing, and a gradient of tracer away from the luminal opening of the cleft (Fig. 27.3). Faint periodicities or bars were seen spanning the cleft, and grazing sections showed an apparent meshwork of randomly orientated extracellular fibrils. In sections from tissue without tracer but treated with tannic acid as mordant, the cleft mouth appeared to be filled with an electron-dense material.

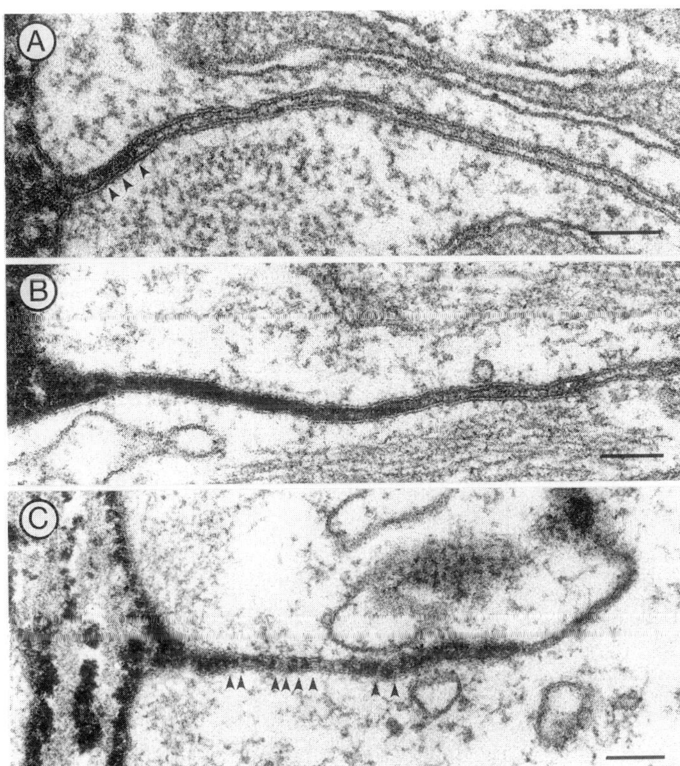

Fig. 27.3. High power views of the restricting zone at the blood–brain interface. (a) and (b) show glial end feet in venous vessels, (c) shows junction between arteriolar pericytes. (a) No tracer; (b) horse-radish peroxidase; (c) lanthanum. Tracer in (b) and (c) shows a gradient of concentration along the cleft, being highest at the luminal end (left) and declining to undetectable levels at the tissue end (right). Electron-lucent periodicities can be seen spanning the cleft in the restricting zone (arrowheads). Scale bars: 0.1 μm

Freeze-fracture studies

Lane and Abbott (1992) conducted a freeze-fracture study of the *Sepia* blood–brain interface. In extensive searches of replicas from freeze-fractured vessels, no linear arrays of intramembranous particles (IMPs) characteristic of vertebrate or invertebrate tight junctions or septate junctions were seen in endothelial, pericyte, or perivascular glial cell membranes (Fig. 27.4); however, isolated IMPs and apparent gap junctional IMP aggregates between glial cells were observed.

Studies in other cephalopod species

Studies of other cephalopod species have been conducted to establish the generality of the results in *Sepia*. Large molecular weight tracers (ferritin, HRP) were not seen in the interstitium in octopus or squid brain following intravascular injection and circulation (Browning and Abbott, in preparation; Brown, in preparation), indicating that a BBB tight to protein is likely to be a universal feature in cephalopod molluscs.

Discussion

Comparison with other *Sepia* tissues

The zone of restricted penetration of electron-dense tracers in *Sepia* is a feature of the brain microvasculature, but not of vessels in other tissues. In tissue taken from the same animals as used in the study of the BBB, HRP was seen to leak from the vascular spaces into the interstitium in the retinal nerves (Abbott *et al.* 1986*a,b*), muscle, and salivary gland (Bundgaard and Abbott, in preparation). In these tissues there was a continuous gradient of tracer density from the vascular lumen to the deeper parts of the tissue, with no abrupt zones of tracer restriction. The different pattern of tracer distribution in brain and non-brain argues strongly against the possibility that the tracer distribution is an artefact, generated for example by localized binding coupled with tracer washout from the tissue during electron microscopic preparation.

The nature of the occluding junction

The characteristic features of the *Sepia* junction are the relatively wide intercellular cleft, the presence of dense extracellular material in the cleft, bars, or 'linkers' spanning the cleft with a faint but not rigid periodicity, the gradual rather than abrupt block of tracer penetration (the location of the restriction being the same for both lanthanum and HRP), and the apparent absence of

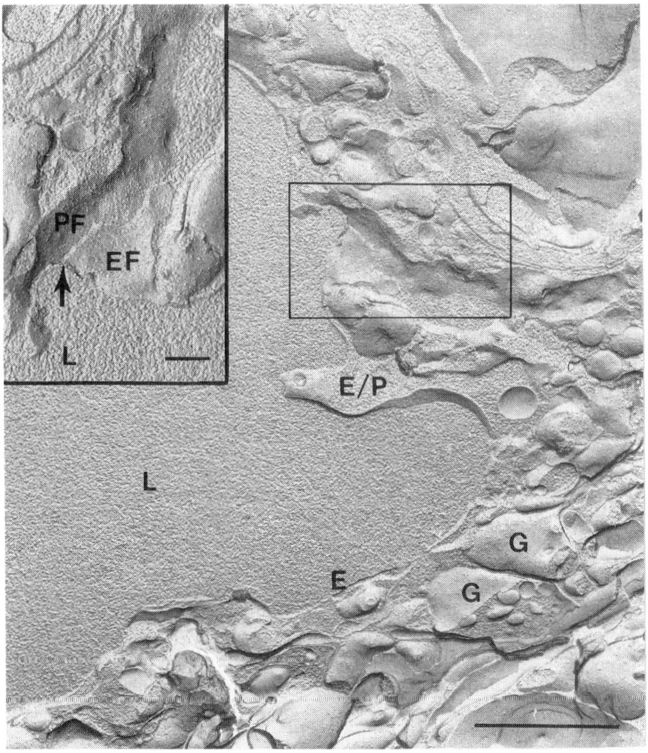

Fig. 27.4. Freeze-fracture replica showing cells surrounding the lumen (L) of a microvessel. No intramembranous particle arrays characteristic of any known occluding junction were detected in endothelial (E), pericyte (P), or perivascular glial (G) membranes, in over 150 vessel profiles examined in 20 replicas from eight brains. The area outlined (at higher power in inset) shows a junctional zone between two presumed endothelial/pericytes. The inset shows the EF (E face) and PF (P face) either side of the intercellular cleft (arrow). Scale bar: 1 μm, inset 0.2 μm.

any corresponding particle arrays in freeze-fracture. Although each of these features has similarities with known vertebrate and invertebrate junctions, no junction previously described demonstrates all five. In particular, the ability of a membrane apposition to restrict the entry of tracers as small as ionic lanthanum without a tight-junctional strand or ridge is unusual.

A fibre-matrix model for the occluding junction

The mechanism of restriction in the *Sepia* junctions is of particular interest, and may throw light on the restricting properties of vertebrate junctions. In vertebrate systemic capillaries, the endothelial glycocalyx extends into the intercellular cleft (Luft 1966; Turner *et al.* 1983). Serum albumin adsorbed on to this matrix is proposed to cross-link its structure, so producing a finer mesh (Curry and Michel 1980; Schneeberger and Hamelin 1984); it has been argued that the mesh rather than the narrow zone of the tight junction is the primary permeability filter (Curry *et al.* 1987). However, the pore size achieved is approximately 5–6 nm radius, too large to block penetration of HRP or lanthanum.

In cephalopods the major plasma protein is haemocyanin, 3.7 MDa molecular weight, with approximate molecular dimensions of 35 × 17 nm (Wichertjes *et al.* 1986); in brain fixed for electron microscopic examination, it is occasionally seen as a regular paracrystalline deposit in the vascular lumen (Barber and Graziadei 1965). Molluscan haemocyanins can be readily dissociated into subcomponents by mild proteolysis, and by changing pH and divalent cation concentration (Préaux and Gieler 1984). It is possible that the dimensions and charge structure of the haemocyanin or its subcomponents, together with the particular features of the glycocalyx at the mouth of the intercellular clefts, enable the protein to cross-link the matrix and plug the fine mesh, forming a much tighter filter than in vertebrate systemic capillaries [Abbott *et al.* 1992; Fig. 27.5(c)]. The restricting properties of the junction would then depend on the extracellular cleft material, and would not require a zone of fusion between adjacent membranes. The fact that the restriction to HRP and ionic lanthanum occurs at a similar site suggests that charge alone is not the critical factor, but rather indicates a localized molecular filter.

If this model is correct, it has several wider implications. One is that *zonulae occludentes* are not the only way to make an effective barrier. The mechanisms by which vertebrate tight junctions restrict permeability are still unclear. According to the simplest model, the anastomosing junctional strands holding cells tightly together are assumed to be responsible for the barrier, by increasing the path-length for diffusion; this model predicts an inverse correlation between the strand number or network complexity and the permeability (Claude and Goodenough 1973; Nagy *et al.* 1984; Bundgaard 1984; Gumbiner 1987). Alternative models are based on the observation that epithelia and endothelia with similar numbers and complexity of junctional strands can show large differences in permeability and electrical resistance (Stevenson *et al.* 1988), suggesting that focal discontinuities within or between the strands, and their spacing, charge and dynamics, may determine the filtering properties of the junction (Karnovsky 1968; Sibley *et al.* 1983; Bundgaard 1984; Ward *et al.* 1988; Stevenson *et al.* 1989; Cereijido *et al.* 1989). It is possible that even in vertebrate junctions, the nature of the extracellular matrix and its adsorbed substances plays a key role in governing permeability. The *Sepia* junction highlights the barrier-forming potential of extracellular matrices.

It is already known that the extracellular material of invertebrate nervous systems may have interesting ion-exchange behaviour (Treherne *et al.* 1982), and that regional differences in extracellular matrix properties may occur

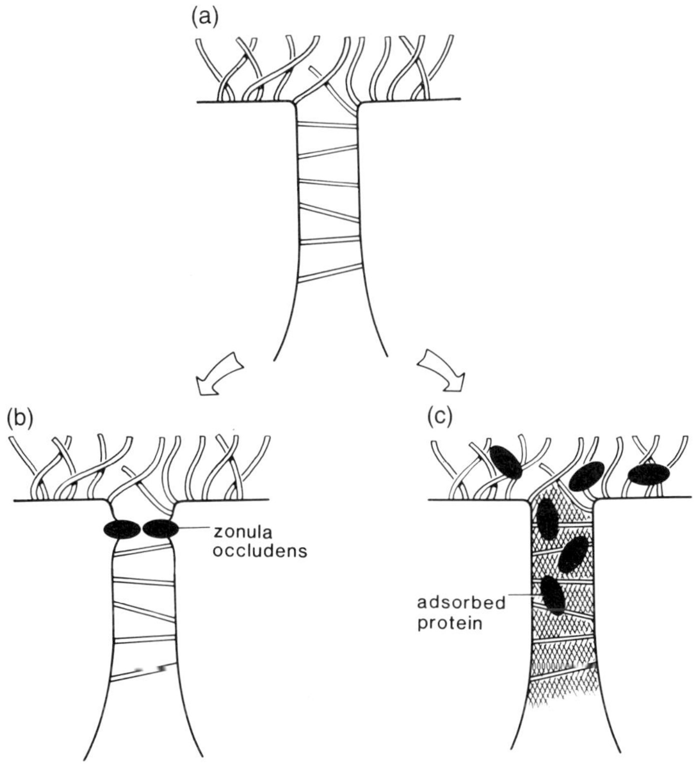

Fig. 27.5. Model illustrating possible evolutionary sequence of restricting junction structure. (a) Hypothetical 'ancestral skeleton' of periodic structures linking the adjacent membranes. (b) One line of evolution leads to insertion of membrane proteins in strands forming the *zonula occludens*. (c) An alternative route elaborates a matrix within the extracellular cleft, cross-linked by adsorbed protein, as suggested for the junctions at the cephalopod blood–brain barrier. The variations in junctional permeability seen in different tissues and animal groups may be explained by differences in the membrane components (b) and in the extracellular matrices (c). (From Abbott *et al.* 1992, with permission.)

within a single tissue (e.g. between neural lamella on the outer side of the perineurium and the interstitium, in insect CNS; Ashhurst and Costin 1971). In the crayfish CNS, the extracellular matrix of the glial perineurium appears to contribute to the diffusion barrier and selective ionic permeability of the perineurium (Abbott and Pichon 1987; Butt 1991). A combination of scattered punctate tight junctions (Lane and Abbott 1975) and dense extracellular matrix within the junctional zone may explain the diffusional restriction. Villegas and Sanchez (1991) have shown that the multilayered sheath surrounding the giant axons of lobster walking legs is able to restrict penetration of ionic lanthanum, although individual tight junctions show incomplete networks of junctional strands in freeze-fracture replicas. The cephalopod findings add weight to the developing view that local specializations of the extracellular matrix can contribute to diffusional restriction in barrier layers.

Evolution of occluding junctions

The cephalopod junctions may also hold clues to junction evolution. The relatively wide intercellular cleft in the junctional region, and the presence of bars or 'linkers' spanning the cleft, is strikingly reminiscent of the wide part of endothelial tight junctions in placental capillaries (Firth *et al.* 1983), and of the occluding junction forming the blood–testis barrier in the locust (Baldwin *et al.* 1987). The similarity, covering three widely separated animal groups, could indicate that this represents a primitive type of junctional 'skeleton' [Fig. 27.5(a)]. During evolution, the basic structure may have been refined and elaborated in two principal ways, either by the addition of a network of proteinaceous strands joining the two membranes [Fig. 27.5(b)], or by decreased permeability of the cleft matrix [Fig. 27.5(c)]. The restricting junction may or may not involve narrowing of the cleft. There is considerable variability in the structure and permeability of junctional zones in different tissues and species (Lane 1984; Gumbiner 1987). This may reflect the combinations possible by subtle alterations in the two major components: the membranous elements of the junction and the material packing the intercellular cleft.

Acknowledgements

This work was supported by grants from the Wellcome Trust and the Danish Natural Sciences Research Council. We thank the Director of the Marine Biological Association Laboratory, Plymouth, for facilities, and Pia Hagman and Bill Lee for technical assistance.

References

Abbott, N. J. and Bundgaard, M. (1987). Microvessel surface area, density and dimensions in brain and muscle of the cephalopod *Sepia officinalis*. *Proceedings of the Royal Society of London, Series B*, **230**, 459–82.

Abbott, N. J. and Bundgaard, M. (1992). Electron-dense tracer evidence for a blood–brain barrier in the cuttlefish *Sepia officinalis*. *Journal of Neurocytology*, **21**, 276–94.

Abbott, N. J. and Pichon, Y. (1987). The glial blood–brain barrier of Crustacea and cephalopods: a review. *Journal de Physiologie*, **82**, 346–56.

Abbott, N. J., Bundgaard, M., and Cserr, H. F. (1985*a*). Brain vascular volume, electrolytes and blood–brain interface in the cuttlefish *Sepia officinalis*. *Journal of Physiology*, **368**, 197–212.

Abbott, N. J., Bundgaard, M., and Cserr, H. F. (1985*b*). Tightness of the blood–brain barrier and evidence for brain interstitial fluid flow in the cuttlefish *Sepia officinalis*. *Journal of Physiology*, **368**, 213–26.

Abbott, N. J., Bundgaard, M., and Cserr, H. F. (1986*a*). Comparative physiology of the blood–brain barrier. In *The blood–brain barrier in health and disease* (ed. A. J. Suckling, M. G. Rumsby, and M. W. B. Bradbury), pp. 52–72. Ellis Horwood, Chichester.

Abbott, N. J., Lane, N. J., and Bundgaard, M. (1986*b*). The blood–brain interface in invertebrates. *Annals of the New York Academy of Sciences*, **481**, 20–41.

Abbott, N. J., Lane, N. J., and Bundgaard, M. (1992). A fibre matrix model for the restricting junction of the blood–brain barrier in a cephalopod mollusc: implications for capillary and epithelial permeability. *Journal of Neurocytology*, **21**, 304–11.

Ashhurst, D. E. and Costin, N. M. (1971). Insect mucosubstances. II. The mucosubstances of the central nervous system. *Histochemical Journal*, **3**, 297–310.

Baldwin, K. M., Loeb, M. J., and Riemann, J. G. (1987). A novel occluding junction which lacks membrane fusion in insect testis. *Tissue and Cell*, **19**, 413–21.

Barber, V. C. and Graziadei, P. (1965). The fine structure of cephalopod blood vessels. I. Some smaller peripheral vessels. *Zeitschrift für Zellforschung und Mikroskopische Anatomie*, **66**, 765–81.

Bundgaard, M. (1984). The three-dimensional organisation of tight junctions in a capillary endothelium revealed by serial-section electron microscopy. *Journal of Ultrastructure Research*, **88**, 1–17.

Bundgaard, M. and Abbott, N. J. (1992). Fine structure of the blood–brain interface in the cuttlefish *Sepia officinalis* (Mollusca, Cephalopoda). *Journal of Neurocytology*, **21**, 260–75.

Butt, A. M. (1991). Modulation of a glial blood–brain barrier. *Annals of the New York Academy of Sciences*, **633**, 363–77.

Cereijido, M., González-Mariscal, L., and Contreras, G. (1989). Tight junction: barrier between higher organisms and environment. *News in Physiological Sciences*, **4**, 72–5.

Claude, P. and Goodenough, D. A. (1973). Fracture faces of *zonulae occludentes* from 'tight' and 'leaky' epithelia. *Journal of Cell Biology*, **58**, 390–400.

Cserr, H. F. and Bundgaard, M. (1984). Blood–brain interfaces in vertebrates—a comparative approach. *American Journal of Physiology*, **246**, R277–88.

Curry, F. E. and Michel, C. C. (1980). A fibre matrix theory of capillary permeability. *Microvascular Research*, **20**, 96–9.

Curry, F. E., Michel, C. C., and Phillips, M. E. (1987). Effect of albumin on the osmotic pressure exerted by myoglobin across capillary walls in frog mesentery. *Journal of Physiology*, **387**, 69–82.

Farquhar, M. G. and Palade, G. E. (1963). Junctional complexes in various epithelia. *Journal of Cell Biology*, **17**, 375–412.

Firth, J. A., Bauman, K. F., and Sibley, C. P. (1983). The intercellular junctions of guinea-pig placental capillaries: a possible structural basis for endothelial solute permeability. *Journal of Ultrastructure Research*, **85**, 45–57.

Gumbiner, B. (1987). The structure, biochemistry, and assembly of epithelial tight junctions. *American Journal of Physiology*, **253**, C749–58.

Karnovsky, M. J. (1968). The ultrastructural basis of transcapillary exchanges. *Journal of General Physiology*, **52**, 64–93s.

Lane, N. J. (1984). A comparison of the construction of intercellular junctions in the CNS of vertebrates and invertebrates. *Trends in Neurosciences*, **7**, 95–9.

Lane, N. J. and Abbott, N. J. (1975). The organization of the nervous system in the crayfish *Procambarus clarkii*, with emphasis on the blood–brain interface. *Journal of Experimental Biology*, **63**, 85–115.

Lane, N. J. and Abbott, N. J. (1992). Freeze-fracture evidence for a novel restricting junction at the blood–brain barrier of the cuttlefish *Sepia officinalis*. *Journal of Neurocytology*, **21**, 295–303.

Lane, N. J. and Chandler, H. J. (1980). Definitive evidence for the existence of tight junctions in invertebrates. *Journal of Cell Biology*, **86**, 765–74.

Luft, J. H. (1966). Fine structure of capillary and endocapillary layer as revealed by ruthenium red. *Federation Proceedings*, **25**, 1773–83.

Nagy, Z., Peters, H., and Hüttner, I. (1984). Fracture faces of cell junctions in cerebral endothelium during normal and hyperosmotic conditions. *Laboratory Investigation*, **50**, 313–22.

Nixon, M. and Messenger, J. (eds) (1977). *The biology of cephalopods. Symposia of the Zoological Society of London*, **38**. Academic Press, London.

Préaux, G. and Gielens, C. (1984). Hemocyanins. In *Copper proteins and copper enzymes*, Vol. II (ed. R. Lontie), pp. 159–205. CRC Press, Florida.

Reese, T. S. and Karnovsky, M. J. (1967). Fine-structural localization of a blood–brain barrier to exogenous peroxidase. *Journal of Cell Biology*, **34**, 207–17.

Schneeberger, E. E. and Hamelin, M. (1984). Interaction of serum proteins with lung endothelial glycocalyx: its effect on endothelial permeability. *American Journal of Physiology*, **247**, H206–17.

Sibley, C. P., Bauman, K. F., and Firth, J. A. (1983). Molecular charge as a determinant of macromolecular permeability across the fetal capillary endothelium of the guinea-pig placenta. *Cell and Tissue Research*, **229**, 365–77.

Staehelin, L. A. (1974). Structure and function of intercellular junctions. *International Review of Cytology*, **39**, 191–283.

Stevenson, B. R., Anderson, J. M., Goodenough, D. A., and Mooseker, M. S. (1988). Tight junction structure and ZO-1 content are identical in two strains of Madin–Darby canine cells which differ in transepithelial resistance. *Journal of Cell Biology*, **107**, 2401–8.

Stevenson, B. R., Anderson, J. M., Braun, I. D., and Mooseker, M. S. (1989). Phosphorylation of the tight-junction protein ZO-1 in two strains of Madin–Darby canine kidney cells which differ in transepithelial resistance. *Biochemical Journal*, **263**, 597–9.

Treherne, J. E., Schofield, P. K., and Lane, N. J. (1982). Physiological and ultrastructural evidence for an extracellular anion matrix in the central nervous system of an insect (*Periplaneta americana*). *Brain Research*, **247**, 255–67.

Turner, M. R., Clough, G., and Michel, C. C. (1983). The effet of cationised ferritin and native ferritin upon the filtration of single frog capillaries. Evidence that proteins in the endothelial coat influence permeability. *Microvascular Research*, **25**, 205–22.

Villegas, G. M. and Sanchez, F. (1991). Periaxonal ensheathment of lobster giant nerve fibres as revealed by freeze-fracture and lanthanum penetration. *Journal of Neurocytology*, **20**, 504–17.

Ward, B. J., Bauman, K. F., and Firth, J. A. (1988). Interendothelial junctions of cardiac capillaries in rats: their structure and permeability properties. *Cell and Tissue Research*, **252**, 57–66.

Wells, M. J. (1978). *Octopus. Physiology and behaviour of an advanced invertebrate.* Chapman & Hall, London.

Wichertjes, T., Gielens, C., Schutter, W. G. Préaux, G., Lontie, R., and Van Bruggen, E. F. J. (1986). The quaternary structure of *Sepia officinalis* haemocyanin. *Biochimica Biophysica Acta*, **872**, 183–94.

Young, J. Z. (1971). *The anatomy of the nervous sytem of* Octopus vulgaris. Clarendon Press, Oxford.

28

Cerebrovascular organization and dynamics in cephalopods

N. JOAN ABBOTT and J. A. MIYAN

Summary

It has proved extremely difficult to do electrophysiological recording from exposed preparations of cephalopod brain, with the result that our understanding of the cell physiology of the cephalopod CNS has lagged behind that of other vertebrate and invertebrate groups. Many of the problems can be traced to features of cerebrovascular dynamics in cephalopods. This chapter reviews the anatomy and physiology of the cerebral circulation in cephalopods, and its behaviour in experimental preparations. Cephalopods have a closed vasculature, and the arrangement of cerebral arteries, arterioles, capillaries and veins follows a similar pattern to that of vertebrates, with some differences at the electron microscopic level. 'Gliovascular' channels may act as routes for flow of interstitial fluid while 'lymphoid' channels return the fluid to the venous system. Careful measurements of vascularity in *Sepia* show that the microvessel density in the brain is similar to that of mammals, while in muscle it is generally lower. There is a good correlation between tissue vascularity and aerobic status as judged from the ratio of oxidative to glycolytic enzymes. The control of the cerebral circulation in cephalopods is poorly understood, but appears to be mainly neurogenic. In larger cephalopods, surgical exposure of the brain, and probing with electrodes, cause ischaemia, hypoxia, cell swelling and eventually cell death, by a combination of a drop in perfusion pressure, local vasoconstriction, and accumulation of toxic metabolites. Experimental strategies to overcome these problems and develop useful preparations for electrophysiology are outlined.

Cephalopod molluscs are the only invertebrate animal group to possess a fully closed vascular system in which a substantial hydrostatic pressure is generated by the heart (Wells 1978). This has enabled coleoid cephalopods (octopus, squid, cuttlefish) to sustain high levels of tissue perfusion and metabolic activity, so that they compete successfully in many benthic and littoral marine environments (Packard 1972; O'Dor 1982). However, the relative inefficiency of cephalopod haemocyanin compared with vertebrate haemoglobin as blood oxygen transport protein means that the cephalopod vascular system is forced to work close to its physiological limits (Houlihan *et al.* 1982; Lykkeboe and

Johansen 1982; Johansen *et al.* 1982*a*,*b*; Brix *et al.* 1989). The cardiovascular specializations that are necessary to maintain tissue oxygen supply in cephalopods give instructive insights into basic principles of cardiovascular physiology.

Cephalopod neural tissue, in common with neural tissues in most animal groups, maintains a high metabolic rate and a predominantly aerobic metabolism, requiring efficient supply of oxygen and aerobic substrates from the blood. Comparison of the vascularity of brain and non-brain tissues and study of regional differences within the brain can highlight important aspects of neural metabolism.

Several chapters in this volume (Chapters 23–26) have mentioned the difficulties encountered in developing a good experimental preparation of cephalopod CNS suitable for intracellular and extracellular electrophysiological recording, and suggested that a failure of vascular perfusion is responsible. This chapter reviews earlier work on cerebrovascular anatomy and physiology in cephalopods, and describes recent experiments to determine the state of cerebrovascular perfusion in the octopus *Eledone* and cuttlefish *Sepia*, as an aid to devising better preparations for electrophysiological study.

Anatomy and fine-structure of brain vasculature

Blood vessels

Figure 28.1 shows the arterial supply to the supra-oesophageal brain lobes of *Octopus vulgaris* (Young 1971). The major input vessels are branches from the cephalic aorta. Each lobe of the brain receives arterial vessels from a single source, with few anastomoses. The arteries enter the centre of each lobe, then divide like the branches of a tree spreading towards the outer zones. The finest vessels (=capillaries) connect with venules and veins, which either run on the outer surface of the lobe (as in the supra-oesophageal lobes; Fig. 28.2), or form a system of venous channels leading back into the centre of the lobe (as in the optic lobe).

At the electron microscopic level, the structure of the brain blood vessel wall follows a similar pattern to that of vertebrates, with arteries containing prominent muscular layers and arterioles characterized by pericytes containing myofilaments (Fig. 28.3; Barber and Graziadei 1965, 1967*a*; Abbott and Bundgaard 1992; Bundgaard and Abbott 1992). The larger capillary vessels are lined by endothelial cells on a basal lamina, and a further layer of pericytes with a second basal lamina; either or both cell layers may be discontinuous. In the finest capillaries the distinction between endothelial cells and pericytes breaks down, and only a single basal lamina may be detected (Figs 28.1–28.3). Venous channels are similar in fine-structure to large capillaries, but with a greater luminal diameter. Arteries, arterioles, and veins are innervated (Alexandrowicz 1928; Barber and Graziadei 1967*b*). Although the earlier literature on the subject is confused, a blood–brain

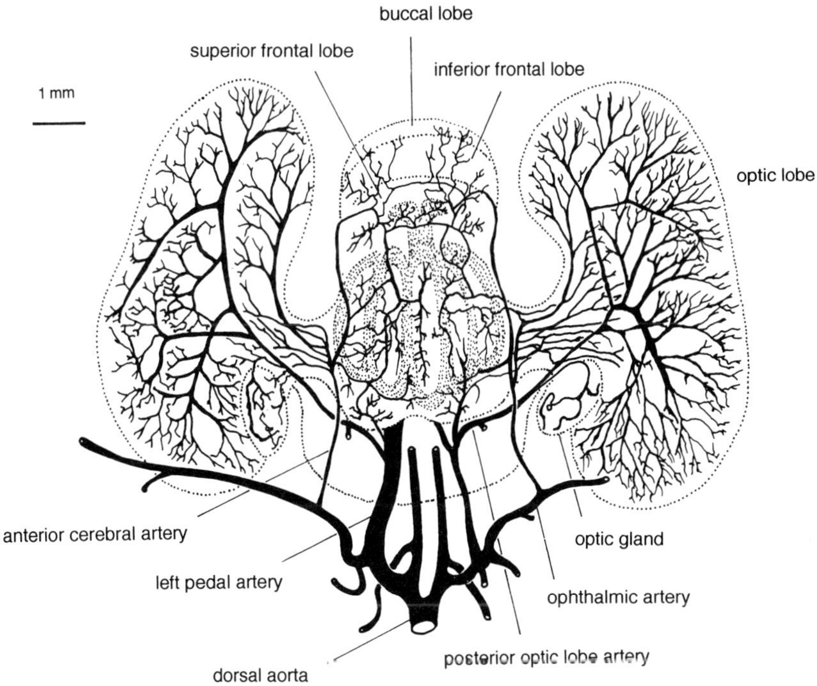

Fig. 28.1. The arterial supply to the supra-oesophageal part of the brain and optic lobes of *Octopus vulgaris*. The outlines of the vertical and subfrontal lobes are shown dotted. Tracing from a photograph of a cleared specimen seen from above after perfusion with Indian ink. The left pedal artery is regularly the larger of the two in *Octopus*. In the optic lobe the main arteries have been drawn as precisely as possible, but the finer branches are diagrammatic. (After Young 1971.)

barrier has been demonstrated in cuttlefish, squid, and octopus, formed by perivascular glial cells in capillaries and venules (Abbott and Bundgaard 1992; see also Chapter 27). The most detailed study, in the cuttlefish *Sepia*, shows that at the arteriolar level, the barrier is formed by junctions between pericytes. The occluding junction has several unusual features, which distinguish it from restricting junctions in other animal groups (Abbott *et al.* 1992; see also Chapter 27).

'Gliovascular' and 'lymphoid' channels

In many cephalopod tissues, there is a collagen-filled space beyond the outermost vessel wall layer, particularly in arterioles and arteries. The analogous space in vertebrate brain appears to be a preferential route for mixing of interstitial fluid, massaged by the fluctuations in arterial pressure (Rennels *et al.* 1985; Ichimura *et al.* 1991), and in neurosecretory zones it is continuous

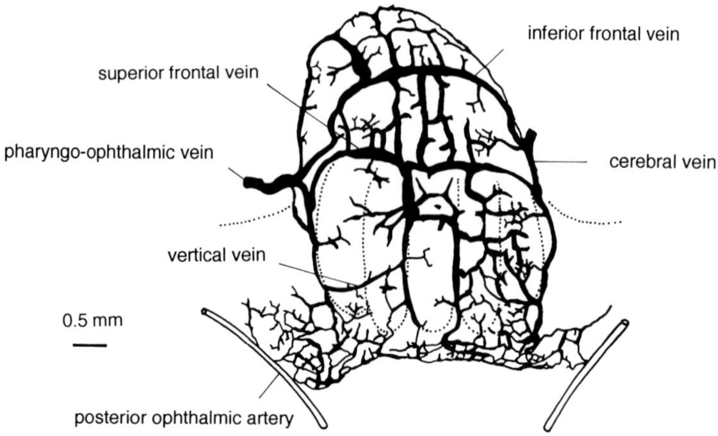

Fig. 28.2. Dorsal view of the venous drainage from the supra-oesophageal lobes of *Octopus vulgaris*, drawn from a specimen injected with Indian ink in which ink filled both arteries and veins. The outlines of the vertical lobes are shown dotted. (After Young 1971.)

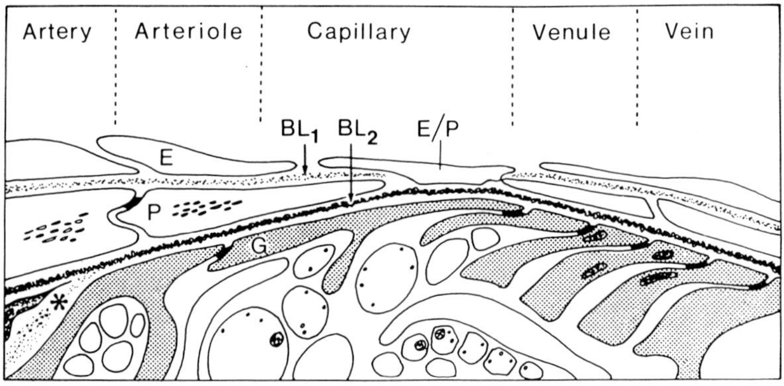

Fig. 28.3. Schematic drawing summarizing the cell layers present at the blood–brain interface in *Sepia* from the arterial to the venous end of the microvasculature, not to scale. E, endothelial cell; P, pericyte; BL_1 and BL_2, basal laminae; G (stippled), perivascular glial layer; (*) collagen-filled extracellular space. In the smaller vessels, only a single basal lamina can be resolved; endothelial cells and pericytes here become indistinguishable, and are shown as a merged class of endothelial/pericytes (E/P). Restricting junctions (black bars) are formed between pericytes in arterial vessels, and between perivascular glial cells in microvessels (capillaries) and venous vessels. A large proportion of the glial sheath covering the microvessel surface is 'seamless' (no intercellular clefts), while in the venous vessels glial end feet make a lamellated layer with extensive overlap between adjacent cells. (From Abbott and Bundgaard 1992.)

with spaces penetrating into the interstitium (Abraham *et al.* 1976). In certain parts of the octopus brain, particularly the optic and sub-oesophageal lobes, the system of perivascular spaces is particularly prominent, and extends as a series of straight projections into the surrounding neural tissue (Stephens and Young 1969; Gray 1969). The spaces accompany arterioles and all but the smallest microvessels. The spaces contain collagen, fibroblasts, and muscle cells, and are lined by glial cells; they have been termed the 'gliovascular system'.

If Indian ink is injected into *Octopus* aorta under high pressure, it fills both arterial and venous vessels in the brain, and in addition a system of grape-like clusters of tissue spaces around the venous vessels. These are distinct from the gliovascular system, and have been termed 'lymphoid channels' (Stephens and Young 1969; Young 1971). Haemocyanin has occasionally been reported in an apparently perivascular location around large (venous) vessels (Gray 1969); if this is not an artefact, it is likely to represent haemocyanin backflux into lymphoid channels. Haemocyanin is notably absent from the gliovascular system and indeed from any perivascular space in well-fixed cephalopod central neural tissue.

In *Sepia* brain it has been calculated that interstitial fluid is being produced at a rate of about 0.1 μl/g per min (Abbott *et al.* 1985*b*), and since there are no specialized secretory structures analogous to the choroid plexus of vertebrates, the fluid must be produced by filtration and/or secretion across the microvascular walls. The gliovascular system with its muscle cells would then provide an efficient system for the circulation of this fluid. As the cephalopod blood–brain barrier has been shown to be as tight as that of mammals, efficiently excluding proteins (Abbott *et al.* 1985*b*; Abbott and Bundgaard 1992), the fluid in gliovascular channels is likely to be essentially protein-free. It could well trickle back into the veins via the lymphoid channels. A flowing interstitial fluid system provides more effective homeostasis of the brain microenvironment than a static system, for example for supply of nutrient substrates, removal of toxic waste-products, and distribution of neuroactive substances. However, the fact that the gliovascular system is poorly developed in some cephalopod species and brain lobes, for example being virtually absent in the vertical lobe of *Sepia* (Bundgaard and Abbott 1992), means it is unlikely to be an obligatory component in blood–brain exchange.

Measurements of tissue vascularity

Young's (1971) photomicrographs of ink-injected *Octopus* brain showed dense networks of microvessels, but no quantitation was attempted. Browning (1982) made a stereometric analysis in the muscle and neural tissue of *Octopus* arms, and showed a relatively low microvessel density, with 45 exchange microvessel profiles per mm^2 section area, corresponding to a maximum blood-to-cell diffusion distance of 60–85 μm. This figure has sometimes been assumed to be representative of all cephalopod tissues, but as the tissues vary greatly in their aerobic/anaerobic activities (Mommsen *et al.* 1981; Baldwin 1982; Storey and Storey 1983; Hochachka *et al.* 1983), differences in vascularity between tissues would be expected.

In a detailed morphometric study of vessels in the vertical and optic lobes of *Sepia* brain, with valve and tentacle muscle for comparison, Abbott and Bundgaard (1987) showed that the brain vascularity was much higher than previously thought, the density of microvessel profiles being 211 and 395/mm^2 in vertical and optic lobe, respectively (Fig. 28.4), compared with 157 and 43/mm^2 in valve muscle and tentacle muscle. These and other morphometric indices are given in Table 28.1. It is clear that the vascularity of cephalopod brain is similar to that of mammals, while valve muscle is somewhat less vascular than slow mammalian muscle. *Sepia* tentacle muscle has a lower vascularity than fast (white) mammalian muscle. The *Sepia* study also showed marked regional differences in vascularity within individual brain lobes, with the ratio of microvessel density in neuropil to cell body regions being 1.55 in vertical lobe and 2.03 in optic lobe. The total vessel volume

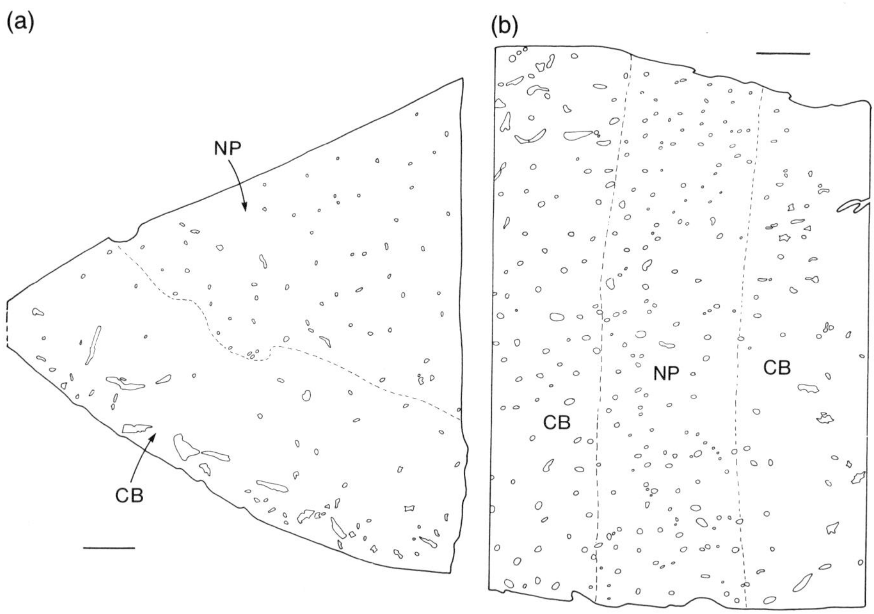

Fig. 28.4. Tracing of vascular profiles from micrographs of 1 μm sections of *Sepia* brain. (a) Vertical lobe; (b) optic lobe. Microvessels are more homogeneous in diameter and more uniformly distributed in neuropil (NP) than in the cell body zones (CB). Morphometric analysis of many such sections shows that the microvascular density is higher in the neuropil than in the cell body zone. Scale bars: 100 μm. (From Abbott and Bundgaard 1987.)

Table 28.1 Summary of *Sepia* microvessel morphometric parameters, mean ± SEM, based on measurements of 60–110 vessels from two to six sections in n = 5 animals

	Tissue	Q_A/mm^{-2}	J_V/mm^{-2}	R/μm	S_V/cm^{-2}	V_V/%	d_m/μm
Brain							
Sepia	VL	211±19	422±37	28	134 ±9	2.94±0.20	4.9
	OL	395±48	790±95	20	176 ±9	3.46±0.21	4.1
Rat	cortex	400	800	20	165	2.2	4.6
Muscle							
Sepia	VM	157±13	315±26	32	67.9±3.6	1.69±0.20	3.9
	TM	43	86	61	13.8	0.28	3.9
Rat	slow (red)	3200	4420	9	833	12.5	—
	fast (white)	190	209	41	38	0.55	—

VL and OL, vertical lobe and optic lobe of brain; VM and TM, valve muscle and tentacle muscle. Q_A, number of vessel transections per mm^2; J_V, length density; R, radius of Krogh cylinder; S_V, surface area density; V_V, vessel volume density; d_m, modal diameter. Mammalian values from the literature for comparison. (From Abbott and Bundgaard 1987.)

density calculated from the sections (large and small vessels) was 3.49 and 4.73 per cent in brain vertical and optic lobe, 1.88 and 0.28 per cent in valve and tentacle muscle, in good agreement with studies using radioactive intravascular markers (Abbott *et al.* 1985*a*).

Correlation between vascularity and tissue biochemistry

Cephalopod brain has a predominantly oxidative metabolism, using glucose as substrate, with brain hexokinase activity being 15–48 times that of muscle (Ballantyne *et al.* 1981; Storey and Storey 1983). *Sepia* muscle is specialized for either aerobic, maintained activity (well vascularized, plentiful mitochondria, oxidative enzymes) or anaerobic burst activity (less vascular, fewer mitochondria, glycolytic enzymes) using glycogen and amino acids as substrates (reviewed in Storey and Storey 1983; Abbott and Bundgaard 1987). Key metabolic enzymes have been used as indicators of the aerobic and anaerobic status of cephalopod tissues. Octopine dehydrogenase (ODH) catalyses the (anaerobic) conjugation of arginine and pyruvate to yield octopine (end product of glycolysis in cephalopods), and α-glycerophosphate dehydrogenase (α-GPDH) reflects the activity of the glycerophosphate cycle, which supports aerobic metabolism (Storey and Storey 1983). High α-GPDH/ODH ratios indicate aerobic metabolism. Figure 28.5 shows the relation between the α-GPDH/ODH ratio and vascular volume (per cent) in cephalopod tissues, plotting mean enzyme ratios from several coleoid species (Ballantyne *et al.* 1981; Baldwin 1982) against determinations of *Sepia* tissue vascular

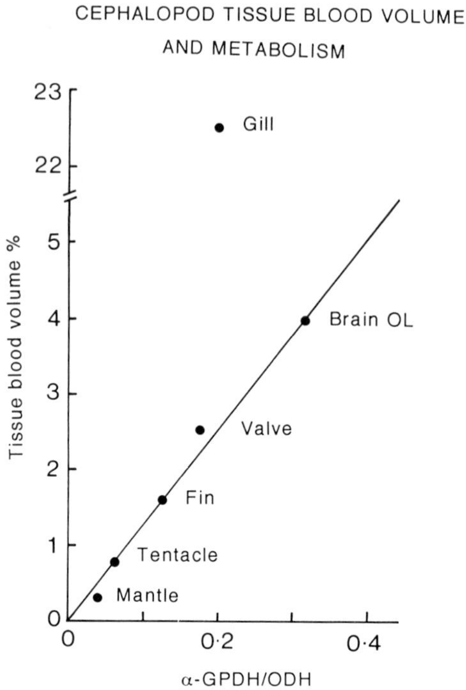

Fig. 28.5. Relation between tissue blood volume (per cent) and metabolic enzyme ratio in cephalopod tissues. Tissue blood volume was measured in *Sepia* using ^{125}I-human serum albumin injected intravenously and allowed to circulate for 10 min (Abbott 1987). α-GPDH/ODH enzyme ratio, a measure of aerobic metabolism of the tissue, was derived from tables in Baldwin (1982) and Storey and Storey (1993). Tissue blood volume is highly correlated with tissue oxidative metabolism, except in the gill (see text).

volumes, measured with ^{125}I-serum albumin (Abbott 1987). In all cases except gill, there is a good correlation between vascularity and aerobic metabolism, expressed as α-GPDH/ODH ratio. Gill tissue has a higher vascularity than predicted from its enzyme status, as would be expected for a tissue whose vascularity subserves the oxygen exchange needs of the whole body, not just the metabolic requirements of the tissue itself.

Control of cerebrovascular tone

All higher vertebrates have mechanisms by which cerebral blood flow is maintained adequate for the needs of the tissue. In mammals, the cerebral circulation exhibits strong 'autoregulation', by which flow is kept relatively constant in the face of variations in blood pressure in the range 60–150 mmHg (Paulson *et al.* 1990). The mechanism appears to involve a combination of myogenic and (local) metabolic processes, with some modulation by sympathetic nervous activity. Removal of the endothelium of cerebral arteries impairs autoregulation, suggesting that the endothelial cells are involved in the myogenic response, for example by sensing changes in pressure and flow and releasing relaxing and constrictor factors that mediate the smooth muscle response. The observed acetylcholine-induced vasodilatation of cerebral vessels is endothelium dependent. There is good evidence that all capillaries are perfused at any one time in brain, so that unlike the situation in skeletal muscle, recruitment is not used as a means of increasing tissue blood flow (Kuschinsky and Paulson 1992). There is a good correlation between capillary density on the one hand, and glucose utilization and leucine uptake on the other, showing that the capillary density is well matched to the physiological needs of each brain region. 'Physiological recruitment', an apparent increase in flow per unit weight of brain tissue, is apparently achieved by increasing the flow rate of the more slowly perfused microvessels, there being some heterogeneity in flow rate in different vessels at rest.

Vasoconstriction of the larger mammalian cerebral arteries, mediated by sympathetic innervation, tends to be accompanied by an autoregulatory vasodilatation of the smaller resistance vessels downstream to maintain a constant cerebral blood flow (Paulson *et al.* 1990). In conditions of mild hypotension, the combination of constricted large resistance vessels and dilated small resistance vessels can lead to inhomogeneous perfusion with patchy tissue ischaemia (Hamar *et al.* 1979). In severe hypotension the vessels collapse leading to widespread ischaemia. Loss of autoregulation in pathological conditions can lead to breakdown of the blood–brain barrier and oedema.

Much less is known about the control of cerebral blood flow in cephalopods. The cephalic aorta behaves as a 'Windkessel' vessel, modulating by its elasticity the pulse generated by the heart (Shadwick and Gosline 1985; Shadwick *et al.* 1987; Schipp 1987*a*). The vessel has a three-layered wall (Schipp 1987*b*). The inner *tunica intima* consists of an endothelium, frequently incomplete, and prominent continuous basal lamina, with underlying pericytes containing myofilaments. The *tunica media* is composed of loose connective tissue and circular and longitudinally arranged obliquely striated muscle cells. The *tunica adventitia* is a collagenous network, with numerous nerve fibres ending at the boundary between the *t. adventitia* and *t. media* (Schipp 1987*b*).

The cephalic aorta can produce weak autonomous contractions in response to stretch, originating in the *tunica media* and mesangium since removal of these layers blocks contraction (Schipp *et al.* 1991). The neural regulation of the highly specialized circular muscle cells of the multilayered *t. media* appears to involve a catecholamine-mediated vasoconstriction, with agonist potency in the order dopamine > noradrenaline > adrenaline (Schipp *et al.* 1991). The response is blocked by phentolamine, evidence for action via an α-adrenergic receptor. The constriction is antagonized by serotonin (5-hydroxytryptamine, 5HT) and FMRFamide, and FMRFamide can be detected in nerve terminals in the *t. media*. Acetylcholine can induce dilatation of the dopamine-precontracted aorta (Schipp 1987*b*), but it is not known whether this is synergistic with the action of FMRFamide. As FMRFamide and analogues are also released into the vascular system by the cardiac nerve and the neurosecretory vena cava system, these agents are also capable of exerting a modulatory role from the blood side.

The pharmacology of the cerebral arterioles is not known, nor is it known whether vascular autoregulation or endothelium-dependent vasomotor responses are present. However, the presence of nerve fibres associated with arteriolar vessels (Barber and Graziadei 1967*b*) and veins (Alexandrowicz 1928), and the evidence for localized vasoconstriction in the microvasculature (see below) suggests that the vascular tone of arteries, arterioles, and veins can be modulated by neural influences.

Tissue perfusion in experimental preparations

In spite of numerous attempts to record neural activity from cephalopod brain using extracellular and intracellular electrodes, these efforts have been relatively unsuccessful (Boycott *et al.* 1965; see also Chapter 24). Boycott and Maynard attempted to maintain neural activity of octopus brain with saline vascular perfusion, previously used successfully in Crustacea (Maynard 1966; Sandeman 1969), but again found the blood vessels shut down (Boycott 1988). Laverack (1980) developed a semi-isolated perfused preparation of octopus brain which was reasonably successful, but neuronal activity remained hard to detect, and the method has not been widely used. Saline perfusion of small cephalopod brains such as that of the Bay squid *Lolliguncula brevis* has been more successful (Dubas *et al.* 1986*a,b*). Implanted electrodes in unanaesthetized freely-moving animals were successful in stimulation experiments in octopus (Boycott 1988; Chicery and Chanlet 1976; Chicery 1983) and more recently for recording in *Sepia* (see also Chapter 24). This historical survey suggests that failure of vascular perfusion in exposed brain preparations results in hypoxia and neuronal death. Recent investigators have side-stepped the problems by working on brain slices, where vascular perfusion is eliminated. However, there will always be physiological questions where recording from semi-intact, exposed preparations is required to give a complete picture of the normal situation, so it is important to establish why cephalopod brain is so difficult to keep functional in experimental preparations. Here we identify several key factors concerning cephalopod neural tissue and its blood supply.

High oxygen requirement of cephalopod neural tissue

Cephalopod neural tissue needs to be kept well oxygenated to remain excitable; its metabolic rate is geared to oxygen supply by haemocyanin-containing blood (4.5 volume per cent oxygen; Altman and Dittmer 1971, p. 655, table 211) flowing to within less than 30 μm of active neurones, that is similar to the dimensions of a single neurone cell body. Thus whereas the oxygen needs of crustacean brain (*in vivo* blood supply contains 1–2 volume per cent oxygen, Altman and Dittmer 1971, p. 654, table 211) can be met by perfusion with oxygenated saline, this is generally inadequate for larger cephalopods.

Cell damage in oxygenated non-perfused cephalopod brain

Some lower vertebrate brains (e.g. fish, frog, turtle) can be excised from the animal, and when incubated in cold oxygenated saline, maintain stable cellular ionic concentrations and size of interstitial space, indicating a lack of cell swelling and good ionic homeostasis (Bradbury *et al.* 1968; Friede and Hu 1971; Nicholson and Rice 1986). Isolated vertical lobes of *Sepia* and *Alloteuthis* when incubated in artificial sea water (ASW) progressively lose K^+ and gain Na^+, and if gassed with oxygen or air, show increased membrane permeability after 1–2 h, as demonstrated by an increase in the distribution space for the normally extracellular tracer ^{51}Cr-EDTA (Abbott 1985). Isolated optic lobe survives better, but membrane damage is demonstrable by 3 h. Gassing with nitrogen results in less membrane damage, suggesting that the accumulation of toxic metabolites in non-perfused brain lobes is responsible for the damage in the presence of oxygen and air. Neuronal death in saline-perfused preparations is therefore likely to be a result of the combined effects of inadequate supply of oxygen and inadequate removal of toxic metabolites.

Injury-mediated vasoconstriction

As described previously (p. 467), cephalopod arteries have an extensive innervation associated with the *t. media*. At the electron microscopic level, nerve endings are also seen on the pericyte layer of arterioles and veins. As cephalopod blood does not clot, alternative mechanisms have evolved to prevent bleeding following vessel damage, and powerful vasoconstriction is one such mechanism. Under light chloral hydrate anaesthesia, a band of vasoconstriction about 2 mm wide can be demonstrated either side of a ventral mid-line cut through the mantle muscle in *Sepia*, persisting for the duration of the experiment (up to 90 min; Abbott *et al.* 1985*a*). In *Octopus vulgaris* and *Eledone cirrhosa*, severing the arms under light ethanol anaesthesia induces immediate vasoconstriction of the vessels of the stump, and cessation of bleeding within seconds (see Chapter 25). Deeper anaesthesia is associated with lack of the vasoconstrictor response, leading to excessive blood loss and death.

A serious problem for normal brain stimulation and recording in cephalopods is that vasoconstriction of cerebral vessels can apparently be triggered simply by electrode placement. B. B. Boycott (personal communication) describes an experiment in the optic lobe of *O. vulgaris*, normally translucent when well perfused. Under light urethane anaesthesia, the outer cell layers of the optic lobe cortex were probed with a 150 μm tungsten electrode; the electrode was withdrawn, and the skin wound sutured. Reopening the exposed area on the following day revealed a white opaque zone around the site of electrode penetration, while the rest of the optic lobe remained translucent.

Circulating leucocytes in stress and surgery

In *Sepia* anaesthetized lightly in chloral hydrate or ethanol under conditions designed to minimize pre-anaesthetic stress, the blood leucocyte (white cell) content is less than 1–2 per cent in blood samples taken from the cephalic vein (Abbott *et al.* 1985*a*). If the animals are agitated and hypoxic before anaesthesia (e.g. if catching involves a chase around the holding tank), the white cell content of blood samples is markedly elevated. Since cephalopod leucocytes have some functions analogous to blood platelets in vertebrates (plugging sites of vessel damage; Wells 1978), it is likely that the circulating leucocytes will tend to adhere to the walls of damaged vessels in sites of neurosurgical intervention, causing microemboli and consequent local zones of ischaemia.

Effect of perivascular connective tissue

Most of the problems of local ischaemia have been reported in the more accessible outer (cortical) regions of the vertical and optic lobes, where the lack of perivascular connective tissue may make the microvessels more prone to irreversible collapse and occlusion.

Effect of anaesthesia on brain in peripheral vascular perfusion

Studies with ion-selective microelectrodes in the outer zones of *Sepia* vertical lobe provided evidence for hypoxia and cell swelling (small values recorded for interstitial volume fraction) in animals anaesthetized with chloral hydrate or ethanol, indicating inadequate tissue perfusion and oxygenation (see Chapter 23). Results were similar whether the preparations were autoperfused (own blood) or saline perfused. Some improvement was seen in the semi-intact preparation of *Eledone*, where anaesthesia was used only for the initial surgery (see Chapter 25).

Studies with ^{14}C-sucrose as intravascular tracer in anaesthetized *Sepia*

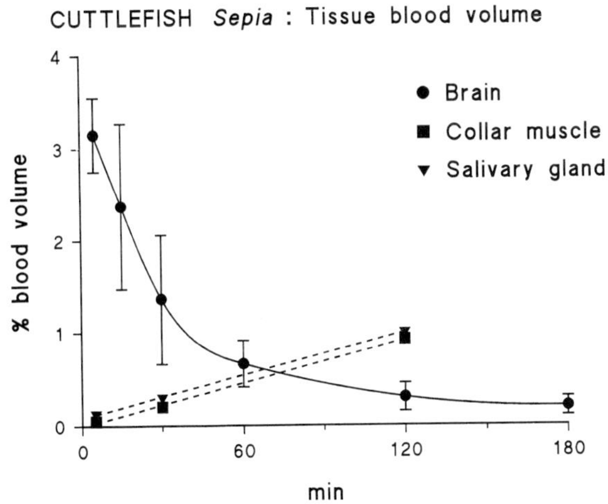

Fig. 28.6. Vascular volume in exposed *Sepia* brain, and in reference tissues, in an anaesthetized preparation. Animals were anaesthetized in 1.5 per cent ethanol, the vertical lobe was exposed by removing dorsal cartilage, and the animals were returned to 0.5 per cent ethanol in sea water. At the end of the experiment, ^{14}C-sucrose was injected into the aorta over 10 s, and tissue blood volume measured as ^{14}C-sucrose dpm per mg tissue divided by dpm per μl injectate × 100 per cent. (From Abbott and Pearson, in preparation.)

showed a progressive decline in cerebral blood volume over a 3 h period (Abbott and Pearson, in preparation; Fig. 28.6). In preparations of *Eledone*, anaesthetized only for surgery, brain blood volume was maintained relatively constant (Fig. 28.7), although somewhat lower than seen in *Sepia in vivo* (Miyan and Abbott, in preparation). If the optic sinuses were not opened, the cerebral blood volume in *Eledone* was similar to that of intact *Sepia* (Abbott and Miyan, in preparation), indicating that blood loss through the orbital sinuses leads to the lower brain blood volume recorded in the semi-intact preparation (Fig. 28.7). The decline in cerebral perfusion in anaesthetized *Sepia* was accompanied by a corresponding increase in vascular volume of peripheral tissues such as muscle and salivary gland (Fig. 28.6), suggesting that the failure of brain perfusion was associated with anaesthesia-induced peripheral vasodilatation, possibly leaving insufficient perfusion pressure to drive blood through the finely divided microvascular network of the brain. Taken together, the studies in *Sepia* and *Eledone* show that if cerebral perfusion is compromised because of blood loss or peripheral vasodilatation, the brain becomes hypoxic and neuronal swelling and death rapidly result.

Future directions

The studies reviewed here point to experimental strategies for the future.

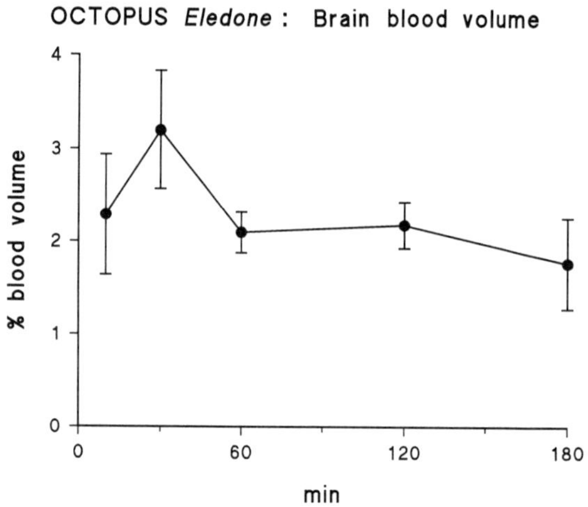

Fig. 28.7. Vascular volume in exposed *Eledone* brain, semi-intact preparation. Under 1.5 per cent ethanol anaesthesia the arms were removed, orbits opened, retinae destroyed, and vertical lobe exposed, to give a decerebrate preparation (see Chapter 25), which was then returned to sea water. ^{125}I-human serum albumin was injected into the cephalic vein 10 min before the end of the experiment and allowed to circulate. Brain vascular volume was measured from ^{125}I-albumin dpm per mg brain divided by dpm per μl blood × 100 per cent, a terminal blood sample having been obtained from the dorsal aorta. (From Miyan and Abbott, in preparation.)

Autoperfused preparations

These should work well providing they are unanaesthetized or lightly anaesthetized so that cerebral perfusion pressure is maintained and blood is well oxygenated in the gills. Neural recording in intact preparations is possible, but relies heavily on stereotaxic methods for localization. *Sepia* so treated remain calm with a normal repertoire of reflexes (see Chapter 24) showing that careful insertion of microelectrodes can avoid trauma and vasoconstriction.

Saline perfusion

For large species this is likely to work only if an oxygen carrier is included, and if cerebral vasodilatation is achieved by pharmacological means. More information is needed on the relation between arterial and arteriolar vasoconstriction and dilatation, in order to devise the best drug cocktail to preserve microvascular perfusion. However, for pharmacological studies of CNS neurones, interference with neurotransmitter actions on the vessel wall could be a complication, especially if the blood–brain barrier is breached by an elevated perfusion pressure. If the heart is normally releasing vasodilator factors, perfusion techniques including the heart in the input side may be more successful than perfusion directly via the cephalic aorta.

Slice preparations

Diffusion distances must be kept short, slices need to be cooled to reduce metabolic rate, and superfused to wash away toxic metabolites. The choice of incubation medium may be more critical than in vertebrates, and should be tested for its ability to maintain cellular and interstitial ionic homeostasis.

Acknowledgements

We are grateful to the Directors and staff of the Marine Biological Association Laboratory, Plymouth, and the Gatty Marine Laboratory, St Andrews, for experimental facilities and supply of animals. We thank the the Wellcome Trust, Royal Society, and Physiological Society for support, and Ben Pearson for technical assistance.

References

Abbott, N. J. (1985). Size of extracellular space in isolated brain of cuttlefish *Sepia* and squid *Alloteuthis*. *Journal of Physiology*, **369**, 57P.

Abbott, N. J. (1987). Vascular volume estimates in *Sepia* tissues, and evidence for vessel recruitment. *Journal of Physiology*, **394**, 165P.

Abbott, N. J. and Bundgaard, M. (1987). Microvessel surface area, density and dimensions in brain and muscle of the cephalopod *Sepia officinalis*. *Proceedings of the Royal Society of London, Series B*, **230**, 459–82.

Abbott, N. J. and Bundgaard, M. (1992). Electron-dense tracer evidence for a blood–brain barrier in the cuttlefish *Sepia officinalis*. *Journal of Neurocytology*, **21**, 276–94.

Abbott, N. J., Bundgaard, M., and Cserr, H. F. (1985*a*). Brain vascular volume, electrolytes and blood–brain interface in the cuttlefish *Sepia officinalis* (Cephalopoda). *Journal of Physiology*, **368**, 197–212.

Abbott, N. J., Bundgaard, M., and Cserr, H. F. (1985*b*). Tightness of the blood–brain barrier and evidence for brain interstitial fluid flow in the cuttlefish, *Sepia officinalis*. *Journal of Physiology*, **368**, 213–26.

Abbott, N. J., Lane, N. J., and Bundgaard, M. (1992). A fibre-matrix model for the restricting junction of the blood–brain barrier in a cephalopod mollusc: implications for capillary and epithelial permeability. *Journal of Neurocytology*, **21**, 304–11.

Abraham, M., Kieselstein, M., and Lisson-Begon, S. (1976). The extravascular channel system in the rostral pituitary of *Mugil cephalus* (Teleostei) as revealed by use of horseradish peroxidase. *Cell and Tissue Research*, **167**, 289–96.

Alexandrowicz, J. S. (1928). Sur l'innervation des vaissaux sanguin des céphalopodes. *Comptes Rendus des Séances de la Société de Biologie et de ses Filiales*, **99**, 1015–17.

Altman, P. L. and Dittmer, D. S. (1971). *Biological handbook of respiration and circulation*. FASEB, Bethesda, MD.

Baldwin, J. (1982). Correlations between enzyme profiles in cephalopod muscle and swimming behaviour. *Pacific Science*, **36**, 349–56.

Ballantyne, J. S., Hochachka, P. W., and Mommsen, T. P. (1981). Studies on the metabolism of the migratory squid, *Loligo opalescens*: enzymes of tissues and heart mitochondria. *Marine Biology Letters*, **2**, 75–85.

Barber, V. C. and Graziadei, P. (1965). The fine structure of cephalopod blood vessels. I. Some smaller peripheral vessels. *Zeitschrift für Zellforschung und Mikroskopische Anatomie*, **66**, 765–81.

Barber, V. C. and Graziadei, P. (1967*a*). The fine structure of cephalopod blood vessels. II. The vessels of the nervous system. *Zeitschrift für Zellforschung und Mikroskopische Anatomie*, **77**, 147–61.

Barber, V. C. and Graziadei, P. (1967*b*). The fine structure of cephalopod blood vessels. III. Vessel innervation. *Zeitschrift für Zellforschung und Mikroskopische Anatomie*, **77**, 162–74.

Boycott, B. B. (1988). Cephalopods, memory, neurofilaments, mammalian hibernation, the vertebrate retina: an autobiographical research note for John H. Welsh in his 85th year. *Comparative Biochemistry and Physiology*, **91C**, 25–9.

Boycott, B. B., Lettvin, J. Y., Maturana, H. R., and Wall, P. D. (1965). Octopus optic responses. *Experimental Neurology*, **12**, 247–56.

Bradbury, M. W. B., Villamil, M., and Kleeman, C. R. (1968). Extracellular fluid, ionic distribution and exchange in isolated frog brain. *American Journal of Physiology*, **214**, 643–51.

Brix, O., Bardgård, A., Cau, A., Colosimo, A., Condó, S. G., and Giardina, B. (1989). Oxygen-binding properties of cephalopod blood with special reference to environmental temperatures and ecological distribution. *Journal of Experimental Zoology*, **252**, 34–42.

Browning, J. (1982). The density and dimensions of exchange vessels in *Octopus pallidus*. *Journal of Zoology*, **196**, 569–79.

Bundgaard, M. and Abbott, N. J. (1992). Fine structure of the blood–brain interface in the cuttlefish *Sepia officinalis* (Mollusca, Cephalopoda). *Journal of Neurocytology*, **21**, 260–75.

Chicery, R. (1983). Motor and behavioral responses obtained by electrical stimulation of peduncle and basal lobes: the control of visual-static centres on oculomotor reaction and locomotion in the cuttlefish *Sepia officinalis*. *Fortschritte der Zoologie*, **28**, 231–40.

Chicery, R. and Chanelet, J. (1976). Motor and behavioral responses obtained by stimulation with chronic electrodes of the optic lobe of *Sepia officinalis*. *Brain Research*, **105**, 525–32.

Dubas, F., Leonard, R. B., and Hanlon, R. T. (1986*a*). Chromatophore motoneurons in the brain of the squid *Lolliguncula brevis*: an HRP study. *Brain Research*, **374**, 21–9.

Dubas, F., Hanlon, R. T., Ferguson, G. P., and Pinsker, H. M. (1986*b*). Localisation and stimulation of chromatophore motoneurons in the brain of the squid, *Lolliguncula brevis*. *Journal of Experimental Biology*, **121**, 1–25.

Friede, R. C. and Hu, K. H. (1971). A new approach to determining the volume of cerebral extracellular fluid and demonstration of its connection with cerebrospinal fluid. *Journal of Physiology*, **218**, 477–93.

Gray, E. G. (1969). Electron microscopy of the glio-vascular organization of the brain of *Octopus*. *Philosophical Transactions of the Royal Society of London, Series B*, **255**, 13–32.

Hamar, J., Kovach, A. B., Reivich, M., Nyary, I., and Durity, F. (1979). Effect of phenoxybenzamine on cerebral blood flow and metabolism in the baboon during haemorrhagic shock. *Stroke*, **10**, 401–7.

Hochachka, P. W., Fields, J. H. A., and Mommsen, T. P. (1983). Metabolic and enzyme regulation during rest-to-work transition: a mammal versus mollusc comparison. In *The Mollusca* (ed. K. M. Wilbur), Vol. 1, *Metabolic biochemistry and molecular biomechanics*, pp. 55–89. Academic Press, New York.

Houlihan, D. F., Innes, A. J., Wells, M. J., and Wells, J. (1982). Oxygen consumption and blood gases of *Octopus vulgaris* in hypoxic conditions. *Journal of Comparative Physiology*, **148**, 35–40.

Ichimura, T., Fraser, P. A., and Cserr, H. F. (1991). Distribution of extracellular tracers in perivascular spaces of the rat brain. *Brain Research*, **545**, 103–13.

Johansen, K., Brix, O., and Lykkeboe, G. (1982*a*). Blood gas transport in the cephalopod *Sepia officinalis*. *Journal of Experimental Biology*, **99**, 331–8.

Johansen, K., Brix, O., Kornerup, S., and Lykkeboe, G. (1982*b*). Factors affecting O_2-uptake in the cuttlefish, *Sepia officinalis*. *Journal of the Marine Biological Association of the United Kingdom*, **62**, 187–91.

Kuschinsky, W. and Paulson, O. B. (1992). Capillary circulation in the brain. *Cerebrovascular and Brain Metabolism Reviews,* **4**, 261–86.

Laverack, M. S. (1980). Electrophysiology of the isolated central nervous system of the Northern octopus *Eledone cirrhosa*. *Marine Behaviour and Physiology*, **7**, 155–69.

Lykkeboe, G. and Johansen, K. (1982). A cephalopod approach to rethinking about the importance of Bohr and Haldane effects. *Pacific Science*, **36**, 305–13.

Maynard, D. M. (1966). Integration in crustacean ganglia. *Symposium of the Society for Experimental Biology*, **20**, 111–49.

Mommsen, T. P., Ballantyne, J., MacDonald, D., Gosline, J., and Hochachka, P. W. (1981). Analogues of red and white muscle in squid mantle. *Proceedings of the National Academy of Sciences, USA*, **78**, 3274–8.

Nicholson, C. and Rice, M. E. (1986). The migration of substances in the neuronal microenvironment. *Annals of the New York Academy of Sciences*, **481**, 55–71.

O'Dor, R. K. (1982). Respiratory metabolism and swimming performance of the squid, *Loligo opalescens*. *Canadian Journal of Fisheries and Aquatic Science*, **39**, 580–7.

Packard, A. (1972). Cephalopods and fish: the limits of convergence. *Biological Reviews*, **467**, 241–307.

Paulson, O. B., Strandgaard, S., and Edvinsson, L. (1990). Cerebral autoregulation. *Cerebrovascular and Brain Metabolism Reviews*, **2**, 161–92.

Rennels, M., Gregory, T. F., Blaumanis, O. R., Fujimoto, K., and Grady, P. A. (1985). Evidence for a 'paravascular' fluid circulation in the mammalian central nervous system, provided by the rapid distribution of tracer protein throughout the brain from the subarachnoid space. *Brain Research*, **326**, 47–63.

Sandeman, D. C. (1969). The synaptic link between the sensory and motoneurones in the eye-withdrawal reflex of the crab. *Journal of Experimental Biology*, **50**, 87–98.

Schipp, R. (1987*a*). General morphological and functional characteristics of the cephalopod circulatory system. An introduction. *Experientia*, **43**, 474–7.

Schipp, R. (1987*b*). The blood vessels of cephalopods. A comparative morphological and functional survey. *Experientia*, **43**, 525–37.

Schipp, R., Jakobs, P. M., and Fiedler, A. (1991). Monoaminergic–peptidergic interactions in neuroregulatory control of the cephalic aorta in *Sepia officinalis* L. (Cephalopoda). *Comparative Biochemistry and Physiology*, **99C**, 421–9.

Shadwick, R. E. and Gosline, J. M. (1985). Mechanical properties of the octopus aorta. *Journal of Experimental Biology*, **114**, 259–84.

Shadwick, R. E., Gosline, J. M., and Milson, W. K. (1987). Arterial haemodynamics in the cephalopod mollusc, *Octopus dofleini*. *Journal of Experimental Biology*, **130**, 87–106.

Stephens, P. R. and Young, J. Z. (1969). The glio-vascular system of cephalopods. *Philosophical Transactions of the Royal Society of London, Series B*, **255**, 1–12.

Storey, K. B. and Storey, J. M. (1983). Carbohydrate metabolism in cephalopod molluscs. In *The Mollusca* (ed. K. M. Wilbur), Vol. 1, *Metabolic biochemistry and molecular biomechanics*, pp. 91–136. Academic Press, New York.

Wells, M. J. (1978). *Octopus. Physiology and behaviour of an advanced invertebrate.* Chapman & Hall, London.

Young, J. Z. (1971). *The anatomy of the nervous system of* Octopus vulgaris. Clarendon Press, Oxford.

PART VII

Sensory function

29

Squid rhodopsin

H. R. SAIBIL, K. A. LANGMACK, C. VENIEN-BRYAN, and J. R. WILKINSON

Summary

The retina of the squid contains highly ordered arrays of photoreceptor microvilli, aligned in two orthogonal directions in the plane normal to the optic axis. The microvillar membranes themselves have ordered membrane–membrane and membrane–cytoskeleton contacts. Their major protein is the visual pigment, rhodopsin. This arrangement aligns the rhodopsin molecules, thereby conferring sensitivity to the plane of polarization of light absorbed by the retinula cells. The biological function of polarization sensitivity in squid is unknown, but could be related to detection of prey or navigation. The photostable metarhodopsin is regenerated to rhodopsin by a chromophore exchange system involving the retinal isomerase retinochrome, and a soluble retinal binding protein. Cephalopod rhodopsins contain a unique C-terminal sequence of proline repeats, which may be involved in the structural organization. Once light is absorbed by rhodopsin, it functions as the activator of a second messenger cascade, which includes, in those invertebrates studied, the GTP-binding protein Gq and the effector enzyme phospholipase C. In this chapter, molecular aspects of squid rhodopsin structure and function are reviewed, and some new results on its purification and reconstitution into phospholipid membranes are presented.

The highly developed eyes of cephalopods have long attracted the attention of neuroscientists (Hubbard and St George 1958; Young 1962). More recently, molecular biologists have developed an interest in this system. Since cephalopod vision (Messenger 1991) and the structure and biochemistry of the squid eye (Saibil 1990*a*) have been reviewed elsewhere, this chapter will concentrate on the structure and function of its visual pigment, rhodopsin. Vertebrate photoreceptor function is now understood in considerable detail, and there are molecular models of rhodopsin stucture based on its sequence and on physical and chemical properties, and initial results of an electron microscopy structure have been published (Schertler *et al.* 1993). However, the three-dimensional structure of rhodopsin and other G-protein-coupled receptors have yet to be determined by X-ray or electron crystallography. Although the collection of suitable material requires specialist marine station facilities, squid rhodopsin is abundant and has structural qualities, unique to cephalopod rhodopsins, that make it a suitable candidate for crystallization experiments. Squid photoreceptors also provide an advantageous preparation for biochemical analysis of the phospholipase C signalling system, a pathway of widespread importance in eukaryotic cells. In addition, the cellular organization of squid photoreceptor microvilli provides a preparation with a very high proportion of membrane–cytoskeleton linkages. The latter two properties are also found in other invertebrate photoreceptors, for example in *Drosophila*, but only cephalopod photoreceptors provide an abundant source of pure photoreceptor microvilli for biochemical and structural studies.

Rhodopsins absorb light by virtue of their 11-*cis*-retinal chromophore, covalently bound to and enclosed by the protein. The wavelength of maximal sensitivity is determined mainly by the amino acid residues in the region of the chromophore, but also by the variant of retinal present (see Chapter 30). For the species we have studied (*Loligo forbesi* and *Alloteuthis subulata*) only one major photochemical form of rhodopsin has been detected.

Molecular components of visual transduction in invertebrate photoreceptors

Light absorbed by the photoreceptor membranes of the retina is transformed by a series of biochemical steps into changes in receptor potential. Rhodopsin and associated enzymes, which relay the visual signal, belong to a widespread family of receptor–enzyme complexes that transmit sensory, hormonal, and neurotransmitter signals into cells (Saibil 1990*b*). In the visual system of several invertebrates, the signal transduction pathway is partly identified. In *Drosophila* and squid, rhodopsin activates the Gq class of G-protein (Lee *et al.* 1990; Pottinger *et al.* 1991; Nobes *et al.* 1992), which is directed to the effector enzyme phospholipase C. The resulting light-induced increase in cytoplasmic calcium plays a major role in visual excitation and light adaptation. Analogues of the vertebrate protein arrestin are present and undergo calcium-stimulated phosphorylation (Yamada *et al.* 1990). Work on *Drosophila* mutants has revealed that the *inaC* locus codes for protein kinase C, and its mutation is deficient in deactivation (Smith *et al.* 1991). The *trp* protein, whose mutation causes the phenotype of transient receptor potential, is a light-sensitive calcium channel involved in inositide-mediated calcium entry (Hardie and Minke 1992). However, light-dependent channels in excised patches of *Limulus* ventral photoreceptors are opened by cyclic guanosine monophosphate (GMP), the other main transmitter candidate, but not calcium (Bacigalupo *et al.* 1991).

Cephalopod photoreceptors have a highly ordered structure

In vertebrate rod photoreceptors, rhodopsin, the G-protein transducin and the GMP phosphodiesterase are free to diffuse in the plane of the photoreceptor membrane. Activation proceeds by sequential encounters between these components. Signal transduction in other cells is more complex, in that many different receptors and G-protein subtypes may be present in the same cell. It is thought that they may be segregated in different subcellular compartments to avoid cross-talk between different pathways.

In the squid retina, the signalling components are found in a structurally ordered array of microvillar membranes. There are regularly arranged junctions linking the extracellular surfaces of the microvilli and also between the cytoplasmic surfaces of the membranes and the actin-based cytoskeletal filament (Saibil and Hewat 1987). The link between membranes and actin filaments is likely to be a form of myosin I (Langmack *et al.*, in preparation), analogous to *ninaC* in *Drosophila* (Montell and Rubin 1988). Some of these organized structures may be specialized forms of those involved in compartmentalization of signalling components in other cell types. Rhodopsin in the microvillar membranes must be aligned and immobilized in order to give the high polarization sensitivity of squid photoreceptor cells (Saidel *et al.* 1983). The Gq, a peripheral membrane protein like other G-proteins, is likely to be freely diffusing on the membrane surface, enabling many Gqs to contact individual photoexcited rhodopsins. Vertebrate rhodopsin is free to diffuse in the plane of the membrane, but the peripheral membrane components transducin and cGMP phosphodiesterase have higher diffusion coefficients (Chabre and Deterre 1989). Unlike bovine rhodopsin, isolated squid rhodopsin tends to spontaneously form organized structures (see below).

Biochemical properties and function of squid rhodopsin

Although cephalopod rhodopsins are related to other rhodopsins and to other members of the family of G-protein-activating receptors, they are unique in possessing a proline-rich repeat at their C-terminus (Ovchinnikov *et al.* 1988; Hall *et al.* 1991). The topology of squid rhodopsin, predicted on the basis of amino acid sequence deduced from cDNA cloning, is shown in Fig. 29.1. Residues that are conserved among most members of the G-protein receptor family are likely to be important in the structure and mechanism of receptor activation of G-protein. These include the helix-disrupting prolines within the transmembrane regions, perhaps lending conformational flexibility to the helices, a sequence including a charge pair at the top of helix 3, important in G-protein activation, and an extracellular disulfide bond, important in rhodopsin structure.

The photochemical properties of squid rhodopsin are different from vertebrate pigments in the behaviour of the metarhodopsin photoproducts. Unlike vertebrate rhodopsin photoproducts, squid metarhodopsin, the all-*trans*-retinal form, is stable and does not release its chromophore. This photostable product absorbs maximally at 500 nm at acid and neutral pH and converts to

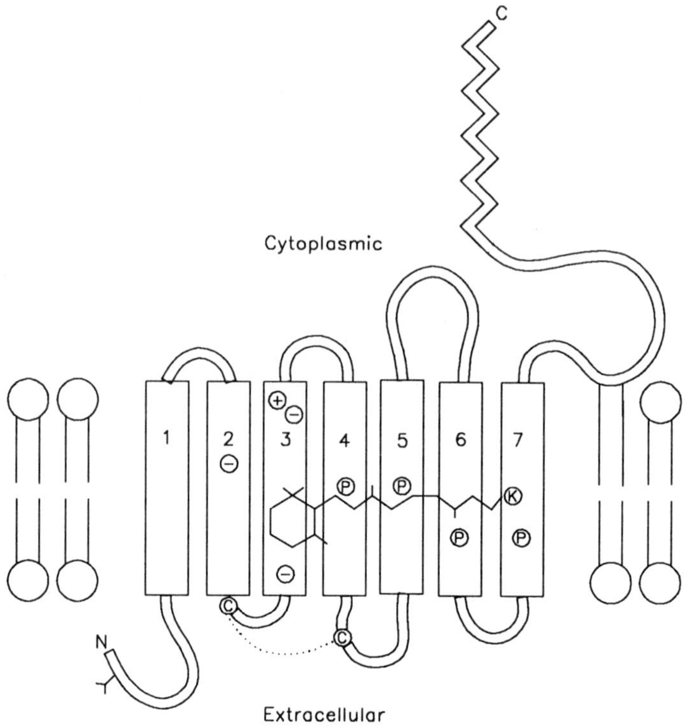

Fig. 29.1. Squid rhodopsin topology. Schematic diagram of squid rhodopsin topology showing seven α-helical segments thought to form a cage around the 11-*cis*-retinal chromophore, covalently linked to lysine 305 (K). The C-terminal proline-rich repeats are shown as an extended zigzag. Their consensus sequence is Pro-Pro-Gln-Gly-Tyr. Residues widely conserved among members of the G-protein coupled receptor family are indicated, including charged residues and prolines (P) in the transmembrane region. The extracellular disulphide bond is shown as a dotted line (C ··· C).

an alkaline form absorbing at 380 nm at high pH. An interesting feature in comparison to vertebrate rhodopsins is the lack of a negative charge at the position equivalent to glu113 in bovine rhodopsin, which is tyr111. This residue was proposed to be an important counterion causing the red shift of retinal absorption in the protein-bound form as compared to free solution. However, squid rhodopsin has very nearly the same red shift as the bovine pigment. It is interesting to note that mutations of this residue in bovine rhodopsin produce a metarhodopsin sensitive to pH and ionic strength. This suggests that glu113 mediates the hydrolysis of the Schiff's base bond and release of the chromophore, rather than acting as a counterion (Hall *et al.* 1991).

The consequence of having a photostable metarhodopsin in squid is seen in the regeneration pathway of rhodopsin. The high proportion of rhodopsin to metarhodopsin, even in light adaptation, does not reflect the photoequilibrium between the two forms (Hara and Hara 1990). Instead, 11-*cis*-retinal is produced by a remarkable photoisomerase, retinochrome, which converts all-*trans*-retinal to the 11-*cis* form upon illumination. This protein is surprisingly homologous to bovine rhodopsin and other G-protein-coupled receptors (Hara-Nishimura *et al.* 1990), considering that it is unlikely to be involved in G-protein activation. The 11-*cis*-retinal is specifically exchanged with all-*trans* by the action of a soluble retinal-binding protein shuttling between rhodopsin and retinochrome, which is found in the inner segments of the retinula cells (Hara and Hara 1990). The retinal-binding protein interacts strongly to exchange chromophore with both metarhodopsin (all-*trans*) and meta-retinochrome (11-*cis*), facilitating chromophore recycling. The existence of this pathway in the photoreceptor cells could be a consequence of the non-inverted orientation of the retina. In the inverted retina of vertebrates, the photoreceptors are at the back of the neural tissue and the pigment epithelium is behind them, in close apposition to the rod and cone outer segments. The pigment epithelium performs the metabolic roles of chromophore metabolism and photoreceptor membrane turnover, and is thus full of strongly absorbing material. This is not possible in a retina where the photoreceptors face the incoming light. Cephalopod photoreceptor cells perform these metabolic tasks in addition to generating the primary light response.

Squid rhodopsin in isolated membranes spontaneously forms clusters and partially ordered arrays (Ryba *et al.* 1993). Negative stain electron micrographs show pronounced extensions that are too large to be single rhodopsins (Fig. 29.2). The C-terminal sequence can be removed by protease treatment, and this is accompanied by the loss of visible extensions (Ryba *et al.* 1993). This structural feature may be involved in ordered interactions between rhodopsins or with the cytoskeleton. However, its presence does not seem to affect the receptor function of squid rhodopsin, which activates G-proteins in response to light. It must therefore constitute a cytoplasmic domain separate from the G-protein activating-domain common to all rhodopsin-like receptors.

Previously published work showed activation of two or more minor G-proteins by squid rhodopsin. These were substrates for modification by bacterial toxins, cholera, and pertussis toxins, reactions diagnostic of two major classes of G-protein (Fyles *et al.* 1991; Robinson *et al.* 1990). It has recently been found that the major G-protein of squid photoreceptors is a member of the Gq class (Pottinger *et al.* 1991; Nobes *et al.* 1992). This G-protein is not a toxin substrate and is a specific activator of phospholipase C (Smrcka *et al.* 1991). Although the minor G-proteins can be activated by squid rhodopsin *in vitro*, they are unlikely to play a major role in the light response, and may be contaminants of the microvillar preparation.

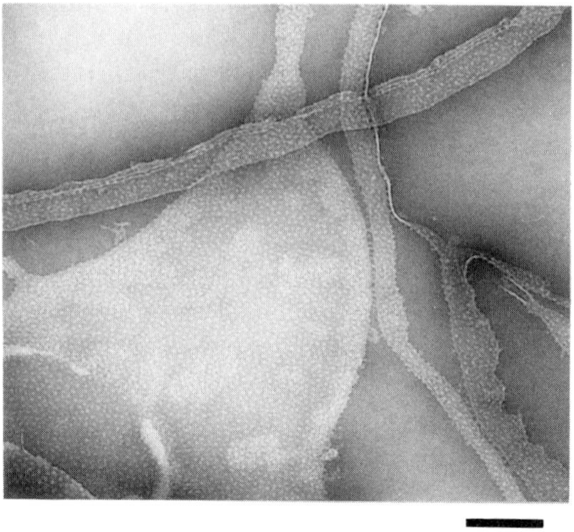

Fig. 29.2. Negative stain EM of isolated, native microvillar membranes, showing structures projecting from the membrane surface (light dots). The projections represent the cytoplasmic regions of small clusters of rhodopsin molecules. Adjacent membranes appear to interact via the projections. Scale bar: 200 nm.

Purification and reconstitution of squid rhodopsin

Rhodopsin is the major protein of the isolated photoreceptor membranes and it is readily purified by chromatography in detergent solution, from which it can be reconstituted into phospholipid bilayers by dialysis (Vandenberg and Montal 1984). About 25 per cent of the microvillar protein is insoluble in detergent, including a significant amount of glycoprotein (expected to have transmembrane regions), but almost all the native phospholipid (96 per cent) is found in the detergent supernatant. Figures 29.3 and 29.4 show SDS gels of rhodopsin purification by chromatography on DE52 cellulose and concanavalin A (conA)-sepharose, respectively, along with plots of the elution profiles of the native phospholipids. Rhodopsin does not bind to DE52 and elutes with the phospholipid, whereas the conA column binds the oligosaccharide residues and rhodopsin is only eluted by the addition of α-methyl mannoside, after the phospholipid and non-glycosylated proteins have been eluted. Rhodopsin purified by these two methods was reconstituted into egg lecithin bilayers by dialysis, and negative stain electron micrographs of the reconstituted samples obtained (Fig. 29.5). The projections are present in both cases, showing that the association of squid rhodopsin into clusters is spontaneous and does not require other proteins.

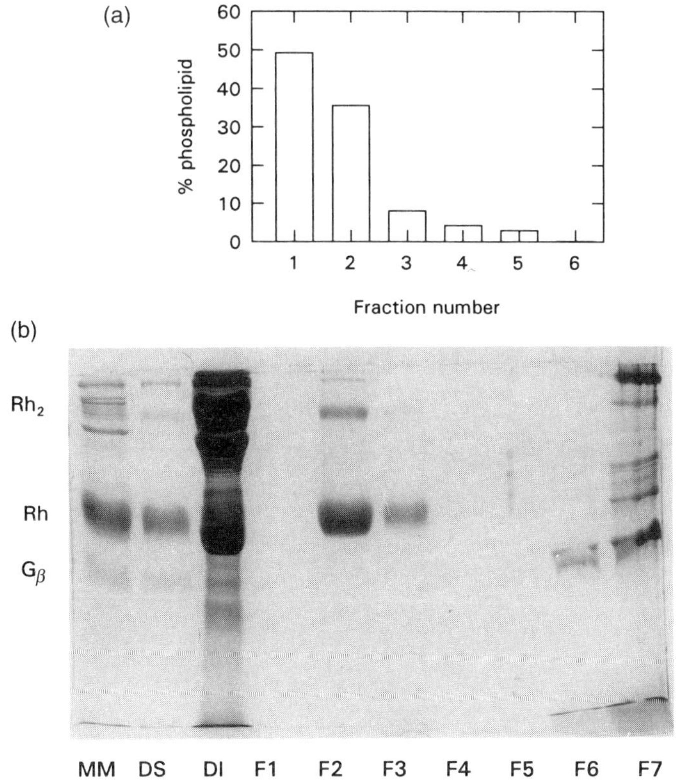

Fig. 29.3. (a) Plot of phospholipid elution from a DE52 ion exchange column, showing the percentage of total phospholipid eluted with fractions 1–7, as measured by a phosphate determination. The values are averages of two measurements that were in close agreement. (b) SDS polyacrylamide gel of protein elution from a DE52 ion exchange column. MM, total microvillar membrane protein; DS, detergent-soluble fraction (sample loaded on column); DI, detergent-insoluble fraction; F1–F5, fractions 1–5 eluted from the column in 50 mM Tris pH 7.2; F6–7, proteins eluted with 500 mM NaCl, 50 mM Tris. Rh, rhodopsin; Rh_2, rhodopsin dimer; A, actin; G_β, β-subunit of G-protein. The protein loadings in the different lanes are not equivalent quantities on this gel.

Conclusions

Squid photoreceptors are interesting because they have a more ordered structure than other photoreceptors, along with the same degree of biochemical specialization. The role of the extended, proline-rich C-terminus of cephalopod rhodopsins is likely to be a structural one, related to the high polarization sensitivity in cephalopod retinas. By analogy with other membrane proteins possessing this type of sequence, it is expected to mediate linkage to

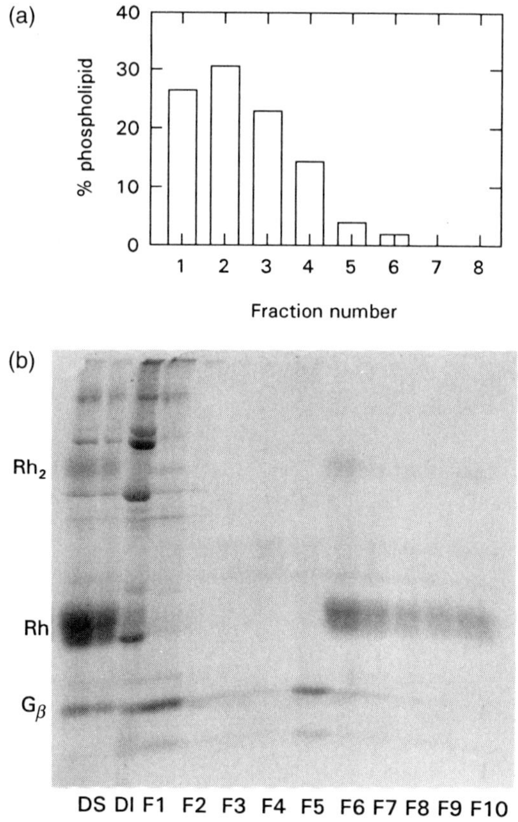

Fig. 29.4. (a) Plot of phospholipid elution from a concanavalin A-sepharose column. (b) SDS gel of protein elution from a concanavalin A-sepharose column run. The labels are as for Fig. 29.3, except that more eluted fractions are shown. Two lanes are shown of the detergent-soluble fraction (DS). Rhodopsin elutes between fractions 5 and 9, after 15 min incubation with 200 mM α-methylmannoside.

cytoskeleton components. Cytoskeletal involvement in phototransduction is suggested by the reversible disruption of the cytoskeleton seen after illumination and rapid freezing (Tsukita *et al.* 1988). The membrane–cytoskeleton interactions in these photoreceptor microvilli provide a model for molecular interactions in the compartmentalization of signal transduction components in other eukaryotic cells. The photochemical properties of the visual pigment may be related to the cell biology of these invertebrate photoreceptor cells and the retinochrome/retinal-binding protein regeneration system. This is the most specialized example known of a receptor—Gq—phospholipase C signal transduction system, and should provide a valuable source of biochemical information of the molecular mechanisms of this enzyme cascade.

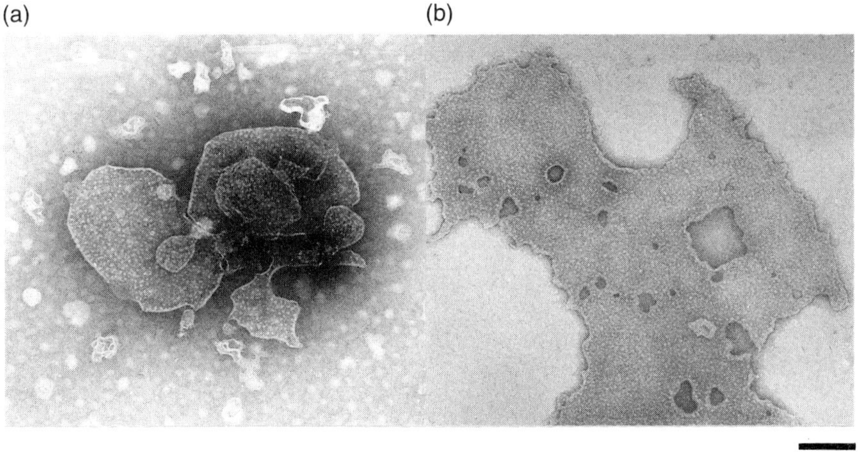

Fig. 29.5. Negative stain EM of purified, reconstituted rhodopsin. (a) DE52 purified; (b) concanavalin A purified. The projections are present in both preparations. Scale bar: 200 nm.

Acknowledgements

We are grateful to the staff of the Marine Biological Association Laboratory, Plymouth, UK, for the excellent supply of live squid.

References

Bacigalupo, J., Johnson, E. C., Vergara, C., and Lisman, J. E. (1991). Light-dependent channels from excised patches of *Limulus* ventral photoreceptor are opened by cGMP. *Proceedings of the National Academy of Sciences, USA*, **88**, 7938–42.

Chabre, M. and Deterre, P. (1989). Molecular mechanism of visual transduction. *European Journal of Biochemistry*, **179**, 255–66.

Hall, M. D., Hoon, M. A., Ryba, N. J. P., Pottinger, J. D. D., Keen, J. N., Saibil, H. R., and Findlay, J. B. C. (1991). Molecular cloning and primary structure of squid (*Loligo forbesi*) rhodopsin, a phospholipase C-directed G-protein-linked receptor. *Biochemical Journal*, **274**, 35–40.

Fyles, J. M., Baverstock, J., Baer, K. M., and Saibil, H. R. (1991). Effects of calcium on light-activated GTP-binding proteins in squid photoreceptor membranes. *Comparative Biochemistry and Physiology B*, **98**, 215–21.

Hara, T. and Hara, R. (1990). Retinal-binding protein: Function in a chromophore exchange system in the squid visual cell. In *Progress in retinal research*, Vol. 10 (ed. N. Osborne and J. Chader), pp. 179–206. Pergamon Press, Oxford.

Hara-Nishimura, I., Matsumoto, T., Mori, H., Nishimura, M., Hara, R., and Hara, T. (1990). Cloning and nucleotide sequence of cDNA for retinochrome retinal photoisomerase from the squid retina. *Federation of European Biochemical Societies Letters*, **271**, 106–10.

Hardie, R. C. and Minke, B. (1992). The *trp* gene is essential for a light-activated Ca^{2+} channel in adult *Drosophila* photoreceptors. *Neuron*, **8**, 643–51.

Hubbard, R. and St George, R. C. C. (1958). The rhodopsin system of the squid. *Journal of General Physiology*, **41**, 501–28.

Lee, Y.-J., Dobbs, M. B., Verardi, M. L., and Hyde, D. R. (1990). *dgq*: A *Drosophila* gene encoding a visual system-specific Gα molecule. *Neuron*, **5**, 889–98.

Messenger, J. B. (1991). Photoreception and vision in molluscs. In *Evolution of the eye and visual systems* (ed. J. R. Cronly-Dillon and R. L. Gregory), pp. 364–97. Macmillan, London.

Montell, C. and Rubin, G. (1988). The *Drosophila ninaC* locus encodes two photoreceptor cell specific proteins with domains homologus to protein kinases and the myosin heavy chain head. *Cell*, **52**, 757–72.

Nobes, C., Baverstock, J., and Saibil, H. (1992). Activation of the GTP-binding protein Gq by rhodopsin in squid photoreceptors. *Biochemical Journal*, **287**, 545–8.

Ovchinnikov, Yu. A., Abdulaev, N. G., Zolotarev, A. S., Artamonov, I. D., Bespalov, I. A., Dergachev, A. E., and Tsuda, M. (1988). Octopus rhodopsin. Amino acid sequence deduced from cDNA. *Federation of European Biochemical Societies Letters*, **232**, 69–72.

Pottinger, J. D. D., Ryba, N. J. P., Keen, J. N., and Findlay, J. B. C. (1991). The identification and purification of the heterotrimeric GTP-binding protein from squid (*Loligo forbesi*) photoreceptors. *Biochemical Journal*, **279**, 323–6.

Robinson, P. R., Wood, S. F., Szuts, E. Z., Fein, A., and Hamm, H. E. (1990). Light dependent GTP-binding proteins in squid photoreceptors. *Biochemical Journal*, **272**, 79–85.

Ryba, N., Hoon, M., Findlay, J. B. C., Saibil, H., Wilkinson, J. R., Heimburg, T., and Marsh, D. (1993). Rhodopsin mobility, structure and lipid–protein interaction in squid photoreceptor membranes. *Biochemistry*, **32**, 3298–305.

Saibil, H. R. (1990*a*). Structure and function of the squid eye. In *Squid as experimental animals* (ed. W. Adelman, D. Gilbert, and J. Arnold), pp. 371–97. Plenum Publishing, New York.

Saibil, H. R. (1990*b*). Cell and molecular biology of photoreceptors. In *Seminars in the neurosciences*, 2 (ed. J. F. Ashmore and H. R. Saibil), pp. 15–23. W. B. Saunders, London.

Saibil, H. R. and Hewat, E. A. (1987). Ordered transmembrane and extracellular structure in squid photoreceptor microvilli. *Journal of Cell Biology*, **105**, 19–28.

Saidel, W. M., Lettvin, J. Y., and MacNicol, E. F., Jr. (1983). Processing of polarized light by squid photoreceptors. *Nature*, **304**, 534–6.

Schertler, G. F. X., Villa, C., and Henderson, R. (1993). Projection structure of rhodopsin. *Nature*, **362**, 770–2.

Smith, D. P., Ranganathan, R., Hardy, R., Marx, J., Tsuchida, T., and Zucker, C. (1991). Photoreceptor deactivation and retinal degeneration mediated by photoreceptor-specific protein kinase C. *Science*, **254**, 1478–84.

Smrcka, A. V., Helper, J. R., Brown, K. O., and Sternweis, P. C. (1991). Regulation of polyphosphoinositide-specific phospholipase C activity by purified Gq. *Science*, **251**, 804–7.

Tsukita S., Tsukita S., and Matsumoto G. (1988). Light-induced structural changes of cytoskeleton in squid photoreceptor microvilli detected by rapid-freeze method. *Journal of Cell Biology*, **106**, 1151–60.

Vandenberg, C. A. and Montal, M. (1984). Purification of squid rhodopsin and reassembly into lipid bilayers. *Biochimica et Biophysica Acta*, **771**, 74–81.

Yamada, T., Takeuchi, Y., Komori, N., Kobayashi, H., Sakai, Y., Hotta, Y., and Matsumoto, H. (1990). A 49-kilodalton phosphoprotein in the *Drosophila* photoreceptor is an arrestin homolog. *Science*, **248**, 483–6.

Young, J. Z. (1962). The retina of cephalopods and its degeneration after optic nerve section. *Philosophical Transactions of the Royal Society of London, Series B*, **245**, 1–18.

30

The firefly squid, *Watasenia scintillans*, has three visual pigments

MASATSUGU SEIDOU, KINYA NARITA, MASANAO MICHINOMAE, and YUJI KITO

Summary

Watasenia scintillans has three visual pigments which have absorbance maxima at 471 nm, 484 nm, and 501 nm. The 484 nm pigment is based on retinal, the 501 nm pigment on 3-dehydroretinal and the 471 nm pigment is based on the novel chromophore 4-hydroxyretinal. In the dorsal retina, the rhabdomes are 200 μm thick and consist of photoreceptor outer segments containing the visual pigment with retinal (A_1 pigment). In the specific region of the ventral retina receiving downwelling light, the rhabdomes, the photoreceptor outer segments are 600 μm thick. The distal outer segment layer, 400 μm thick from the retinal surface, contains the yellowish visual pigment based on 4-hydroxyretinal (A_4 pigment) and the proximal outer segment layer contains the pinkish visual pigment based on 3-dehydroretinal (A_2 pigment). It was observed that one type of photoreceptor had rhabdomes only in the proximal outer segment layer, and the other types had rhabdomes only in the distal outer segment layer. From these results, it was concluded that these visual pigments seem to be contained in separate photoreceptor cells. Due to the structure of the stratified outer segment layer, the A_4 pigment layer acts as a yellowish filter on the photoreceptor cells containing A_2 pigment and this filtering effect shifts the wavelength of maximal photosensitivity of the photoreceptors to 550 nm. The various small photophores on the ventral surface of the body emit light, peaking at 470 nm at 4°C and at 535 nm at 15°C. The A_2 pigment photoreceptor cell with photosensitivity shifted towards longer wavelength is useful for detecting this green light.

Introduction

Watasenia scintillans lives in the open ocean around Japan. The mantle length of mature squid is about 6 cm. The squid is known as the firefly squid in Japan due to its intense bioluminescence from three large photophores at the tips of the fourth arm pair. The squid has five relatively large photophores forming an array at the ventral surface of the eye and numerous small photophores distributed mainly over the ventral surface of the body. Tsuji reviewed in detail many earlier reports about the distribution, structure of the photophores, and mechanism of their bioluminescence (Tsuji 1983). In the daytime the firefly squids have never been caught at a depth less than 250 m, while at night they have been found at 50 m depth. Thus the firefly squids seem to move vertically during the day. Large groups of the squids, mostly egg-bearing females, come to Toyama bay of the Japan Sea in spring to spawn. These squids were captured at night and used in our experiments.

It is known that cephalopods can change their skin colour in order to camouflage, threaten, and display sexuality. Since they can well match their background colour, it has been expected that cephalopods are able to discriminate colours. Although Wells (1978) reviewed many experiments using discrimination learning, electroretinogram studies, and examination of the optomotor responses, he could find no evidence precisely showing the existence of the ability to discriminate colours.

We have investigated cephalopod visual pigments and found that *W. scintillans* has three visual pigments different from each other in absorbance maxima. The existence of the three visual pigments suggests that the squid may have the ability of colour discrimination. In this chapter we describe the three visual pigments and their distribution in the retina.

The visual pigments of *W. scintillans*

The visual pigment of *W. scintillans* had an absorbance maximum at 482 nm, when it was extracted from whole retina and purified (Nashima *et al.* 1979). The visual pigment chromophore was extracted and analysed, using high performance liquid chromatography (HPLC). It was found that the extract contained not only retinal but also 3-dehydroretinal amounting to about 15 per cent of the total chromophores and that decrease in absorbance at 501 nm was accompanied by the conversion of 11-*cis*-3-dehydroretinal to the all-*trans* form. This fact suggested that the *W. scintillans* retina contained some 3-dehydroretinal-based visual pigment (A_2 pigment). In order to investigate the localization of the A_2 pigment, the retina was dissected into small parts and the chromophore composition of each part was analysed. The 3-dehydroretinal was located in the ventral retina receiving downwelling light (Kito *et al.* 1986).

Visual pigment was extracted from the specific ventral retina where the 3-dehydroretinal was concentrated. The visual pigment was purified by column chromatography with DEAE-cellulose and concanavalin A-sepharose. The 3-dehydroretinal comprised more than 90 per cent of total chromophores in the purified pigment solution by analysis with HPLC. We expected that the visual pigment solution would contain the 501 nm (A_2) pigment amounting to more than 90 per cent and have an absorbance maximum at about 501 nm; nevertheless, the visual pigment solution had an absorbance maximum at 473 nm.

The visual pigment solution of ventral retina was further analysed by partial bleaching and it was found that the solution contained two visual pigments with absorbance maxima at 471 and 501 nm, respectively (Matsui *et al.* 1988*a*). The visual pigment extracted from dorsal retina had only retinal and an absorbance maximum at 484 nm (A_1 pigment).

The 501 nm pigment was based on 3-dehydroretinal as described above. If the 471 nm pigment was based on retinal, the 471 nm pigment would have to exhibit an abnormally high molar extinction coefficient because the visual pigment solution contained so little retinal. It was considered that the chromophore of this visual pigment might be neither retinal nor 3-dehydroretinal but some more polar retinal derivative, which was not eluted from the HPLC column with the eluent used for retinal and 3-dehydroretinal. After both retinal and 3-dehydroretinal were eluted from the HPLC column, the column was washed with a more polar solvent system. The unknown chromophore was eluted from the HPLC column. It was analysed by nuclear magnetic resonance, infra-red, and mass spectroscopy, and identified as 4-hydroxyretinal. The 4-hydroxyretinal extracted from *W. scintillans* had negative ellipticity in circular dichroism spectrum. From an empirical rule, the 4-hydroxyretinal was possibly the S-enantiomer because of the negative ellipticity (Matsui *et al.* 1988*b*). Thus the 473 visual pigment extracted from the ventral retina was a mixture of A_4 pigment and A_2 pigment contaminated with a residual amount of A_1 pigment.

The A_4 pigment was located together with A_2 pigment in the ventral retina of *W. scintillans*. The length of the outer segments of the dorsal retina was 200 μm while that of the ventral retina was 600 μm. The ventral retina was sliced perpendicular to its surface using a guillotine equipped with a razor blade. The distal two-thirds of the outer segment layer were yellowish and the proximal one-third was pinkish (Matsui *et al.* 1988*a*). The pinkish and yellowish layers were separately dissected out under the microscope for chromophore analysis. As shown in the HPLC chromatograms (2) and (3) of Fig. 30.1, 3-dehydroretinal was contained in the pinkish proximal layer and 4-hydroxyretinal was contained in the yellowish distal layer.

The ratio of retinal to 3-dehydroretinal to 4-hydroxyretinal in whole retina of *W. scintillans* was 67:8:25 (Seidou *et al.* 1990). Seasonal variation in the ratio of 3-dehydroretinal to retinal has been found in eyes of several kinds of fish and crayfish (Dartnall and Lythgoe 1965; Munz 1965; Suzuki *et al.* 1985), so we expected that only the *W. scintillans* coming to Toyama Bay to spawn in spring might yield 3-dehydroretinal and 4-hydroxyretinal. However, squid captured in other seasons had the same chromophore composition. Most of the squids captured at Toyama Bay in spring were female. The retina of male squid also had the same composition (Seidou *et al.* 1990).

In order to compare the A_4 pigment with A_1 pigment, a visual pigment was synthesized from 4-hydroxyretinal and bovine opsin. The synthesized pigment had an absorbance maximum at 486 nm, 15 nm shorter than that of

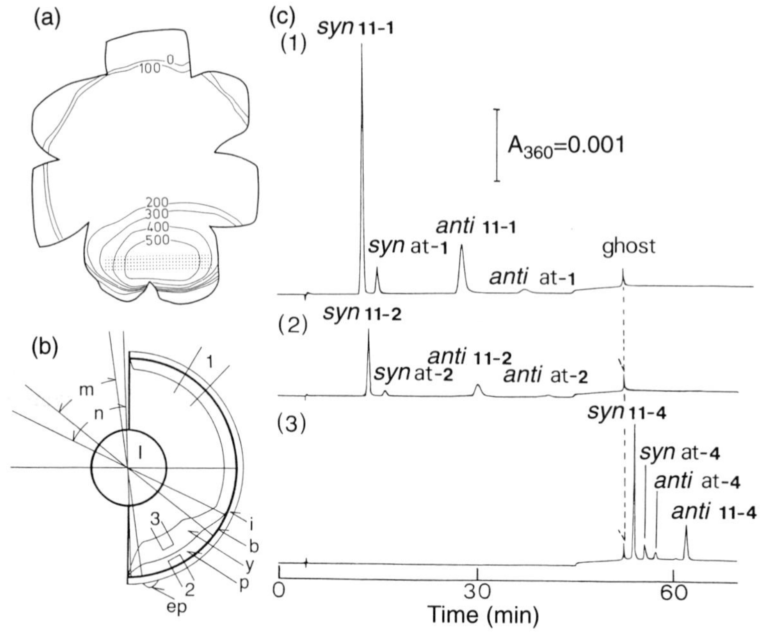

Fig. 30.1. Topology of the outer segment length of the photoreceptors and distribution of visual pigment chromophores in *W. scintillans* retina. (**a**) Diagram of unfolded retina. The retina was sliced with a guillotine. The length from the black pigment layer to the tip of the outer segments was measured and the points with the same length were connected by a line in the diagram. The figures in the diagram are lengths in μm. Eye photophores are behind the dotted area. (**b**) Diagram of vertical view of *W. scintillans* eye. m: the direction of the incident light on outer segments longer than 400 μm, n: the direction of the incident light on outer segments longer than 200 μm, l: lens, y: yellowish outer segment layer; p: pinkish outer segment layer; ep: eye photophores; b: black pigment layer; i: inner segment layer. (**c**) HPLC profiles of the chromophores extracted as oxime derivatives from the eye shown in (b). Curve (1): extract of block (1) in the dorsal retina. Curve (2): extract of segment (2) from the pinkish stratum of the ventral retina. Curve (3): extract of segment (3) from the yellowish stratum of the ventral retina. The peaks were assigned by comparison with the respective authentic compounds. The labels on the peaks, *syn* 11–1, *anti* at-1, *syn* 11–2, *syn* 11–4 and *anti* 11–4 should be read as *syn* 11-*cis*-retinal oxime, *anti* all-*trans*-retinal oxime, *syn* 11-*cis*-3-dehydroretinal oxime, *syn* 11-*cis*-4-hydroxyretinal oxime and *anti* 11-*cis*-4-hydroxyretinal oxime, respectively. HPLC was carried out using the following step elution: the initial solvent system for retinal oximes until the appearance of *anti* all-*trans*-3-dehydroretinal oxime and then the second solvent system for 4-hydroxyretinal oximes. 'Ghost' is a peak due to the solvent exchange. Flow rate: 1 ml/min.

the parent bovine A_1 pigment. The difference in the wavelength of absorbance maxima between the synthetic pigment and bovine pigment was comparable with the difference between A_4 pigment and A_2 pigment of *W. scintillans* (Kito *et al.* 1991). This suggested that *W. scintillans* uses 4-

hydroxyretinal to shift the absorbance spectrum of the visual pigment towards shorter wavelengths without changing the apoprotein. Visual pigment synthesized from 3-hydroxyretinal and bovine opsin had an absorbance maximum at 492 nm. 4-hydroxyretinal could shift the absorbance spectrum to shorter wavelengths more effectively than 3-hydroxyretinal.

The three visual pigments of *W. scintillans* could not be separated by column chromatographies utilizing affinity to concanavalin A, isoelectric point, and/or molecular size (Matsui *et al.* 1988*a*). From the similarity in those properties and the same amino acid sequence near the chromophore binding site, it was concluded that the three visual pigments were composed of the same apoprotein, but different chromophores (Seidou *et al.* 1988).

The visual pigment synthesized from 4-hydroxyretinal and bovine opsin was very similar in absorbance spectrum and photosensitivity to bovine visual pigment (Kito *et al.* 1991). This suggested that the A_4 pigment and A_1 pigment of *W. scintillans* were probably very similar in absorbance spectrum and photosensitivity. The difference in the absorbance maxima of the two pigments was 13 nm. It may be due to these similarities and the small (13 nm) difference of absorbance maximum that the A_4 pigment, amounting to 25 per cent of total visual pigments, could not be easily separated by partial bleaching.

Watasenia scintillans had a considerable amount of 11-*cis*-retinylester in the hepatopancreas, but neither 3-dehydroretinoids nor 4-hydroxyretinoids were found there. A small amount of 4-hydroxyretinol was found exceptionally in the eggs (Seidou *et al.* 1990). It was reported that the squid contained neither 4-hydroxycarotenoids nor 3-dehydrocarotenoids (Maoka *et al.* 1989). It has been reported that brine shrimp contain 4-hydroxy-β-carotene (Underwood 1984). It is difficult to rule out the possibility that 4-hydroxyretinal in *W. scintillans* came from food such as the brine shrimp. Existence of 3-dehydroretinal and 4-hydroxyretinal, respectively, in the different photoreceptor cells of *W. scintillans* strongly suggests that these retinoids may be produced in each cell from the storage of large amounts of 11-*cis*-retinylester in the hepatopancreas. To test this hypothesis, we examined the conversion of retinol to 4-hydroxyretinol and/or to 3-dehydroretinol in the retinal homogenate. By incubation of 11-*cis*-retinol with a homogenate of ventral retina, it was found that 3 per cent of the added 11-*cis*-retinol was converted to 11-*cis*-4-hydroxyretinol. The dorsal retina did not produce any 4-hydroxyretinoid. The synthesis of 3-dehydroretinol could not be detected.

The structure of rhabdomes in the ventral retina of the *W. scintillans*

The outer segment layer of the ventral retina of *W. scintillans* was yellowish in the distal two-thirds with A_4 pigment and pinkish in the proximal one-third with A_2 pigment. If A_4 pigment and A_2 pigment are contained in different photoreceptor cells, this could form the basis of colour discrimination. In order to provide further information on this problem, the outer segment layers were observed in the electron microscope. Figure 30.2(a) shows a cross-section at 50 μm depth from the retinal surface. It can be seen that each rhabdome consists of microvilli from four identical photoreceptor cells (labelled

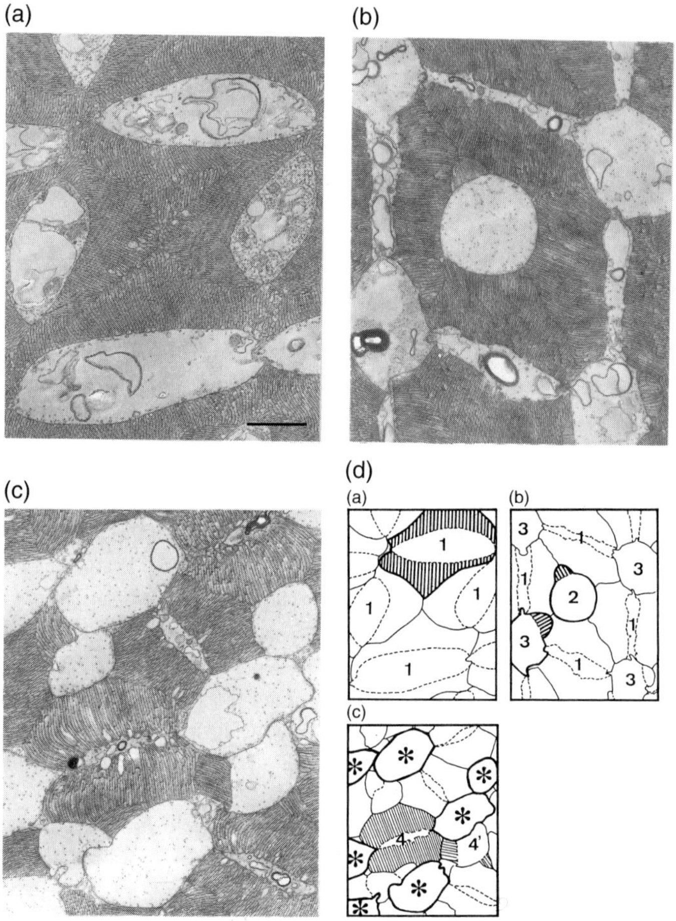

Fig. 30.2. Electron micrographs of cross-sections of rhabdomes in the ventral retina. (a) Cross-section sliced at 50 μm depth from the surface of the retina. (b) Cross-section in the middle of the yellowish outer segment layer. (c) Cross-section in the pinkish layer. (d) Drawings of the photographs (a), (b), and (c) to show all types and microvilli. (1) The four identical photoreceptor cells make a square rhabdome; (2) the centre cell; (3) the corner cells; (4) and (4′) photoreceptor cells with microvilli in the pinkish layer. *photoreceptor cells which did not extend their microvilli into the pinkish layer. Scale bar: 2 μm.

1). A similar structure has generally been observed in the outer segment layer of other cephalopods (Wolken 1958; Yamamoto *et al.* 1965) and in that of the dorsal retina of *W. scintillans*. Figure 30.2(b) shows a cross-section in the middle part of the yellowish layer. Two different types of outer segments now appear at the centre (2) and at the corner (3) of each square rhabdome. The microvillar orientation of the new cells differs from the others (type 1) by 45°. Figure 30.2(c) shows a cross-section through the pinkish outer segment layer. There are two types of cell in this section. One of them (4 and 4′) has microvilli in this layer, while the other cells labelled with asterisks) do not extend their microvilli in this layer. Thus the former photoreceptor cells extend their microvilli only in the pinkish layer while the others have microvilli in the yellowish layer. These results show that *W. scintillans* has at least four kinds of photoreceptor cells, and one of them, in the pinkish layer, contains A_2 pigment, while the other three kinds of photoreceptor cells have A_4 pigment. These results lead us to expect that the squid may discriminate colours.

Discussion

As shown in Fig. 30.1(b), if the retina of *W. scintillans* extends around the hemisphere of the eye, the specific ventral region, where the rhabdomes are longer than 400 μm, will receive light with an incident angle between 38 and 72° from the horizontal plane, as marked 'm' in the diagram. The area having rhabdomes longer than 200 μm will receive light with an incident angle between 25 and 88° as marked 'n' in the diagram. Thus the specific ventral retinal region will receive downwelling light. The intensity of the downwelling light is much stronger than that of the light in other directions in the deep sea (Denton and Locket 1989). The outer segments of the dorsal retina of *W. scintillans* are as long as that of other cephalopods that live successfully in the photic environment of the deep sea. Therefore the presence of longer rhabdomes in the ventral retina of *W. scintillans* is not merely to compensate for weakness of intensity of the environmental light because the specific region of the ventral retina is the region receiving the strongest incident light in the eye.

Denton proposed that 'multibank' retinae in deep sea fish enable the outer segments to discriminate colour with even one visual pigment, because the layered outer segments act as a filter to change the photosensitivity curve of underlying photoreceptors. With this assumption, two or more visual pigments are more effective (Denton and Locket 1989). The ventral retina of *W. scintillans* consists of, at least, a two bank retina with two visual pigments.

We measured the absorbance spectra of the visual pigments in the fresh retinae of *W. scintillans*, as shown in Fig. 30.3. The absorbance of the A_4 pigment was 0.7 OD at 470 nm in a 100 μm thick slice and that of the A_2 pigment was 0.49 OD at 500 nm in the same slice. The yellowish layer was

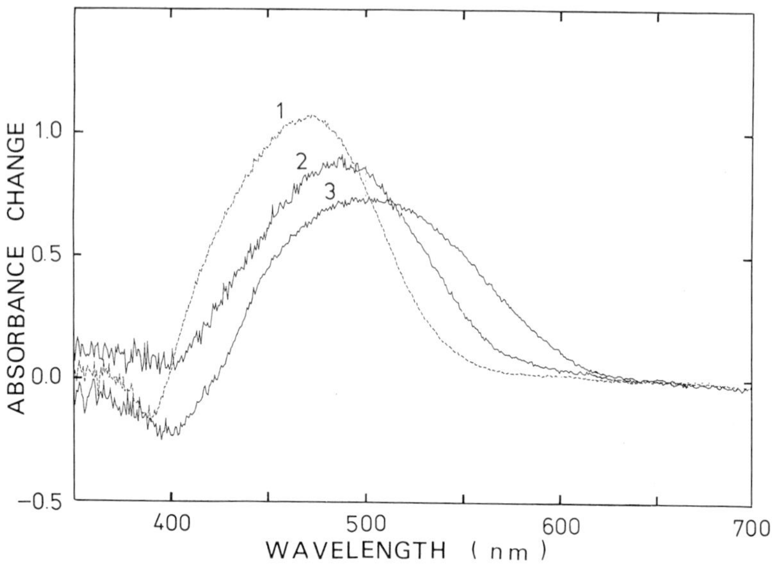

Fig. 30.3. Absorption spectra of the outer segment layers of *W. scintillans*. Fresh retina was sliced perpendicular to its surface without fixation and bathed in artificial squid Ringer's solution containing 100 mM hydroxylamine. The difference spectrum was measured before and after irradiation with light of $\lambda > 560$ nm. Curve 1: the absorbance spectrum of the A_4 pigment in the yellowish distal layer of the ventral retina. Curve 2: the absorbance spectrum in the dorsal retina. Curve 3: the absorbance spectrum of A_2 pigment in the pinkish proximal layer of the ventral retina. The spectra 1 and 3 were measured in the same slice. The measurements were not accurate below 400 nm because of the large absorbance of the slices at wavelengths shorter than 400 nm.

400 μm thick so the absorbance of this layer was 2.8 OD. The incident light is first absorbed by the rhabdome layer of A_4 pigment and the rhabdomes of A_2 pigment therefore receive light absorbed through the A_4 pigment which acts as a filter. As a result of this filter effect, the visual sensitivity curve of the photoreceptor cells in the pinkish layer of A_2 pigment is shifted towards longer wavelengths, correlating well with the 550 nm absorbance maximum of the visual pigment (Fig. 30.4). One type of small photophore on the ventral surface of these squid emits light with intensity maximum at 470 nm at 4°C and the other emits light with an intensity maximum at 535 nm at 15°C, as shown in Figure 30.4. *Watasenia scintillans* are likely to benefit from the red shift in the sensitivity curve of the photoreceptor cells with A_2 pigment when they see each other's bioluminescence.

Vertebrate cone pigments have only retinal as chromophore and different apoproteins. *Watasenia scintillans* varies the absorbance maximum of their visual pigment by using three chromophores, retinal, 3-dehydroretinal, and

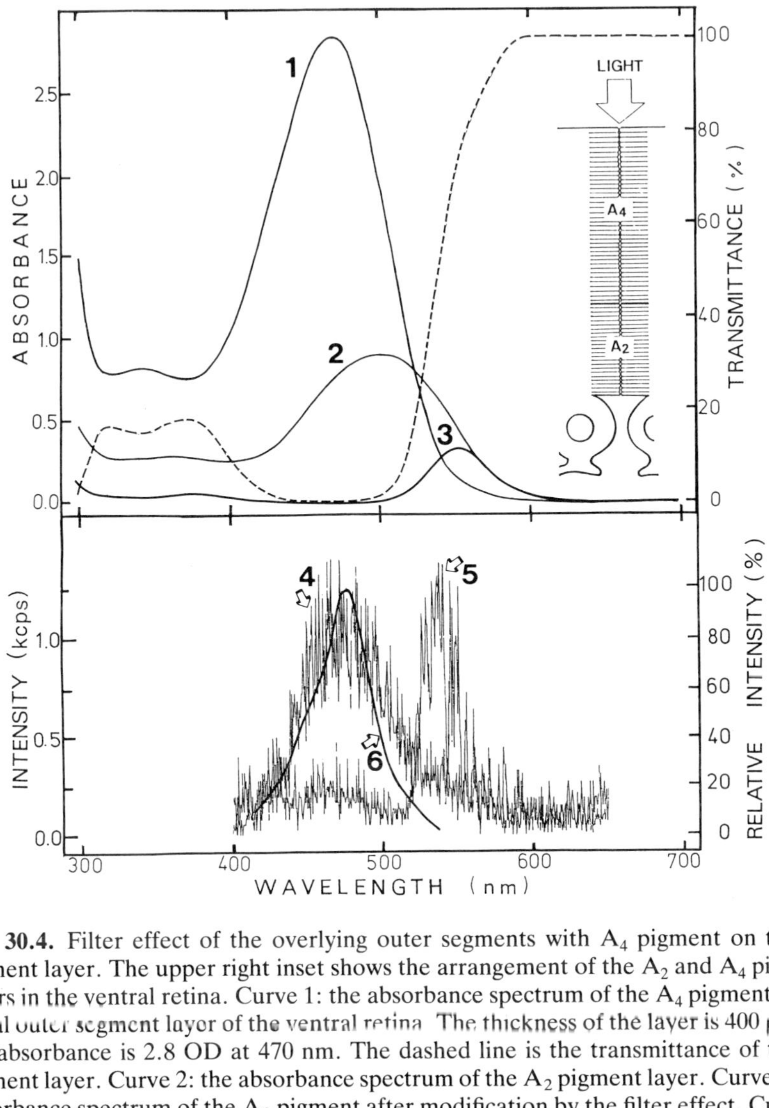

Fig. 30.4. Filter effect of the overlying outer segments with A_4 pigment on the A_2 pigment layer. The upper right inset shows the arrangement of the A_2 and A_4 pigment layers in the ventral retina. Curve 1: the absorbance spectrum of the A_4 pigment in the distal outer segment layer of the ventral retina. The thickness of the layer is 400 μm, so the absorbance is 2.8 OD at 470 nm. The dashed line is the transmittance of the A_4 pigment layer. Curve 2: the absorbance spectrum of the A_2 pigment layer. Curve 3: the absorbance spectrum of the A_2 pigment after modification by the filter effect. Curve 4: the emission spectrum of the bioluminescence of the ventral photophores at 4°C. Curve 5: the emission spectrum of bioluminescence at 15°C. Curve 6: the spectral distribution of the intensity of the light at depths of about 500 m in clear oceanic waters (Kampa 1970).

4-hydroxyretinal. In order to examine whether other deep sea cephalopods or fish generally use different chromophores, we analysed the chromophore composition of various species. Neither 3-dehydroretinal nor 4-hydroxyretinal could be found in other cephalopods (Seidou *et al.* 1990). The ventral retina of *W. scintillans* was made of two banks, the A_2 pigment layer and A_4 pigment layer. In electron microscopic observation, there were two kinds of long and intermediate photoreceptor outer segments in the distal yellowish layer. The photosensitivity of the intermediate receptors in the yellowish layer could be modified by the overlying long yellowish outer segments. This may suggest that the receptor system for colour discrimination of *W. scintillans* is even more complex and delicate than the two banked retina.

References

Dartnall, H. J. A. and Lythgoe, J. N. (1965). The clustering of fish visual pigments around discrete spectral positions and its bearing on chemical structure. In *Color vision* (ed. A. V. S. de Reuck and J. Knight), pp. 3–26. J. and A. Churchill, Ltd, London.

Denton, E. J. and Locket, N. A. (1989). Possible wavelength discrimination by multibank retinae in deep-sea fishes. *Journal of the Marine Biological Association of the United Kingdom*, **69**, 409–35.

Kampa, E. (1970). Underwater daylight and moonlight measurements in the eastern North Atlantic. *Journal of the Marine Biological Association of the United Kingdom*, **50**, 397–420.

Kito, Y., Seki, T., Suzuki, T., and Uchiyama, I. (1986). 3-Dehydroretinal in the eye of a bioluminescent squid, *Watasenia scintillans. Vision Research*, **26**, 275–9.

Kito, Y., Partridge, J. C., Seidou, M., Narita, K., Hamanaka, T., Michinomae, M., *et al.* (1992). The absorbance spectrum and photosensitivity of a new synthetic 'visual pigment' based on 4-hydroxyretinal. *Vision Research*, **32**, 3–10.

Maoka, T., Yokoi, S., and Matsuno, T. (1989). Comparative biochemical studies of carotenoids in nine species of cephalopoda. *Comparative Biochemistry and Physiology*, **92B**, 247–50.

Matsui, S., Seidou, M., Horiuchi, S., Uchiyama, I., and Kito, Y. (1988*a*). Adaptation of a deep-sea cephalopod to the photic environment. *Journal of General Physiology*, **92**, 55–66.

Matsui, S., Seidou, M., Uchiyama, I., Sekiya, N., Hiraki, K., Yoshihara, K., and Kito, Y. (1988*b*). 4-Hydroxyretinal, a new visual pigment chromophore found in the bioluminescent squid, *Watasenia scintillans. Biochemica et Biophysica Acta*, **966**, 370–4.

Munz, F. W. (1965). Adaptation of visual pigments to the photic environment. In *Color vision* (ed. A. V. S. de Reuck and J. Knight), pp. 27–51. J. & A. Churchill Ltd, London.

Nashima, K., Mitsudo, M., and Kito, Y. (1979). Molecular weight and structural studies on cephalopod rhodopsin. *Biochemica et Biophysica Acta*, **579**, 155–68.

Seidou, M., Kubota, I., Hiraki, K., and Kito, Y. (1988). Amino acid sequence of the retinal binding site of squid visual pigment. *Biochemica et Biophysica Acta*, **957**, 318–21.

Seidou, M., Sugahara, M., Uchiyama, H., Hiraki, K., Hamanaka, T., Michinomae, M., *et al.* (1990). On the three visual pigments in the retina of the firefly squid, *Watasenia scintillans. Journal of Comparative Physiology A*, **166**, 769–73.

Suzuki, T., Arikawa, K., and Eguchi, E. (1985). The effects of light and temperature on the rhodopsin-porphyropsin visual system of the crayfish, *Procambarus clarkii. Zoological Science*, **2**, 455–61.

Tsuji, F. I. (1983). Molluscan bioluminescence. In *Environmental biochemistry and physiology* (ed. P. W. Hochachka), pp. 257–79. Academic Press, New York and London.

Underwood, B. A. (1984). Vitamin A in human and animal nutrition. In *The retinoids*, Vol. 1 (ed. M. B. Sporn, A. B. Roberts, and D. S. Goodman), pp. 281–392. Academic Press, New York and London.

Wells, M. J. (1978). *Octopus*. Chapman & Hall Ltd, London.

Wolken, J. J. (1958). Retinal structure, Mollusc Cephalopods: *Octopus, Sepia. Journal of Biophysics and Biochemistry Cytology*, **4**, 835–8.

Yamamoto, T., Tasaki, K., Sugawara, Y. and Tonosaki, A. (1965). Fine structure of the octopus retina. *Journal of Cell Biology*, 345–59.

31

The statocysts of cephalopods

RODDY WILLIAMSON

Summary

The statocysts of most cephalopods are sense organs of great sophistication with macula/statolith systems that respond to gravity and crista/cupula systems that operate as angular velocity transducers. These systems show strong parallels with their vertebrate counterparts in the vestibular organ of the inner ear, both in terms of their afferent response characteristics and in the mechano-electric transduction mechanism of the sensory hair cells. The cephalopod statocysts, unlike those of most other invertebrates, have both primary and secondary sensory hair cells. These hair cells have a unidirectional morphological and physiological polarization. The primary afferent neurones of the secondary sensory hair cells have their somata in the statocyst sensory epithelium. The statocyst sensory epithelia receive a very large efferent innervation from the brain that can excite or inhibit the afferent activity. These two efferent effects appear to use separate neurotransmitters, acetylcholine for the inhibition and a catecholamine for excitation. Within the central nervous system, the sensory input from the statocysts is integrated into a variety of behaviours, including locomotion, posture, control of eye movements, and control of body colour pattern. The morphological and physiological basis of these processes are beginning to be understood. The sensory solutions to the problems of balance and orientation that cephalopods and vertebrates have evolved show close parallels. It is anticipated that a comparison of these two systems will give a greater understanding of the basic mechanisms that are important for the operation of all such sensory systems.

From the early anatomical work on cephalopod statocysts (Owsjannikow and Kowalewsky 1867; Hamlyn-Harris 1903) it was recognized that they were sense organs capable of responding to gravity and possibly capable of detecting rotational movements of the animal. This view was strengthened by behavioural work, particularly ablation experiments showing that removal of one or both statocysts had profound effects on the ability of the animal to orientate, maintain an upright posture, or to perform compensatory eye movements to rotational stimuli (Muskens 1904; Boycott 1960; Dijkgraff 1961). Recent research has not only confirmed the statocysts as detectors of linear and angular accelerations, but has shown that their level of performance and sophistication rivals other such systems in the animal kingdom, including our own vertebrate vestibular system.

Morphology

Types of statocyst

The structure of the cephalopod statocysts has been extensively studied and the early work reviewed by Barber (1968) and Budelmann (1976, 1990). Within the living cephalopod groups there are three main types of statocyst.

The simplest are those found in the nautiloids [Fig. 31.1(a)]. Here the statocysts are a pair of simple oval sacs, embedded in the central nervous

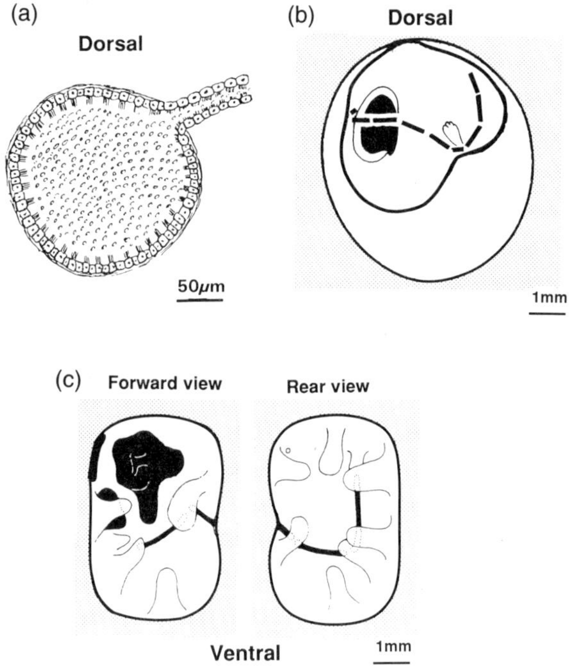

Fig. 31.1. (a) Diagram of the statocyst of *Nautilus* showing the sensory hair cells, the small statoconia within the sac, and the open Kölliker's canal. Note that this statocyst is much smaller than those of the octopods and decapods. (After Barber 1968.) (b) *Octopus* statocyst viewed from the side. The statocyst sac is suspended within the statocyst cavity by fibrous strands. There are two areas of receptor epithelium: a single, oval shaped macula with an attached statolith, and a crista strip that passes around the inside of the sac, such that it runs in all three axes. (c) Forward and rear view of the statocyst of *Sepia*. There are three maculae (shown in black), arranged in three different planes, and the crista strip is divided into four segments. Anticristae and hamuli project into the cavity of the statocyst. (After Budelmann 1980.)

system and open to the exterior via Kölliker's canal (Young 1965). The insides of these ectodermal sacs are lined with numerous sensory hair cells whose cilia project into the lumen of the sac. The sensory hair cells appear to be primary hair cells, such as are found in other non-cephalopod molluscs, with axons that gather together to form the static nerve and pass into the brain. The sac itself is partially filled with small statoconia that press down on the cilia and hence stimulate the underlying sensory hair cells. The hair cells have not been reported to be grouped into clusters, but to be fairly evenly spread along the sac wall; thus there is no differentiation into cristae and maculae (Barber 1967). The basic form of this statocyst is similar to those found in other groups of non-cephalopod molluscs and may represent the ancestral form.

The second type of cephalopod statocyst is found in the octopods [Fig. 31.1(b)]. The right and left statocysts are in separate cavities in the cranial cartilage, just below and to the side of the brain. Here, however, there is an almost spherical inner sac, filled with endolymph, and suspended in perilymph within the cavity by fibrous strands containing blood vessels (Young 1960). The sensory hair cells lie on the inside wall of the inner sac and are grouped into two main areas of sensory epithelium. First, an oval plate containing about 5100 sensory hair cells underlying a calcareous stone (Budelmann *et al.* 1973); this is the macula/statolith system, and from basic physical principles, is likely to respond to linear accelerations, particularly gravity. The second area of sensory epithelium is a narrow strip of sensory hair cells that winds around the inside of the statocyst sac such that it runs in three almost orthogonal planes. The strip is usually divided into nine segments or cristae. Each crista segment consists of up to nine rows of sensory hair cells, with up to 100 hair cells per row; the precise hair cell make-up of the crista segments varies between species and sometimes even within a single statocyst (Budelmann *et al.* 1987). Overlying each crista segment is a sail-like gelatinous cupula, which is deflected by the flow of endolymph during movements of the animal. Between crista segments 6 and 7 there is a single cartilaginous projection into the statocyst cavity (Young 1960). This anticrista is thought to influence the endolymph flow within the statocyst sac. The cupulae in some octopods vary in size and shape for alternate crista segments (Budelmann *et al.* 1987); thus the first crista segment from the macula side has a small, wide cupula whereas the next has a tall, thin cupula, and so on. It is believed that this changes the sensitivity of the responses from the individual segments, effectively fractionating the response range of the system (Williamson and Budelmann 1985*a*,*b*).

The third type of cephalopod statocyst is found in decapod cephalopods [Fig. 31.1(c)], the squid and cuttlefish (Stephens and Young 1982, Budelmann 1990). This is similar to the octopod statocyst but here the sac remains in contact with the cartilage wall (i.e. there is no perilymphatic space), there are four crista segments instead of nine, again made up of rows of sensory hair

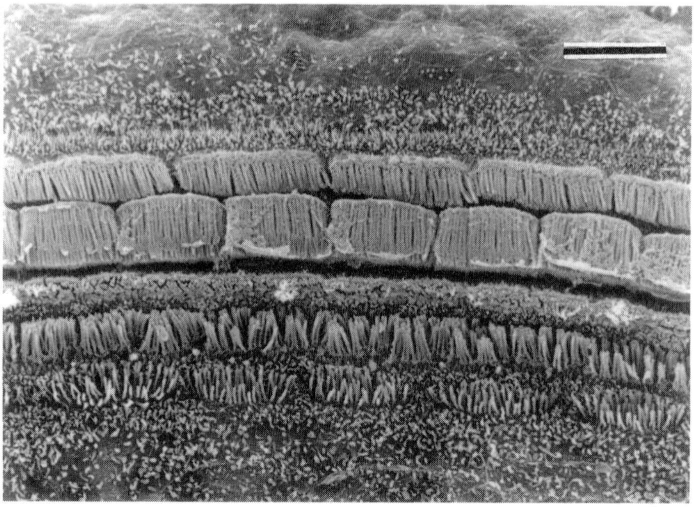

Fig. 31.2. Scanning electron micrograph of part of a cuttlefish crista segment showing the rows of hair cells. The individual hair cells can be identified from the groups of kinocilia protruding from their apexes. The hair cells towards the bottom of the picture are primary sensory hair cells and those towards the top are secondary hair cells. Scale bar: 10 μm.

cells (Fig. 31.2). In decapods there are two additional small maculae, these having overlying statoconia instead of the more dense statolith; as before the crista segments, and this time also the maculae, are arranged to cover the three axes of movement. The statocysts are less regularly shaped than in the octopods and can bulge outwards into the cartilage to form small sac-like projections, or have many fingers of cartilage (the anticristae and hamuli) projecting into the main cavity. These features vary for different species and are believed to modify the pattern of flow of endolymph within the cavity and hence change the response characteristics of the system (Maddock and Young 1984; Young 1989).

An exception to this classification of statocysts is found in *Vampyroteuthis* (Stephens and Young 1976). This animal has a statocyst that is intermediate between the octopod and decapod types. It has an inner sac within the statocyst cavity, like *Octopus*, but has a partly divided macula as in decapods. The crista has four parts, as in decapods, but these are not sharply separated. Other unusual characteristics are the presence of a very large opening for Kölliker's canal, a large posterior element to the statocyst sac, and the inner sac wall is covered with ciliated cells, which apparently connect with the crista nerve.

The sensory epithelia

The sensory epithelia of both decapod and octopod cephalopods are more complex than those found in nautiloids or other types of mollusc. Here, the

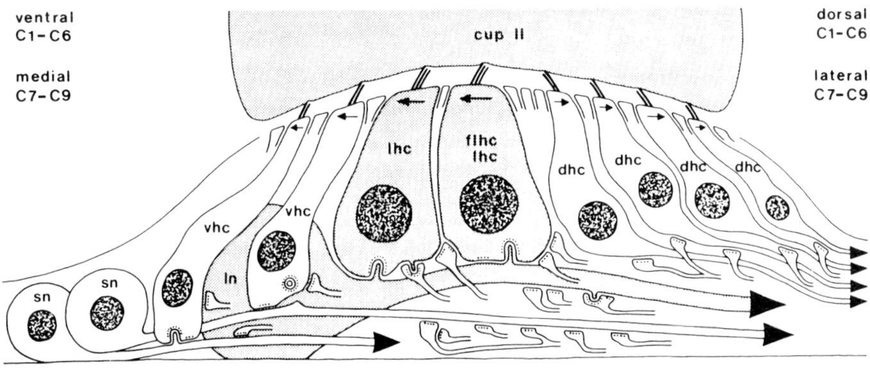

Figure 31.3. Diagram of a cross-section of the crista of *Octopus vulgaris*. It represents both the odd-numbered (C1, C3, C5, C7, C9) and the even-numbered (C2, C4, C6, C8) crista segments, and demonstrates the different types of hair cells and first-order afferent neurones, as well as their afferent and efferent synaptic organization with finger-like and flat synaptic contacts. In the middle (shaded): the large (lhc), and large or fairly large (flhc) secondary hair cells, together with their large first-order afferent neurones (ln). To the left: the rows of small secondary ventral hair cells (vhc), together with their small first-order afferent neurones (sn). To the right: the rows of small dorsal primary hair cells (dhcs). Some afferent and efferent synapses are represented and the arrows in the distal ends of the hair cells indicate the direction of their morphological polarization. (From Budelmann *et al.* 1987.)

statocysts contain both primary and secondary sensory hair cells; the secondary sensory hair cells have no centripetal axons, but make synaptic contact with first-order afferent neurones whose axons pass to the central nervous system (Fig. 31.3). In addition, each hair cell is morphologically polarized. The direction of polarization can be determined from features such as the location of the basal foot at the base of the kinocilium or the orientation of the internal tubuli structure of each kinocilium (Budelmann 1979). A single hair cell can have up to 200 kinocilia protruding from its apical surface (Fig. 31.2) and in a polarized cell these all have their basal feet pointing in the same direction. We shall see later that this morphological polarization corresponds with a physiological polarization. The polarized hair cells are arranged within the sensory epithelia in distinct patterns. For example, in the maculae, there is a radial pattern of polarization whereas in the cristae the cells are polarized in the direction perpendicular to the line of the crista segment (Budelmann 1979). In the species so far examined in detail, the primary hair cells in the cristae segments are polarized in the opposite direction to that of the secondary hair cells (Budelmann *et al.* 1987).

The sensory epithelia of the maculae and cristae receive an efferent innervation from the central nervous system. In *Octopus*, this efferent innervation consists of many thousands of small nerve fibres which form a plexus beneath the sensory epithelium. Axon counts have indicated that more than 70 per cent of the axons in the statocyst nerves are efferent fibres; this compares with a figure of about 10 per cent for vertebrate systems (Budelmann *et al.* 1987). The efferent fibres make at least two morphologically different kinds of synaptic contacts with the hair cells and afferent neurones of the system, perhaps implying that they have two different functions (Budelmann *et al.* 1987). We shall see later that this efferent system exerts a powerful control over the output from the sense organ.

Development

Although the early embryologists clearly recognized the developing statocysts or otocysts as they were then called (e.g. Kölliker 1844; Lancaster 1875), there have been few specific descriptions of their development (Hamlyn-Harris 1903; Fioroni 1978). As far as is known, the early development of decapod and octopod statocysts is similar (Bülow and Fioroni 1989) and has been described for *Loligo* (Fig. 31.4) by Meister (1972). The statocysts in the developing embryo can first be identified as two small circular depressions of ectodermal cells at about stage VII (after Naef 1928), roughly 25 days old—at 15°C (Boletzky 1979). These depressions form into pockets, which then close over by stage XI to produce the statocyst cavities, leaving only a small opening to the exterior via Kölliker's canal. By stage XII, Kölliker's canal has closed and the statocyst sensory epithelia and maculae statoliths can be recognized.

At hatching (stage XX—about 70 days at 15°C), *Loligo* has formed only a few of its final 12 cartilaginous projections into the statocyst cavity (anticristae and hamuli) and none of the invaginations into the surrounding cartilage; it thus resembles the *Octopus* statocyst, which at this stage has not yet detached itself from the surrounding cartilage to form the free statocyst sac (Bülow and Fioroni 1989). Similarly, the squid macula has not yet divided into its three parts and so resembles the octopus condition. The maculae, at this stage, are probably partially functional, since the hair cells and parts of the statoliths have developed. However, the cristae of both squid and octopus do not yet have cupulae and so cannot be expected to be fully functional (Bülow and Fioroni 1989). In *Sepia*, however, the crista cupulae are present before hatching (Glanfield *et al.* 1991) and so the angular acceleration receptor system is probably functional at this stage.

After hatching, the statocysts rapidly develop their specific features, that is anticristae and hamuli; however, in terms of their overall size, they grow at a slower rate than the rest of the body. Thus, although at hatching the squid statocysts are relatively large, about one-quarter the length of the animal's mantle, in the adult the statocyst are less than 1/20th of the mantle length (Maddock and Young 1984). The relatively large initial size of the statocysts

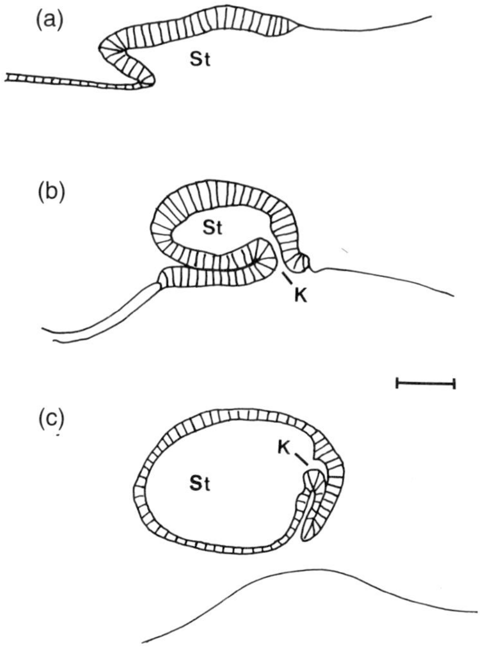

Fig. 31.4. Diagram showing the early development of the statocyst of *Loligo* at (a) stage IX—where the statocyst (St) is just a depression in the ectodermal cells, (b) stage X—where the statocyst cavity has formed but is still open to the exterior through Kölliker's canal (K); and (c) stage XII—where Kölliker's canal has closed and the statocyst is entirely encased in cartilage. All transverse sections at the level of Kölliker's canal. Scale bar: 50 μm. (After Meister 1972.)

probably reflects a hydrodynamic constraint on the minimum physical size that permits the angular acceleration receptor system to function usefully. As the sensory epithelia grow they add new sensory cells with up to five times more cells in the adult than the hatchling (Budelmann *et al.* 1987). The cristae add new hair cells, both by increasing the number of cells per row and also by initially increasing the number of rows; the actual sizes of the hair cells, however, does not appear to change with animal size (Glanfield *et al.* 1991).

Physiology

Despite the wealth of data on the statocyst morphology of many species of cephalopods (Ishikawa 1929; Young 1989), there is physiological information from only a handful of species. In the main, most of the neurophysiological studies have been performed on the statocysts of *Octopus vulgaris*, with some work on *Eledone, Sepia, Alloteuthis*, and *Loligo*. So far no great differences have been found in the responses of these species, but this is likely to be due to the dearth of detailed comparative data rather than a uniformity in their responses.

Afferent activity

The first electrophysiological evidence for the function of cephalopod statocysts was obtained by Maturana and Sperling (1963) who recorded from an *Octopus* crista nerve and obtained afferent excitatory responses when the animal was rotated. These results were extended by Budelmann and Wolff (1973) who also showed that the afferents from the crista/cupula system not only responded to angular acceleration, but were also sensitive to linear accelerations (i.e. gravity). The cupulae, therefore, are probably not neutrally buoyant.

A more complete analysis of the response characteristics of the *Octopus* cristae system was performed by Williamson and Budelmann (1985*a*,*b*). Their main finding was that the statocyst crista afferents are part of an angular acceleration detector system, but that over a middle range of frequencies (0.1–1 Hz) the receptors code for angular velocity (Fig. 31.5). The *Octopus* response characteristics are thus very similar to those of the vertebrate semicircular canal system. This work also showed that although each afferent has a unidirectional response to stimulation (Fig. 31.5), as had been predicted from

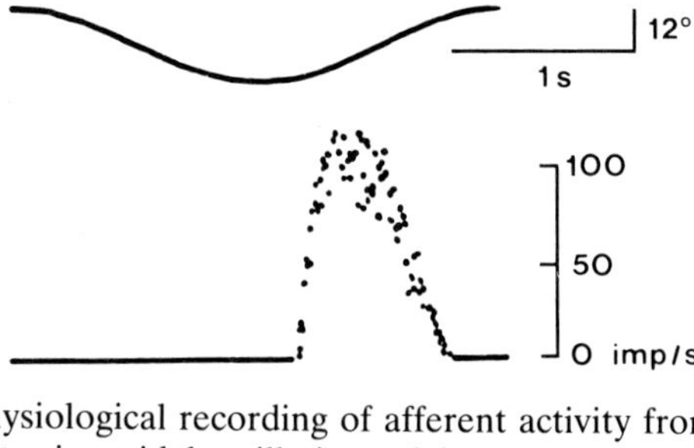

Fig. 31.5. Electrophysiological recording of afferent activity from the crista nerve of *Octopus* in response to sinusoidal oscillations of the statocyst in the pitch plane. Upper trace extracellular suction electrode recording from crista segment C2, middle trace indicates the position of the statocyst, lower trace is an instantaneous frequency plot of the spike activity in the upper trace; this clearly shows that the response is unidirectional and that the peak of nerve activity coincides with the time of peak velocity of the movement. (From Williamson and Budelmann 1985*a*.)

the morphology, there are two populations of afferents in each crista and these have opposing polarities of responses. Thus, in contrast to vertebrate systems, groups of afferents will be excited regardless of the direction of movement. For *Octopus*, the primary sensory hair cells in a crista segment are polarized in one direction while the secondary sensory hair cells are polarized in the opposite direction (Budelmann *et al*. 1987). By recording from afferent units from different crista segments it was shown that afferents from crista segments with a large, narrow cupula are more sensitive to rotational stimuli than those from segments with a small, wide cupula. This was interpreted as being a mechanism for fractionating the response range of the receptor system, possibly correlated with the two different forms of locomotion employed by the animals, slow crawling and rapid jet-propelled swimming (Williamson and Budelmann 1985*a*,*b*).

The results obtained from extracellular recordings from the statocyst afferents have been supplemented by intracellular recordings (Williamson 1990, 1991) from squid and cuttlefish sensory hair cells while they were stimulated by controlled displacements of the overlying cupula (Fig. 31.6). These experiments confirmed the response polarization of the primary and secondary sensory hair cells, the latter so far only inferred from the response of the afferents, and showed that the secondary hair cell's displacement/response curve has a sigmoidal shape similar to that already found for vertebrate hair cells. It should be borne in mind that, although the response characteristics of cephalopod and vertebrate hair cells appear similar, and hence the mechano-electric transduction processes may also be similar, the cephalopod hair cells use true cilia, kinocilia, in this process, whereas vertebrate cells use stereocilia.

One unexpected finding from these intracellular recording studies (Williamson 1989*a*) was that some groups of secondary sensory hair cells in the crista segments are quite strongly electrically coupled (Fig. 31.7), although, as far as yet seen, they are not dye-coupled. Electrical coupling is known to occur between the sensory hair cells in the statocysts of some non-cephalopod molluscs (Detwiler and Alkon 1973) but is not usually found between secondary sensory hair cells in vertebrates, although there may be some exceptions (Miller and Beck 1990). The function of this coupling is not yet clear but it may act to improve the signal-to-noise ratio, and hence the sensitivity, of the system at some response frequencies.

Efferents

There is considerable morphological evidence indicating the presence of a very large, dual efferent innervation of the cephalopod statocysts. Efferent systems are present in many sense organs but their function is poorly understood. In cephalopods, it is possible to record efferent activity and, to some extent, determine what drives it and when it is active. The efferent activity has been examined by recording from the central end of the cut crista nerve as

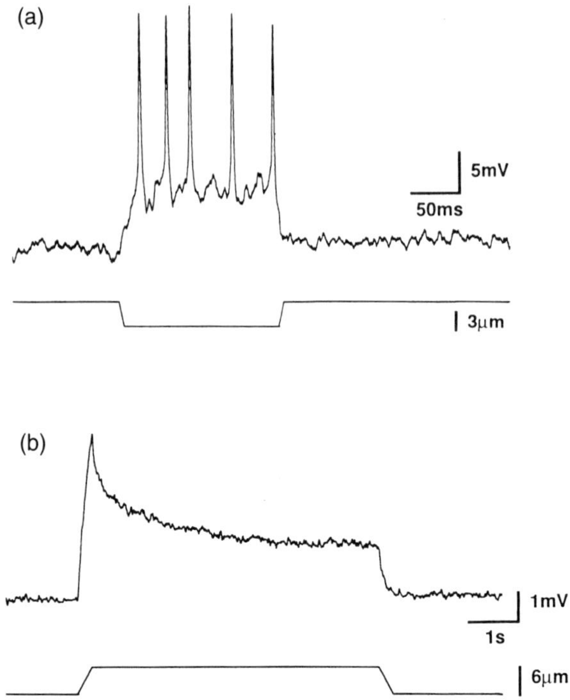

Fig. 31.6. Intracellular recordings from (a) a primary sensory hair cell and (b) a secondary sensory hair cell in the crista of the squid showing their responses to small mechanical displacements of the overlying cupula; displacements shown in the lower traces. Note that the cells are depolarized by displacements in opposite directions and that action potentials are seen only in the primary hair cells. (From Williamson 1990.)

it leaves the brain in an isolated CNS/statocyst preparation (Williamson 1986). These recordings, from *Octopus*, demonstrated that many of the efferents have a resting level of activity and that the activity could be entrained and increased by mechanical stimulation of the contralateral statocyst. Efferents showing this entrained response (84 per cent of all efferents seen) could be divided into two populations on the basis of when they were active with respect to the imposed stimulus oscillation, that is both populations were entrained by a sinusoidal mechanical stimulus but were 180° out of phase with each other. This difference in the phase of their activities could be due to the efferents having two different targets, for example sensory cells polarized in opposite directions, or to the efferents having different effects on the same targets, for example one efferent population excitatory and the other inhibitory. The phase position of the peak efferent activity in these experiments coincided with the phase of the peak afferent activity; one

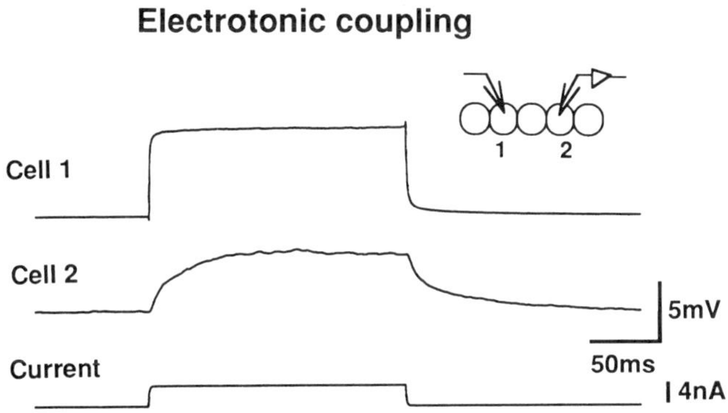

Fig. 31.7. Electrical coupling between hair cells. Intracellular recordings from two secondary sensory hair cells in the crista of *Sepia* showing their responses to injection of a small current into one cell. The hair cells were in the outer ventral row of hair cells and were close neighbours. (From Williamson 1991.)

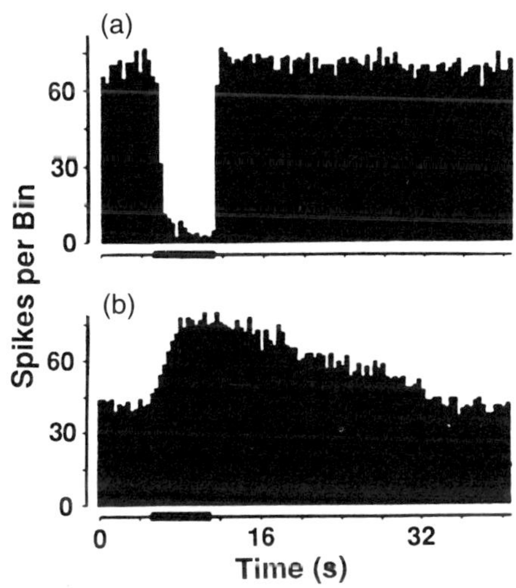

Fig. 31.8. Efferent (a) inhibition and (b) excitation of the afferent activity. Time histograms of statocyst crista activity showing the two different responses to electrical activation of the statocyst efferents. Extracellular recording from *Octopus*. (After Williamson 1985*d*.)

interpretation of this is that the brain can precisely time its use of the efferents for maximal impact on the afferent activity.

The effect of the efferents on the statocyst afferent activity can be observed by recording the afferent activity from the *Octopus* medial crista nerves and electrically stimulating the anterior crista nerve in order to activate the efferents (Fig. 31.8). It was found that the afferent activity could be enhanced or depressed, depending on the stimulus site and stimulation parameters (Williamson 1985). These experiments provided good evidence that there are indeed two populations of efferents from the brain to the statocyst with opposing effects. This was confirmed by later work from squid statocysts where intracellular recordings were made from hair cells and afferent neurones while electrically stimulating the crista nerve to activate the efferents (Williamson 1989*b*). Recordings from the secondary sensory hair cells showed that they all responded to efferent stimulation (Fig. 31.9): 48 per cent showed a depolarization, 15 per cent hyperpolarization, and 37 per cent a mixed response of a depolarization followed by a hyperpolarization. Each hair cell probably receives both inhibitory and excitatory efferent inputs; the observed different responses probably reflect the ease of electrically activating different efferent axons in the nerve, for example the inhibitory axons may be larger or more numerous and therefore their effects more often seen. Further experiments, looking at ways of blocking synaptic transmission, and the results of varying the cell membrane potential, indicated that these effects were due to

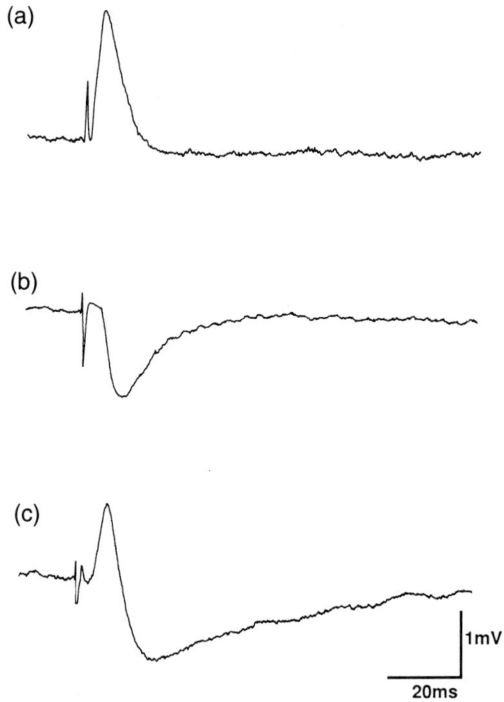

Fig. 31.9. Intracellular recordings from hair cells in the squid statocyst crista showing the different efferent responses to electrical stimulation of the crista nerve: (a) depolarization; (b) hyperpolarization; (c) mixed depolarization and hyperpolarization. (From Williamson 1989*b*.)

chemical synapses and that two different transmitters were probably involved (Williamson 1989*c*). These results show that the cephalopod CNS has extensive control over the information it receives from the statocyst. It can enhance or depress the input at both the hair cell and the afferent neurone level. This could be used in a variety of ways, for example, to increase the sensitivity of the input during slow gentle movements, or decrease it during jet-propelled movements that may well otherwise saturate the input system.

Pharmacology

The neurotransmitter between the secondary hair cells and the primary afferent neurones in the cephalopod statocyst is probably glutamergic (Tu and Budelmann 1994). Glutamate is a well established neurotransmitter in cephalopods, occurring at the squid giant synapse (De Santis and Messenger 1989) and at some neuromuscular junctions (Bone and Howarth 1980). Glutamate is also considered to be the afferent neurotransmitter in the vertebrate vestibular system (Blesdoe *et al.* 1988).

In the statocyst efferent system, there is morphological evidence indicating that there are two efferent transmitters, one acetylcholine and the other a catecholamine. First, Budelmann and Bonn (1982) used the Falck–Hillarp fluorescence technique to demonstrate the presence of catecholamines in the neuronal plexus beneath the sensory epithelium of the *Octopus* statocyst macula and crista. From their data they proposed that either dopamine and/or noradrenaline is a likely efferent transmitter. This was supplemented by later work (Auerbach and Budelmann 1986) which used a combination of histochemical, lesion/degeneration, and bioassay techniques, to support the hypothesis that acetylcholine is also a statocyst efferent transmitter.

In addition, electrophysiological recordings from crista nerves in the isolated *Octopus* statocyst have shown (Fig. 31.10) that the resting level of afferent activity is depressed by the bath application of acetylcholine and its agonists, and increased by the bath application of dopamine or noradrenaline (Williamson 1989*d*). By recording from the afferents, while electrically stimulating the efferent fibres it was also shown that the efferent-induced inhibition of the afferent activity could be blocked by cholinergic antagonists such as gallamine, and the efferent-induced excitation could be blocked by catecholaminergic antagonists such as phentolamine. These data consistently support the view that there are at least two efferent neurotransmitters employed in the statocyst: acetylcholine which is inhibitory and a catecholamine, dopamine, or noradrenaline, which are excitatory. There is evidence for acetylcholine as an efferent transmitter in the vertebrate vestibular system (Bledsdoe *et al.* 1988) and noradrenaline has been proposed as an efferent neurotransmitter in the statocysts of gastropod molluscs (Janse *et al.* 1988).

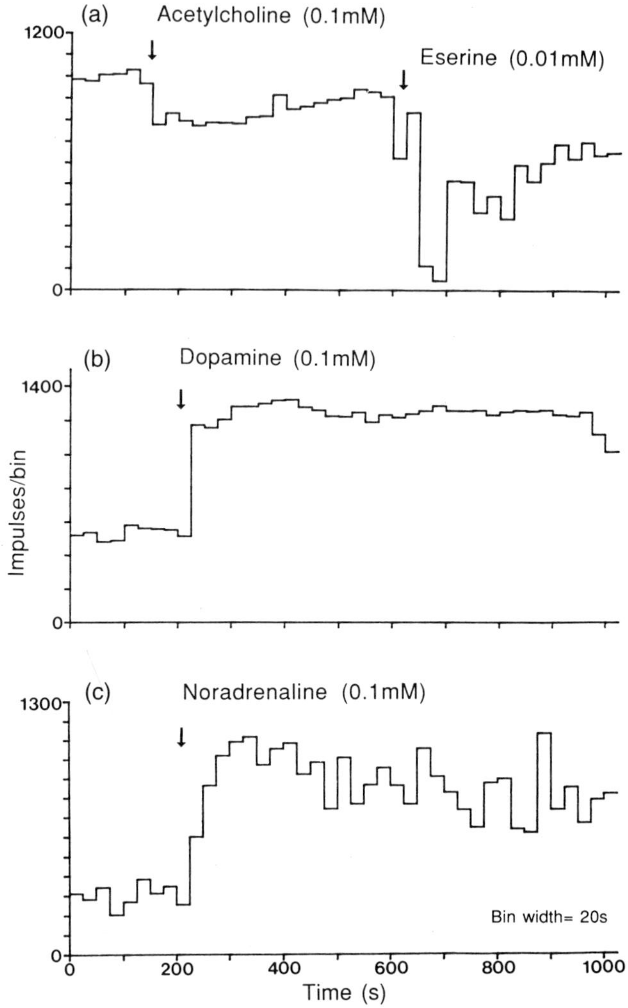

Fig. 31.10. Time histograms showing the effect on the afferent resting activity from the *Octopus* statocyst crista nerve of bath application of: (a) acetylcholine and eserine; (b) dopamine; and (c) noradrenaline. These results support the hypothesis that acetylcholine is employed as an efferent inhibitory transmitter in the statocyst and that there is also a catecholaminergic efferent excitatory transmitter. (From Williamson 1989*d*.)

Central connections

From behavioural work, it is clear that the sensory input from the statocysts is used in many of the central nervous control systems of the cephalopods, for example oculomotor responses, control of posture and swimming, and even control of body shading (Messenger 1970; Ferguson and Messenger 1991). Lesion/degeneration experiments (Young 1971) and direct dye backfills of statocyst nerves (Colmers 1982; Budelmann and Young 1984) have shown that statocyst afferents enter the brain and then some proceed ipsilaterally to the anterior and posterior lateral pedal lobes, some to the median basal lobes, the peduncle lobes, and in both directions along the brachial to palliovisceral lobe connective. Contralaterally, fibres pass to the lateral pedal and median basal lobes.

Efferent fibres travelling to the *Octopus* statocyst come from three main areas of the brain: the anterior lateral pedal lobe, the posterior pedal lobe, and the magnocellular lobes (Colmers 1982; Budelmann and Young 1984). It is not known whether these different sources of efferent fibres are correlated with the different actions of the efferents on the statocyst afferent activity, or with different target cells within the statocyst sensory epithelium.

By combining behavioural and morphological studies we can surmise that the statocyst afferents contact the efferents in, for example, the magnocellular lobes, and that the statocyst input is integrated with the visual input in the peduncle lobe (the cephalopod 'cerebellum') and the basal lobes. However, direct physiological evidence for statocyst central connections has, as yet, only been shown for oculomotor neurones in the anterior lateral pedal lobe (Williamson and Budelmann 1991).

Concluding remarks

These data have shown the degree of sophistication attained by the cephalopod statocysts and their close parallels with the vertebrate vestibular system, particularly in terms of afferent response characteristics, hair cell mechanoelectric transduction process, efferent control, and central nervous systems influenced by the statocysts. Since the cephalopods have been separated from the vertebrate line for some 500 million years, we can only conclude that these similarities are due to the process of convergent evolution (Packard 1972), with direct competition between cephalopods and vertebrates in almost all marine niches, resulting in sensory systems of comparable performance. There is still much to be understood in the operation of sense organs, particularly in the processes of sensory integration and the function of the efferents. It is anticipated that a comparative study of both cephalopod and vertebrate systems will highlight the fundamental principles that are important in these sense organs, and advance our understanding of all such sensory systems.

Acknowledgements

The support of the Wellcome Trust and the Alexander von Humboldt Stiftung is gratefully acknowledged. R.W. is a Wellcome Trust Senior Research Fellow.

References

Auerbach, B. and Budelmann, B. U. (1986). Evidence for acetylcholine as a neurotransmitter in the statocyst of *Octopus vulgaris*. *Cell and Tissue Research*, **243**, 429–36.

Barber, V. C. (1967). The sense organs of *Nautilus*: preliminary observations on their fine structure. *Journal of Microscopy*, **6**, 1067–72.

Barber, V. C. (1968). The structure of mollusc statocysts, with particular reference to cephalopods. *Symposia of the Zoological Society (London)*, **23**, 37–62.

Bledsdoe, S. C., Bobbin, R. P., and Puel, J. L. (1988). Neurotransmission in the inner ear. In *Physiology of the ear* (ed. A. F. Jahn and J. Santos-Sacchi), pp. 385–406. Raven Press, New York.

Boletzky, S. (1979). Observations on the early post-embryonic development of *Loligo vulgaris* (Mollusca, Cephalopoda). *Rapport. Commission Internationale pour la Mer Méditerrannée. Paris*, **25/26**(10), 159–68.

Bone, Q. and Howarth, J. V. (1980). The role of L-glutamate in neuromuscular transmission in some molluscs. *Journal of the Marine Biological Association of the United Kingdom*, **60**, 619–26.

Boycott, B. B. (1960). The functioning of the statocysts of *Octopus vulgaris*. *Proceedings of the Royal Society of London, Series B*, **152**, 78–87.

Budelmann, B. U. (1976). Equilibrium receptor systems in molluscs. In *Structure and function of proprioceptors in the invertebrates* (ed. P. J. Mill), pp. 529–66. Chapman & Hall, London.

Budelmann, B. U. (1979). Hair cell polarization in the gravity receptor systems of the statocysts of the cephalopods *Sepia officinalis* and *Loligo vulgaris*. *Brain Research*, **160**, 261–70.

Budelmann, B. U. (1980). Equilibrium and orientation in cephalopods. *Oceanus*, **23**, 34–43.

Budelmann, B. U. (1990). The statocysts of squid. In *Squid as experimental animals* (ed. D. L. Gilbert, W. J Adelman, and J. M. Arnold), pp. 421–42. Plenum Press, New York.

Budelmann, B. U., Barber, V. C., and West, S. (1973). Scanning electron microscopical studies of the arrangements and numbers of hair cells in the statocysts of *Octopus vulgaris, Sepia officinalis* and *Loligo vulgaris*. *Brain Research*, **56**, 25–41.

Budelmann, B. U. and Bonn, U. (1982). Histochemical evidence for catecholamines as neurotransmitters in the statocyst of *Octopus vulgaris*. *Cell and Tissue Research*, **227**, 475–83.

Budelmann, B. U., Sachse, M., and Staudigl, M. (1987). The angular acceleration receptor system of the statocyst of *Octopus vulgaris*: morphometry, ultrastructure, and neuronal and synaptic organization. *Philosophical Transactions of the Royal Society of London, Series B*, **315**, 305–43.

Budelmann, B. U. and Wolff, H. G. (1973). Gravity response from angular acceleration receptors in *Octopus vulgaris*. *Journal of Comparative Physiology*, **85**, 283–90.

Budelmann, B. U. and Young, J. Z. (1984). The statocyst–oculomotor system of *Octopus vulgaris*: extraocular eye muscles, eye muscle nerves, statocyst nerves and the oculomotor centre in the central nervous system. *Philosophical Transactions of the Royal Society of London, Series B*, **306**, 159–89.

Bülow, B. and Fioroni, P. (1989). Der Bau der Statocyste im Schlüpfstadium der Cephalopoden *Loligo vulgaris* Lam. und *Octopus vulgaris* Lam. *Zoologische Jahrbücher. Abteilung für Anatomie*, **119**, 89–105.

Colmers, W. F. (1982). The central afferent and efferent organisation of the gravity receptor system of the statocyst of *Octopus vulgaris*. *Neuroscience*, **7**, 461–76.
De Santis, A. and Messenger, J. B. (1989). New evidence that L-glutamate is a transmitter at the squid giant synapse. *Quarterly Journal of Experimental Physiology*, **74**, 219–22.
Detwiler, P. B. and Alkon, D. L. (1973). Hair cell interactions in the statocyst of *Hermissenda*. *Journal of General Physiology*, **62**, 618–42.
Dijkgraff, S. (1961). The statocyst of *Octopus vulgaris* as a rotation receptor. *Pubblicazioni della Stazione Zoologica di Napoli*, **32**, 64–87.
Ferguson, G. B. and Messenger, J. B. (1991). A countershading reflex in cephalopods. *Proceedings of the Royal Society of London, Series B*, **243**, 63–1.
Fioroni, P. (1978). Cephalopoda, Tintenfische. In *Morphogenese der Tiere*, Vol. 2 (ed. F. Seidel), p. 112. Gustav Fischer, Jena.
Glanfield, W., Witkam, A., and Williamson, R. (1991). Can swimming performance be predicted from statocyst morphology in cephalopods? *Journal of the Marine Biological Association of the United Kingdom*, **71**, 736–7.
Hamlyn-Harris, R. (1903). Die Statocysten der Cephalopoden. *Zoologische Jahrbücher*, **18**, 327–58.
Ishikawa, M. (1929). On the statocysts of the American Cephalopod genera. *Journal of Morphology*, **48**, 563–84.
Janse, C., van der Wilt, G. J., van der Roest, M., and Pieneman, A. W. (1988). Intracellularly recorded responses to tilt and efferent input of statocyst sensory cells in the pulmonate snail *Lymnaea stagnalis*. *Journal of Comparative Biochemistry and Physiology*, **90A**, 269–78.
Kölliker, A. (1844). *Entwicklungsgeschichte der Cephalopoden*. Zürich.
Lancaster, E. R. (1875). Observations on the development of the Cephalopoda. *Quarterly Journal of the Microscopy Society*, **15**, 37–47.
Maddock, L. and Young, J. Z. (1984). Some dimensions of the angular acceleration receptor system of cephalopods. *Journal of the Marine Biological Association of the United Kingdom*, **64**, 55–79.
Maturana, H. R. and Sperling, S. (1963). Unidirectional response to angular acceleration recorded from the middle crista nerve in the statocyst of *Octopus vulgaris*. *Nature*, **197**, 815–16.
Meister, G. (1972). Organogenese von *Loligo vulgaris* Lam. (Mollusca, Cephalopoda, Teuthoidea, Myopsida, Loliginidae). *Zoologische Jahrbücher. Abteilung für Anatomie*, **89**, 247–300.
Messenger, J. B. (1970). Optomotor responses and nystagmus in intact, blinded and statocystless cuttlefish (*Sepia officinalis* L.). *Journal of Experimental Biology*, **53**, 789–96.
Miller, M. R. and Beck, J. (1990). Innervation of a lizard auditory organ having gap junctions between most hair cells: a serial transmission electron microscopy study. *Journal of Comparative Neurology*, **293**, 223–35.
Muskens, L. J. J. (1904). Über eine eigenthümliche kompensatorische Augenbewegung der Octopoden mit Bemerkungen über deren Zwangsbewegung. *Archiv für Anatomie und Physiologie*, **28**, 49–56.
Naef, A. (1928). *Die Cephalopoden*. Monographie 35. Fauna e Flora del Golfo di Napoli, Volume 1.
Owsjannikow, P. and Kowalewsky, A. (1867). Über das Centralnervensystem und das Gehörorgan der Cephalopoden. *Mémoires de l'Académie Impériale des Sciences de St. Petersbourg*, **11**, 36.

Packard, A. (1972). Cephalopod and fish: the limits of convergence. *Biological Reviews*, **47**, 241–307.

Stephens, P. R. and Young, J. Z. (1976). The statocyst of *Vampyroteuthis infernalis* (Mollusca: Cephalopoda). *Journal of Zoology*, **180**, 565–88.

Stephens, P. R. and Young, J. Z. (1982). The statocysts of the squid *Loligo*. *Journal of Zoology*, **197**, 241–66.

Tu, Y. and Budelmann, B. U. (1994). The effect of L-glutamate on the afferent resting activity in the cephalopod statocyst. *Brain Research*, **642**, 47–58.

Williamson, R. (1985). Efferent influences on the afferent activity from the *Octopus* angular acceleration receptor system. *Journal of Experimental Biology*, **119**, 251–64.

Williamson, R. (1986). Efferent activity in the *Octopus* statocyst nerves. *Journal of Comparative Physiology*, **158A**, 125–32.

Williamson, R. (1989*a*). Electrical coupling between secondary hair cells in the statocyst of the squid, *Alloteuthis subulata*. *Brain Research*, **486**, 67–72.

Williamson, R. (1989*b*). Secondary hair cells and afferent neurones of the squid statocyst receive both inhibitory and excitatory efferent inputs. *Journal of Comparative Physiology*, **165A**, 847–60.

Williamson, R. (1989*c*). Efferent neurotransmitters onto the hair cells in the statocyst of the squid. *Society for Neuroscience Abstracts*, **15**, 738.

Williamson, R. (1989*d*). Electrophysiological evidence for cholinergic and catecholaminergic efferent transmitters in the statocyst of *Octopus*. *Journal of Comparative Biochemistry and Physiology*, **93C**, 23–7.

Williamson, R. (1990). The responses of primary and secondary sensory hair cells in the squid statocyst to mechanical stimulation. *Journal of Comparative Physiology*, **167A**, 655–64.

Williamson, R. (1991). The responses of the sensory hair cells in the statocyst of *Sepia*. In *The cuttlefish* (ed. E. Boucaud-Camou), pp. 211–22. Centre de Publication de l'Université de Caen.

Williamson, R. and Budelmann, B. U. (1985*a*). The response of the *Octopus* angular acceleration receptor system to sinusoidal stimulation. *Journal of Comparative Physiology*, **156**, 403–12.

Williamson, R. and Budelmann, B. U. (1985*b*). An angular acceleration receptor system of dual sensitivity in the statocysts of *Octopus vulgaris*. *Experientia*, **41**, 1321–3.

Williamson, R. and Budelmann, B. U. (1991). Convergent inputs to *Octopus* oculomotor neurones demonstrated in a brain slice preparation. *Neuroscience Letters*, **121**, 215–18.

Young, J. Z. (1960). The statocysts of *Octopus vulgaris*. *Proceedings of the Royal Society of London, Series B*, **52**, 3–29.

Young, J. Z. (1965). The central nervous system of *Nautilus*. *Philosophical Transactions of the Royal Society of London, Series B*, **249**, 1–25.

Young, J. Z. (1971). *The anatomy of the nervous system of* Octopus vulgaris. Clarendon Press, Oxford.

Young, J. Z. (1989). The angular acceleration receptor system of diverse cephalopods. *Philosophical Transactions of the Royal Society of London, Series B*, **325**, 189–238.

32

Physiology of squid olfaction

MARY T. LUCERO and WM. F. GILLY

Summary

The work presented in this report focuses on the characterization of the squid olfactory system in terms of morphology, behaviour, and electrophysiology. Although squid and other cephalopod olfactory systems have been described anatomically and anecdotally for many years, only recently have behavioural and electrophysiological experiments examined potential odorants and mechanisms of signal transduction in squid olfaction. Thus the questions addressed herein reflect the basic level of our current understanding of this complex and interesting sensory system.

Chemoreceptive capabilities are present to varying degrees in organisms ranging from bacteria to humans (Finger and Silver 1987). Both vertebrates and invertebrates have evolved specialized chemosensory structures that are appropriate to sample the chemical environment. Information obtained by chemosensory structures is strongly linked to behavioural responses including feeding, avoidance, arousal, withdrawal, alarm, homing, mating, and settling of larvae (Bardach 1975). Thus neuroscientists have been attracted to the study of chemoreception both to discover how different animals have solved the problem of collecting relevant chemical signals from their surroundings and also to form a model neural system for examining receptor–ligand interactions, signal transduction, neural coding, integration, development, cell death, and nerve regeneration.

There have been many discussions regarding usage of the specific vertebrate-oriented labels of 'taste' and 'olfaction' in relation to the various chemosensory systems of invertebrates, especially marine and aquatic species (Kohn 1961; Bardach 1975; Atema 1980). One classification scheme suggests that contact chemoreception, which can often be mediated by receptors distributed over a large portion of the body (Bardach and Atema 1971), can be considered the aquatic equivalent to taste, whereas chemoreception of dissolved chemicals originating from a distant source falls into the category of olfaction (Kohn 1961; Atema 1980). The focus of this chapter will be on the chemosensory system associated with distance chemoreception in squid and hence is referred to as squid olfaction.

Squid are predatory animals with highly developed visual (Packard 1972), vestibular (Budelmann 1990), auditory (Hanlon and Budelmann 1987; Budelmann and Bleckmann 1988), and tactile (Wells 1964; Wells and Young 1975) systems. They are also prey for fish and marine mammals. To avoid predation, squid have developed a very effective escape jetting response. Although escape jetting is an alarm response, it is not clear to what degree the squid perceive danger by visual, tactile, auditory, and chemical cues. Although anatomical descriptions of a putative olfactory organ in cephalopods first appeared in the literature in 1869 (Zernoff 1869), it has only recently been shown that squid have chemosensory capabilities (Gilly and Lucero 1992). Behavioural experiments in the latter study used an assay based on escape jetting responses, and chemical stimuli that the animals responded to included the alarm substance in squid ink and its precursor L-DOPA as well as quaternary ammonium ions and other potassium (K^+) channel blockers.

Anatomy of the squid olfactory system

The olfactory organ is a highly specialized sensory structure found in all coleoid cephalopods (Messenger 1967; Woodhams and Messenger 1974). In many squids, the organ is situated on a muscular ridge of tissue posterior and ventral to each eye (Watkinson 1909; Emery 1975; Wildenburg and Fioroni 1990; Gilly and Lucero 1992) where it can sample the water that is pulled into the mantle cavity during each inhalation [see Fig. 32.1(a) and (b)]. The organ is composed of several types of sensory cells with apical dendritic processes and basal unmyelinated axons that form the olfactory nerve and make synapses in the olfactory lobe, optic lobe, optic gland, and higher motor centres in the brain (Messenger 1979). There are also mucus-secreting cells (Barber and Wright 1969), and epithelial support cells with long motile cilia that form a dense mat over the surface of the organ as shown in Fig. 32.1(c) and (d).

When olfactory organs from the squid *Loligo opalescens* Berry were excised, treated with non-specific protease, rinsed in fresh filtered sea water and placed in tissue culture medium, the sensory cells maintained their bipolar morphology for several days in culture (Lucero *et al.* 1992). Figure 32.2(a)–(c) show examples of olfactory organ cells that have been freshly dissociated. The most common chemoreceptor cell type [Fig. 32.2(a)] is termed pyriform and has a long axonal process (a) arising from the basal pole containing the nucleus (n), and a spike-like process that projects from the apical pole (arrow). The bulk of the cell's 'interior' is actually an invaginating extracellular cavity filled with cilia (Emery 1975). A low-power scanning electron micrograph of a fixed pyriform receptor is shown in Fig. 32.2(e).

A second receptor cell type is termed floriform [Fig. 32.2(b)]. This type also has a basal axon (a), but in this case a long, thin neck extends from the cell body and is crowned by a swelling covered with non-motile cilia arranged in a petaloid fashion (arrow). Figure 32.2(d) and (f) are scanning electron

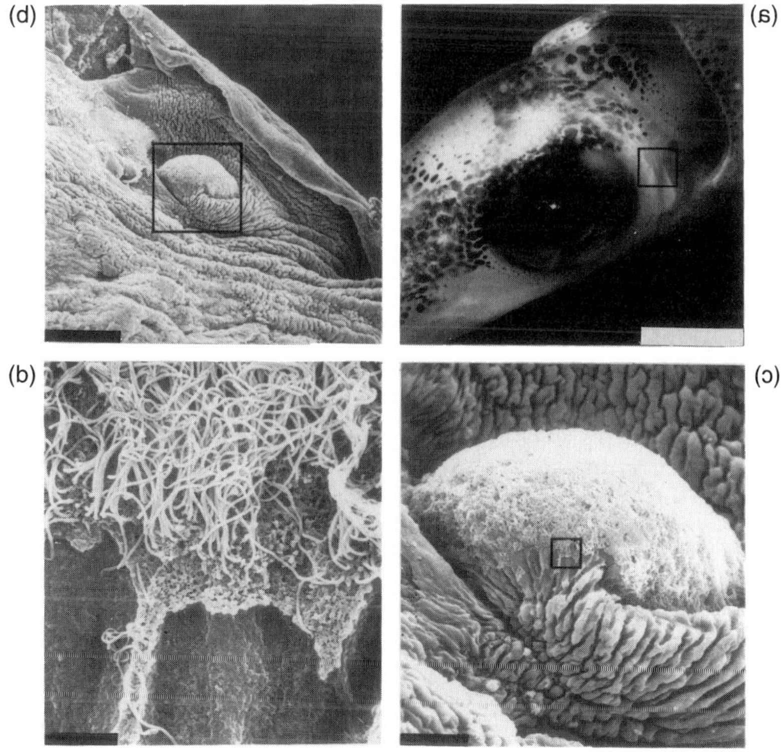

Fig. 32.1. Location and appearance of the olfactory organ in *Loligo opalescens*. (a) The olfactory organ in a living squid is indicated by the square posterior to the eye. (b)–(d) Scanning electron micrographs of each of the boxed areas at progressively higher magnifications. Scale bars: (a) 1 cm; (b) 250 μm; (c) 50 μm; and (d) 5 μm. (From Gilly and Lucero, 1992, with permission.)

micrographs of a fixed floriform receptor. Pyriform and floriform receptor cells were the most distinctly identifiable and common types of cells. Electrophysiological experiments were conducted on both.

A third, less common receptor may be related to the pyriform-type cell based on the invaginating cup-like structure on the apical pole [Fig. 32.2(c)]. In this type of cell, however, the portion of the cell body containing the nucleus is separated by a thin neck similar to the floriform receptor. These cells were relatively rare and were not studied electrically.

Behavioural experiments

Screening potential odorants

Behavioural and physiological experiments were performed on the squid *L. opalescens* Berry. Healthy adult animals were secured to a plexiglass support

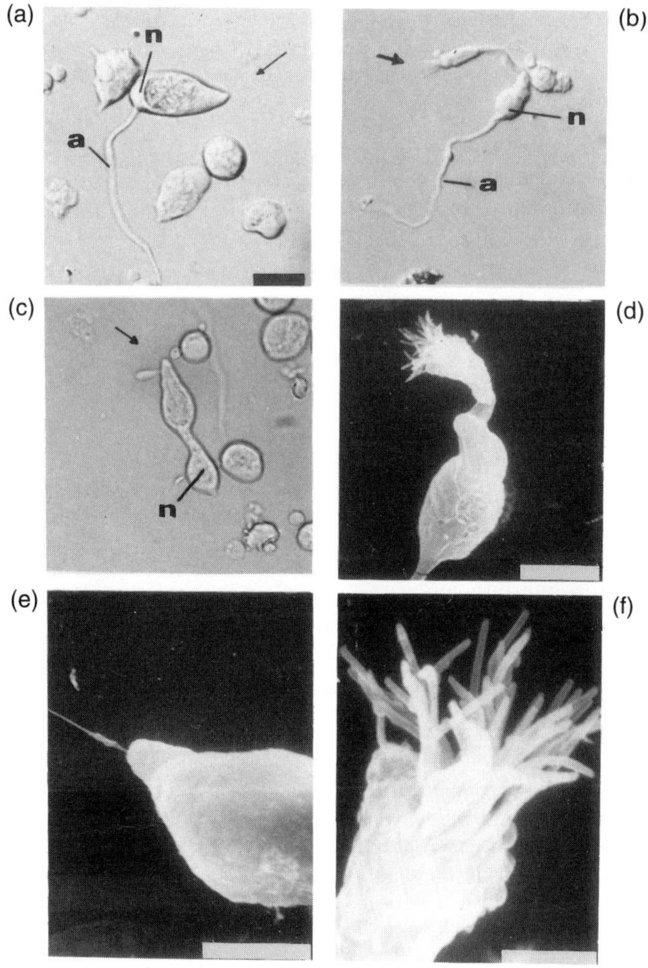

Fig. 32.2. Primary sensory neurones retain normal morphology when dissociated from the olfactory organ. (a)–(c) Light micrographs of living pyriform, floriform and intermediate receptor cells, respectively. (d)–(f) SEM micrographs of a pyriform cell, a floriform cell, and an enlarged floriform knob, respectively. Scale bars: (a)–(c) 10 μm; (d)–(e) 5 μm; (f) 1 μm. (From Lucero *et al.* 1992, with permission.)

in a 30 litre tank containing slowly flowing sea water. A hypodermic needle attached to a pressure transducer was inserted into the animal's mantle cavity and measured intramantle pressure transients associated with high pressure escape jetting in response to the application of small volumes (100–300 μl) of chemical stimuli (Gilly and Lucero 1992).

Under these conditions, squid were capable of chemically detecting a 1:20 dilution (in sea water) of squid ink, 1 mM L-DOPA, and various K^+ channel blockers including 20 mM tetrabutyl ammonium (TBA), 20 mM tetraethyl ammonium (TEA), 10–250 μM propyl paraben, 0.5–1 mM methadone, and 5–10 mM 4-aminopyridine (4AP) (Gilly and Lucero 1992). Stimuli that did not produce escape responses included 20 mM tetramethyl ammonium (TMA), 1 mM tetrodotoxin (TTX), 100 μM Brilliant Blue, 100 μM Fast Green, 5 mM sodium nitrite, 20 mM methionine, 1 mM proline, 1 mM taurine, 5 mM menthol, 20 mM trimethylamine N-oxide, pH 5 sea water, and filtered homogenates of squid mantle, gonads, and head (Gilly and Lucero 1992).

Mapping experiments

Detailed mapping experiments clearly indicated that the olfactory organ is the site of highest chemosensitivity and that escape jetting responses were not due to a generalized irritation of the skin on the head or mantle. A high concentration of chemosensory cells are also found on the buccal lobes surrounding the beak ('lips') of squids (Emery 1975) and cuttlefishes (Graziadei 1965). These are morphologically distinct from those in the olfactory organ and have not yet been studied physiologically.

Figure 32.3 shows one type of mapping experiment in which a potent stimulus was applied at different locations on the squid's head, and at each site the duration of the pressure pulse was varied in order to identify a threshold value (the shortest pulse needed to elicit a high-pressure jet). Because the spread of a stimulus diminished as pulse duration decreased, it was assumed that the stimulation site with the minimum threshold corresponded to the site of highest chemosensitivity. In such experiments, minimum threshold was consistently observed at the site of the olfactory organ (Gilly and Lucero 1992).

The second type of mapping experiment used a smaller stimulating pipette containing a chemical stimulus (10 mM TBA) and a tracking dye (5 mM Brilliant Blue). Video observations identified the exact site of impact for stimuli that produced escape jets. The animals only responded when the stimulus enveloped the olfactory organ, again suggesting that the organ is a specific site for chemoreception (data not shown).

Chemical 'ablation'

Localization of chemosensitivity to the olfactory organ was also achieved by reversibly blocking the function of the receptor cells and/or afferent transmission with a local anaesthetic. Figure 32.4 shows the data from an experiment in which high pressure escape jets were produced by stimulation with 100 μM propyl paraben on the olfactory organs of both sides of the head. The animal was then removed from the tank and 10 μl of 0.5 mM dibucaine (in sea water) was applied to the olfactory organ on the test side of the head. Subsequent

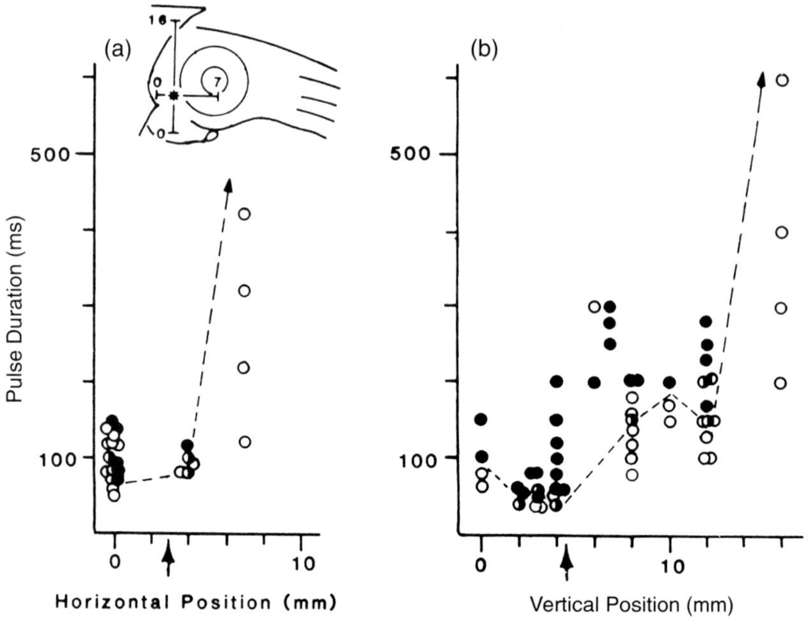

Fig. 32.3. The region of highest chemosensitivity maps to the olfactory organ. A chemical stimulus (20 mM TBA and 1:25 blue food colouring in sea water) was delivered at different positions along two axes intersecting at the olfactory organ (*), as indicated on the inset (distances in mm). The stimulus pulse duration was varied at each location. Pulses that elicited escape responses (●); showed no response (○). The threshold duration for a positive response (– – – –). For both horizontal (a) and vertical (b) positions the minimum threshold duration mapped to the olfactory organ as indicated by the arrows. (From Gilly and Lucero 1992, with permission.)

stimulation with propyl paraben on the anaesthetized olfactory organ failed to produce an escape response, but 'control' stimulation of the contralateral, untreated side reliably produced escape jets. Function of the dibucaine-treated organ recovered approximately 20 min after anaesthetic application.

Thus, results from mapping and ablation experiments led to the conclusion that the olfactory organ is the site of chemosensitivity and that the detection of chemical stimuli is linked to motor control pathways involved in escape jetting behaviour (Gilly and Lucero 1992).

Electrophysiological experiments

Both whole cell (Hamill *et al.* 1981) and nystatin patch recordings (Horn and Marty 1988) were performed on sensory cells isolated from squid olfactory organs. The advantage of whole cell recording is that both the internal and

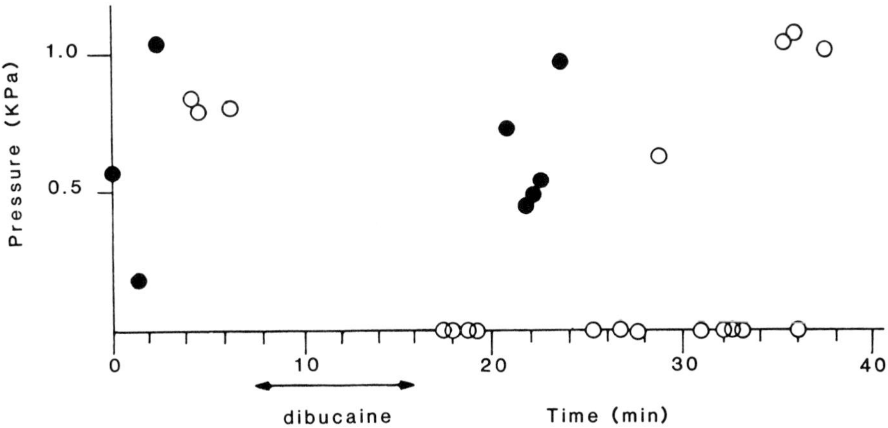

Fig. 32.4. Chemoreceptive capability is reversibly impaired by treating the olfactory organ with a local anaesthetic. A chemical stimulus (propyl paraben) produced escape responses in either the left (control) or right (test) olfactory organ, and amplitudes of the resultant pressure transients (●, control and ○, test) are plotted as a function of time during the experiment. Initially, both sides were responsive. Following application of anaesthetic, the test side became unresponsive while the control side maintained responsiveness. The test side recovered function after approximately 20 min. (From Gilly and Lucero 1992, with permission.)

external solutions can be controlled. The advantage of nystatin patch recording is that second messenger and calcium (Ca^{2+}) buffering systems remain relatively intact. Thus, whole cell recordings were used to examine the membrane properties of the olfactory receptor cells in voltage-clamp experiments and nystatin patch recordings (in current-clamp) were used to study the effects of previously identified chemicals that cause escape jetting.

Passive properties of floriform and pyriform receptor cells

Resting membrane potentials of pyriform receptor cells, measured immediately after obtaining the whole cell configuration, averaged −58 ± 1 mV and are similar to resting potentials in isolated olfactory cells from both vertebrates and invertebrates (Firestein and Werblin 1987, −54 mV; Schmiedel-Jakob *et al.* 1989, −55.8 mV). The resting potential of floriform receptor cells was less negative, varying even in a single cell, from −50 to −25 mV. The variation was probably due to the high input resistance of the floriform cells, so that opening or closing of relatively few ion channels caused marked changes in membrane voltage.

Resting potential data, input resistance, input capacitance, and time constants for the capacity current transients measured for both pyriform and floriform cells are listed in Table 32.1

Table 32.1 Passive membrane properties

Cell type	Resting potential (mV)	Input resistance (MΩ)	Input capacitance (pF)	Capacity current Time constant	
				Fast (μs)	Slow (ms)
Pyriform	−58.2 ± 1.2	406 ± 77	11.9 ± 1.0	78 ± 6	1.7 ± 0.2
(*n*)	(91)	(30)	(40)	(16)	(16)
Floriform	−25 to − 50	1700 ± 400	5.7 ± 0.5	88 ± 8	—
(*n*)	(5)	(14)	(11)	(14)	

Passive membrane properties were obtained for pyriform and floriform receptor cells. (*n*) indicates the number of cells used to generate each mean ± 1 SE. Input capacitance refers to the fast component of the capacitive current. (See Lucero *et al.* 1992 for additional details.) The resting potentials of floriform cells were very unstable, and are indicated by a range of values rather than a mean.

Voltage-dependent properties

Both pyriform and floriform receptor cells contain TTX-sensitive sodium (Na^+) currents, delayed rectifier type K^+ currents and small labile Ca^{2+} currents, all of which resemble currents found in other squid neurones (Brismar and Gilly 1987; Gilly and Brismar 1989). The selectivity, voltage dependence, and kinetics of Na^+ currents are similar in pyriform and floriform receptor cells. K^+ currents in each cell type are also basically alike; however, both Na^+ and K^+ currents in pyriform cells are more than twice as large as in floriform cells when normalized to input capacitance (see Table 32.1).

Effects of channel blocking compounds on receptor cells

Isolated olfactory receptor cells are capable of generating action potentials with rates of spontaneous and induced firing ranging from 0.08 to 17.00 Hz. A single action potential elicited from a pyriform receptor cell by current injection is shown in Fig. 32.5(a). Action potentials are generated by voltage-dependent Na^+ channels and are blocked by 100 nM TTX. The recovery phase of the action potential is due to both Na^+ channel inactivation and the opening of K^+ channels. The importance of K^+ channels in determining spike duration and influencing cell excitability is demonstrated in Fig. 32.5(b). In the presence of 20 mM TBA, an identical current pulse elicited a broadened spike and a second action potential. Such changes in the cell's excitability would presumably alter information encoded by the pattern of impulses passing from an olfactory receptor cell to the brain of a living squid and lead to the escape jetting observed in behavioural experiments.

The mechanism of TBAs effect on excitability was examined with whole cell voltage-clamp. Twenty millimoles of TBA (and 20 mM TEA) selectively blocked approximately 20 per cent of the peak K^+ current at +40 mV in

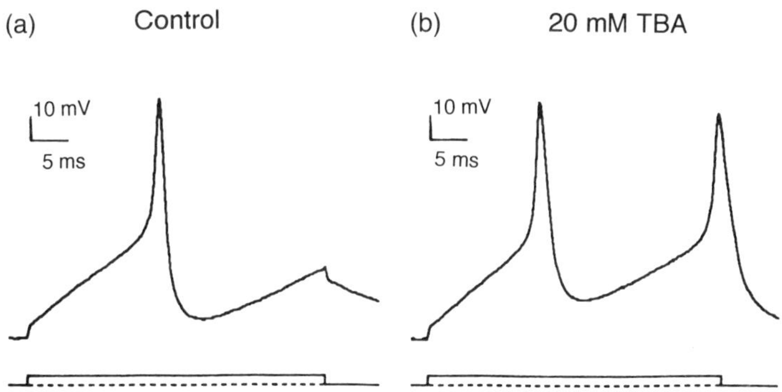

Fig. 32.5. 20 mM TBA increases excitability of current-clamped olfactory cells. (a) A single action potential was elicited from a pyriform receptor cell during a 30 ms current injection of 200 pA. (b) Two action potentials were elicited during an identical pulse in the presence of 20 mM TBA. Bath solution: artificial sea water (ASW) 470 mM NaCl, 10 mM KCl, 50 mM $MgCl_2$, 10 mM $CaCl_2$, 10 mM HEPES, and 100 nM TTX tetrodotoxin (drugs were added to this solution). 400 K internal solution: 20 mM KCl, 355 mM K-glutamate, 25 mM KF, 10 mM EGTA and 15 mM HEPES. (From Lucero *et al.* 1992, with permission.)

isolated pyriform receptor cells without affecting inward Na^+ currents [Fig. 32.6(a) and (b)] A related quaternary ammonium ion, TMA^+, had no behavioural effects and did not block K^+ channels [see Fig. 32.6(c)].

Experiments with TBA and TEA indicate that selective block of K^+ channels is important for behavioural responses. Further support is provided by experiments using 4AP, another behaviourally active chemical that selectively blocks K^+ channels in both pyriform and floriform cells.

Thus, all of the channel blocking compounds that have thus far proven to be active in behavioural experiments act to block K^+ channels when employed at the same concentrations. This is also true for chemicals (methadone and propyl paraben) that block both K^+ and Na^+ channels; however, TTX, which selectively blocks Na^+ currents, does not produce behavioural effects. Therefore it appears that K^+ channel block is one of the mechanisms involved in signal transduction and results in increased excitability in the receptor cell. Squid do not respond behaviourally to impermeant Na^+ channel blockers and this may be due to the low density of apical Na^+ channels observed in preliminary channel distribution studies.

Effects of naturally occurring substances on receptor cells

In the search for physiologically relevant chemical stimuli using behavioural assays, it was found that a 1:20 dilution of squid ink caused escape jetting (Gilly and Lucero 1992). Behavioural observations were followed up with electrophysiological studies of action potentials in isolated receptor cells.

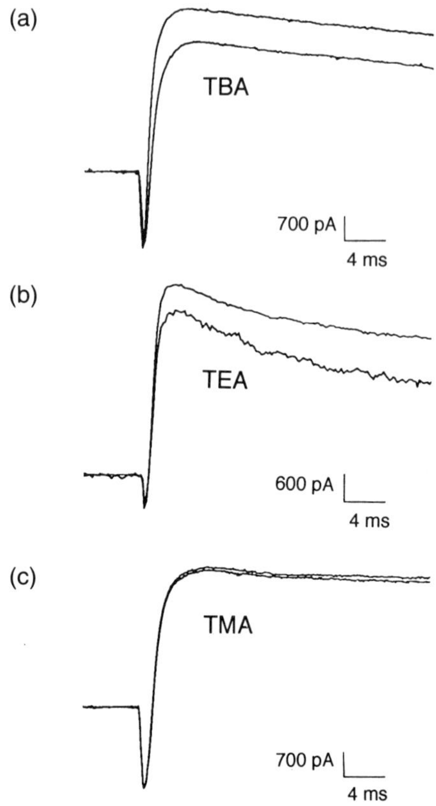

Fig. 32.6. Effects of quaternary ammonium compounds on ionic currents in isolated pyriform cells. Holding potential − 70 mV; voltage step to + 40 mV. (a) 20 mM TBA blocks ~ 20 per cent of the outward K^+ current without affecting inward N^+ current. (b) 20 mM TEA blocks ~ 15 per cent of the outward K^+ current, but has no effect on inward Na^+ current. (c) 20 mM TMA has very little effect on either the K^+ currents or the Na^+ currents. Test substances were applied via a perfusion pipette for a few seconds before, during, and after each voltage pulse. Same solutions as in Fig. 32.5. (From Lucero *et al.* 1992, with permission.)

Control application of artificial sea water (ASW) to the pyriform receptor cell had no effect on the action potential firing rate as shown in Fig. 32.7(a). In contrast, diluted squid ink reversibly hyperpolarized the membrane potential by ~ 5 mV and inhibited firing in 65 per cent of the cells tested [Fig. 32.7(b)]. This hyperpolarization was associated with a significant decrease in membrane resistance (data not illustrated). Thus it was suggested that K^+, or possibly Cl^-, channels open in response to squid ink and hyperpolarize the cell.

Melanin is the major component in squid ink and its precursor L-DOPA is also present in the ink (Jimbow *et al.* 1984). One millimol of L-DOPA, like

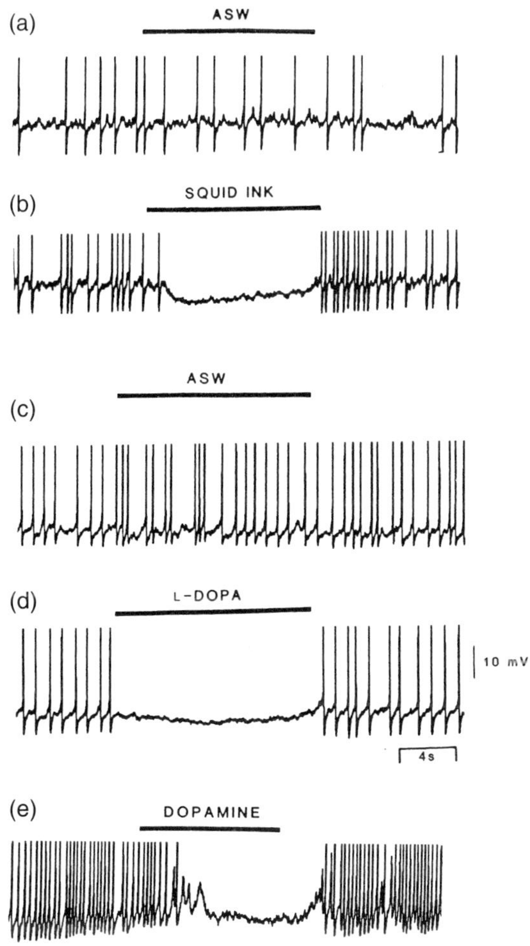

Fig. 32.7. Action potentials are blocked during the application of squid ink, L-DOPA, and dopamine. ASW (a) or 1:20 dilution of squid ink (b) were applied to the same cell with a small perfusion pipette (solid bar). (c) and (d) Show similar responses in a different cell during application of ASW and 10 mM L-DOPA, respectively. (e) 50 μM dopamine also inhibits action potentials. (From Lucero *et al.* 1992, with permission.)

squid ink, elicited escape jetting in behavioural experiments (Gilly and Lucero 1992). Similarly, 1–10 mM L-DOPA mimicked the effect of squid ink in isolated pyriform cells with a hyperpolarization and a decrease in membrane resistance [Fig. 32.7(d)]. Control application of ASW had no effect as shown in Fig. 32.7 (c). Fifty millimoles of dopamine also hyperpolarized the membrane potential and inhibited action potential firing [Fig. 32.7(e)].

The hyperpolarization associated with the application of squid ink, L-DOPA or dopamine on to olfactory receptor cells apparently requires the 'intact' second messenger and Ca^{2+} buffering systems that are preserved in nystatin patch recordings. When similar experiments were performed on internally dialysed cells using conventional whole cell recording, there were no responses. Thus L-DOPA or a closely related compound such as dopamine, may be the component of squid ink that causes the second messenger-mediated hyperpolarization in olfactory receptor neurones.

Conclusions

Squid are capable of distance chemoreception and the olfactory organ is the site of highest chemosensitivity. Living restrained squid respond with vigorous escape jetting behaviour to squid ink, L-DOPA, and K^+ channel blockers including TBA, TEA, 4AP, methadone, and propyl paraben. Mapping and reversible chemical ablation experiments show that chemosensitivity is directly correlated with a stimulus contacting a functional olfactory organ.

Electrophysiological experiments on isolated olfactory receptor cells (Lucero *et al.* 1992) show that the two most common sensory cell types, the pyriform and floriform receptor cells, are excitable and contain voltage gated Na^+, K^+, and Ca^{2+} channels. Channel blockers that cause behavioural responses suppress K^+ currents in the receptor cells resulting in an increase in excitability and a broadening of the action potential. Natural substances such as squid ink, dopamine, and its precursor L-DOPA increase membrane conductance, hyperpolarize the cell, and inhibit the firing of action potentials. It thus appears that when olfactory receptor cells are chemically stimulated by either natural alarm substances or K^+ channel blockers, change in the ongoing firing rate of the receptor cells is linked to the activation of the motor pathways that control the escape jetting response.

It is interesting that squid are capable of chemically detecting their ink. Generally, squid ink is thought of as a visual signal used to confuse predators and indicate danger. The present work suggests that these often nocturnal schooling animals, are able to chemically sense and avoid the ink released by recently alarmed conspecifics.

Acknowledgements

We would like to thank the Wellcome Trust for providing a Wellcome Research Travel Grant to Dr M. Lucero. This work was supported by the U.S. Office of Naval Research Grant N0014–89–J–1744 and a grant from the Whitehall Foundation.

UNIVERSITY
COLLEGE LONDON
LIBRARY

References

Atema, J. (1980). Smelling and tasting underwater. *Oceanus*, **23**, 4–18.

Barber, V. C. and Wright, D. E. (1969). The fine structure of the sense organs of the cephalopod mollusc *Nautilus. Zeitschrift für Zellforschung und Mikroskopische Anatomie*, **102**, 293–312.

Bardach, J. E. (1975). Chemoreception of aquatic animals. In *Olfaction and taste*, V (ed. D. A. Denton and J. P. Coghlan), pp. 121–32. Academic Press, New York.

Bardach, J. E. and Atema, J. (1971). Taste in Fishes. In *Handbook of sensory physiology* (ed. L. Beidler), Vol. IV, pp. 293–31. Springer-Verlag, Berlin.

Brismar, T. and Gilly, W. F. (1987). Synthesis of sodium channels in the cell bodies of squid giant axons. *Proceedings of the National Academy of Sciences USA*, **84**, 1459–63.

Budelmann, B. U. (1990). The statocysts of squid. In *Squid as experimental animals* (ed. D. L. Gilbert, W. J. Adelman, and J. M. Arnold), pp. 421–39. Plenum Press, New York.

Budelmann, B. U. and Bleckmann, H. (1988). A lateral line analogue in cephalopods: water waves generate microphonic potentials in the epidermal head lines of *Sepia* and *Lolliguncula. Journal of Comparative Physiology*, **164**, 1–5.

Emery, D. G. (1975). The histology and fine structure of the olfactory organ of the squid *Lolliguncula brevis. Tissue and Cell*, **7**, 357–67.

Finger, T. E. and Silver, W. L. (ed.) (1987). *Neurobiology of taste and smell.* Wiley Series in Neurobiology, John Wiley & Sons, New York.

Firestein, S. and Werblin, F. S. (1987). Gated currents in isolated olfactory receptor neurons of the larval tiger salamander. *Proceedings of the National Academy of Sciences, USA*, **84**, 6292–6.

Gilly, W. F. and Brismar, T. (1989). Properties of appropriately and inappropriately expressed sodium channels in squid giant axon and its somata. *Journal of Neuroscience*, **9**, 1362–74.

Gilly, W. F. and Lucero, M. T. (1992). Behavioral responses to chemical stimulation of the olfactory organ in the squid, *Loligo opalescens. Journal of Experimental Biology*, **162**, 209–29.

Graziadei, P. (1965). Sensory receptor cells and related neurons in cephalopods. *Cold Spring Harbor Symposium*, **30**, 45–57.

Hamill, O. P., Marty, A., Neher, E., Sakman, B., and Sigworth, F. J. (1981). Improved patch-clamp techniques for high-resolution current recording from cells and cell-free membrane patches. *Pflügers Archiv*, **391**, 85–100.

Horn, R. and Marty, A. (1988). Muscarinic activation of ionic currents by a new whole cell recording method. *Journal of General Physiology*, **92**, 145–59.

Hanlon, R. T. and Budelmann, B.-U. (1987). Why cephalopods are probably not 'deaf'. *American Naturalist*, **129**, 312–17.

Jimbow, K., Miyake, Y., Homma, K., Yasuda, K., Izumi, Y., Tsutsumi, A., and Ito, S. (1984). Characterization of melanogenesis and morphogenesis of melanosomes by physicochemical properties of melanin and melanosomes in malignant melanoma. *Cancer Research*, **44**, 1128–34.

Kohn, A. J. (1961). Chemoreception in gastropod molluscs. *American Zoologist*, **1**, 291–308.

Lucero, M. T., Horrigan, F. T., and Gilly, W. F. (1992). Electrical responses to chemical stimulation of squid olfactory receptor cells. *Journal of Experimental Biology*, **162**, 231–49.

Messenger, J. B. (1967). The peduncle lobe: a visuo-motor centre in *Octopus*. *Proceedings of the Royal Society of London*, **167**, 225–51.

Messenger, J. B. (1979). The nervous system of *Loligo*. IV. The peduncle and olfactory lobes. *Philosophical Transactions of the Royal Society of London*, **285**, 275–309.

Packard, A. (1972). Cephalopods and fish: the limits of convergence. *Biological Reviews*, **47**, 241–307.

Schmiedel-Jakob, I., Anderson, P. A. V., and Ache B. W. (1989). Whole cell recording from lobster olfactory receptor cells: responses to current and odor stimulation. *Journal of Neurophysiology*, **61**, 994–1000.

Watkinson, G. B. (1909). Untersuchungen über die sog. Geruchsorgane der Cephalopoden. *Jenaische Zeitschrift für Naturwissenschaft*, **44**, 353–414.

Wells, M. J. (1964). Tactile discrimination of surface curvature and shape by octopuses. *Journal of Experimental Biology*, **41**, 435–45.

Wells, M. J. and Young, J. Z. (1975). The subfrontal lobe and touch learning in the octopus. *Brain Research*, **92**, 103–21.

Wildenburg, G. and Fioroni, P. (1990). Ultrastructure of the olfactory organ during embryonic development and at the hatching stage of *Loligo vulgaris* Lam. (Cephalopoda). *Journal of Cephalopod Biology*, **1**, 56–70.

Woodhams, P. L. and Messenger J. B. (1974). A note on the ultrastructure of the *Octopus* olfactory organ. *Cell and Tissue Research*, **152**, 253–8.

Zernoff, D. (1869). Über das Geruschsorgane der Cephalopoden. *Bulletin de la Societe des Naturalistes de Moscou*, **42**, 71–90.

Index

Italic numbers denote reference to illustrations.

UNIVERSITY
COLLEGE LONDON
LIBRARY